D0161625

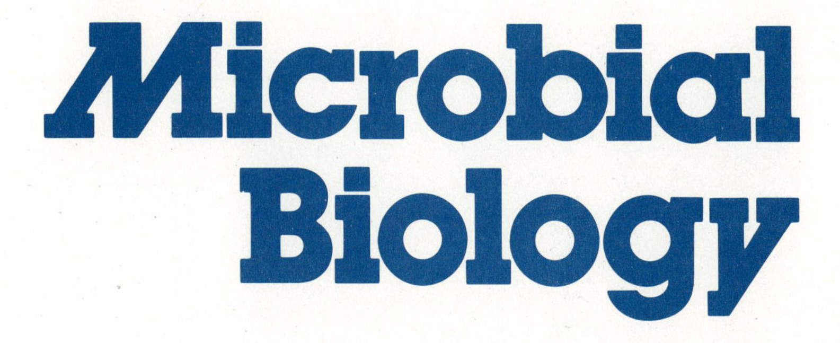
Microbial
Biology

Microbial Biology

Eugene Rosenberg
Professor of Microbiology
Tel-Aviv University

Irun R. Cohen
Professor of Cell Biology
Weizmann Institute of Science

SAUNDERS COLLEGE PUBLISHING

Philadelphia New York Chicago
San Francisco Montreal Toronto
London Sydney Tokyo Mexico City
Rio de Janeiro Madrid

Address orders to:
383 Madison Avenue
New York, NY 10017

Address editorial correspondence to:
West Washington Square
Philadelphia, PA 19105

This book was set in Melior by The Clarinda Company.
The editors were Michael Brown, Janis Moore, Lynne Gery, and Cate Barnett Rzasa.
The art & design director was Richard L. Moore.
The text design was done by Nancy E. J. Grossman.
The cover design was done by Richard L. Moore.
The artwork was drawn by Vantage Art, Inc.
The production manager was Tom O'Connor.
This book was printed by Fairfield Graphics.

Cover: Chains of blue-green algae (bacteria). These microorganisms perform both photosynthesis and nitrogen fixation. See page 105 and Chapter 9 for more details.

MICROBIAL BIOLOGY ISBN 0-03-085658-2

Library of Congress catalog card number 81-53096.

3 4 5 016 9 8 7 6 5 4 3 2

CBS COLLEGE PUBLISHING
Saunders College Publishing
Holt, Rinehart and Winston
The Dryden Press

This book is dedicated to the memory of
Miriam Petlak
and
Michal Cohen

PREFACE

This book is motivated by a desire to share with others our own feelings of excitement and intellectual joy in important discoveries in the field of scientific endeavor called microbiology and its rapidly growing offspring, immunology. Our approach has been to select topics of interest and relevance, such as the *origin of life, genes and genetic engineering, ecology,* the contributions of microbes to our *foods and beverages,* the *ecology of man* and his *microbial parasites, antibiotics, the immune response, infectious diseases,* and *cancer,* and to explain these subjects in a way that will be intelligible and satisfying both to the reader who has some knowledge of the natural sciences and to the reader who has no specific training in microbiology or medicine.

Many of the consequential discoveries in biology have been related, to a greater or lesser degree, to microbiology, either as a science or as a research tool. Hence, a survey of microbiology is an excellent means of grasping the significance of factors in the biological world that are of concern to most people, whether or not they have a professional interest in microbiology.

We have used different vehicles to convey the reader into the field of microbiology. The historical or case-study approach is prominent in Chapters 1, 3, and 5 through 8. We have outlined a number of seminal ideas from their earliest conception to their latest development, emphasizing the observations and reasoning behind each concept and the critical experiments that were performed to test them. In this way we intend to stimulate the reader's innate sense of scientific inquiry. For it is precisely in the substitution of evidence for dogma, as a basis for belief, that science has made its greatest offering.

An additional approach has been to emphasize the relevance of microbiology to human welfare. This is evident in the chapters dealing with microbial ecology, foods and beverages, antibiotics, and other valuable microbial products.

We have developed what we believe to be a unique presentation of medical microbiology and immunology in Chapters 12 through 16. Rather than confront the reader with a traditional list of pathogenic microbes and the diseases they "cause," we have attempted to outline a fundamental concept of microbial disease as an incident in an ongoing ecological relationship between humans and microbial parasites. We have emphasized the mutual adaptation and benefits of the normal host–parasite accommodation. Disease is seen to result when the balanced relationship is upset by changes in the parasite and/or host. The immune response is analyzed as a factor in the mutual communication between human and microbe. Human behavior, culture, and history are viewed as elements with both positive and negative effects on health, parasitism, and disease.

Throughout the book, we have attempted to strike a balance between theoretical and applied aspects of microbiology. *Scientific progress is motivated by both a desire to know and a desire to use this knowledge*. Nowhere is this more evident than in microbiology. We hope this book stimulates interest in microbiology and provides the necessary backbone for greater appreciation of the increasing number of microbiological discoveries that are making an impact on our way of life.

Acknowledgements

We take pleasure in expressing our appreciation to the many individuals who helped us in the preparation of this book. We are especially grateful to R. J. Martinez for constructive criticism of the entire text. Dina Silverman and Malvine Baer gave us invaluable help with the typing of the manuscript. We thank the people of Saunders College Publishing, particularly Janis Moore and Michael J. Brown, for helping to convert our manuscript into this book. We also acknowledge the diligent reviewers of the manuscript: J. C. Lara, University of Washington; D. W. Grant, Colorado State University; James Fuchs, University of Minnesota; D. E. Duggan, University of Florida; W. S. Riggsby, University of Tennessee; L. J. Berry, University of Texas; and R. E. Kallio, University of Illinois.

The authors are grateful to their teachers: I. R. Cohen to Gene H. Stollerman for introducing him to science and stimulating him to think about the relationship between microbes and humans, and to Moshe Prywes for introducing him to the responsibility of teaching others; E. Rosenberg to Stephen Zamenhof and Sydney C. Rittenberg for more than 20 years of encouragement, guidance, and inspiration.

Our wives, Leah and Yael, and children, Robin, Stephenie, and Denise, and Tammy, Ruthy, and Yonatan, deserve special thanks for inspiration, encouragement, and patient understanding.

E. R.
I. R. C.

CONTENTS

CHAPTER 9: ENVIRONMENTAL MICROBIOLOGY 231

CHAPTER 10: MICROBIOLOGY OF FOODS AND BEVERAGES: ART AND SCIENCE 255

CHAPTER 11: ANTIBIOTICS AND OTHER VALUABLE MICROBIAL CHEMICALS 280

The Origin of Life and the Origin of Microbiology

CHAPTER 1

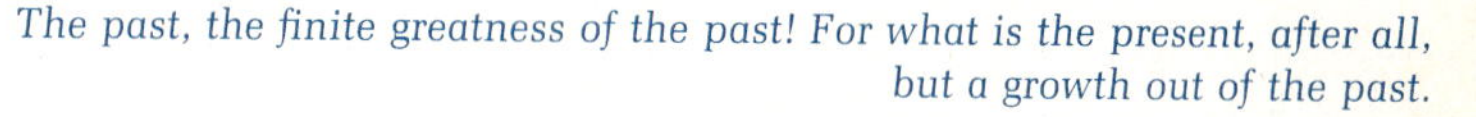

The past, the finite greatness of the past! For what is the present, after all, but a growth out of the past.

Walt Whitman

The opening chapter describes the discovery of microorganisms by Antony van Leeuwenhoek and discusses how the early development of microbiology was closely associated with the spontaneous generation controversy over the origin of life. The experiments of Spallanzani, Pasteur, and Tyndall not only refuted the age-old theory of spontaneous generation, but more importantly, in doing so, converted the science of microbiology from one solely of observation to one of controlled experimentation. In the latter part of this chapter, the important chemicals of life are presented in the context of the Oparin-Haldane theory of the origin of life through chemical evolution. In this "natural" way, we attempt to provide the necessary chemical background for subsequent chapters.

Genesis has an obvious and compelling fascination for all humanity. The question of how life came to be is fundamental to theology and philosophy as well as to the natural sciences, for it is only possible to understand the essence of life when we understand its origin. Recent progress in science and technology makes this question more timely than ever. Especially significant is the more detailed understanding of the essential features of existing life that has been obtained during the last 30 years. Also, as the human race begins its adventures into space, the possible existence of extraterrestrial life brings new significance to this age-old question.

Rather than present only the current views on the origin of life, several of the explanations that have been put forth in the past are also discussed. This is being done for two reasons. First, the essence of the scientific method is learning from experience. Second, it provides an opportunity to demonstrate how clear thinking and experimentation on non-practical problems can lead to discoveries of great potential significance.

The following theories that are proposed to account for the origin of life are analyzed in historical sequence: **spontaneous generation, continuous life, panspermia,** and **chemical evolution.** Emphasis is placed on observation and experiments that were performed to test each of these theories. In this way, students can obtain what they need most—an attitude of scientific inquiry. The substitution of evidence for dogma, as a basis for belief, is the greatest offering that science has bestowed on man. In this tradition, students are asked to accept nothing on faith. Students should understand the evidence, analyze the hypothesis, and be willing to change their ideas as new evidence is presented.

1.1 The Theory of Spontaneous Generation of Animals and Plants

Although the theory of spontaneous generation (abiogenesis) can be traced back at least to the Ionian school (600 B.C.), it was Aristotle (384–322 B.C.) who presented the most complete arguments for and the clearest statement of this theory. In his *On the Origin of Animals,* Aristotle states not only that animals originate from other similar animals, but also that **living things do arise and always have arisen from lifeless matter.** Aristotle's theory of spontaneous generation was adopted by the Romans and Neo-Platonic philosophers and, through them, by the early fathers of the Christian Church. With only minor modifications, these philosophers' ideas on the origin of life, supported by the full force of Christian dogma, dominated the mind of mankind for more that 2000 years.

According to this theory, a great variety of organisms could arise from lifeless matter. For example, worms, fireflies, and other insects arose from morning dew or from decaying slime and manure, and earthworms originated from soil, rainwater, and humus. Even higher forms of life could originate spontaneously according to Aristotle. Eels and other kinds of fish came from the wet ooze, sand, slime, and rotting seaweed; frogs and salamanders came from slime.

Rather than examining the claims of spontaneous generation more closely, Aristotle's followers concerned themselves with the production of even more remarkable recipes. Probably the most famous of these was van Helmont's (1577–1644) recipe for mice. By placing a dirty shirt into a bin containing wheat germ and allowing it to stand 21 days, live mice could be obtained. Another example was the slightly more complicated but equally "foolproof" recipe for bees. By killing a young bullock with a knock on the head, burying him in a standing position with his horns sticking out of the ground, and finally sawing off his horns one month later, out will fly a swarm of bees.

The more exact methods of observation that were developed during the seventeenth century soon led to a realization of the complex nature of the anatomy and life cycles of certain living organisms. Equipped with this better understanding of the complexity of living organisms, it became more difficult for some to accept the theory of spontaneous generation. This skepticism signaled the beginning of three centuries of heated controversy over a theory that had gone unchallenged for the previous 2000 years. What is more significant is that the controversy was to be resolved not by powerful arguments but by ingeniously designed, simple experiments.

1.2 The Experiment of Redi

To Francisco Redi (1626–1698), an Italian physician, goes the honor of being the first to test the theory of spontaneous generation by using carefully controlled experimental techniques. He put some meat in two jars. One he left open to the air (the control); the other he covered securely with gauze. At that time it was well recognized that white worms would arise from decaying meat or fish. Sure enough, in a few weeks, the meat was infested with the white worms *but* only in the control jar which was not covered. This experiment was repeated several times, using either meat or fish, with the same result. On closer examination he noted that common houseflies went down into the meat in the open jar, later the white worms appeared, and then new flies. Redi reported that he had observed the flies deposit their eggs on the gauze; however, worms developed in the meat only when the eggs got to the meat. He therefore concluded from his observations that the white worms did not arise from the putrid meat. The worms developed from the eggs that the flies had deposited. The white worm then was the larva of the fly, and the meat served only as food for the developing insect.

Redi's experiment provided the impetus for testing other well-established recipes. In all cases that were examined carefully, it was demonstrated that the living organism arose not by spontaneous generation, but from a parent. Thus it was shown that the theory of spontaneous generation was based on a combination of the weakness of the human eye and bits and snatches of information gathered by accidental observation. The early biologists had seen earthworms coming out of the soil and frogs emerging from the slime of pond water, but they had not been able to see the tiny eggs from which these organisms arose. Because their observations had not been systematic, they had not seen how the mice invaded the grain bin in search of food, so they thought that the grain produced the mice. Based on the more exact methods of observation, the evidence that supported the theory of spontaneous generation of animals and plants was largely demolished by the end of the seventeenth century.

1.3 The Discovery of Microorganisms

Just as the theory of the abiogenesis of higher organisms was being refuted, the controversy was reopened, more heated than ever, because of the discovery of microorganisms by Antony van Leeuwenhoek. Before proceeding to the controversy over the spontaneous generation of these microorganisms, a brief account of the discovery of microorganisms is presented. Much of the material in this section is based on a fascinating biography of Leeuwenhoek by Clifford Dobell. (See Suggested Readings at the end of this chapter.)

As a young bacteriologist, Dobell was especially interested in studying the microbial flora of the mouth. However, each time he presented his professor with what he thought was the discovery of a new type of microbe, his professor would shake his head and respond, "No, no, Leeuwenhoek already discovered that one." Finally, motivated by a mixture of curiosity and skepticism, he decided to find out more about this man Leeuwenhoek. After 25 years of painstaking research, Dobell published a truly inspiring biography of Leeuwenhoek in 1932. Those students who find time to read Dobell's book will be treated to a masterpiece of English biography and rewarded with an insight into the true meaning of scientific research.

Leeuwenhoek was born in Delft, Holland, in 1632 (Figure 1.1). After leaving school at 16, he moved to Amsterdam and was an apprentice haberdasher for five years. He also became the bookkeeper and cashier in the shop where he worked. After returning to Delft, Leeuwenhoek married and opened a dry goods store. At the age of 28, he was appointed Chamberlain of the Council of Delft with the main task of keeping the council chambers clean and warm. Between 1653 and 1673, he developed the curious hobby of constructing microscopes. Although he was not the first to build a microscope, his instruments were the finest of that time. Equally important, however, were his almost child-like curiosity and great skills as an objective observer of nature. Leeuwenhoek patiently improved his microscopes and developed his techniques of observation for 20 years before he reported any of his results.

Finally, in 1673, Leeuwenhoek sent a letter to Henry Oldenburg, the first secretary of the Royal Society of England, describing the mouth and eye of the bee as viewed through his simple microscope. This letter was followed by several hundred more over the next 50 years, each written in "Nether-Dutch" (Leeuwenhoek knew neither Latin nor English), sent to England, and after translation published in the *Philosophical Transactions of the Royal Society*. Although these letters describe in great detail, for the first time, many of the parts of higher animals and plants, Leeuwenhoek's greatest recognition comes from his discovery of the previously invisible world of

Figure 1.1
Antony van Leeuwenhoek (1632–1723). (Courtesy of Rijksmuseum, Amsterdam.)

microorganisms. His discovery of protozoa in fresh water is described in his sixth letter, dated from Delft, September 7, 1674. In his eighteenth letter, dated October 9, 1676, he described bacteria for the first time. One of his most famous letters (number 39, September 17, 1683), which described bacteria in the human mouth (Figure 1.2), is an excellent illustration of his charming style and accurate observational ability.

'Tis my wont of a morning to rub my teeth with salt, and then swill my mouth out with water; and often, after eating, to clean my back teeth with a toothpick, as well as rubbing them hard with a cloth: therefore my teeth, back and front, remain as clean and white as falleth to the lot of few men of my years, and my gums (no matter how hard the salt be that I rub them with) never start bleeding. Yet notwithstanding, my teeth are not so cleaned thereby, but what there sticketh or groweth between some of my front ones and my grinders (whenever I inspected them with a magnifying mirror), a little white matter, which is as thick as if 'twere batter. On examining this, I judged (albeit I could discern nought a-moving in it) that there yet were living animalcules therein. I have therefore mixed it, at divers times, with clean rainwater (in which there were no animalcules), and also with spittle that I took out of my mouth, after ridding it of airbubbles (lest the bubbles should make any motion in the spittle): and I then most always saw, with great wonder, that in the said matter there were many very little living animalcules, very prettily a-moving. The biggest sort had the shape of Fig. A. These had a very strong and nimble motion, and they shot through the water (or spittle) like a pike does through the water. These were most always few in number.

The second sort had the shape of Fig. B. These oft-times spun around like a top, and every now and then took a course like that shown between C and D: and these were far more in number. The third sort I could assign no figure: for at times they seemed to be oblong, while anon they looked perfectly round. These were so small that I could see them no bigger than Fig. E: yet therewithal they went ahead so nimbly, and hovered so together, that you might imagine them to be a big swarm of gnats or flies; flying in and out among one another, these last seemed to me e'en as if there were, in my judgement, several thousand of 'em in an amount of water or spittle (mixed with the matter that I took from betwixt my front teeth, or my grinders).

Furthermore, the most part of this matter consisted of a huge number of little streaks, some greatly differing from others in their length, but of one and same thickness withal; one being bent crooked, another straight,

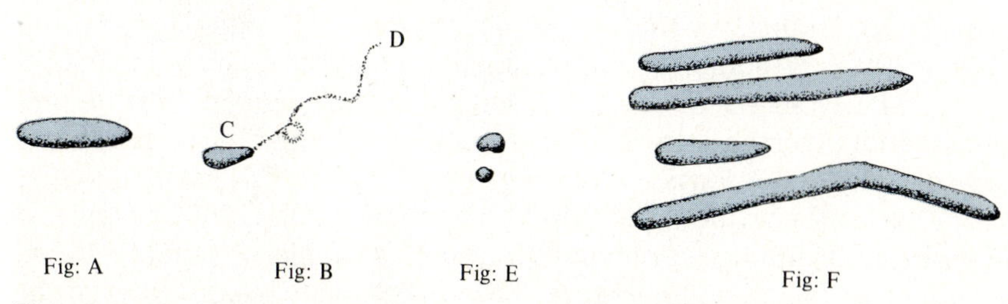

Figure 1.2
Leeuwenhoek's figures of bacteria from the human mouth. (Taken from Letter 39, September 17, 1683.)

> like Fig. F, and which lay disorderly raveled together. And because I had formerly seen, in water, living animalcules that had the same figure, I did make every endeavor to see if there was any life like anything alive, in any of 'em.

The most significant feature of Leeuwenhoek's work and the reason there is no question that, in fact, he did see bacteria are that his experiments are described in sufficient detail so that they can be repeated today. He tells where he obtained his material, how he handled it, and what he observed. For example, if you take the white matter between your teeth and examine it under a modern microscope, you will observe essentially what Leeuwenhoek described in 1683. Since his clear observation and correct interpretations were responsible for the discovery of bacteria and protozoa, Leeuwenhoek is recognized justly as the father of bacteriology and protozoology.

He continued until he was 91 years old, making better lenses and watching what he called his "little beasties." Although Leeuwenhoek had little formal education, studied under no distinguished professor, owed nothing to any university, and knew no language but his own, he belongs to that small group of gifted people who truly penetrated and discovered some of the great secrets of nature. He eventually was honored by being elected a foreign member of the Royal Society of London, and was visited by such famous personages as Peter the Great of Russia, Frederick I of Prussia, and Queen Mary II of England.

1.4 Needham versus Spallanzani

As soon as the discoveries of Leeuwenhoek became known, the proponents of spontaneous generation turned their attention to these microscopic organisms and suggested that surely they most have formed by spontaneous generation. Finally, experimental "proof" for this notion was published in 1748 by an Irish priest, John Tuberville Needham (1713–1781). Needham reported that he had taken mutton gravy fresh from the fire, transferred it to a flask, heated it to boiling, stoppered it tightly with a cork, and then set it aside. Despite boiling, the liquid became turbid in a few days. When examined under a microscope, it was teeming with microorganisms of all types. The experiments were repeated by and gained the support of the famous French naturalist, Georges Louis Leclerc, Comte de Buffon (1707–1788). Needham's demonstration of spontaneous generation was generally accepted as a great scientific achievement, and he was immediately elected into the Royal Society of England and the Academy of Sciences of Paris.

Meanwhile in Italy, Lazzaro Spallanzani (1729–1799) performed a series of brilliantly designed experiments of his own that refuted Needham's conclusions (Figure 1.3). Spallanzani found that if he boiled the food for *one hour* and hermetically sealed the flasks (by fusing the glass so that no gas could enter or escape), no microorga-

Figure 1.3
Lazzaro Spallanzani (1729–1799). (Courtesy of Tel Aviv University, Medical History Library.)

nisms would appear in the flasks. If, however, he boiled the food for only a few minutes, or if he closed the flask with a cork, he obtained the same results that Needham reported. Thus he wrote that Needham's conclusions were invalid because (1) he had not heated the gravy hot enough or long enough to kill the microorganisms, and (2) he had not closed the flask sufficiently to prevent other microbes from entering.

Count Buffon and Father Needham immediately responded that, of course, Spallanzani did not generate microorganisms in his flasks because his extreme heating procedures destroyed the **vegetative force** in the food and the **elasticity** of the air. Regarding Spallanzani's experiments, Needham wrote, "from the way he has treated and tortured his vegetable infusions, it is obvious that he has not only much weakened, and maybe even destroyed, the vegetative force of the infused substances, but also that he has completely degraded . . . the small amount of air which was left in his vials. It is not surprising, thus, that his infusions did not show any sign of life."

Rather than engage in theoretical arguments over the possible existence of these mystical forces, Spallanzani returned to the laboratory and performed another set of ingenious experiments. This time he heated the sealed flasks to boiling not for one hour, but for three hours, and even longer. If Needham was right, this treatment should certainly have destroyed the vegetative force. As Spallanzani had previously observed, nothing grew in these heated, sealed flasks. However, when the seal was broken and replaced with a cork, the broth soon became turbid with microbes. Since even three hours of boiling did not destroy anything in the food necessary for the production of microbes, Needham could no longer argue that he had killed the vegetative force by the heat treatment.

Spallanzani continued to perform experiments that led him to the conclusion that properly heated and hermetically sealed flasks containing broth would remain permanently lifeless. He was, however, unable to answer adequately the criticism that by sealing the flasks he had excluded the "vital forces" in the air that Needham claimed were also necessary ingredients for spontaneous generation. With the discovery of oxygen gas in 1774 and the realization that this gas is essential for the growth of most organisms, the possibility that spontaneous generation could occur, but only in the presence of air (oxygen), gained additional support.

The situation was brought to a crisis in 1859 when Félix Archimède Pouchet (1800–1872), a distinguished scientist and director of the Museum of Natural History in Rouen, France, reported his experiments on spontaneous generation. Pouchet claimed to have accomplished spontaneous generation using hermetically sealed flasks and pure oxygen gas. These experiments, he argued, demonstrated that "animals and plants could be generated in a medium absolutely free from atmospheric air and in which therefore no germ of organic bodies could have been brought by air."

1.5 The Experiments of Pasteur and Tyndall

The impact of Pouchet's experiments on his contemporaries was so great that the French Academy of Sciences offered the Alhumpert Prize in 1860 for exact and convincing experiments that would end this controversy once and for all. The prize eventually went to one whom many consider the greatest biological scientists of the nineteenth century, Louis Pasteur (1822–1895). Pasteur by this time was already famous for his experiments on the crystals of tartaric acid, diseases of silkworms, "diseases of wine" (sufficient reason in itself for French citizens to consider him a savior), and studies on fermentation. Once more, the persuasive genius of Pasteur was to exert itself. Pasteur first demonstrated that air could contain numerous microorganisms.

> My first problem was to develop a method which would permit me to collect in all seasons the solid particles that float in the air and exam-

> ine them under the microscope. It was at first necessary to eliminate if possible the objections which the proponents of spontaneous generation have raised to the age-old hypothesis of aerial dissemination of germs.
>
> The procedure which I followed for collecting the suspended dust in air and examining it under the microscope is very simple. A volume of the air to be examined is filtered through guncotton which is soluble in a mixtire of alcohol and ether. The fibers of the guncotton stop the solid particles. The cotton is then treated with the solvent until it is completely dissolved. All of the particles fall to the bottom of the liquid. After they have been washed several times, they are placed on the microscope stage where they are easily examined.

From his microscopic observation, Pasteur concluded that there are large numbers of organized bodies suspended in the atmosphere (Figure 1.4). Furthermore, some of these organized bodies are indistinguishable by shape, size, and structure from microorganisms found in contaminated broths. Later he showed that these organized bodies that collected on the cotton fibers not only looked like microorganisms, but when placed in a sterile broth were capable of growth!

Pasteur's second series of experiments provided further circumstantial evidence that it was the microbes on floating dust particles and not the so-called vital forces that were responsible for sterilized broth's becoming contaminated. In these experiments, Pasteur carried sterile-sealed flasks to a wide variety of locations in France. At the various sites, he would break the seal, allowing air to enter the flask. The flask was immediately resealed and brought back to Paris for incubation. The conclusion from these numerous experiments was that where considerable dust existed, all the flasks would become turbid. For example, if he opened sterile flasks in the city, even for a brief period, they all became turbid, whereas in the mountainous regions, especially at high altitudes, a large proportion of the flasks remained sterile.

His third and most conclusive experiment utilized the now famous swan-neck flask (Figure 1.5). As a result of the experiments described, Pasteur hypothesized that the source of contamination was

Figure 1.4
Louis Pasteur (1822–1895). (Courtesy of Institut Pasteur, Paris.)

Figure 1.5
Pasteur's swan-neck flask. (A) Unbroken neck; nutrient medium in flask uncontaminated. (B) Flask with broken neck; contents of flask contaminated.

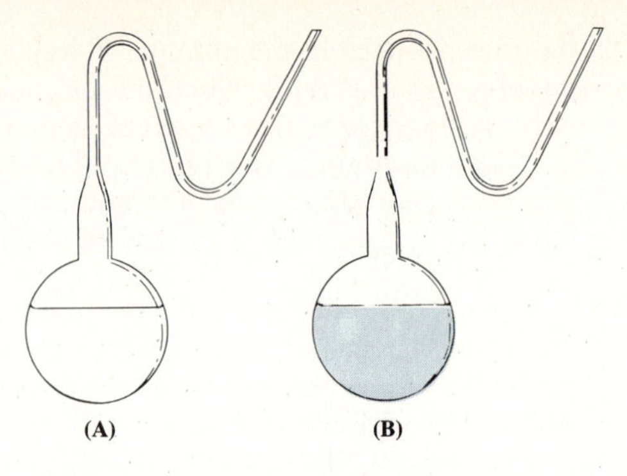

dust. If true, then it should be possible to keep a broth sterile even in the presence of air as long as the dust is kept out. In order to test this hypothesis, Pasteur constructed several bent-neck flasks such as the swan-neck flasks shown in Figure 1.5. After placing broth into the flask, he boiled the liquid for a few minutes, driving the air from the orifice of the flask. As the flask cooled, fresh air entered the flask. Despite the fact that the broth was in contact with the gases of the air, the fluid in the swan-neck flask always remained sterile. Pasteur reasoned correctly that the dust particles that entered the flask were absorbed onto the walls of the neck and never penetrated into the liquid. As an experimental control, Pasteur demonstrated that nothing was wrong with the broth. If he broke the neck off the flask or tipped liquid into the neck (in both cases dust would enter the broth), the fluid soon became turbid with microorganisms.

With these simple, ingenious experiments, Pasteur not only overcame the criticism that air was necessary for spontaneous generation but he was also able to explain satisfactorily many of the sources (dust) of the contradictory findings of other investigators. Although Pasteur's conclusions gained wide support in both the scientific and the lay communities, they did not convince all the proponents of spontaneous generation.

Pouchet and his followers continued to publish reports of spontaneous generation. They claimed their techniques were as rigorous as those of Pasteur. Where Pasteur failed to obtain spontaneous generation they succeeded in every case. For example, they carefully opened 100 flasks at the edge of the Maladetta Glacier in the Pyrenees Mountains at an elevation of 10,850 feet. In this region which Pasteur had found to be almost dust free, all 100 of Pouchet's flasks became turbid after a brief exposure to the air. Even when Pouchet used swan-neck flasks, there was growth. To Pasteur, this disagreement no longer revolved around the interpretation of experiments; rather, either Pouchet was lying or his techniques were faulty.

Pasteur had complete faith in his own procedures and results and had no respect for those of his opponents. Thus he challenged Pouchet to a contest in which both of them would repeat their experiments in front of their esteemed colleagues of the Academy of Science. Pouchet accepted the challenge with the added statement, "If a single one of our flasks remains unaltered, we shall loyally acknowledge our defeat." A date was set, and the place was to be the laboratory in the Museum of Natural History at the Jardin des Plantes, Paris. Pasteur arrived early with the necessary apparatus for demonstrating his techniques. Newspaper photographers and reporters were also on hand for this event of great public interest. However, Pouchet did not show up, and Pasteur won by default. It is difficult to ascertain whether Pouchet was intimidated by Pasteur's confidence or, as he later stated, he refused to take part in the "circus" atmosphere that Pasteur had created, and that their scientific findings should instead be reported in the reputable scientific journals. At any rate, in Pouchet's absence, Pasteur repeated his experiments in front of the referees with the same results he had previously obtained. As far as the scientific community was concerned, the matter was settled. The law *Omne vivium ex vivo* (All life from life) also applied to microorganisms.

In retrospect, however, the most ironic aspect of this extraordinary contest was not that Pouchet failed to show up, but rather that is he had appeared, *he would have won!* Pouchet's experiments are reproducible. Pouchet performed his experiments in the following manner: He filled swan-neck flasks with a broth made from hay, boiled them for one hour, and then allowed the flasks to cool. He obtained growth in every flask. Pasteur's experiments differed in only two respects. Pasteur used a mixture of sugar and yeast extract for broth and boiled it for just a few minutes. Pasteur never obtained growth in his swan-neck flasks. The reason for their contradictory results was not understood until 1877, 17 years later.

Mainly because of the careful work of the English physicist John Tyndall (1820–1893), Pouchet's experiments could be explained without invoking spontaneous generation. Tyndall found that foods vary considerably in the length of boiling time required to sterilize them. For example, the yeast extract and sugar broth of Pasteur could be sterilized with just a few minutes of boiling, whereas the hay medium of Pouchet required heating for several hours to accomplish sterilization. Tyndall postulated that certain microorganisms can exist in heat-resistant forms, which are now referred to as spores. Furthermore, studies by Tyndall and the French bacteriologist Ferdinand Cohen revealed that hay media contain a large number of such spores. Thus the contradictory results of Pasteur and Pouchet were due to differences in the broths they used.

Tyndall went on to demonstrate that nutrient medium containing spores can be sterilized easily by boiling for one-half hour on three successive days. This procedure of discontinuous heating, now called **Tyndallization,** works as follows: The first heating kills all the cells that are not spores and induces the spores to germinate (in the process of germination, the spores lose their heat resistance as they begin to grow); on the second day, the spores have germinated and are thus susceptible to the heating. The third day heating "catches" any late germinating spores. Thus, with the publication of Tyndall's work, *all* the evidence that supported the theory of spontaneous generation was destroyed. Since that time, there has been no serious attempt to revive this theory.

It should be pointed out, however, that by its very nature, the theory of spontaneous generation cannot be disproved. One can always argue that the conditions necessary for spontaneous generation have not yet been discovered. Pasteur was well aware of the difficulty of a negative proof, and in his concluding remarks on the controversy, he merely showed that spontaneous generation had never been demonstrated.

> There is no known circumstance in which it can be affirmed that microscopic beings came into the world without germs, without parents similar to themselves. Those who affirm it have been duped by illusions, by ill-conducted experiments, by errors that they either did not perceive, or did not know how to avoid.

1.6 Techniques for Sterilization

Although Leeuwenhoek, working alone, was able to discover and describe numerous species of microbes, it took the combined skill and imagination of many minds to convert the science of microbiology from one solely of observation to one of controlled experimentation. What was necessary for this development was the discovery of new techniques for handling the microorganisms. In later chapters it will be shown that the ability to perform a wide variety of controlled experiments with microorganisms is a powerful tool for probing the secrets of biology. Since many of the microbiological techniques came about as practical consequences of the controversy over spontaneous generation, we shall digress in Sections 1.6 to 1.8 from the main theme of the chapter, the origin of life, in order to describe some of the more important techniques. The most fundamental of these were techniques for sterilization.

The central issue in the controversy over the spontaneous generation of microorganisms was sterilization and subsequent contamination. Sterilization can be defined as the *complete* destruction or removal of all living organisms. Furthermore, if a substance is to remain sterile, contamination from the outside must be avoided. Primarily from the experiences gained in the last half of the nineteenth century, the following reliable techniques were developed for sterilization:

1. Steam heat: 120°C for 20 minutes;
2. Tyndallization: 100°C for 30 minutes on each of three successive days;
3. Filtration: a process of excluding microorganisms by use of filtering agents;
4. Chemical treatment;
5. Radiation.

Figure 1.6
A research autoclave. Also shown is a flask containing nutrient medium which has a cotton plug to prevent dust from entering.

In the first technique, steam at 120°C is obtained by using a modified pressure cooker, called an **autoclave** (Figure 1.6). This technique is the most common, not only in research laboratories but also in hospitals and industry. Tyndallization has already been discussed in connection with the experiments of Tyndall. Filtration is widely used for sterilizing solutions that contain heat-sensitive substances such as certain vitamins. In this process, the solution is passed through a filter that has the appropriate pore size to prevent the passage of microorganisms. The filtered solution can then be transferred to a previously sterilized (autoclaved) flask. With this method, direct heating of the solution can be avoided. The technique of filtration was suggested by Pasteur in 1872 when he noted that water from deep wells, which had undergone slow filtration through sandy soil, was essentially free of microorganisms. In general, viruses (see Chapter 4) were not removed by filtration. The last two techniques have specialized applications in sterilization of confined areas (e.g., germ-free rooms) and solid objects that are sensitive to heat (e.g., plastic materials). The most effective chemical sterilization technique is exposure to ethylene oxide gas. It had been known for some time that solar radiation is lethal to many microorganisms. Later, it was found that the shorter wavelength ultraviolet radiation was primarily responsible for the killing effect of sunlight. Today, ultraviolet radiation is produced in germicidal lamps for the purpose of sterilizing air and the surface of objects, such as surgical instruments.

Regardless of which sterilization technique is used, it is necessary to avoid recontamination if sterility is to be maintained. As Pasteur had so clearly demonstrated, dust particles in the air or on instruments contain microbes, and if these microbes enter a sterile solution they may initiate growth. Thus, any material that comes in contact with the solution must itself be sterilized. The chief source of contamination is the air. Although swan-neck flasks prevent air contamination, they are troublesome to prepare and inconvenient for removing or adding samples. Today the most common method of preventing microbes from entering flasks or tubes that must be kept open to the air is simply to plug the opening with cotton. The cotton plug filters out the dust, and thus the microbes, but allows the oxygen in the air to enter. It this way, a flask containing broth or any other solution when plugged with cotton and autoclaved for 20 minutes will remain sterile indefinitely.

As soon as these techniques for sterilization were introduced into research laboratories, they were immediately adopted for many practical operations in public health and industry. For example, the use of sterilized surgical instruments, rubber gloves, bandages, and so on, revolutionized surgery and made possible many of the dramatic advances in medicine of the past century; the entire canning industry developed as a direct result of the sterilization technique.

Many foods, such as milk, cheese, beer, and wine, cannot be sterilized without destroying their tastes. In these instances, a heat treatment called **pasteurization** is commonly used. Pasteurization kills many of the microorganisms that are responsible for spoilage or for causing diseases without destroying the taste of the food. For example, the pasteurization of milk is carried out at 72°C for 15 seconds; this treatment kills the disease-producing germs in the milk but does not destroy its flavor. It should be emphasized that pasteurization is *not* a method of sterilization, because many of the nondetrimental microorganisms are not killed.

1.7 Pure Culture Technique

The liquid suspensions that Leeuwenhoek and other pioneers in bacteriology examined under their microscopes contained a wide assortment of microbes of varying sizes and shapes. Was this because a particular organism could exist in various forms, or was it the result of a mixture of different organisms, each having a relatively fixed form? To answer this and many other microbiological questions that were being asked toward the end of the nineteenth century, it became necessary to obtain pure cultures. A **pure culture** is one that contains only a single type of microorganism. The problem then was to devise a method in which different types of microbes could be separated.

The first pure culture was obtained by the English surgeon, Lord Lister, in 1878, using a dilution technique. The principle of his method was to take a turbid culture, containing a mixed population of microbes, and to dilute it with a sterile broth until a point of dilution was reached where only *one* microbe occurred in a flask. That microbe then multiplied and gave rise to a population that was derived from the single parent. Lister reasoned in the following manner. Assume that one milliliter (1 ml) of pond water contains 20,000 microbes of various sorts. If 1 ml of this pond water is mixed with 99 ml of sterile broth, a 1:100 dilution is obtained, which now contains approximately 200 microbes per 1 ml. When this procedure is repeated, a 1:100 dilution of the 1:100 dilution or a 1:10,000 dilution is achieved, which now contains on the average only two microbes per 1 ml. One milliliter of this 1:10,000 dilution is then divided into ten equal parts, and each part is added to sterile broth. Only two of the ten broths can become turbid because there are just two microbes in the entire 1 ml. The resulting two turbid broths would likely be pure cultures since the chance that both microbes would fall in the same 0.1-ml aliquot is only one in ten.

From this example we can appreciate some of the practical difficulties of the dilution technique. Not only does it require considerable trial and error until the correct dilution is reached, but more significantly, only the most abundant type of microorganism can be purified. Because of these reasons, the dilution technique is rarely used now as a means of obtaining pure cultures.

The method currently used comes directly from the work of the brilliant German bacteriologist Robert Koch (1843–1910) (Figure 1.7). One day, Koch walked into his laboratory and noticed several colored spots on the flat surface of a slice of boiled potato that he had inadvertently left on his work bench over the weekend. Koch, in his meticulous manner, removed a bit of one of the colored spots with a sterile needle, mixed it with sterile water, and then examined it un-

Figure 1.7
Robert Koch (1843–1910). (Courtesy of Tel Aviv University, Medical History Library.)

der the microscope. It was teeming with microbes. After noting that, for some unexplained reason, all the microbes from the first spot were rods of rather uniform size, he proceeded to remove and examine material from another spot. Again, he noted, it contained numerous microbes, but this time only spherical bacteria. Each spot he examined consisted of a different type of microorganism, but within any one spot all organisms were invariably the same. He reasoned thus: Pasteur had shown that the air contained microbes; if a single microbe from the air landed on the potato, it could use the potato for food and start multiplying; on the solid surface, the microbe could not swim far and would thus give rise to a tightly packed **colony;** since the entire colony arose from a single parent, it must be a pure culture.

1.8 Streak Technique

This chance observation provided the basis for the development by Koch and his co-workers of the streak technique of obtaining pure cultures. First, they prepared a series of potato slices, placed them each in a covered glass jar, and then sterilized them. These remained sterile as long as they were covered. A sterile needle was then dipped into a turbid broth culture that contained a mixture of different microorganisms. The jar was opened, and the needle containing a small droplet of the broth was streaked lightly over the surface of the potato as shown in Figure 1.8.

The organisms on the needle are thus deposited onto the solid surface. On the initial section of the streak, many microbes are deposited close together so that contiguous areas of growth result. As the streaking progresses, fewer and fewer organisms are deposited, until at the end of the streaking process only occasional and well-separated microbes are deposited. The process is analogous to dipping a paintbrush in a bucket of paint and then drawing the brush across a wall; at first a large quantity of paint is deposited; toward the end of the brush stroke, only a few drops of paint are deposited. On incubation, the microbes multiply and give rise to colonies. These mi-

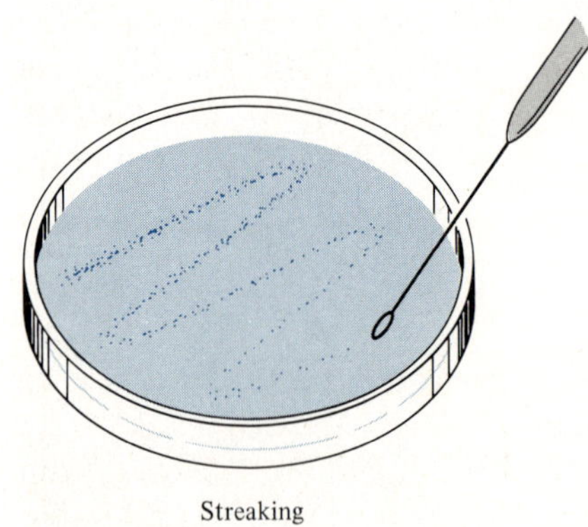

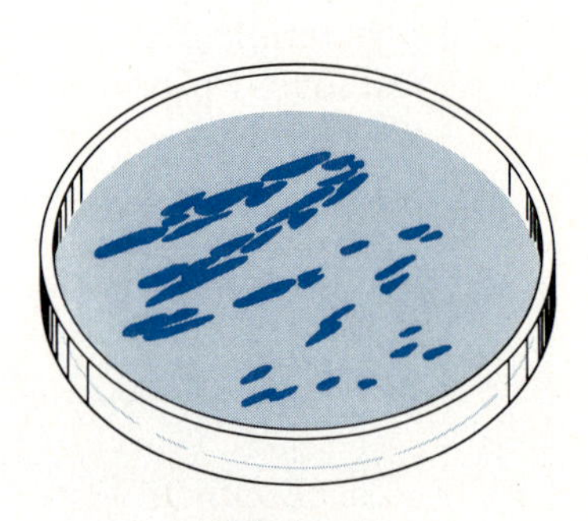

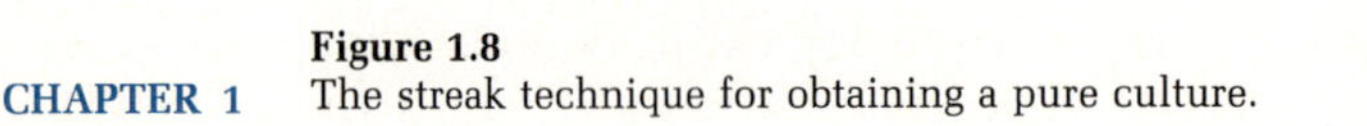

Figure 1.8
The streak technique for obtaining a pure culture.

crobes near the end of the streak give rise to isolated colonies that arise from a single cell and are thus pure cultures, that is, uncontaminated by other types. These can be easily transferred to sterile broth or other potato slices for further study.

Potato slices have several serious disadvantages when used for obtaining pure cultures of bacteria. The colonies tend to merge when the surface is moist; it is difficult to see colorless colonies because the potato surface is opaque; and most significant, potato is not a proper food for all bacteria. What was needed was a solid, transparent, sterile medium to which ingredients could be added as desired. Koch was able to meet these requirements by using gelatin as a solidifying agent. He simply added gelatin to the desired broth, either the yeast extract and sugar mixture of Pasteur or the hay infusion of Pouchet, autoclaved it, and then transferred the liquid into sterile dishes. When it cooled, the "Jello" of bacteriologists was obtained. The gelatin medium of Koch was a major improvement over potatoes, but it also had serious defects. Gelatin is a protein, and many microbes will digest the gelatin and thus liquify the medium. In addition, gelatin melts above 28°C and many bacteria grow best at temperatures above 30°C. Both objections to gelatin were overcome by introducing **agar** as the solidifying agent. Agar, an inert substance that can be extracted from red algae (seaweed), is widely used as a solidifying agent for cooking in the Orient. Once agar solidifies, it will not melt until the temperature reaches nearly 100°C, thus making it an ideal agent for bacteriological work. Nutrients can be mixed into agar to produce various "media." By utilizing the simple technique of streaking onto a solidified agar medium, the isolation of bacteria in pure culture is a routine procedure in modern laboratories.

The digression from the main theme of this chapter, the origin of life, was undertaken for two reasons. First, the sterilization and pure culture techniques are basic to many of the discoveries that are discussed in later chapters. Without these procedures, it would have been impossible to uncover causative agents of diseases or discover methods for combatting them, such as immunization and antibiotic treatments. The use of microbes as tools for unraveling the mysteries of genetics and biochemistry depends absolutely on the development of these bacteriological procedures. Second, the development of these techniques as by-products of pure research should be emphasized. Without technology, there is no science; without science there is no technology. They are mutually dependent. Through understanding we not only conquer to a large extent our fear, but also feed our stomachs and cure our diseases.

1.9 The Theory that Life Is Continuous

The abandonment of the theory of spontaneous generation by the end of the nineteenth century left biologists in a rather uncomfortable position. Apparently, the only alternative to spontaneous generation was the belief that life is eternal. If life does not originate from non-

TABLE 1.1 Chronology of Physical and Biological Evolution

Event	Approximate Time (billions of years ago)
Origin of the universe	12
Origin of our solar system	5.0
Formation of the earth with its present size and composition	4.5
Formation of the earth's crust	4.0
Age of oldest minerals	3.6
Earliest manifestation of life (microfossils)	3.2
Age of blue-green algae	1–3
Formation of oxygen-containing atmosphere	1.0
First hard-shell animals	0.6
Age of the dinosaurs	0.151
Earliest appearance of man	0.001

living material, then it must have always been present. If this is so, the question of its origin has no meaning.

The theory that life is continuous clearly indicates that although living organisms can change their forms, they can never be created from lifeless substances. This theory eventually came into serious question because it conflicted with certain astronomical and geochemical data. Table 1.1 summarizes the approximate time of some important events to consider in the origin and evolution of life. Most significant is that the earth's crust and the oldest minerals were formed considerably later than the earth. Both of these processes require such extreme temperatures that life in that period could not exist. At the temperatures required to melt rock in order to form the earth's crust, not only would living organisms as we now know them be unthinkable, but *any* form of life based on carbon compounds would be impossible. Furthermore, the earliest manifestation of life, as evidenced by microfossils in shales of the upper Onverwacht in the eastern Transvaal, South Africa, has been dated at 1.2 billion years after the earth was formed. These and other facts strongly suggest that there was a time when the earth was sterile, thus indicating that the theory that life on earth has been continuous is incorrect.

1.10 Panspermia Theory

The evidence that contradicts the theory that life on our planet was continuous does not exclude, however, the more general theory of continuous life in the universe. Thus in 1908, the Swedish physical chemist Arrenhius revived the theory of panspermia. This theory hold that the earth is constantly being bombarded by spores from interstellar space. Once the earth had cooled sufficiently, these invading spores from other celestial bodies found the conditions favorable for growth and gave rise to living organisms on this planet.

Arrenhius made detailed calculations demonstrating how small particles could be carried upward by powerful air currents and then shot into space by electric discharges and light pressure. Once in space, the spores could move with great speed. He estimated that it could take a spore "only" 9000 years to reach us from our closest neighboring solar system, Alpha Centauri. According to Arrenhius, the cold temperature and lack of oxygen and moisture would allow the spores to survive their long trip.

Many biologists today believe that it is unlikely that spores could survive the heavy radiation of space or the frictional heat once they came in contact with the earth's atmosphere. One noted biochemist has recently suggested that, to survive such a trip, a spore would need to be fitted with a lead shield for the flight and a ceramic nose-cone for the landing! The panspermia theory has been further criticized because it only dodges the question. If the earth was infected by spores, we would still have to explain how these spores came into existence on their native planet. For these reasons, the theory of panspermia, although not rigorously excluded, is highly unlikely and has few advocates.

1.11 The Theory of the Origin of Life by Chemical Evolution

In 1924, Alexander I. Oparin, a Russian biochemist, published a booklet that outlined his views on the origin of living matter. Later, a more detailed account of Oparin's theory, documented by various investigations, appeared in his book, *The Origin of Life* (1936). These two scholarly works revolutionized thinking on the subject and immediately provided a new experimental approach to the problem. Academician Oparin's theory of the **origin of life by chemical evolution** is described in the next few paragraphs, followed by a discussion of some recent experiments that test his theory.*

Oparin begins by rejecting both the theory of spontaneous generation and the theory that life is continuous. Although these theories appear to be contradictory, he clearly points out that both are based on the same dualistic outlook on nature. Both theories suggest that there is something special about life, that living organisms obey different laws from inanimate objects. The continuous life theory denotes an absolute barrier between living and non-living; only living organisms can give rise to more life. In the theory of spontaneous generation, it was necessary to invoke a "vital force" or "vegetative force" to convert non-living to living matter. Such a "vital force" must, by its very definition, be a force distinct from the usual chemical and physical forces. Oparin's theory is fundamentally different from previous theories, because it **requires no special laws for the origin of life.**

---

*The English chemist J. B. S. Haldane proposed a theory very similar to that of Oparin, which he arrived at independently, but a few years later.

The phrase "chemical evolution" is used to emphasize the gradual rather than the spontaneous appearance of living things. Oparin postulates a long series of chemical changes as a prerequisite to the formation of life. In this respect, his theory is much like Darwin's story of the *Origin of Species*. Both suggest a gradual transition from simple to more complex structures. Oparin's description of the chem-

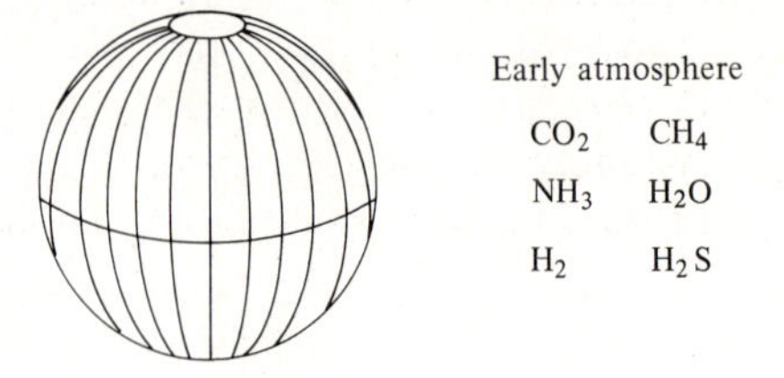

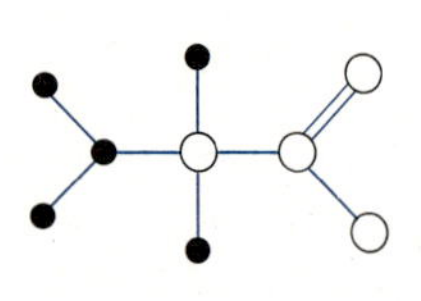

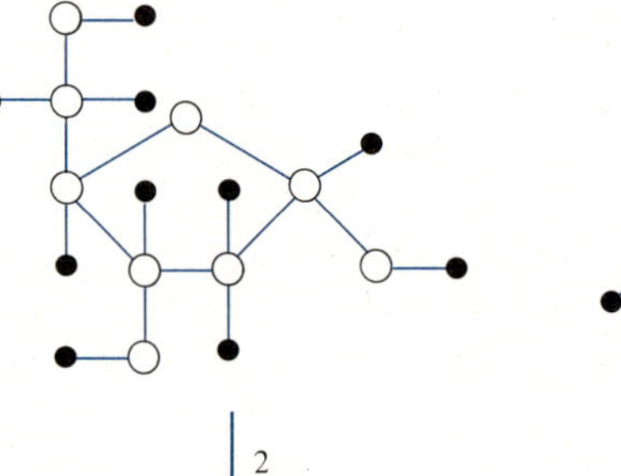

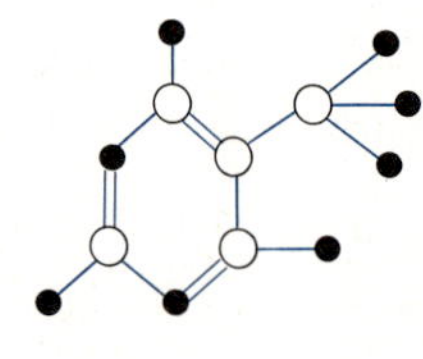

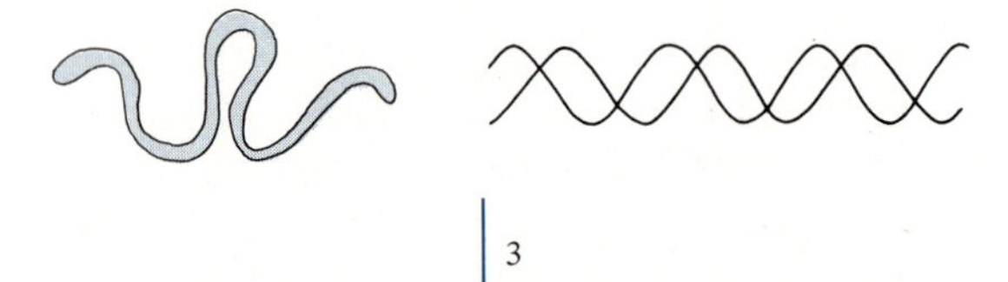

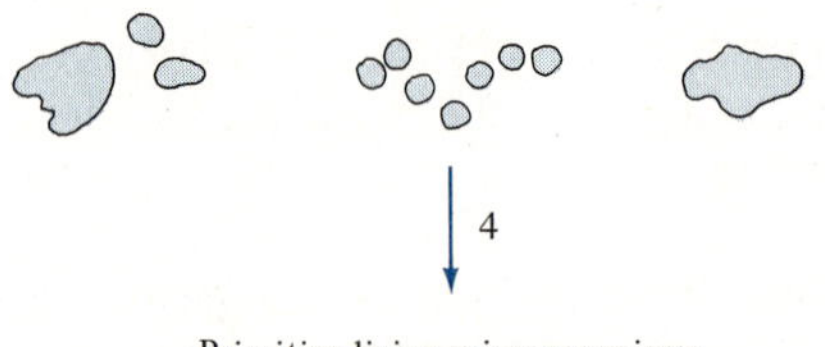

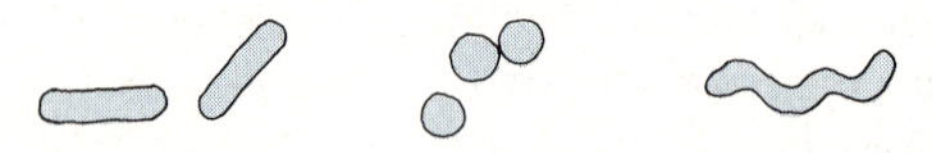

Figure 1.9
Evolution of matter and origin of life according to the Oparin-Haldane theory of the Origin of Life.

ical evolution leading to the origination of life can be considered in four phases (Figure 1.9).

In order to discuss these four transitions, it is necessary to introduce certain basic chemical structures with each transition. This introduction to the important chemicals of life will not only allow the reader to obtain a better understanding of Oparin's theory but will also serve as the necessary chemical background for subsequent chapters. Such a mode of introducing chemistry is a "natural" one in the sense that it attempts to follow the historical development of the chemical substances. If Oparin's theory is correct, then the chemicals will be introduced in the order of their appearance on earth.

Phase One: Production of Small Organic Molecules

The first step in Oparin's theory is the production of simple organic compounds from the inorganic gases present in the primitive atmosphere. Although we now know of 106 different elements, we need to consider here only six elements for our discussion of the origin and basic processes of living organisms: carbon (C), hydrogen (H), oxygen (O), nitrogen (N), sulfur (S), and phosphorus (P). These six elements comprise more than 99% of the weight of living matter. The smallest unit in which an element can exist is called an **atom.** However, in nature, matter usually exists in the form of **molecules,** combinations of atoms held together by chemical bonds. Recently Harold Urey, a Nobel Prize–winning chemist, calculated that two billion years ago, when life may have arisen, the atoms of these six elements existed primarily in the following molecules:

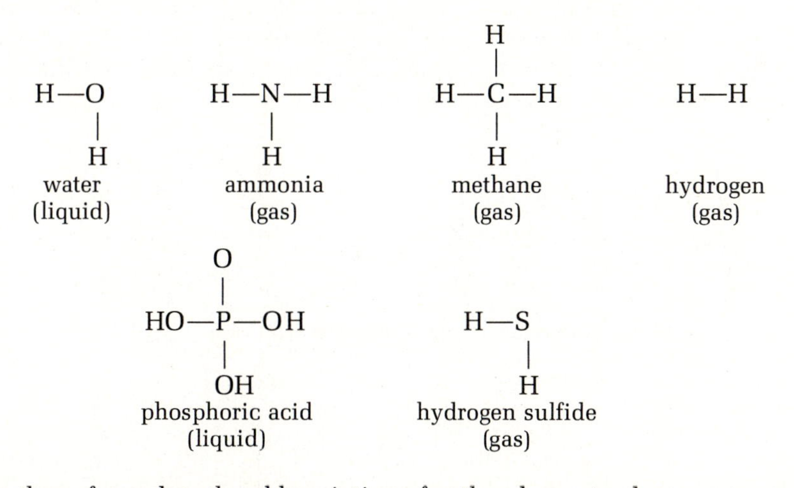

In these formulas, the abbreviations for the elements shown are used, and the lines connecting the atoms represent attractive forces between atoms called **chemical bonds.** The number of bonds that a particular element forms is characteristic of that element; for example, carbon generally forms four bonds, nitrogen three bonds, oxygen two bonds, and hydrogen one bond. The chemical bonds that hold the atoms together are breakable and, if subjected to sufficient energy, can be ruptured and the atoms separated. Once some of the molecules are disrupted, they can recombine to form more complex molecules. For example, if one of the hydrogens of methane is removed, it can combine with another such molecule to form the more complex organic compound, ethane gas.

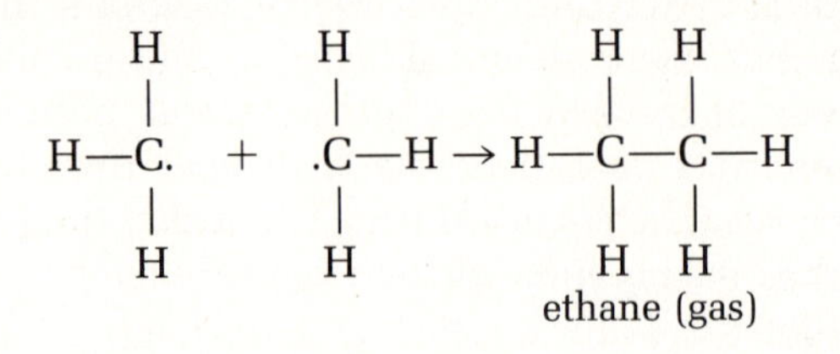

Molecules that contain carbon are called **organic molecules** because they are characteristic of living organisms.

Urey did some further calculations which suggested that if the six molecules were present on earth a few billion years ago, they should have formed small organic compounds as Oparin had previously speculated. Urey, at that time a professor at the University of Chicago, gave a lecture describing how various forms of energy, such as electrical storms and solar radiation, could break down these molecules and, as these bonds are rejoined, give rise to a series of more complex organic molecules. Present in his audience was Stanley Miller, a young graduate student from California. Miller was inspired by Urey's lecture and decided to put Urey's theory to a test.

Over the next three years, Miller planned, built, and tested the glass apparatus shown in Figure 1.10. The apparatus was designed to simulate the conditions on the primitive earth. The water was the ocean, the circulating gases were the atmosphere, and the spark was an electic storm. After first sterilizing the entire apparatus, Miller circulated and sparked the gases for several days. The electrodes were

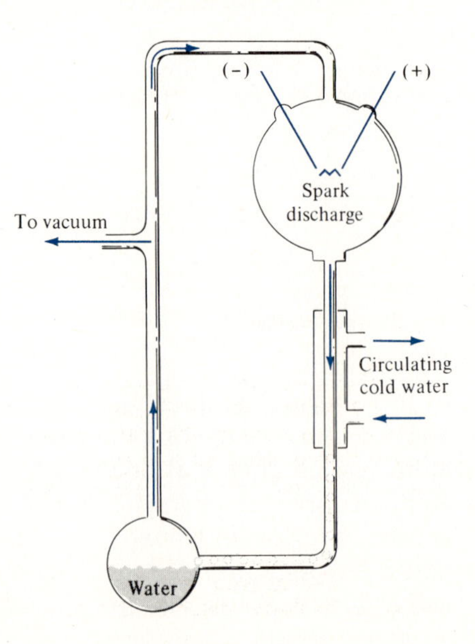

Figure 1.10
Apparatus designed by Stanley Miller to simulate conditions on the primitive earth. Methane, ammonia, hydrogen, and water vapor circulate as indicated by the arrows. After passing through the spark discharge, the gases are cooled by circulating cold water. The condensed steam containing dissolved products of the chemical reaction is collected in the flask.

then disconnected, the gases carefully† removed through the stopcock, and the liquid examined for the presence of organic molecules. The results were exciting! A wide variety of molecules was found in rather substantial quantities. Miller's experiment, based on Urey's theory, was immediately reproduced and extended in several laboratories. The general conclusion from all these experiments was that a large diversity and quantity of organic molecules were produced when a mixture of these gases (or slightly modified mixtures) was exposed to various forms of energy. In addition to electric sparking, ultraviolet rays, visible light, heat, x-rays, and ionizing radiation were successful as energy sources.

Thus the first phase in Oparin's theory now had a solid experimental basis. Small organic molecules could and *should* be formed from the mixture of primitive gases. Scientists have estimated that prior to life, the ocean was a "hot thick soup," containing from 1 to 10% organic molecules. These organic molecules could serve two functions: they could evolve further to more complex structures (phases two to four); and they could provide the first living organisms with an excellent and easily available food supply.

Of the wide variety of organic molecules identified by Miller and others in their "prebiological" syntheses, three important classes will be considered: **amino acids, sugars,** and **nitrogenous bases.** The chemical structures of some examples of these substances are shown in Figures 1.11 to 1.13.

Phase Two: Production of Large Organic Molecules

Oparin postulated that once these small organic molecules were concentrated in the ocean and given millions upon millions of years, some of them would unite to produce larger molecules or polymers. For example, **proteins** would be formed from the union of several

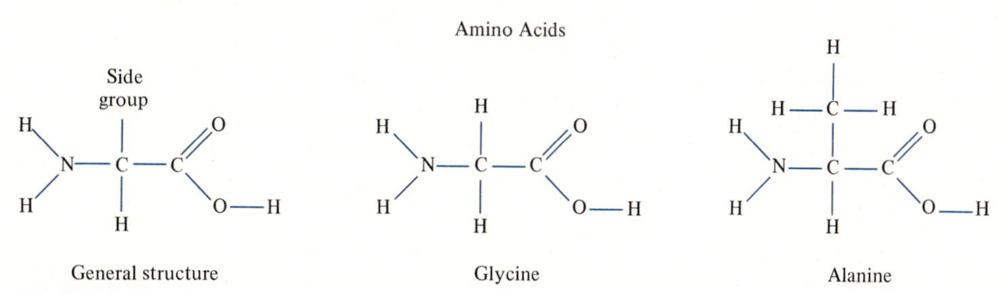

Figure 1.11
The amino acids are composed of carbon, hydrogen, oxygen, and nitrogen atoms. Of the approximately 20 different kinds of amino acids found in proteins, two also contain S atoms. All the amino acids have the same general chemical structure, differing only in their side groups. For example, the simplest amino acids, glycine and alanine, have as their side groups an H and a CH_3 group, respectively. Amino acids are the building blocks from which proteins are constructed.

---

†Any student who attempts to repeat this experiment should be warned of the potentially explosive nature of this gaseous mixture.

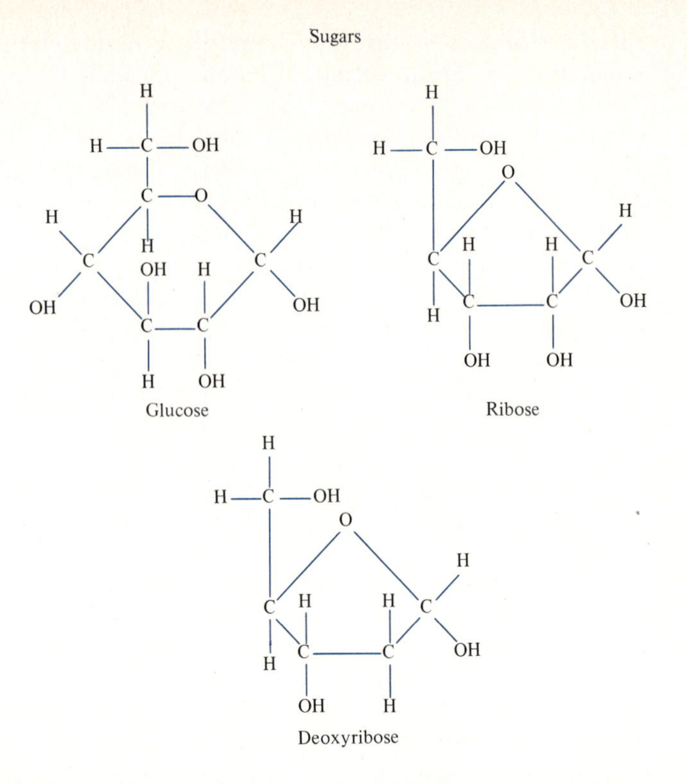

Figure 1.12
The sugars are composed of carbon, hydrogen, and oxygen atoms in a ratio close to 1:2:1. Of the large number of different sugars found in nature, three important examples are shown. Glucose, or dextrose, which contains six carbon atoms, is the most abundant sugar. It occurs in large amounts in many sweet fruits such as grapes and is also the major building block for polysaccharides. The five-carbon sugars ribose and deoxyribose are essential components of the nucleic acids. Deoxyribose differs from ribose by having one less oxygen in each sugar molecule.

amino acids, **polysaccharides** from sugars, and **nucleic acids** from a mixture of sugars, nitrogenous bases, and phosphoric acid. The basic structures of these very important polymeric substances are shown in Figure 1.14.

Proteins are large molecules that contain from 50 to 1000 amino acids joined together in a long chain. The word protein comes from the Greek *proteios,* meaning "of the first rank." These macromolecules comprise not only the major structural material of all cells but also the enzymes that promote and direct all cellular processes. These varied tasks are performed by proteins that have the same basic structure but that **differ in the number and sequence of the amino acids in the chain.** The sequence is crucial. A change in the order of the amino acids may alter the ability of the proteins to perform their varied tasks.

Since there are 20 different kinds of amino acids and a typical protein chain is about 200 amino acids long, the number of possible arrangements of these amino acids is astronomical, 20^{200}. It should be emphasized, however, that the actual number of different proteins found in even a simple organism such as a bacterium is about 1000.

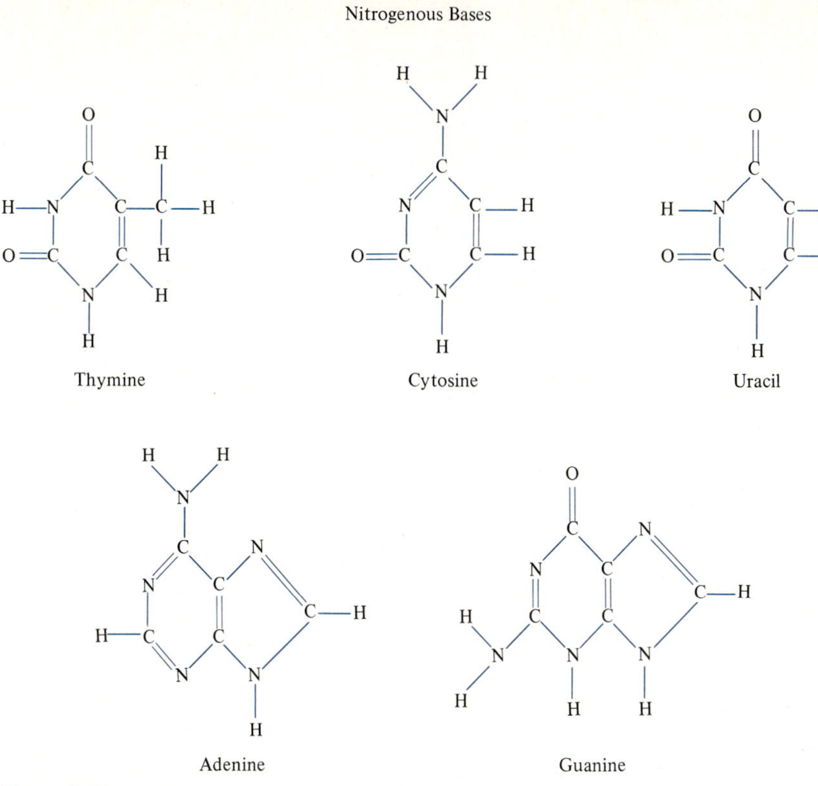

Figure 1.13
The nitrogenous bases are molecular rings, formed by several C and N atoms. The cytosine, uracil, and thymine rings are composed of 4 C and 2 N atoms. The larger and more complicated guanine and adenine are composed of two linked rings. The five nitrogenous bases are found in the nucleic acids, where they play the central role in the hereditary mechanism of all living organisms.

How the cell can produce the 1000 or so proteins that are required for its growth and not produce an enormous number of possible wrong proteins is a major theme of Chapter 7.

The chemical structure of the polysaccharides is simpler than that of either the protein or nucleic acids. The three most important polysaccharides, **starch, glycogen,** and **cellulose,** are composed of a single kind of sugar, glucose. Plants store starch as a reserve fuel supply, whereas animals store glycogen. Cellulose is the main structural material of plants. Although it also consists of glucose units, it is useless to humans as a food. Our bodies cannot break down cellulose to the usable glucose. If an economically feasible method is devised for converting cellulose to glucose, it will be possible to change such materials as paper and cotton to sugar that could be used as food.

Nucleic acids are composed of sugars, nitrogenous bases, and phosphoric acid in equal quantities. These gigantic polymers have a main chain, or backbone, consisting of alternating sugar and phosphoric acid components; connected to each sugar residue is one of the five nitrogenous bases.

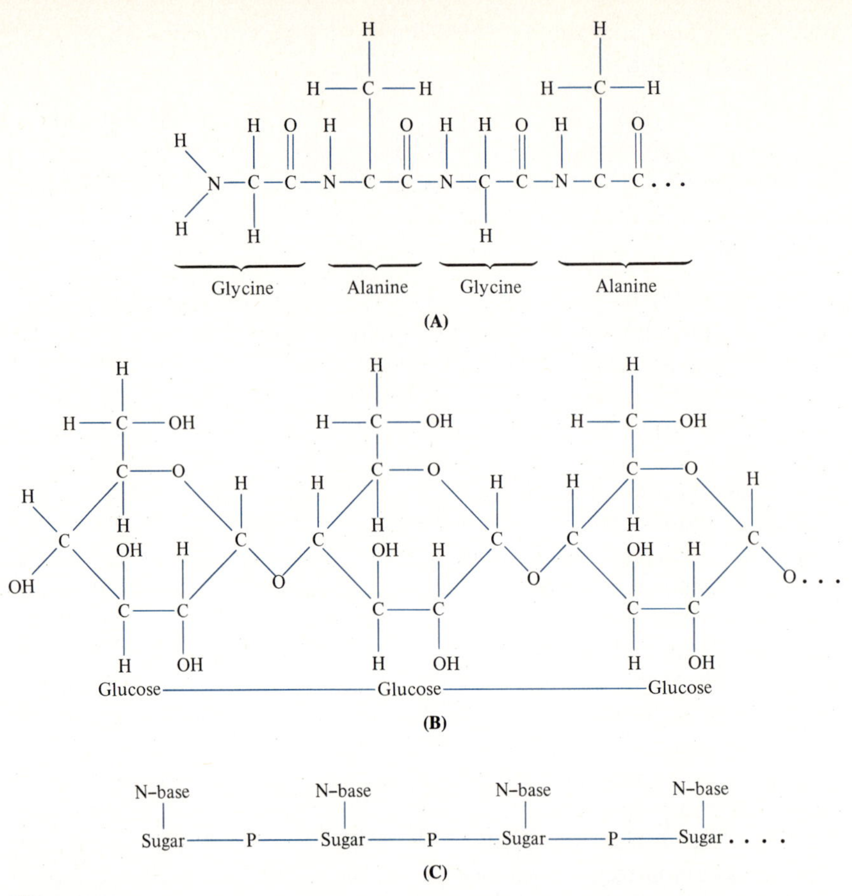

Figure 1.14
The basic structure of (A) protein, (B) polysaccharide, and (C) nucleic acid.

Two kinds of nucleic acids are ribonucleic acid (RNA) when the sugar component is ribose and deoxyribonucleic acid (DNA) when the sugar is deoxyribose. Both types of nucleic acids contain the nitrogenous bases, adenine, guanine, and cytosine, but the base uracil is found only in RNA, whereas the base thymine is found exclusively in DNA. Within each type of nucleic acid, the possibilities for variation are enormous. For example, although DNA molecules from different organisms have the same backbone of deoxyribose and phosphoric acid, they differ in the order in which the four bases are connected to this backbone. The central role of the nucleic acids in the expression and transmission of genetic information is discussed in subsequent chapters.

A few experiments lend some support to phase two of Oparin's theory, that macromolecules can be formed from small molecules prior to the existence of life. These experiments take advantage of the fact that when the small molecules are joined together, water is produced. For example, when two glycine molecules are united, two hydrogen atoms and one oxygen atom (which usually combine to form water) are produced:

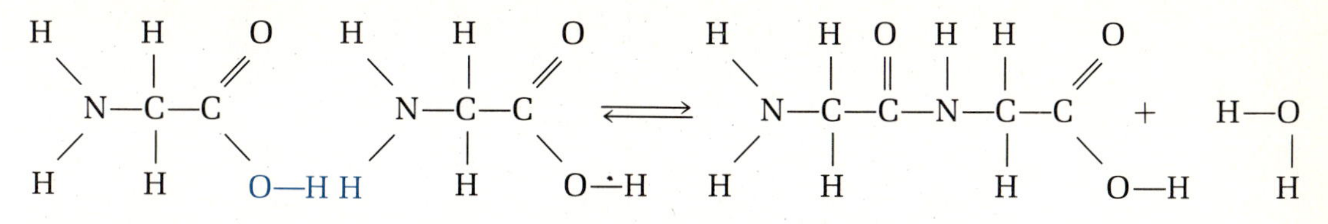

Water must also be produced in the formation of polysaccharides and nucleic acids. The arrows pointing in opposite directions indicate that the reaction can also proceed in reverse. If fact, when any of these macromolecules are exposed to water, they break down to simpler organic molecules much faster than they are produced. To overcome this difficulty, Sidney Fox of Florida State University heated a mixture of *dry* amino acids at 160°C for three hours. The water that was produced when amino acids were thereby united was immediately boiled off, and protein-like molecules accumulated. Although the experiment was criticized by certain scientists because the temperature was too severe, Fox pointed out that there are local "hot spots" on earth, such as hot springs and volcanoes. Furthermore, temporary dehydrating areas containing high concentrations of organic molecules could be produced in tidal pools as the ocean recedes and the water evaporates.

Using a variety of dehydrating conditions, scientists have succeeded recently in producing all the different types of polymers from their respective building blocks. Although this problem is not a simple one, it is reasonable to assume that somewhere on the primitive earth during the millions of years prior to life, conditions were such that the "organic soup" condensed and gave rise to polymeric substances.

Phase Three: Formation of Aggregates

Large organic molecules have a tendency to aggregate and form what Oparin called **coacervates.** Coacervates can be easily prepared in the laboratory by mixing dilute solutions of various polymers. For example, when mixing a solution of gelatin (0.67%) with a solution of gum arabic (0.67 percent) coacervates are formed that are stable below 42°C. Coacervates appear as microscopic droplets of varying size and composition, floating in liquid medium. The formation of coacervates is dynamic; while some molecules are entering into the coacervates others are escaping back into the solution. If molecules enter the aggregate faster than they depart, the coacervate grows in size. The increase in size makes the coacervate more susceptible to breakage by collisions or other mechanical disturbances. The fragments produced have the same composition as the original coacervate and will continue to grow and fragment (Figure 1.15).

One of the most significant features of these coacervates is that,

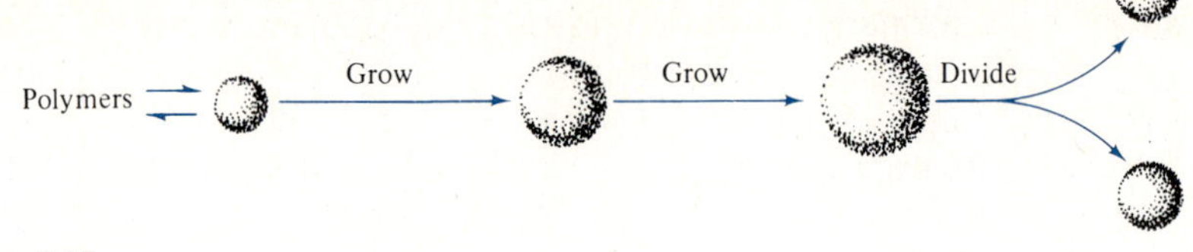

Figure 1.15
The formation and growth of coacervates.

like living organisms, they occupy a definite region in space. The process of coacervate formation causes a sharp boundary to be formed for the first time between the aggregated polymers and the rest of the solution. Prior to this event, the polymers were uniformly distributed in the solution. Oparin made the ingenious suggestion that in the formation and growth of coacervates, a prebiological "natural selection" could operate. The various coacervates formed would compete with one another for the polymers in solution. Since the different coacervates were constantly being formed and broken down, those coacervates that were the stablest and grew the fastest would eventually predominate in the ancient seas (survival of the fittest).

Phase Four: From Coacervates to Primitive Living Organisms

The coacervates have some properties that we associate with living organisms. They are able to take certain (but not all) organic matter from their environment and concentrate it within; thus they slowly become larger and eventually divide. They are, however, at the absolute mercy of the outside world in obtaining the polymers necessary for their growth. Although living organisms are also dependent on their environment for organic matter, they are able to alter the organic matter and thus provide their own building blocks. For example, when you eat a steak, the protein in the meat is broken down into amino acids which then are rebuilt into human protein. It does not matter whether the protein came from a cow, a pig, a fish, a peanut, or any other source; it always ends up as distinctly human protein. The sum of all the closely coordinated, delicately balanced chemical reactions that take place in living cells, entailing the breakdown of different foods, and the syntheses of the myriad of components necessary for growth, is called **metabolism.**

> It's a very odd thing—
> As odd as can be—
> That whatever Miss T. eats
> Turns into Miss T.
>
> *Walter de la Mare*

However, what happens to a coacervate when the supply of polymers in the ocean is depleted? Lacking metabolism, its growth would cease and the process of chemical evolution would come to a halt, short of the formation of living organisms. For the coacervate to cross the twilight zone between living and non-living, it must be able to transform some of the organic material in the ocean into building blocks for its own growth. In short, it needs to develop the ability of organized self-replication through metabolism.

This is the transition step that is least understood and for which no experimental evidence exists. It must be emphasized that there is a great gap between a coacervate which grows and divides by purely physical forces and the fantastically complex nature of a living cell. Nevertheless, Oparin argued that during the millions of years that the coacervates had to evolve, they developed through natural selection: (1) various patterns of metabolism and (2) the ability to transmit the information for these patterns from generation to generation. As these properties became established, a living organism slowly emerged.

Conclusions

The theory of the chemical origin of life has gained wide acceptance by modern biochemists and biologists. There is strong experimental evidence indicating that a variety of organic molecules could be formed on earth prior to the existence of living organisms. The transitions from these molecules to coacervates and to living organisms will require considerable more experimentation before they can be understood in detail. The most appealing feature of the theory is that it invokes no special laws. Life was a natural consequence of chemical evolution. According to Oparin, on a planet such as the earth, the formation of life was not only possible, but given enough time, it was inevitable

If the formation of life is such a natural phenomenon, then it is reasonable to inquire: Is life still emerging from non-living matter on earth? Is there life on other planets?

The formation of living organisms by chemical evolution is much less likely today than it was two billion years ago for at least two reasons: (1) the present atmosphere contains oxygen gas and (2) existing organisms would interrupt the process of chemical evolution by using the necessary organic molecules as sources of food.

As mentioned previously, there is strong evidence that the primitive atmosphere contained little or no oxygen. Oxygen gas was produced on earth as a direct result of photosynthetic organisms. Attempts to produce organic compounds from mixtures of gases containing oxygen have been unsuccessful

According to Oparin's theory, the formation of a living organism from organic molecules is a very slow process. When the earth was sterile, these molecules could slowly accumulate. At the present time, however, the large number of microscopic organisms that populate the earth would digest such molecules and prevent their accumulation. It is an interesting historical point of Oparin's materialistic theory that once life originated, the probability of its happening again was greatly reduced. This problem was realized as long ago as 1871, when Charles Darwin wrote

> It is often said that all the conditions for the first production of a living organism are now present, which could ever have been present. But if (and oh! what a big if) we could conceive in some warm little pond, with all sorts of ammonia and phosphoric salts, light, heat, electricity, etc., present, that a protein compound was chemically formed ready to undergo still more complex changes, at the present day, such matter would be instantly devoured or absorbed, which would not have been the case before living creatures were formed.

The possible existence of extraterrestrial life is an important philosophical and scientific question intimately related to the question of the origin of life. In the next 20 years, we should know whether or not life exists elsewhere within our solar system. The most promising extraterrestrial habitat is Mars. Experiments have indicated that certain terrestrial microbes could survive and multiply under simulated Martian conditions. Even if Mars is sterile, it will be interesting to see how far chemical evolution has proceeded. Are there small organic molecules? Polymers? Coacervates?

Similarly, recent investigations have suggested that chemical evolution should have proceeded and life may have evolved on the Jovian planets, Saturn, Jupiter, and Pluto, as well as on Titan, Saturn's largest satellite. The probability of life outside our solar system is highly speculative. If we assume that life is a natural consequence of a given set of chemical and physical conditions, then life should have originated on any planet whose history of conditions is similar to those of earth. Present estimates of the number of earth-like planets in the universe are completely unreliable. Also, there are the additional interesting possiblilities that life radically different from that on earth exists elsewhere, for example, employing silicon instead of carbon as the central element and existing in an environment of liquid ammonia rather than water.

1.12 On the Nature of Scientific Progress

> The basic texture of research consists of dreams into which the threads of reasoning, measurement, and calculation are woven.
>
> *Albert Szent-Gyorgyi*

In this chapter we have analyzed, in historical sequence, several theories on the origin of life. We have seen how the theory of spontaneous generation went unchallenged for more than 2000 years, whereas in the last 60 years, our ideas on this subject have changed continuously. The mechanism that brought about these rapid changes is often called **the scientific method.** In reality, there is no such thing as *the* method of science. There is no such ideal Platonic form, which if followed, guarantees to lead you down the path of truth. The method of science is not a fixed thing. Just as our ideas on the origin of life have changed with time, so have our methods for investigating the problems. What we have as a mechanism for scientific progress is a series of operations, some manual, some mental, which in the past have proven useful for testing ideas. Many of these operations are simply refinements of the thinking of daily life. These operations can be divided into the following four categories: observations, hypotheses, experiments, and theories.

Observations

Everyone, whether a scientist or not, observes. The important thing is how to observe. In this respect, the trained scientist differs greatly from the creative artist. The artist shapes and colors observations with his or her own feelings within the bounds of the medium. The artist is free to abstract from observation only those items that are essential, and to distort their dimensions in order to emphasize certain features and arouse certain emotions. Scientists, on the other hand, must observe and describe natural phenomena independent of their own sentiments

The science of bacteriology began with observations by a single man, Antony van Leeuwenhoek. His discovery of the microbial world represents an excellent example of how to observe. First, he accurately recorded the source of his material and how he handled it, in sufficient detail so that others could later repeat his observations. Next, he patiently observed the material under a variety of conditions, using the best observational aids which existed at that time. Since these aids themselves can introduce artifacts into the observations, it is essential that the observer be as familiar as possible with the tools of his trade. Remember that Leeuwenhoek studied his microscope and developed his techniques for 20 years before submitting any of his results for publication! Even when using well-established instruments, it is important to understand how they work and what their limitations are. Finally, Leeuwenhoek recorded his observations in explicit language and as free as possible from preconceived notions.

The following short passage from Leeuwenhoek's eighteenth letter to the Royal Society exemplifies some of these points:

> I did now place anew about 1/3 ounce of whole pepper in water, and set it in my closet, with no other design than to soften the pepper, that I could the better study it. This pepper having lain about three weeks in the water, and on two several occasions snow-water having been added thereto, because the water had evaporated away; by chance observing this water on the 24th April, 1676, I saw therein, with great wonder, incredibly many very little animalcules, of divers sorts; and among others, some that were 3 or 4 times as long as broad, though their whole thickness was not, in my judgment, much thicker than one of the hairs wherewith the body of a louse is beset.

Notice the use of measurements by Leeuwenhoek to make his description more explicit. He does not use a pinch of pepper, but 1/3 ounce of pepper; he does not leave it for a while, but for three weeks. Since there was no language to describe the size of microbes in the seventeenth century, Leeuwenhoek relates the size of his microbe relative to that of the diameter of a louse's hair. Leeuwenhoek realized that the hair of a louse, although varied in length, has a relatively constant diameter

When a science such as bacteriology evolves, it continually develops better tools that help the observer to make more accurate measurements and to extend the range of observable phenomena. Whereas Leeuwenhoek's simple microscopes had maximum magnification of 200-fold, we now have compound light microscopes with magnification powers of more than 1000 and electron microscopes with useful magnifications of more than 100,000. Equally important is the development of techniques used to prepare the material for observation. For example, staining procedures have been developed that aid in identifying bacteria; other staining procedures are used to make the appearance of certain parts of cells more obvious. It is important to emphasize that each discipline of science has its own characteristic tools and procedures for making observations. The further development of that discipline will depend to a large degree on how it can adapt the techniques of other disciplines to its own needs, and invent new ones.

Hypotheses

If scientists were merely satisfied with accumulation of observations, then science soon would become unwieldy and as difficult to comprehend as the nature from which it was derived. Observations do not solve problems, but suggest them. When an observation is not satisfactorily accounted for by existing knowledge, it introduces a difficulty or a problem. The scientist then formulates a hypothesis to explain the difficulty. Precisely *how* the human mind is able to originate new thoughts (hypotheses) or combinations of thoughts is certainly not clear. Most likely, guesses, hunches, and intuition play a more important role than deductive reasoning. For this discussion, however, the important point is that a hypothesis is simply a tentative explanation to account for observed phenomena

The discovery of microbes by Leeuwenhoek presented a new problem for biology. Where did the microbes come from? Needham and Buffon hypothesized that microbes arose spontaneously from non-living matter. Such a hypothesis was useful (even though it was later disproven) because it focused on the problem and made certain predictions possible that could then be tested experimentally.

Experiments

Since a hypothesis is only a plausible speculation, it must be tested extensively and much evidence for it adduced before it can be accepted. The experimental process is essentially a means of testing hypotheses under controlled conditions. It is by performing experiments that the scientist attacks the problem directly. We all attempt to learn from our experiences, but the scientist attempts to experience in order to learn. In this regard, it is interesting that the French word *experience* means both experience and experiment.

An excellent example of the experimental process is Pasteur's renowned refutation of the spontaneous generation hypothesis, which predicted that sterile broth would become turbid with microbes if exposed to air. Pasteur was able to design a swan-neck flask in which the broth remained sterile even when it came in contact with air. Thus the prediction (hypothesis) was wrong. Pasteur concluded from this and other studies that the microbes arose not by spontaneous generation but from the dust in the air.

The principle of a control is fundamental in the experimental approach. The control group corresponds to the experimental group at every point except the one at issue. Pasteur's control flask was exactly like the experimental flask except that dust was allowed to enter. Immediately after sterilization, the neck was broken off. Since the control became turbid with growth, it was evident that the broth was suitable culture medium. In general, it is much easier to establish controls in microbiology than in other areas of biology. Not only can one control the environmental conditions more easily, but by using large numbers of organisms derived from the same parent (that is, pure culture), one can reduce the possibility of variations due to the living organisms themselves.

Theories

As generally used, the term **theory** is applied to a hypothesis that has been extensively tested and that ties together and arranges the results of a number of observations and experiments. Theories, however, are not the end; they are also tentative. Whenever a theory is shown to be inconsistent with an experimental result, it is the theory and not the experiment that must be discarded. Consider the following hypothetical case. Theory A exists, which explains a large number of experimantal data. Another theory B is presented, which is also consistent with the experimental facts. A crucial experiment is then designed and performed in order to choose between the two theories. The results of the experiment are found to be contradictory to theory A. Thus theory A is wrong and must be discarded. This does not mean, however, that theory B is right. Some bright young scientist might think of still another theory C that is also consistent with existing facts. It then will be necessary to perform an experiment to distinguish between theories B and C.

This simple hypothetical case exemplifies the two most significant features that characterize the discipline of science. First, science is constantly changing. Since science claims no eternal truths, theories are presented with the expectation that they will need to be modified sooner or later. The method of science is logically incapable of arriving at complete and final theories. It is, so to speak, constantly under repair. Second, science has no authorities other than observations and experiments. The men of science do not ask that a theory be believed because some important authority has said that it is true. Rather, only those doctrines that are based on the facts, that can (and must) be verified by other scientists, should be accepted even temporarily. Connected with the theoretical aspects of science is technology, which can utilize the knowledge of science to produce comforts and luxuries that were impossible in the prescientific era. It is this latter aspect that gives such importance to science even for those who are not scientists.

This discussion of science and the scientific method is not meant to imply that science and its methods are any better than other fields and other methods. When the methods of science can be applied, they have provided powerful tools for understanding nature. In the past when science came into conflict with religious creeds or authoritarian principles, it was science that was victorious. The Copernican Revolution and Evolution are two examples of these conflicts. However, the method of science is severely limited in what kinds of questions it can answer. For example, it cannot answer the question of whether or not sciences should be used for the enhancement or destruction of life. This is why Albert Einstein wrote

> Religion without science is lame;
> Science without religion is blind.

To this we add

> Science without morality is perilous.

QUESTIONS

1.1 What do each of the following terms signify?

Tyndallization	Coacervate
Sterilization	Metabolism
Pasteurization	Polysaccharides
Pure culture	Nucleic acids
Organic chemical	Proteins
Polymer	Macromolecule

1.2 What are the similarities and differences between the theories of spontaneous generation and chemical evolution?

1.3 What were the reasons for

a) Pasteur using a swan-neck flask rather than an ordinary flask?
b) Miller sparking a mixture of gases?
c) Discounting the theory that life has always existed on earth?

1.4 What are the differences between elements, atoms, and molecules?

1.5 Which technique would you use to sterilize a solution of vitamins? a plastic dish?

1.6 Give two reasons why

a) Needham obtained living forms, whereas Spallanzani did not.
b) Agar is better than gelatin as a solidifying agent.
c) The conditions for the slow emergence of life are no longer present.

Suggested Readings

Barghoorn ES: The Oldest Fossils. Scientific American, May 1971.

A beautifully illustrated account of the discovery of Precambrian microfossils.

Bernal JD: The Origin of Life. London: Weidenfeld and Nicolson, 1967.

A recent collection of works dealing with biological, chemical, and physical aspects of the origin of life.

Bulloch W: The History of Bacteriology. London: Oxford University Press, 1938.

This standard history of bacteriology was reprinted in 1960.

Dobell C: Antony van Leeuwenhoek and His "Little Animals." New York: Harcourt, Brace and Company, 1932.

This classic biography is now available in paperback by Dover, New York, 1960.

Dubos R: Louis Pasteur, Free Lance of Science. Boston: Little, Brown and Company, 1950.

An interpretive biography by one of the world's foremost microbiologists and humanists.

Frieden E: The Chemical Elements of Life, Scientific American, July 1972.

A readable presentation of the chemical elements found in living cells.

Gabriel ML and S Fogel, Eds and Translators: Great Experiments in Biology. Englewood Cliffs, NJ: Prentice-Hall, Inc., 1955.

An excellent collection of classic papers, including works by Redi, Spallanzani, Pasteur, Koch, and Leeuwenhoek.

Miller SL: The Atmosphere of the Primitive Earth and the Prebiotic Synthesis of Amino Acids. In: Origins of Life, Vol. 5. Dordrecht, Holland: D. Reidel Publishing Company, 1974.

An international journal devoted to the publication of scientific papers on the origin and evolution of life.

Monod J: An Essay on the Natural Philosophy of Modern Biology. New York: Alfred A. Knopf, 1971.

Few scientists have taken the time and effort to articulate how their science fits into the wider framework of culture and society. Monod has done exactly that.

Oparin AI: The Origin of Life on Earth. New York: Macmillan Company, 1938.

A translation by Sergius Morgulis of Oparin's book first published in Moscow in 1936; the theory of chemical evolution is fully discussed. An exact reprint of the 1938 edition is available in paperback by Dover, New York, 1953.

Pauling L and R Hayward: The Architecture of Molecules. San Francisco; W.H. Freeman and Company, 1964.

An introduction to molecular architecture for the layman, including many beautiful pictures of simple and complex molecules.

CHAPTER 2 Through the Microscope: Cells-Viruses-Molecules

Faith is a fine invention
When Gentlemen can see—
But Microscopes are prudent
In an Emergency.

Emily Dickinson

This chapter offers a pictorial essay of microbial cells, subcellular organelles, viruses, and the macromolecules of which they are composed. After a brief description of the instruments that are used to visualize biological matter, photographs of microbial, animal, and plant cells are presented in order to develop two of the central themes of biology—the cell as the structural unit of life and the fundamental difference between procaryotic cells and eucaryotic cells. The component parts of the cell are then analyzed by high resolution microscopy. The subcellular structures of different microorganisms demonstrate some of the possible variations on the common theme of the cell.

Viruses constitute a special form of life not composed of cells. Their life cycles, exemplified by a series of high magnification electron micrographs of a bacterial virus, demonstrate their dependency on living cells for multiplication. Finally, some examples of the most important macromolecular species for both cells and viruses, the nucleic acids and proteins, are presented.

In order to best appreciate visually this collection of micrographs, representing some of the most outstanding products of the art of photomicroscopy, descriptive material has been kept to a minimum. In later chapters, the nature of the structures shown here will be further amplified and correlated with their biological functions.

2.1 The Microscope—Resolution

Cytology, the study of cells, relies heavily on the use of microscopes to extend the limit of visual observation. Two major types of instruments used in research today are the light microscope (Figure 2.1) and the electron microscope (Figure 2.2). The unaided human eye is capable of distinguishing two points as distinct from each other only if they are separated by at least 0.1 mm. When observing an object whose fine detail has spacings that are less than this distance, the object will appear unclear, for its fine structure will not be resolved.

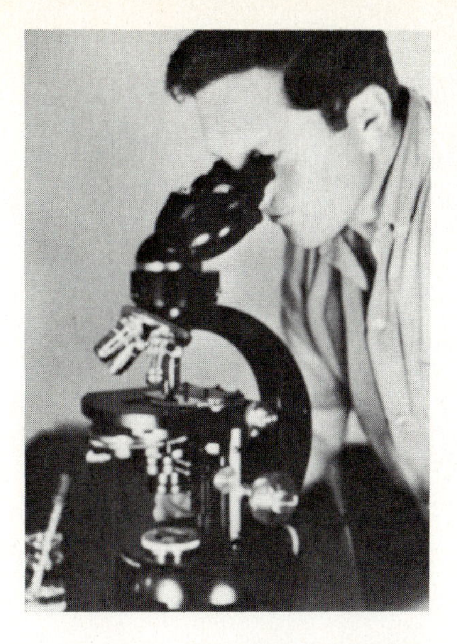

Figure 2.1
Light microscope. Magnifying power: 1500; resolving power: 0.1 micron; source of energy: light; types of specimens: living or preserved; cost: $1000 to $2000.

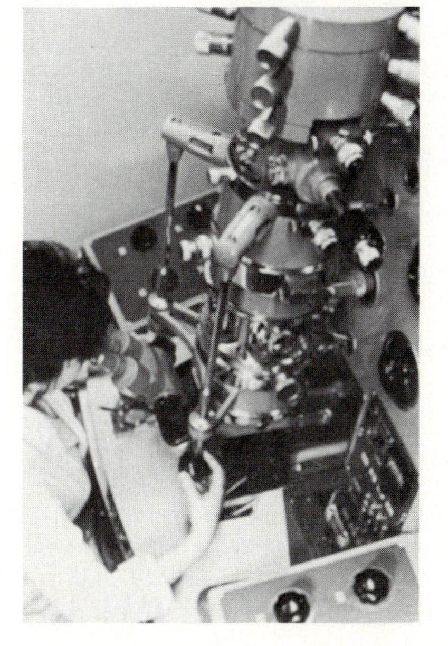

Figure 2.2
Electron microscope. Magnifying power: 200,000; resolving power: 0.0005 micron; source of energy: electrons; types of specimens: preserved only; cost: $50,000 to $100,000.

Merely magnifying an unclear object (for example, by photographic techniques) will result in nothing more than an unclear enlargement—blurring. The major function of the microscope is twofold: It must resolve fine detail, and it must enlarge the resolved image to dimensions that can be perceived by the human eye. These points are illustrated in Figures 2.3 and 2.4.

Recent modifications of the electron microscope have increased its effective depth of focus to several millimeters, thus making possible three-dimensional images of microscopic objects. An example of such an image is seen in Figure 2.5.

Figure 2.3
Sectioned frog muscle in the light microscope (A, 2500×; B, 9400×).

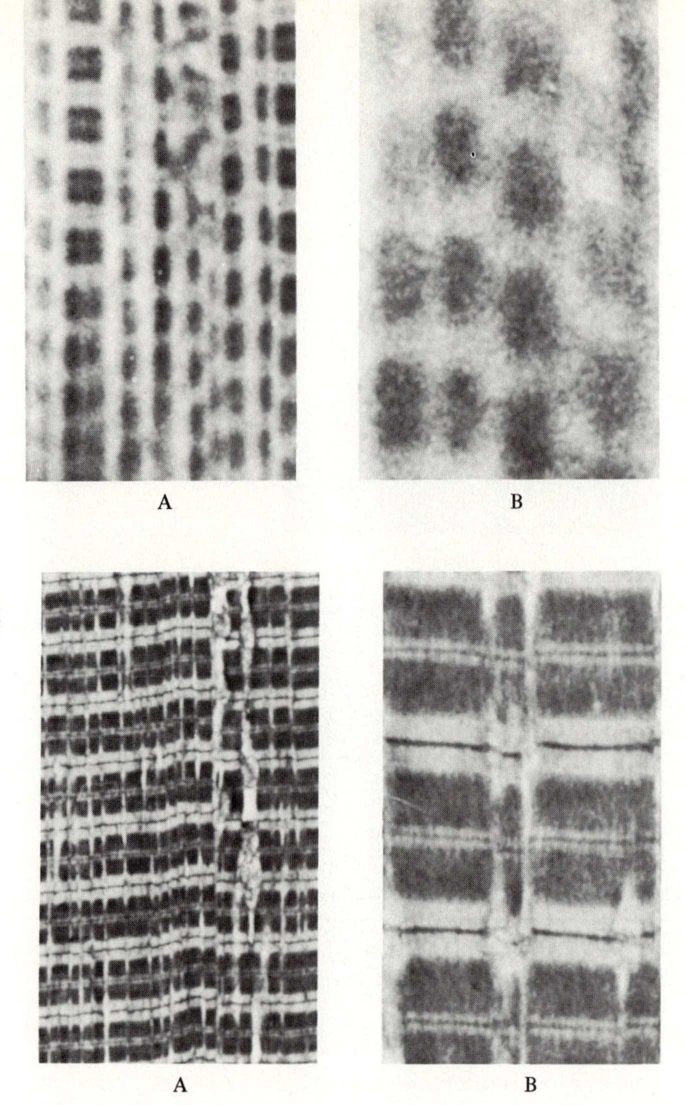

Figure 2.4
Sectioned frog muscle in the electron microscope (A, 2500×; B, 9400×). (Courtesy of M. Reedy.)

2.2 Two Basic Cell Types: Procaryotic and Eucaryotic

Although the resolving power of the light microscope is sufficient to examine the size, shape, and arrangement of living cells, the electron microscope is necessary to observe their internal structures. Electron microscopic studies of cells, beginning around 1950, have made it clear that two basic plans of cellular organization exist, the simpler **procaryotic** cell and the more complex **eucaryotic** cell.

The procaryotic cell, characteristic of bacteria, has a relatively small size and simple organization. The nuclear material is not separated from the cytoplasm by any defined structure. Procaryotes contain only one chromosome (consisting of a single double-stranded

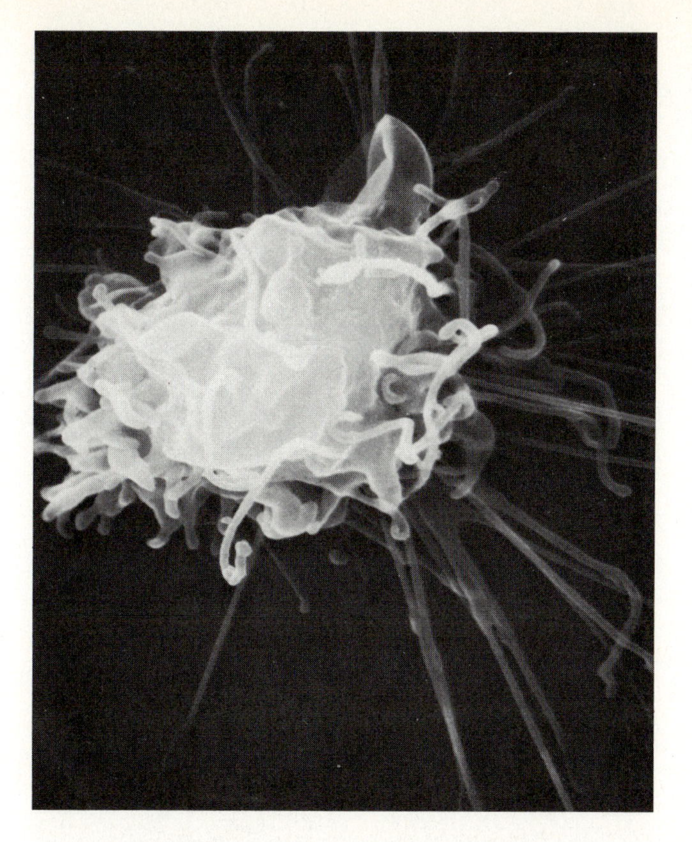

Figure 2.5
Scanning electron micrograph of an embryonic chick cell (10,000×). (Courtesy of Y. Ben Shaul.)

DNA molecule) and lack a special class of proteins called histones found in eucaryotic chromosomes. In addition, bacteria do not have clearly defined membrane-limited organelles, such as mitochondria and chloroplasts. In general, respiration and photosynthesis take place on the cytoplasmic membrane or on extensions of it. An example of a procaryotic cell is shown in Figure 2.6.

The eucaryotic cell is characteristic of plants, animals, and all microorganisms except bacteria. The typical structural features include (1) a nucleus separated from the cytoplasm by a nuclear membrane and (2) cytoplasm containing many structurally differentiated units. For example, respiration is localized in mitochondria and photosynthesis in chloroplasts. The nuclei of eucaryotic cells contain nucleoli and more than one chromosome containing histones in addition to DNA. An example of a eucaryotic cell is shown in Figure 2.7. Additional differences between procaryotic and eucaryotic cells are discussed in Section 2.5, following a more detailed description of cell structure and function.

2.3 The Size, Shape, and Arrangement of Cells

Although both eucaryotic and procaryotic cells occur in a wide variety of sizes (Table 2.1), certain generalizations can be made. The average procaryotic cell is several hundred times smaller than the av-

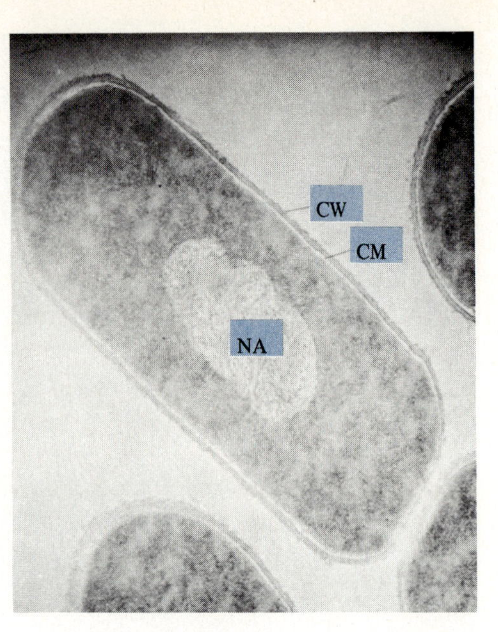

Figure 2.6
Thin section of the bacterium *Bacillus subtilis* in the electron microscope. Note the cell wall (CW), cytoplasmic membrane (CM), and the nuclear area (NA). The darkness of the cytoplasm is the result of large numbers of ribosomes (66,000 ×). (Courtesy of F. Eiserling.)

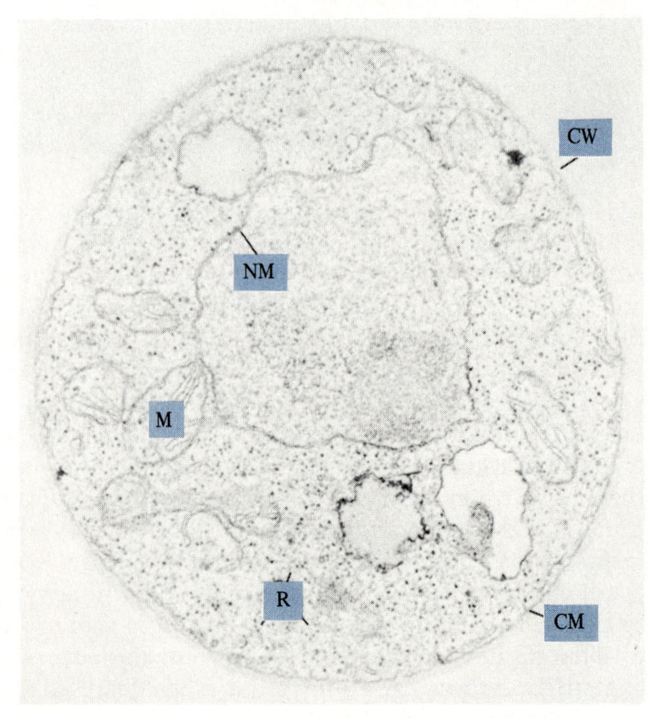

Figure 2.7
Thin section of a fungus in the electron microscope. Note the nuclear membrane (NM), cell wall (CW), cytoplasmic membrane (CM), mitochondria (M), and ribosomes (R) (34,000 ×). (Courtesy of M. R. Edwards.)

erage eucaryotic cell. What is the advantage of small size for a procaryote? Bacteria have no special organelles for taking in nutrients and excreting waste products. Those processes take place at the cell surface. Therefore, for a given cell mass or volume, the larger the surface area, the more rapidly the cell can grow and multiply. The fact that surface area increases as materials are broken into smaller pieces can

TABLE 2.1 Size of Cells and Viruses		
	Volume of Structural Unit (cubic microns[a])	
	Normal Range	*Example*
Eucaryotic cell	20–10,000	*Amoeba proteus:* 2,000
Procaryotic cell	0.01–50	Tuberculosis bacillus: 1.5
Virus	0.00001–0.01	Flu virus: 0.0005

[a]A micron is one millionth of a meter.

readily be appreciated by simply slicing an apple. The mass of apple does not change, but with every slice there is additional surface area. It follows that one of the main advantages of small size for a procaryotic cell is a greater surface-to-volume ratio.

What determines the minimum size for a procaryotic cell? The smallest cells known, bacteria called mycoplasmas, have a volume of 0.01 cubic micron (about the same size as the largest viruses). As discussed in Chapter 1, living independent cells must have (1) specific patterns of metabolism and (2) the ability to transmit the information for these patterns from generation to generation. It is possible to calculate that the volume required for the proteins and nucleic acids needed to carry out these minimum functions would occupy approximately 0.01 cubic micron. It is therefore unlikely that cells much smaller than the mycoplasmas will ever be discovered. Viruses can be much smaller because they rely on the metabolism of their living host cells for much of their self-duplication.

The range of sizes found in eucaryotic cells is much greater than that in procaryotes. The smallest eucaryotic cell, an alga called *Micromonas,* is somewhat smaller than the largest bacterium. Therefore, size alone cannot be used to distinguish the two cell types. Essentially, the entire volume of a *Micromonas* is taken up by its membrane-bound nucleus, a single small chloroplast, and one mitochondrion. Any further reduction in size would require the loss of an essential organelle. Thus, the lower size limit of a eucaryotic cell is the volume necessary to harbor its structural elements. The upper limit in cell size is determined to a large extent by the need for efficient intracellular communication. The nucleus, which contains the genetic information and is the cell's control center, must be responsive to the changing conditions in the cytoplasm. If the time it takes for molecules to diffuse between the nucleus and the various parts of the cell is too great, the cell cannot adapt rapidly enough to its environment. One way that some large cells partially overcome this problem is by having two or more nuclei.

Except in a few cases, the relationship between cell shape and function is not clearly understood. Nevertheless, cell shape or morphology is useful for identification purposes. The following four fundamental shapes of procaryotic cells are exemplified in Figures 2.8 to 2.11: (1) more-or-less spherical organisms known as **cocci** (singular: coccus); (2) cylindrical ones called **bacilli** (singular: bacillus); (3)

Figure 2.8
Thin sections of a coccus in the electron microscope (6500×).

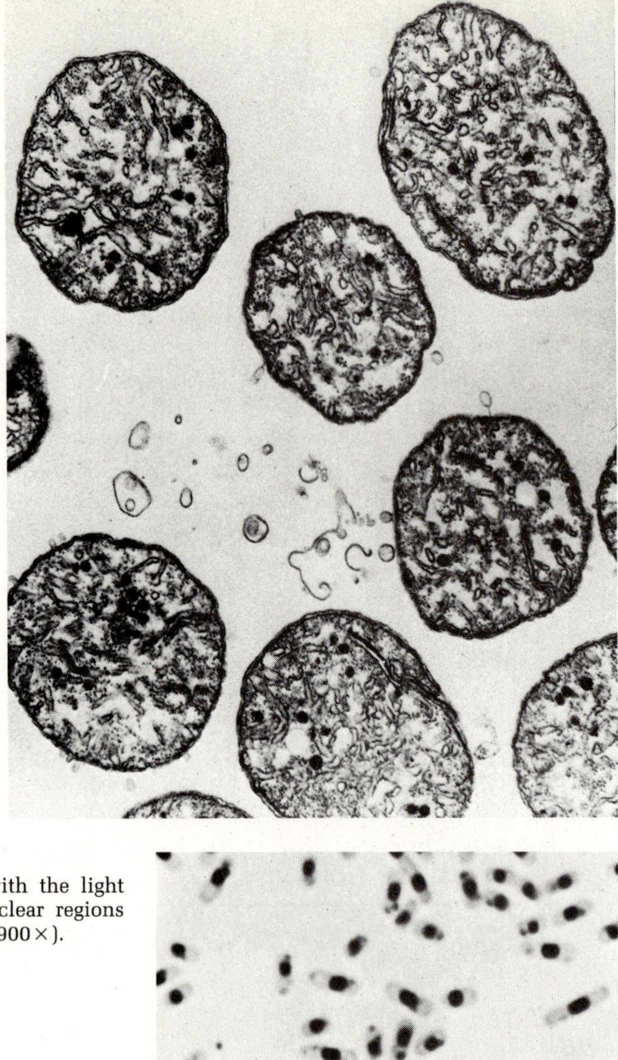

Figure 2.9
Bacilli as viewed with the light microscope. The nuclear regions have been stained (1900×).

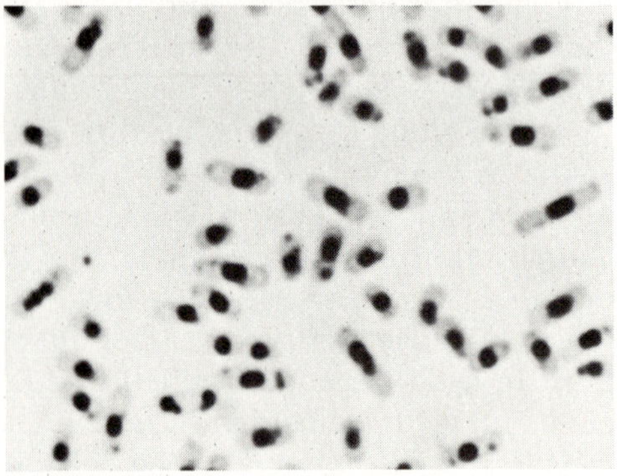

curved rods referred to as **spirilla** (singular: spirillum); and (4) filamentous forms.

Although most types of bacteria exist as single, unattached cells, certain groups form characteristic cell aggregation patterns. This is especially common with cocci (Figure 2.12). These aggregation patterns arise from the fact that cells may adhere to each other following cell division. In the simplest case, cocci divide in a single plane and remain attached. When predominately in pairs, they are called **diplococci;** in chains they are referred to as **streptococci.** Cocci that di-

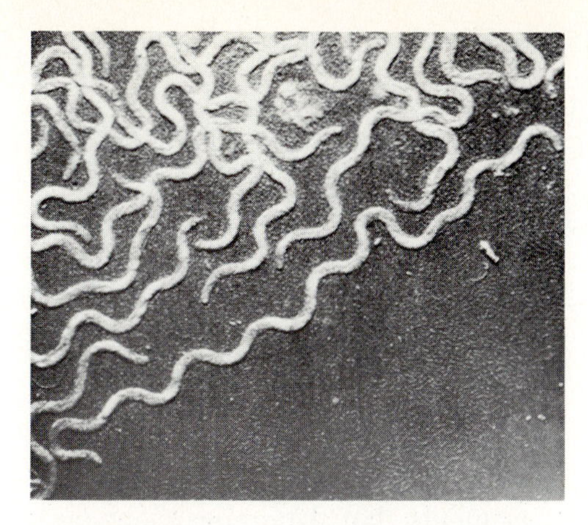

Figure 2.10
Spirilla (23,000×).

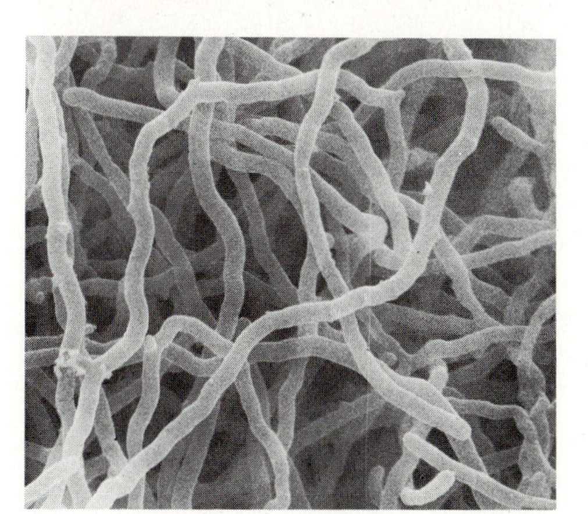

Figure 2.11
Scanning electron micrograph of filamentous *Streptomyces* bacteria (9500×). (Courtesy of Y. Aharonovitz.)

(A) (B)

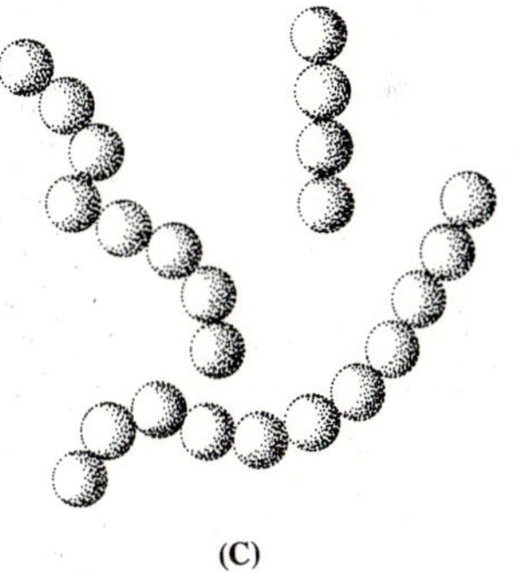

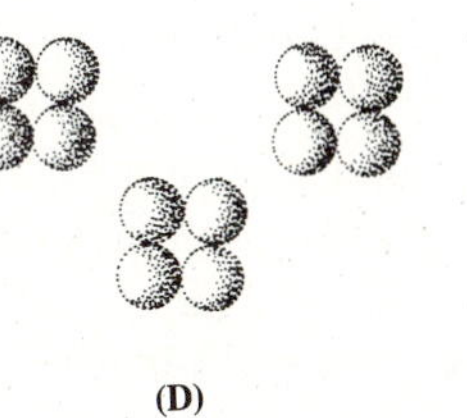

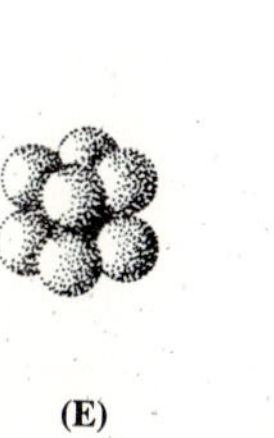

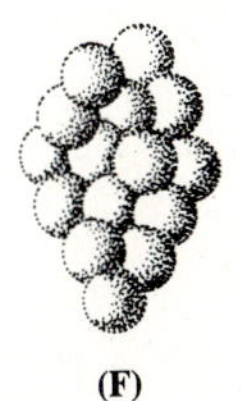

Figure 2.12
Common aggregation patterns of cocci. (A) Single cocci. (B) Diplococci. (C) Streptococci. (D) Tetracocci. (E) Sarcinae. (F) Staphylococci.

vide in two planes at right angles to each other and form tetrads are called **tetracocci.** Finally, some cocci divide successively in three planes and form either cubical masses of cells called **sarcinae** or irregular clusters resembling bunches of grapes referred to as **staphylococci.**

Most eucaryotic cells have distinct shapes. In the case of eucaryotic microorganisms, such as protozoa and algae, cells exhibit a complex geometric beauty unsurpassed in the natural world. In Chapter 4, photomicrographs of unicellular eucaryotes are presented along with a discussion of the microbial world.

2.4 Structure and Function in the Procaryotic Cell

Seven important structural elements that determine the typical architecture of the procaryotic cell are schematically represented in Figure 2.13. Three of the structures—the **nuclear region, cytoplasmic membrane,** and **ribosome—**are found in all procaryotic cells and are absolutely essential for cell viability. An invagination of the cytoplasmic membrane, referred to as a **mesosome,** has been found in some bacteria. Except for a few rare exceptions, procaryotic cells have rigid **cell walls** that exist just outside the cytoplasmic membrane. The presence of one or more **flagella** and/or **storage granules** varies from one group to another. In general, bacteria that are able to swim contain flagella. The existence and type of storage granules in the cytoplasm depend upon the nutritional environment and growth conditions.

The **nuclear region** of a procaryotic cell consists of a single molecule of DNA that is about 1000 times longer than the entire bacterial cell. In order to fit within the cell, the long thread of deoxyribonucleic acid (DNA) is highly twisted into a form that molecular biologists call "supercoiled DNA." The chemical structure of DNA is mentioned briefly in Chapter 1 and is discussed in more detail in Chapter 5. Also in Chapter 5, evidence is presented that proves DNA is the genetic material in all cellular organisms and most viruses.

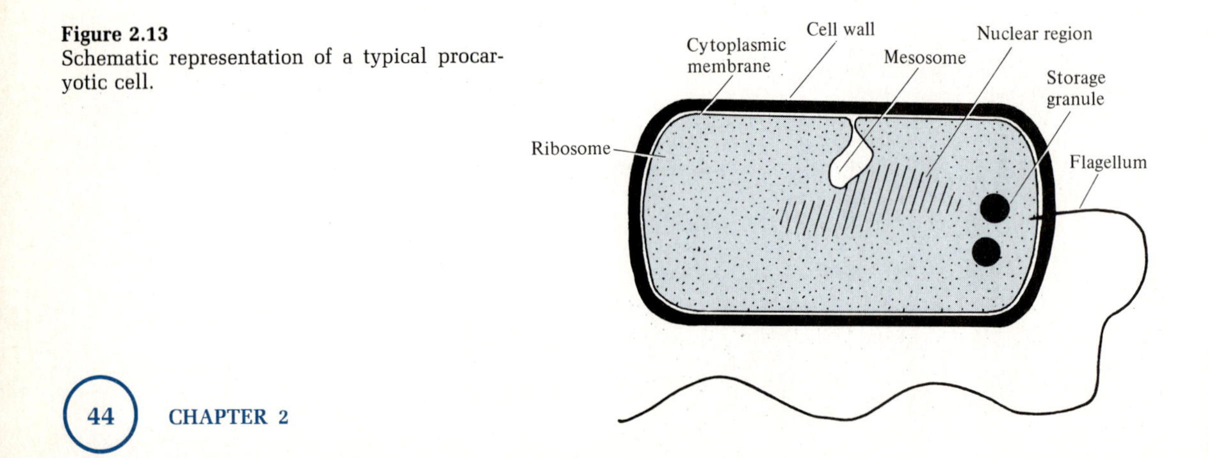

Figure 2.13
Schematic representation of a typical procaryotic cell.

The fact that the DNA of procaryotic cells is located in the nuclear region can be demonstrated by the elegant technique of **autoradiography.** When cells are grown in the presence of radioactive thymine, they incorporate the radioactive molecule into their DNA. The cells are then sliced and covered with photographic emulsion. The radioactive DNA molecule, as a consequence of its emission of ionizing radiation, can expose the photographic emulsion, just as light does. After an appropriate time of exposure, the cell sections are processed as for photography. The resulting picture (Figure 2.14) can then be viewed in the microscope. The dark, reduced silver grains indicate the intracellular location of the radioactive material. Thus, Figure 2.14 demonstrates that DNA is concentrated in the nuclear region of the procaryotic cell. This general technique has been exploited to make visible a variety of different chemical structures within the cell.

The **cytoplasmic membrane** constitutes the most important boundary between the interior of the cell and the external environment. It is composed of about equal amounts of lipid and protein. Since the cytoplasmic membrane is elastic, the internal pressure inside the cell forces the membrane against the rigid cell wall. The cytoplasmic membrane performs several vital functions:

1. The membrane is semipermeable; that is, it allows only certain molecules to enter and leave the cells, thereby controlling carefully the internal composition of the cytoplasm. Only a few very small molecules, such as water, can freely pass through the membrane.
2. The membrane contains an important group of enzymes called **permeases.** These permeases are responsible for transporting various inorganic ions, sugars, amino acids, vitamins, and other substances into the cell. Each permease sys-

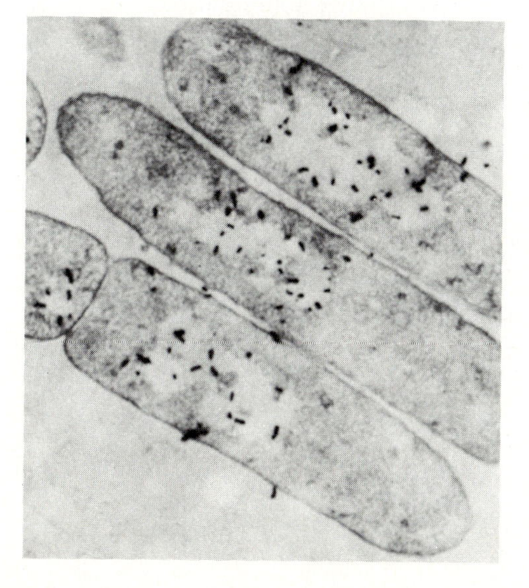

Figure 2.14
Thin section of several *Escherichia coli* cells in the electron microscope. Note the absence of a nuclear membrane. The dark spots on the micrograph are a result of the autoradiographic deposition of silver grains on the deoxyribonucleic acid of the cells (22,000×). (Courtesy of F. Eiserling.)

tem is specific for a given compound or small group of closely related compounds. For example, there are separate permeases for each of the three common sugars: glucose, galactose, and sucrose. In many cases, the transport system not only demonstrates **specificity** but also the capacity to scavenge nutrients from dilute solutions and concentrate these materials inside the cell. This latter process, which requires the expenditure of energy, is called **active transport.** The ability to concentrate nutrients allows bacteria to maintain a relatively constant environment inside the cell, even when the external conditions change dramatically.

3. In many bacteria, the cytoplasmic membrane contains enzymes involved in generating energy. In the case of photosynthetic bacteria, the enzymes and pigments involved in converting light energy into chemical energy are located on the cytoplasmic membrane.

All types of cells contain in their cytoplasm small, almost spherical particles called **ribosomes.** As discussed in Chapter 7, ribosomes are the site of protein synthesis in living cells. The ribosome consists of about equal amounts of ribonucleic acid (RNA) and protein. The number of ribosomes per cell can vary greatly. Generally, rapidly growing cells contain many ribosomes, whereas resting cells have few ribosomes.

The exact function of the **mesosome** is not known. Electron micrographs show that the mesosome is continuous with the cytoplasmic membrane and is probably in contact with DNA in the nuclear region. It has been suggested that the mesosome plays a role in cell division by assuring that each daughter cell receives a copy of the bacterial genome. A fine example of a mesosome can be seen in Figure 2.15.

The **cell wall** of procaryotic cells is extremely interesting for several reasons. First, it has a unique chemical and physical structure not found anywhere else in the biological world. Second, it is responsible for maintaining the rigidity and shape of the cell. Third, the cell wall is the target of an important group of antibiotics. In order to fully

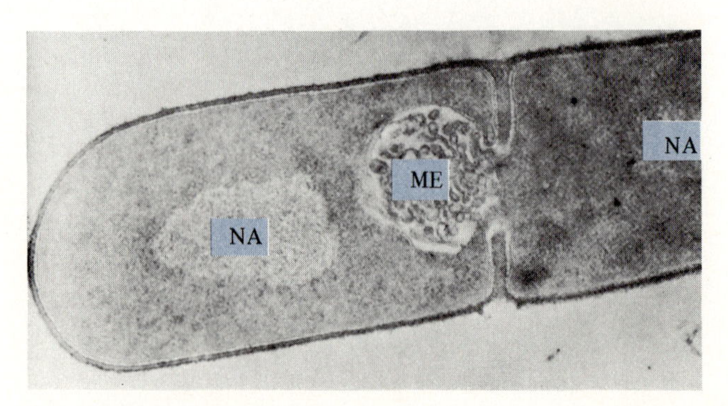

Figure 2.15
A thin section of *Bacillus subtilis* in the process of dividing. Note the deposition of cell wall between the two daughter cells and the presence of two nuclear areas (NA). A mesosome (ME), an invagination of cytoplasmic membrane, can be seen at the region of septum formation (58,000×). (Courtesy of F. Eiserling.)

understand these points, it is necessary to consider how the cell wall is assembled.

Figure 2.16 describes the eight steps in the synthesis of the procaryotic cell wall. In the first step, a sugar (designated M) is combined with an amino acid (designated a). In steps 2 through 5, four other amino acids (designated b, c, d, and e) are sequentially added onto sugar M. Then another sugar (designated G) is linked to sugar M to form the basic building block of the cell wall—the disaccharide pentapeptide. Up to this point, all the steps take place inside the cytoplasm of the cell. In step 7, the disaccharide pentapeptide building block is transported through the cytoplasmic membrane with the help of a special lipid carrier molecule and then is polymerized to form a linear chain of alternating sugar G and sugar M components with short amino-acid side arms. In the final step, this long chain is linked to other chains in the cell wall through the amino-acid side arms. This produces a highly cross-linked molecule. In fact, the entire cell wall can be considered as a single gigantic, bag-shaped macromolecule.

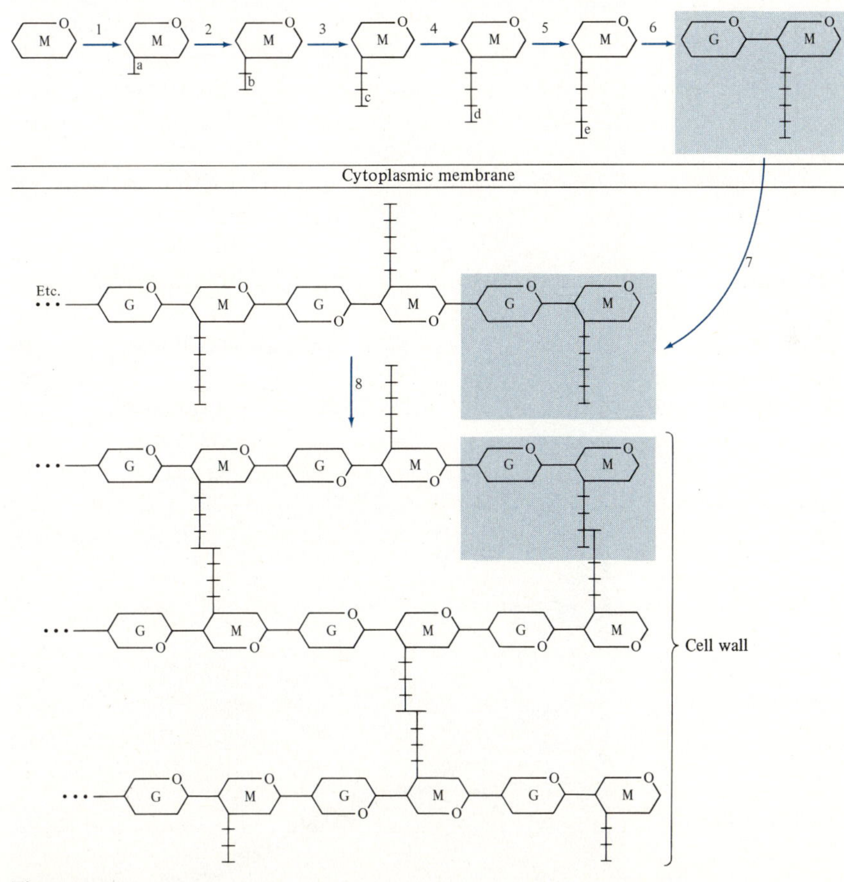

Figure 2.16
Synthesis of the bacterial cell wall. Each of the eight steps is catalyzed by a separate enzyme. The final product is a highly cross-linked polymer containing sugar M, sugar G, and five different amino acids, abbreviated a, b, c, d, and e.

Electron micrographs of purified bacterial cell walls are presented in Figures 2.17 and 2.18.

With an understanding of the architectural design and assembly system for the cell wall, it is now possible to appreciate some of its important properties. The molecular network structure of the cell wall makes it extremely rigid and resistant to mechanical breakage. In that sense, it is like a chain-link fence. Even if you cut a few of the links, the fence maintains its integrity. In order to destroy a chain-link fence or a bacterial cell wall, it is necessary to introduce several breaks in the cross-linked structure.

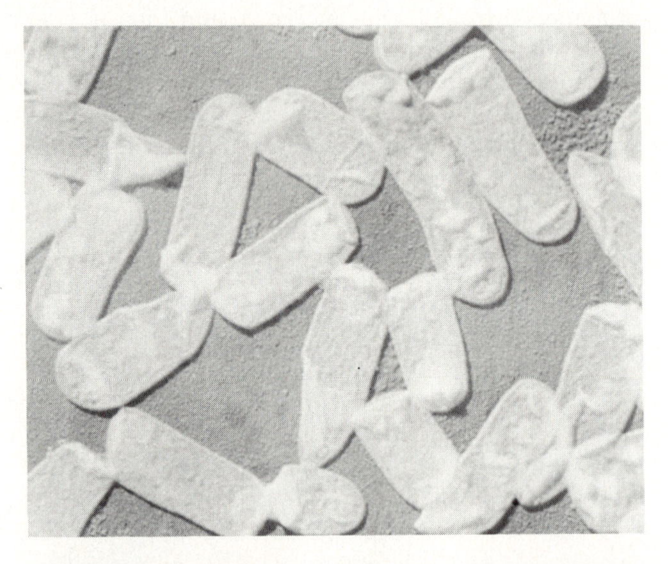

Figure 2.17
Electron micrograph of isolated bacterial cell walls. This rigid structure is responsible for maintaining cell shape (16,000×). (Courtesy of E. Ribi.)

Figure 2.18
High-power magnification of a fragment of a bacterial cell wall (70,000×). (Courtesy of R. G. E. Murray).

The unique structure of the cell wall makes it an ideal target for antibiotics. Since eucaryotes do not have any structures that resemble the bacterial cell wall, any material that specifically poisons or inhibits one of the steps in the cell wall synthesis is a potentially useful antibacterial substance. For example, penicillin blocks the final step in cell wall synthesis (Figure 2.16, step 8). This concept of selective toxicity is the basis of chemotherapy and is discussed in Chapter 10.

A simple experiment that exemplifies several important aspects of the cell wall is shown in Figure 2.19. Normally, cells contain a higher concentration of molecules in the cytoplasm than exists in the outside medium. In an attempt to equalize the concentrations of molecules on both sides of the cytoplasmic membrane, water enters the cell. However, the rigid cell wall prevents the cell from expanding. Therefore, an internal force is developed in the cell, called **osmotic pressure.** When penicillin is added to the bacillus, cell wall synthesis is inhibited, but growth continues. This leads to the development of weak spots around the cell, deficient in cell-wall content. The high internal osmotic pressure forces the cytoplasmic membrane through the weakened wall to form, initially, cytoplasmic bulges or blebs. If the concentration of molecules on the outside remains low, the wall-deficient cell continues to take up water, expands, and rapidly bursts, releasing the cytoplasmic contents into the medium. On the other hand, if enough sucrose is added to medium so that the osmotic pressure inside the cell equals that of the medium, then the wall-deficient cell assumes its most stable shape, a sphere. This simple experiment demonstrates the role of the cell wall in (1) preventing the cell rupture due to osmotic forces and (2) maintaining cell shape.

The most common organelle of motility in bacteria is the **flagellum** (plural: flagella). The number of flagella per cell and their distribution around the cell are characteristic features of different groups of bacteria; one type of bacterium, for example, typically has a single flagellum attached to one end of the cell (Figure 2.20), whereas another type contains many flagella distributed around the cell surface (Figure 2.21). As can be seen in Figure 2.22, the flagellum emanates from just inside the cytoplasmic membrane, passes through both the membrane and the cell wall, and extends for a considerable distance outside the cell. Sometimes the flagellum is many times longer than the entire cell.

It is possible to make swimming bacteria temporarily immobile

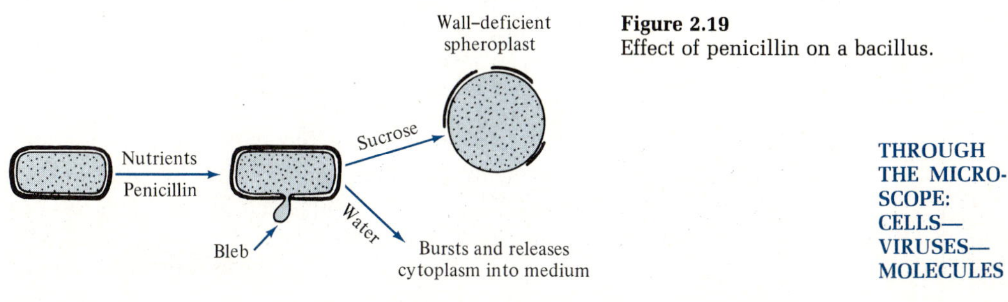

Figure 2.19
Effect of penicillin on a bacillus.

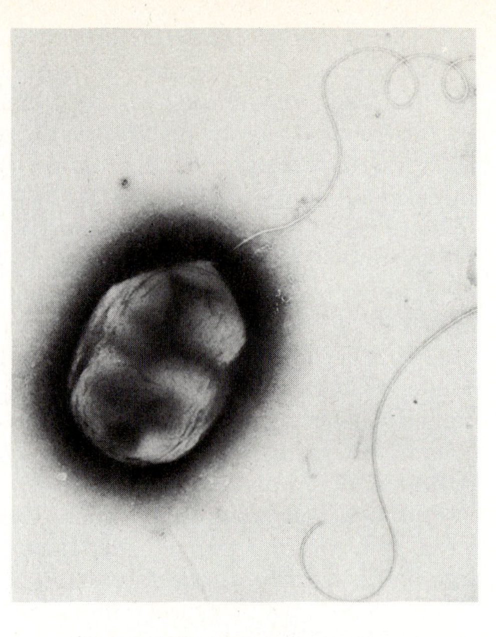

Figure 2.20
Electron micrograph of the bacterium *Pseudomonas aeruginosa* showing a single polar flagellum (26,000×). (Courtesy of M. Kessel.)

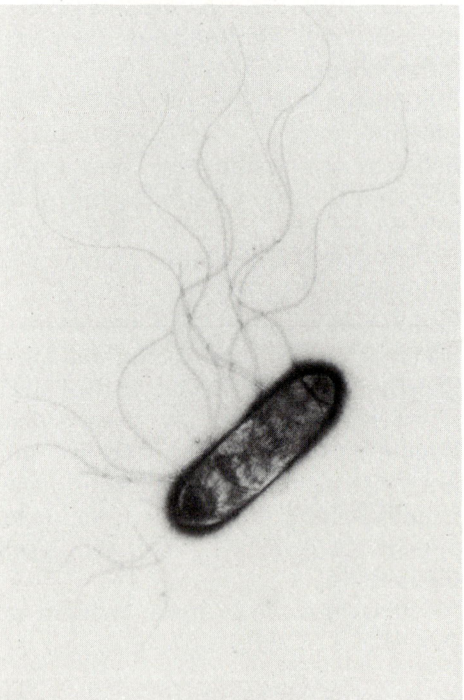

Figure 2.21
Electron micrograph of *Escherichia coli* showing flagella distributed around the cell (18,000×). (Courtesy of M. Kessel.)

by giving them a "shave." Using a common kitchen blender, flagella can be sheared off the cells without killing the bacteria. The purified flagella fraction is composed almost entirely of a single type of protein. The shaved cells slowly grow new flagella. The ability of the bacteria to swim coincides with the production of new flagella, providing strong circumstantial evidence that the proteinaceous flagellum is the organelle of motility.

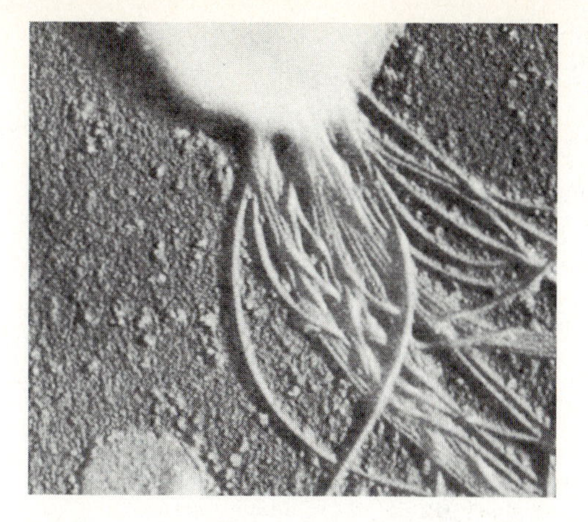

Figure 2.22
Bacterial flagella in the electron microscope. The flagellum, an organelle of locomotion, is composed entirely of protein. Note the flagella originating from within the cytoplasm (214,000×). (Courtesy of R. J. Martinez.)

The mechanism by which flagella propel bacteria is now well established (see reference "How Bacteria Swim" at the end of this chapter). The flagella become oriented in one direction and then rotate as a single helical unit. This type of propeller action pushes the cell in the opposite direction. A typical velocity for a bacterium is about 18 cm per hour. Relative to its size, this is an astounding speed. A bacterium is about one millionth the length of man. If you multiply 18 cm per hour by a million, you arrive at 180 km per hour!

Many procaryotic cells are able to concentrate and store nutrients in the form of **cytoplasmic granules** or inclusions. In addition to glycogen, starch, and fats (storage materials of certain eucaryotic cells), procaryotes store a wide variety of inorganic and organic materials, such as phosphates, sulfur, hydrocarbons, and organic acids. These substances are generally stored in the form of water-insoluble polymers. It is unusual for any particular bacterium to produce more than one type of storage granule. When the cells become starved, specific enzymes break down the polymers into small molecules that can serve as nutrients. A thin section of a bacterium containing polyphosphate granules is shown in Figure 2.23. The function and chemical compo-

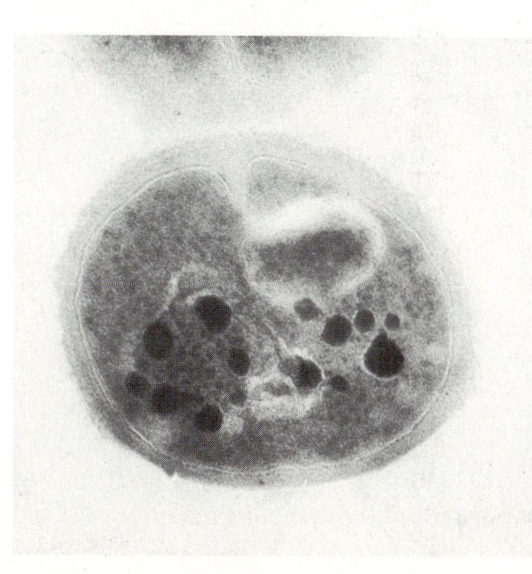

Figure 2.23
Electron micrograph of a thin section of a coccus. The dark cytoplasmic granules are composed of phosphate polymers (60,000×). (Courtesy of I. Friedberg.)

sition of the most common subcellular structures found in procaryotic cells and discussed in this section are summarized in Table 2.2.

TABLE 2.2 Subcellular Structures of Procaryotic Cells

Structure	Principal Function	Chemical Composition
Nuclear region	Contains the genetic information	DNA
Cytoplasmic membrane	Regulates movement of materials in and out of the cell; generates cell energy	Lipid and protein
Ribosome	Site of protein synthesis	RNA and protein
Mesosome	May be involved in cell division	Lipid and protein
Cell wall	Maintains cell shape and rigidity	Polysaccharide–polypeptide
Flagellum	Motility	Protein
Storage granule	Reserve supply of nutrients	Different types

2.5 Structure and Function in the Eucaryotic Cell

The eucaryotic cell is far morc complcx than the procaryotic cell, with regard to both the number of different types of subcellular organelles and the organization of each of the structures. For comparative purposes, it is interesting to ask the following question: What structure in the eucaryotic cell is equivalent to each of the seven structures listed in Table 2.2?

In place of a single nuclear region, the eucaryotic cell has a well-defined nucleus that is separated from the cytoplasm by a nuclear membrane (Figure 2.24). The genetic information of eucaryotes is or-

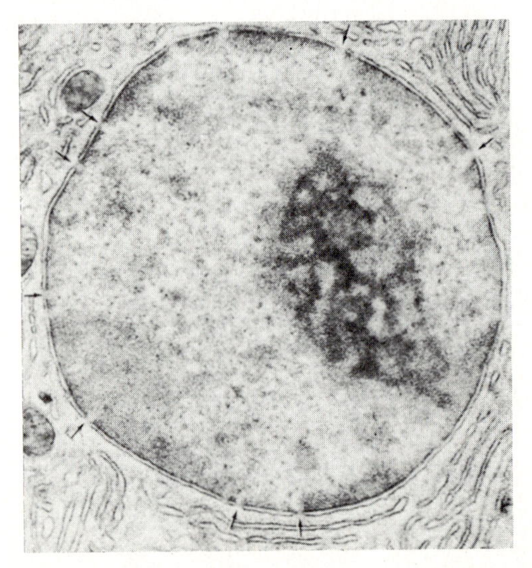

Figure 2.24
Electron micrograph of the nucleus from a pancreas cell of the mouse. Note the pores in the nuclear membrane. The cytoplasm contains extensive membranes (17,500×). (Courtesy of D. W. Fawcett.)

ganized onto chromosomes, which reside inside the nucleus. Prior to cell division, the nucleus separates into two equivalent nuclei by a process called **mitosis.** Some of the different phases of this complex process of mitosis are shown in Figures 2.25 and 2.26.

The membrane system of eucaryotic cells is highly complex. In some cases, the cytoplasmic membrane will increase its surface area by developing elongated slender protrusions, as shown in Figure 2.27. Even more common in eucaryotic cells is an extensive system of membranes found in the cytoplasm called **endoplasmic reticulum** (Figure 2.28). In some cases, the endoplasmic reticulum can be completely covered with ribosomes. As mentioned in Section 2.4, energy-generating reactions in procaryotic cells are carried out by enzymes associated with the cytoplasmic membrane. In eucaryotic cells, special membrane-bound organelles are involved in performing these tasks. Photosynthesis in algae and higher plants is carried out in chloroplasts (Figure 2.29). Respiration in all eucaryotic cells takes place in the mitochondrion (Figures 2.30 and 2.31). Energy-generating processes are discussed in Chapter 4.

The overall design of the procaryotic and eucaryotic ribosomes is similar. Although both structures are composed of about equal amounts of RNA and protein, there are sufficient differences in their size and detailed structure to allow certain antibiotics to selectively

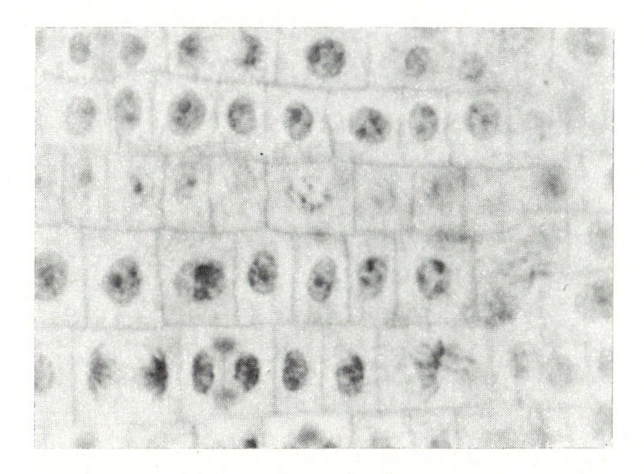

Figure 2.25
Onion root cells at various stages of mitosis (1900×).

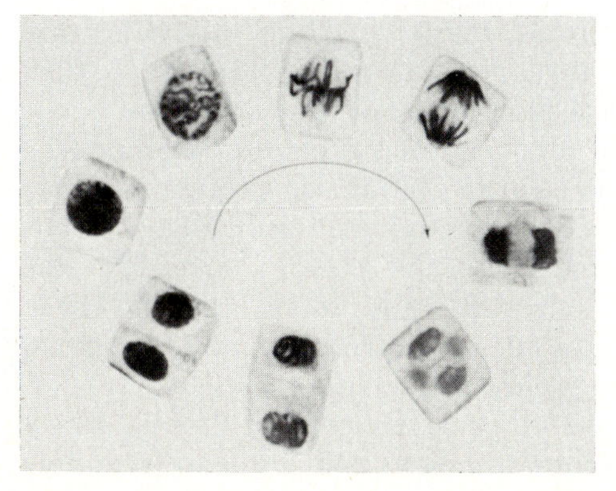

Figure 2.26
The ordered sequence of events during mitosis in the onion root cell. Reconstructed from micrographs like Figure 2.27.

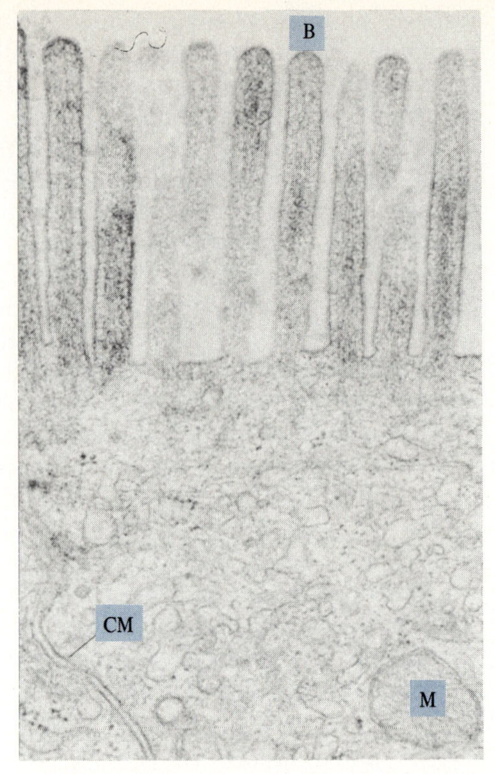

Figure 2.27
Electron micrograph of a thin section of an epithelial cell from a mouse intestine. In addition to the cytoplasmic membrane (CM), a mitochondrion (M) and the brush border (B) are clearly visible (60,000×). (Courtesy of F. Sjostrand.)

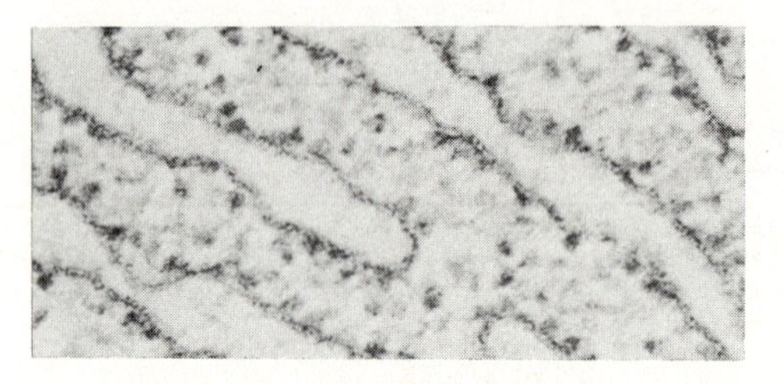

Figure 2.28
Endoplasmic reticulum at high-power magnification from sectioned mouse pancreas. This extensive system of membranes is found in the cytoplasm of eucaryotic cells. In this micrograph, the membranes are coated with small spherical bodies called ribosomes. As discussed in Chapter 7, ribosomes are the sites of protein synthesis (184,000×). (Courtesy of F. Sjostrand.)

inhibit only one type. For example, the tetracycline antibiotics owe their effectiveness to the fact that they poison the procaryotic ribosome without affecting the corresponding eucaryotic structure.

As mentioned previously, the chemical structure of the bacterial cell wall is unique and can be used to distinguish procaryotic from eucaryotic cells. Most animal cells lack a rigid cell wall. Higher plants have cell walls composed of cellulose and pectin. Most algae and fungi also contain polysaccharide cell walls. In no case, however, does the chemical structure of a eucaryotic cell resemble the cross-linked peptidoglycan structure typical of bacteria.

Eucaryotic cells that are motile generally contain flagella. These flagella, however, are much more complex than the simple protein

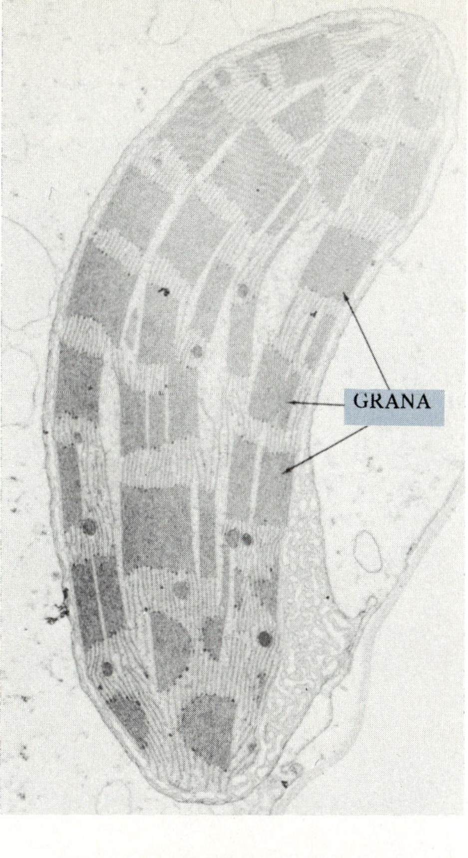

Figure 2.29
A section of a corn leaf chloroplast. Note the complex internal membranes which contain chlorophyll. The heavily concentrated membranes are called grana (50,000×). (Courtesy of L. K. Shumway.)

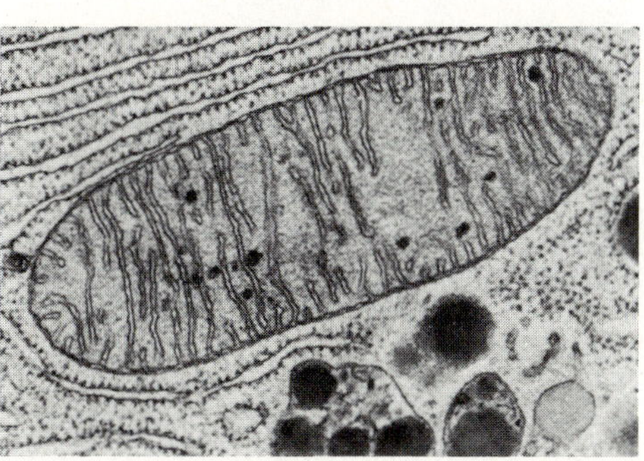

Figure 2.30
A typical mitochondrion from a bat pancreas, showing characteristic projections of membranes into the interior. The mitochondria are the major sites of energy generation in eucaryotic cells. Note the endoplasmic reticulum coated with ribosomes in the upper left corner (40,000×). (Courtesy of K. R. Porter.)

flagella of bacteria. Figure 2.32 shows the characteristic "9 + 2" arrangement of eucaryotic flagella. Each flagellum consists of nine double tubules around the circumference of two single tubules in the center. The entire structure is surrounded by a membrane. Certain protozoa contain cilia as organelles of motility instead of flagella. Cilia are shorter than flagella but contain the same "9 + 2" arrangement. One mode of locomotion that is possible in eucaryotic cells, but not in procaryotes, is amoeboid movement. The rigid bacterial cell wall and lack of directed cytoplasmic streaming preclude amoeboid motion in bacteria.

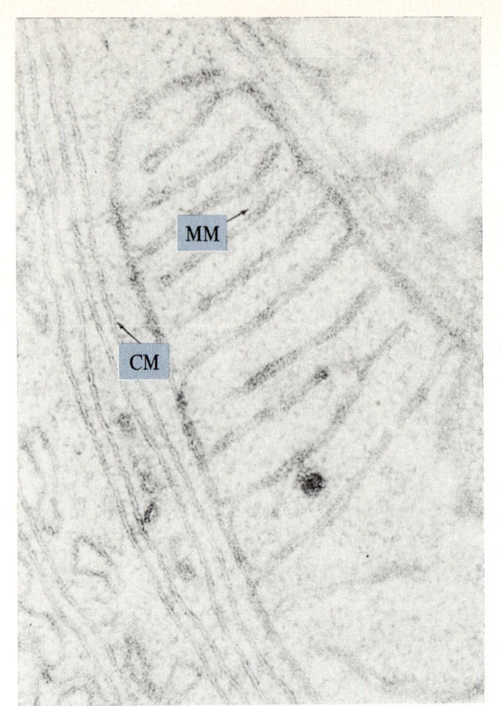

Figure 2.31
A thin section of mouse kidney showing two types of membrane, the layered cytoplasmic membrane (CM) and the globular mitochondrial membrane (MM) (160,000×). (Courtesy of F. Sjostrand.)

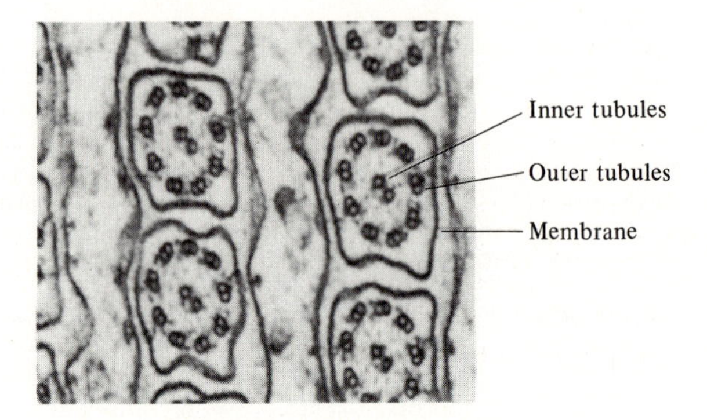

Figure 2.32
Cross-section of a flagellum from a protozoan. Each flagellum consists of nine double tubules around the circumference of two single tubules. The entire structure is enclosed by a membrane (65,000×). (From Nester, E. W., et al.: Microbiology, 2nd ed. New York: Holt, Rinehart and Winston, 1978.) (Courtesy of A. W. Grimstone.)

Storage granules in eucaryotic cells are generally composed of lipid or a polysaccharide: starch in the case of plants and glycogen in the case of animals. A photosynthetic microorganism showing numerous starch granules is shown in Figure 2.33.

2.6 Virus Architecture

Viruses were discovered long before they were seen. In the late nineteenth century, several diseases of plants and animals were shown to be caused by "microbes" so small that they could not be detected in the light microscope. Nevertheless, these submicroscopic germs were purified and studied because of their obvious importance in the disease process. It was not until about 1950 that the first clear

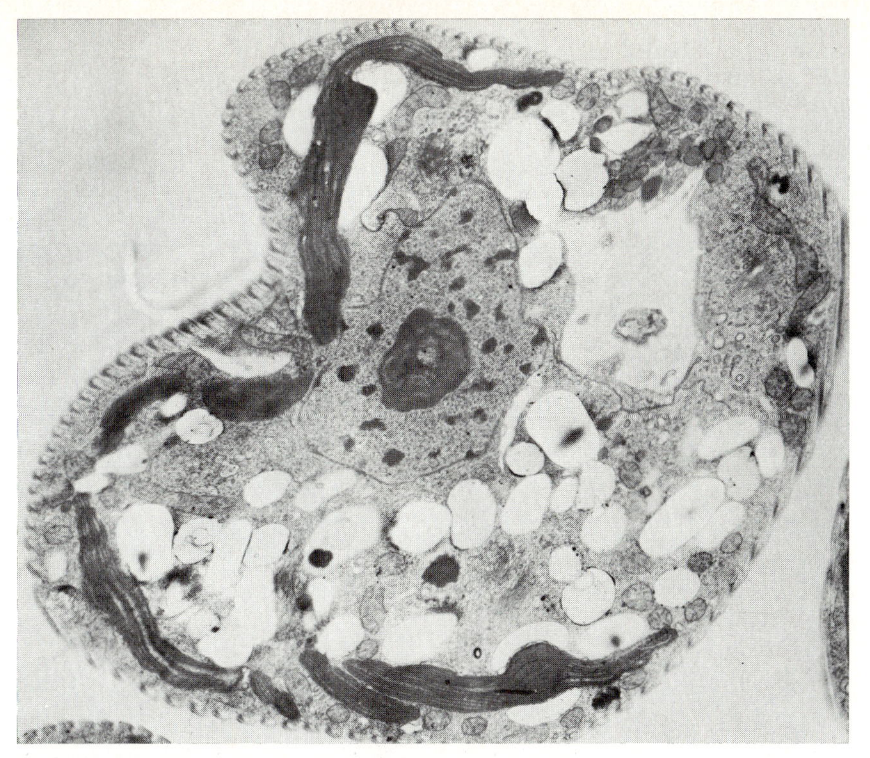

Figure 2.33
Thin section of a photosynthetic microorganism in the electron microscope. The complexity of the eucaryotic cell is well illustrated (16,650×). (Courtesy of Y. Ben Shaul.)

photographs of viruses were published. Resolution of the beautiful symmetry in viral architecture required both the development of the electron microscope and procedures for preparing samples for viewing in such a way that they would not be distorted. This latter point was appreciated by Lewis Carroll in 1876 when he wrote

> You boil it with saw dust; you salt it with glue:
> You condense it with locusts and tape:
> Still keeping one principal object in view—
> To preserve its symmetrical shape.

Some of the fascinating shapes of viruses can be seen in Figures 2.34 to 2.37. These electron micrographs clearly demonstrate two of the four features common to all viruses: (1) Viruses are **very small** (compare the magnifications of these pictures with those of typical eucaryotic and procaryotic cells, e.g., Figures 2.6 and 2.7), and (2) viruses are **acellular.** A third property shared by viruses is that they generally are **composed of only two kinds of macromolecules**—protein and either DNA or RNA. The nucleic acid is surrounded by a protein coat. If the coat is damaged, the tightly coiled nucleic acid molecule is ejected (see Figure 2.38). The fourth characteristic that helps to define the kingdom of viruses is that they are all **obligate intracellular par-**

Figure 2.34
Portrait of the bacterial virus PBSI. The structural features include the head (H), collar (C), tail (T), and tail fibers (TF) (386,000×). The biology of viruses is discussed in Chapter 4. (Courtesy of F. Eiserling.)

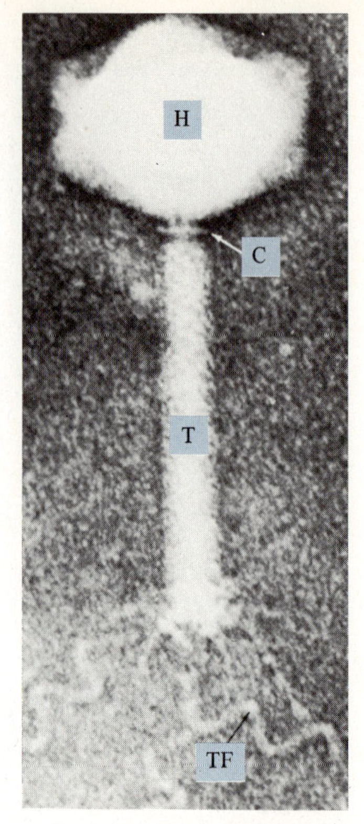

Figure 2.35
A mixture of three different viruses—one that infects tobacco plants, tobacco mosaic virus (TMV), and two that infect the bacterium *Escherichia coli*, bacteriophages T_4 and ϕX174 (213,000×). (Courtesy of F. Eiserling.)

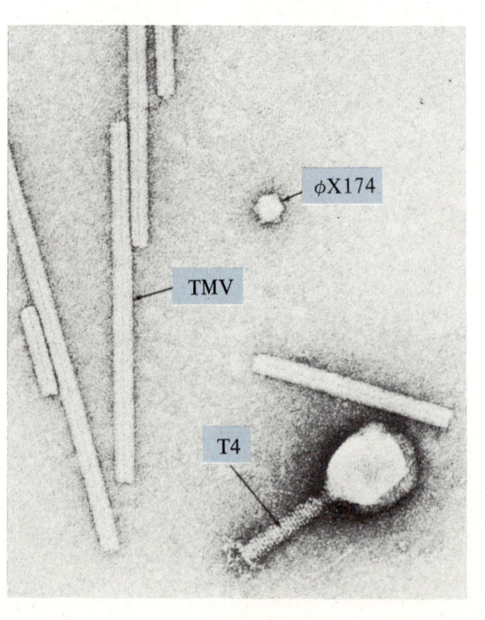

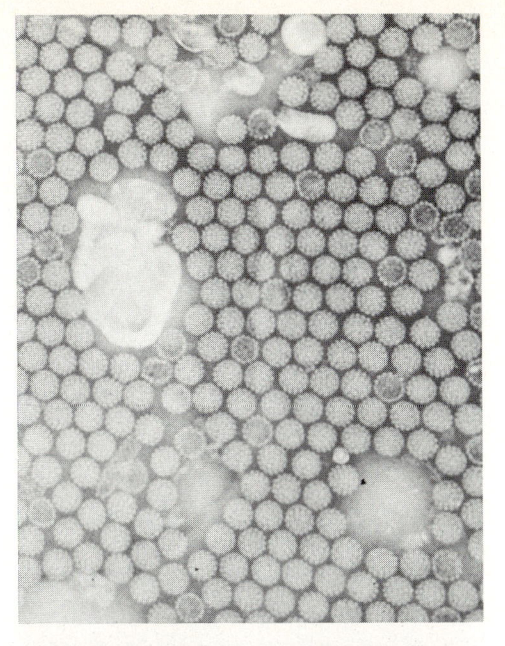

Figure 2.36
Human wart viruses (116,000×). (Courtesy of A. Klug and J. T. Finch.)

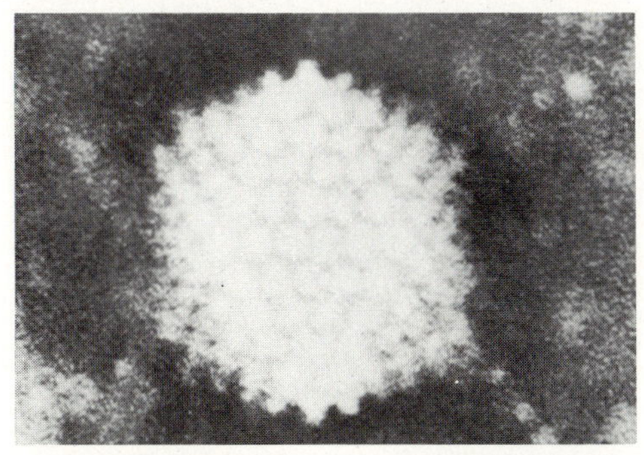

Figure 2.37
Adenovirus (740,000×). (Courtesy of R. C. Valentine and H. G. Pereira.)

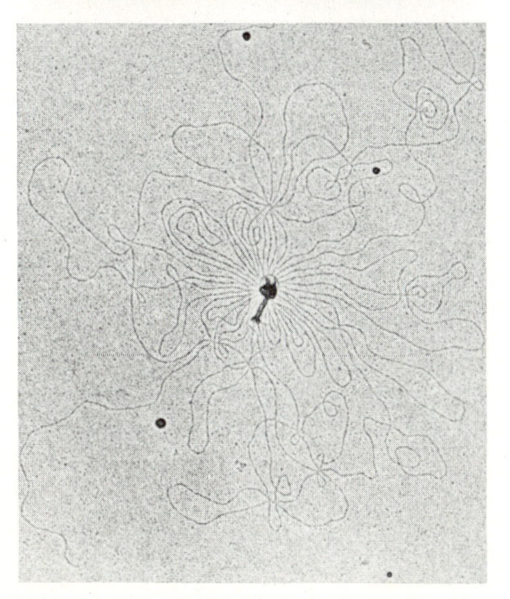

Figure 2.38
Thread-like DNA released from a virus. Note the viral remnant in the center (51,000×). (Courtesy of A. K. Kleinschmidt.)

asites; that is, they multiply only inside living cells. The unique manner in which viruses propagate is discussed in Chapter 5.

2.7 Molecules: Protein, RNA, and DNA

Using special staining techniques, it is possible to observe large molecules in the electron microscope. An example of a protein molecule is the enzyme glutamine synthetase, shown in Figure 2.39. A

Figure 2.39
A doughnut-shaped protein molecule, glutamine synthetase (400,000×). (Courtesy of F. Eiserling.)

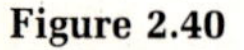

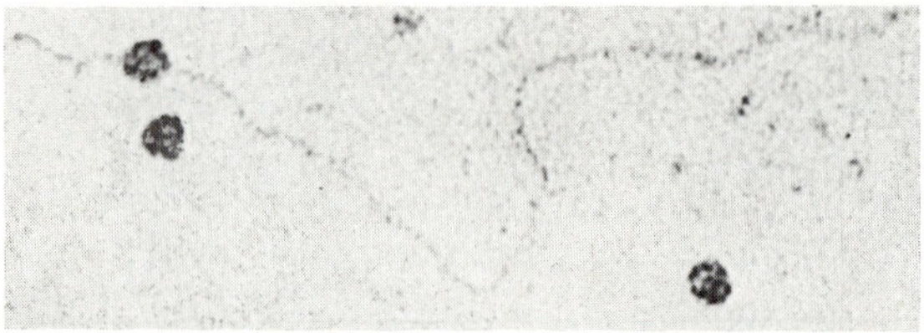

Figure 2.40
High resolution electron micrograph of an RNA molecule. One of the ribosome particles is bound near the end of the RNA (50,000×). (Courtesy of M. Hertzberg.)

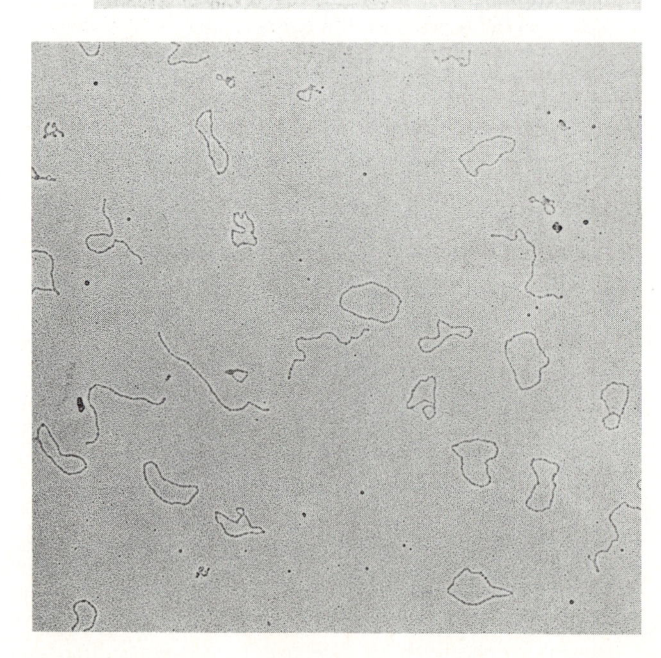

Figure 2.41
Electron micrograph of DNA molecules from bacterial viruses. Each virus contains a single DNA molecule of a defined length and no ends. During preparation for electron microscopy, some of the native circular molecules are broken (70,000×). (Courtesy of Y. Ben Shaul.)

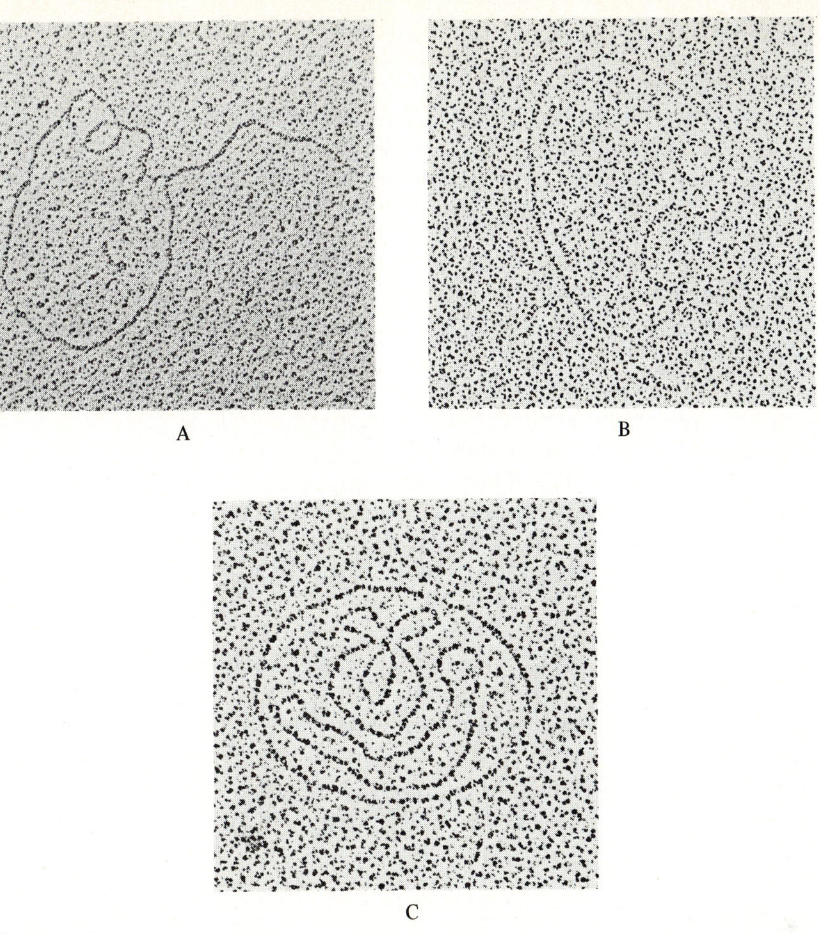

Figure 2.42
The DNA of an animal virus: (A) smoking DNA; (B) shouting DNA; and (C) smiling DNA. These DNA molecules were selected from hundreds of photographs. The "smoking" and "shouting" DNAs are made of single molecules, while the "smiling" DNA is made of three identical molecules. In all cases, the molecule has a length of 1.4 microns and no ends. (Courtesy of M. Hertzberg.)

ribosome attached near the end of an RNA molecule can be seen in Figure 2.40. Finally, one of the favorite subjects of electron microscopists is viral DNA molecules (Figure 2.41). When placed on the supporting medium for examination, the DNA assumes a variety of shapes. To demonstrate that scientists are not (necessarily) without a sense of humor, one electron microscopist put together the assemblage shown in Figure 2.42.

QUESTIONS

2.1 The magnification of most of the photographs in this chapter is indicated in the legend. Using a ruler, carefully measure the diameter of the typical eucaryotic cell shown in Figure 2.7, the procaryotic coccus in Figure 2.25, the mitochondrion in Figure 2.29, a ribosome in Figure 2.30, and the virus ϕ174 in Figure 2.37. Using these measurements and the appropriate magnifications, calculate the following:

a) The *unmagnified diameter* of each structure in centimeters and microns (1 centimeter = 10,000 microns).

b) The *volume* of each structure in cubic microns using the equation $V = 4/3\ \pi r^3$, where V is the volume and r is the radius of the sphere.

c) The *surface area* of the eucaryotic and procaryotic cells, using the equation, $S = 4/3\ \pi r^2$.

2.2 Microscopes are used far more frequently by biologists than by chemists. Explain why.

2.3 Two common bacterial species causing disease in humans are *Streptococcus pyogenes* and *Staphylococcus aureus*. What can you infer about their cell morphology from their names?

2.4 The functions and chemical compositions of seven important subcellular structures found in procaryotic cells are summarized in Table 2.2. What equivalent structure in the eucaryotic cell is involved in performing each of these functions? How do each of these structures differ in procaryotic and eucaryotic cells? What additional subcellular structures are present in eucaryotic cells?

2.5 Describe what would happen if a fresh-water bacterium was placed in seawater.

2.6 Explain the reason why penicillin is not effective in combating fungal or viral infections.

2.7 The length of the virus DNA molecule shown in Figure 2.42 is 1.4 microns. Calculate the magnification of the electron micrograph.

Suggested Readings

Berg H: How Bacteria Swim. Scientific American, August 1975.
A lucid account of how flagella propel bacteria through water.

Bradbury S: The Microscope: Past and Present. Oxford: Pergamon Press, 1968.
A historical account of the development of the microscope from Leeuwenhoek's time to the scanning electron microscope.

Everhart TE and TL Hayes: The Scanning Electron Microscope. Scientific American, January 1972.
A good explanation of how the electron microscope works, with special emphasis on the newer scanning electron microscope.

Fox CF: The Structure of Cell Membranes. Scientific American, February 1972.

Nester EW, CE Roberts, NN Pearsall, and BJ McCarthy: Microbiology, 2nd Edition. New York: Holt, Rinehart and Winston, 1978.
Chapter Three of this excellent textbook has a detailed discussion of functional anatomy of procaryotic and eucaryotic cells.

Sharon N: The Bacterial Cell Wall. Scientific American, May, 1969.

Thomas L: The Lives of a Cell. New York: Bantam Books, Inc., 1974.
This is a rare book, combining wit, style, humanism, and modern biology.

Energy Transformations in Biological Systems

Give me matter and motion and I will make the world.

René Descartes

As seen in Chapter 2, living material at all levels—multicellular, cellular, subcellular, and viral—is in a highly organized state. To maintain and reproduce these complex structures, living organisms require a constant supply of energy. The series of closely coordinated, delicately balanced biochemical reactions by which the cell is able to acquire energy from various nutrients and then utilize the energy for the synthesis of cellular material is called **metabolism.**

3.1 The Fermentation Controversy

The laws of thermodynamics* allow us to predict whether or not a particular chemical reaction is possible and in which direction it will proceed, but thermodynamics cannot tell us *when* or *how fast* the reaction will go. For example, consider the union of hydrogen and oxygen gas to produce water:

$$2\,H_2 + O_2 \rightarrow 2\,H_2O$$

Since the energy of two water molecules is considerably less than the energy of one oxygen and two hydrogen molecules, the reaction must proceed in the direction indicated by the arrow. However, at temperatures below 400°C, water is not formed at a measurable rate. If a small amount of finely divided platinum is added, however, the reaction will then proceed rapidly. Surprisingly, the platinum is not used up, but may be recovered unchanged at the end of the reaction. This is an example of a rather common phenomenon known as **catalysis.** Any substance that speeds up the rate of a chemical reaction,

*The first and second laws of thermodynamics are discussed in Appendix A.

but which itself can be recovered unchanged at the end of the overall process, is known as a **catalyst.**

Most of our knowledge concerning catalysis in living cells can be traced back to the nineteenth-century controversy regarding the causative agent of **fermentation.** The phenomenon of fermentation as seen in the production of wine, beer, cheese, and bread has been known since antiquity. For this discussion, the production of beer will serve to illustrate the basic principles.

Beer is made from cereal seeds, such as barley, wheat, rye, rice, and corn. After the seeds have been steeped for some time in water, they are drained and subjected to sufficient temperature to cause the moist grain to germinate; dried germinated barley seeds are called *malt.* The malt is ground, mashed in warm water, and boiled to extract the sugar and flavor from the broken cells. Hops (the dried flowers of *Humulus lupulus)* are added when the boiling is almost over, both for their bitter flavor and because the extract tends to inhibit bacterial growth. The resulting extract, the *wort,* contains a moderate concentration of sugar which has been set free from the starch in the grain. The wort is immediately cooled, and large quantities of yeast are added to it; the yeast is usually obtained from a previous fermentation. Shortly after addition of the yeast, the wort starts the bubbling and frothing from which fermentation derives its name (*fervere,* to boil). After several days the reaction ceases and a copious precipitate begins to settle to the bottom of the vessel. The clarified liquid (beer) is then stored in vats at low temperatures for several weeks (lagering) prior to bottling. The sediment, which has been called the ferment or yeast, is used to inoculate a fresh wort.

The art of brewing was developed by trial and error over a 6000-year period and practiced without any understanding of the underlying principles. From long experience, the brewer learned the conditions, not the reasons, for success. Only with the advent of experimental science in the eighteenth and nineteenth centuries did man attempt to explain the mysteries of fermentation. Let us, then, from our vantage point in time, trace the observations, experiments, and debates from which evolved our present understanding of fermentation and biological catalysis.

For centuries, fermentation had a significance that was almost equivalent to what we would now call a chemical reaction, an error that probably arose from the vigorous bubbling seen during the process. The conviction that fermentation was strictly a chemical event gained further support during the early part of the nineteenth century, when French chemists led by Lavoisier and Gay-Lussac determined that the alcoholic fermentation process could be expressed chemically by the following equation:

$$\underset{\text{glucose}}{C_6H_{12}O_6} \rightarrow 2\ \underset{\text{ethyl alcohol}}{C_2H_5OH} + 2\ \underset{\text{carbon dioxide}}{CO_2}$$

It was, of course, known that yeast must be added to the wort in order to ensure a reproducible and rapid fermentation. The function of the yeast, according to the chemists, was merely to catalyze the process. All chemists agreed that fermentation was in principle no different from other catalyzed chemical reactions.

Then in 1837, the French physicist Charles Cagniard-Latour and the German physiologist Theodor Schwann independently published studies that indicated yeast was a living microorganism. Prior to their publications, yeast was considered a proteinaceous chemical substance. The reason the two workers came up with the same observations at approximately the same time is most likely due to the production of better microscopes. As Cagniard-Latour stated,

> Twenty-five years ago I first examined fresh yeast under the microscope. However, my instrument was very poor and I concluded that the yeast was like a very fine sand composed of crystalloid particles. These observations were in error. The majority of the microscopic observations indicated in this memoir was performed on a microscope recently constructed by M. Georges Oberhauser. It enabled me to obtain enlargements of 300–400 times.

It should be mentioned that one of the reasons it was difficult to ascertain whether or not yeast is living was because, like most other fungi, yeast is not motile. The organized cellular nature of yeast was discovered only when improved microscopes became available. Schwann and Cagniard-Latour also observed that alcoholic fermentation always began with the first appearance of yeast, progressed only with its multiplication, and ceased as soon as its growth stopped. Both scientists concluded that alcohol is a by-product of the growth process of yeast. As Schwann stated,

> The alcoholic fermentation must be considered to be that decomposition which occurs when the sugar fungus utilizes sugar and nitrogen containing substances for its growth, in the process of which the elements of these substances which do not go into the plant are preferentially converted into alcohol. Most of the observations on the alcoholic fermentation fit quite nicely with this explanation.

The biological theory of fermentation advanced by Cagniard-Latour and Schwann was immediately attacked by the leading chemists of the time. The eminent Swedish physical chemist Jons Jakob Berzelius reviewed the two papers in his *Jahresbericht* for 1839 and concluded that microscopic evidence was of no value in what was obviously a purely chemical problem. According to Berzelius, nothing was living in yeast:

> It was only a chemical substance which precipitated during the fermentation of beer and which had the usual shape of a noncrystalline precipitate.

To the scorn of Berzelius was soon added the sarcasm of two great German chemists, Justus von Liebig and Friedrich Wohler. In 1839 there appeared in the *Annalen der Chemie*, a reputable scientific journal, an anonymous† article entitled "The Riddle of the Alcoholic

†It was later revealed that the article was authored by none other than Liebig and Wohler, the editors of the *Annalen*.

Fermentation Solved." In this scientific travesty, the yeast is viewed through a powerful new microscope (Figure 3.1). When fed a solution of sugar, the yeast devours it, and a stream of alcohol is seen flowing from the anus while carbon dioxide bubbles out of its enormously enlarged genital organs. The bladder of the yeast when full has the shape of a champagne bottle.

In addition to this farce, Liebig published a serious paper, containing several important arguments against the biological theory of fermentation. Liebig's two major points can be summarized as follows:

1. Certain types of fermentation, such as the lactic acid (souring of milk) and acetic acid (formation of vinegar) fermentations, can occur in the complete absence of yeast.
2. Even if yeast is living, it is not necessary to conclude that the alcoholic fermentation is a biological process. The yeast is a remarkably unstable substance which, as a consequence of its own death and decomposition, catalyzes the the splitting of sugar. Thus, fermentation is essentially a chemical change catalyzed by breakdown products of the yeast.

Liebig's views were widely accepted, partly because of his powerful influence in the scientific world and partly because of a desire to avoid seeing an important chemical change relegated to the domain of biology. And so the stage was sct—biology against chemistry—for the entrance once again of Louis Pasteur.

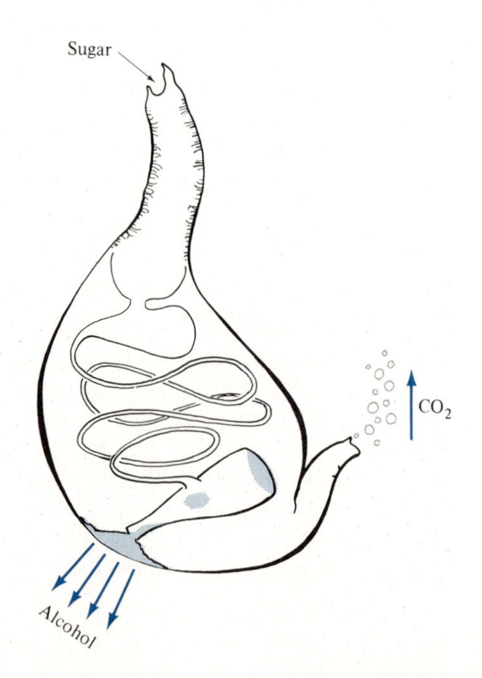

Figure 3.1
The yeast cell as described by Liebig and Wohler. (Original drawing by F. W. Taylor from an English translation.)

In 1857, Pasteur published his first paper on the topic of fermentation. The publication dealt with lactic acid fermentation, not alcoholic fermentation. Utilizing the finest microscopes of the time, Pasteur discovered that souring of milk was correlated with the growth of a microorganism, but one considerably smaller than the beer yeast. During the next few years, Pasteur extended these studies to other fermentative processes, such as the formation of butyric acid as butter turns rancid. In each case he was able to demonstrate the involvement of a specific and characteristic microorganism; alcoholic fermentation was always accompanied by yeasts, lactic acid fermentation by nonmotile bacteria, and butyric acid fermentation by motile rod-shaped bacteria. Thus, Pasteur not only disposed of one of the opposition's strongest arguments, but also provided powerful circumstantial evidence for the biological theory of fermentation.

Now Pasteur was ready to attack the crucial problem, alcoholic fermentation. Liebig had argued that this fermentation was the result of the decay of yeast; the proteinaceous material that is released during this decomposition catalyzes the splitting of sugar. Pasteur countered this argument by developing a protein-free medium for the growth of yeast. He found that yeast could grow in a medium composed of glucose, ammonium salts, and some incinerated yeast. If this medium is kept sterile, neither growth nor fermentation takes place. As soon as the medium is inoculated with even a trace of yeast, growth commences and fermentation ensues. The quantity of alcohol produced parallels the multiplication of the yeast. In this protein-free medium, Pasteur was able to show that fermentation takes place without the decomposition of yeast. In fact, the yeast synthesizes protein at the expense of the sugar and ammonium salts. Thus Pasteur concluded in 1860:

> Fermentation is a biological process, and it is the subvisible organisms which cause the changes in the fermentation process. What's more, there are different kinds of microbes for each kind of fermentation. I am of the opinion that alcoholic fermentation never occurs without simultaneous organization, development and multiplication of cells, or continued life of the cells already formed. The results expressed in this memoir seem to me to be completely opposed to the opinion of Liebig and Berzelius.

Pasteur argued effectively, and more important, all the data were on his side. Thus the vitalistic theory of fermentation predominated until 1897, when an accidental discovery by Eduard Buchner finally resolved the controversy and threw open the door to modern biochemistry.

Buchner, working in Munich with his brother Hans, was attempting to obtain from yeast an extract that might have medicinal value. After several unsuccessful trials, he discovered that yeast could be disrupted and the cell sap released by grinding a mixture of intact cells and fine sand with a mortar and pestle. After filtering the mash to remove the sand and any unbroken cells, a clear yeast juice was obtained. The juice, however, soon became contaminated with bacterial growth. Since the extract was to be used for human consumption, Buchner could not utilize ordinary antiseptics to prevent the spoilage. Therefore, he attempted to preserve the yeast extract by adding large quantities of sugar—much as one uses high concentrations of sugar in preserving jam and jelly. To Buchner's utter amazement, the yeast extract began to bubble soon after the sugar was added. Careful analysis revealed that the sugar was decomposing to carbon dioxide and ethyl alcohol. Fermentation had proceeded **in the absence of living cells.**‡

3.2 Biological Catalysts: The Enzymes

Buchner's achievement began a new era in the study of alcoholic fermentation and other metabolic processes. Reactions that normally take place only in living cells (in vivo) could now be studied in test tubes (in vitro). The agents that are present in cell extracts and that catalyze these reactions (make them go faster) are called **enzymes.** The term enzyme comes from the Greek words *en zyme,* meaning "in yeast."

During the twentieth century, enzymes have been separated and obtained in pure form from a wide variety of organisms. From studies on these purified enzymes, certain generalizations have emerged.

1. **Enzymes are true catalysts.** They control the rate of a chemical reaction, but they themselves are not used up during the process. Thus, they can be used over and over again. Although enzymes can control the speed of reactions, they cannot bring about reactions that otherwise would not occur, that is, reactions involving an increase in energy.
2. **Enzymes are large protein molecules.** Most enzyme molecules are composed of 100 to 1000 amino acids joined together in a specific sequence. In addition to the sequence of amino acids, the three-dimensional structure of the protein also plays a critical role in catalytic activity. For example, heating an enzyme can alter the shape of the molecule, and thus its catalytic activity will be destroyed.
3. **Enzymes are highly specific in their action.** Even the simplest microorganism contains more than 1000 different enzymes, each capable of catalyzing a specific reaction. In general, a different enzyme is used to catalyze each of the different steps in the synthesis and breakdown of organic molecules that take place in living organisms.
4. **Enzymes determine the metabolic pattern of the cells.** All important reactions that take place in living cells are catalyzed by enzymes; thus, species of cells must achieve their individ-

‡In 1907 Eduard Buchner received the Nobel Prize in chemistry "for his biochemical researches and his discovery of cell-less fermentation."

uality from the kind and amount of enzymes they contain. For example, the ability to derive nutrient value from protein depends on the presence of a group of enzymes, called proteinases.§ Absence of these proteinases means that the cell or organism cannot digest protein. Enzymes also determine the synthetic capacity of cells. Dark-skinned individuals have abundant quantities of the series of enzymes that catalyze production of the pigment melanin. The complete absence of any one of these enzymes results in albinism—fair skin, white hair, and pink eyes.

One of the most important areas of current biochemical research deals with the mechanism by which enzymes accelerate reactions. Although at present no theory adequately explains the phenomenon of enzyme catalysis, Figures 3.2 and 3.3 are offered as a visual representation of two possible mechanisms.

Before discussing biochemical aspects of energy metabolism, let us take one final look at the fermentation controversy in the light of

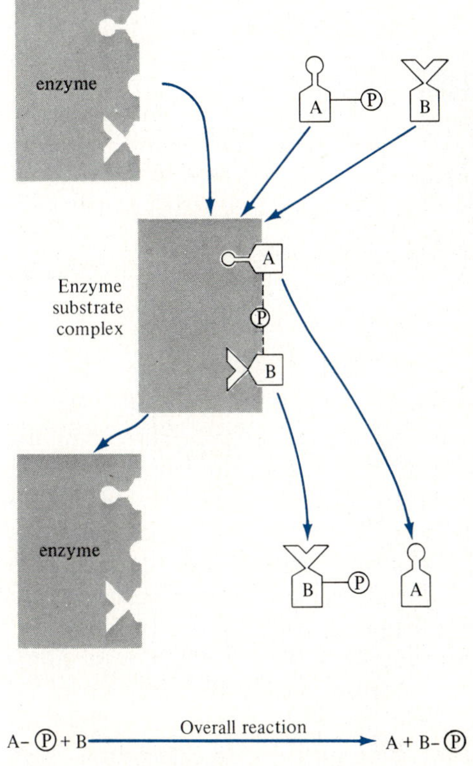

Figure 3.2
A schematic version of one type of enzyme catalysis. The overall reaction catalyzed by this hypothetical enzyme is the transfer of a phosphate from A to B. First, A–P and B are bound to a specific enzyme because of its surface geometry. In the complex that is formed, the phosphate comes into close contact with B, increasing the chance for the phosphate to be transferred. If no enzyme were present, the chance that A–P and B would collide in such a manner that the phosphate could be transferred would be greatly reduced. After the complex has dissociated to form A and B–P, the enzyme is released and can be reutilized. Such catalyzed reactions can occur at fantastic rates, more than one million per minute.

§The suffix *-ase* indicates that the substance is an enzyme. The few enzymes that end with *-in*, such as trypsin and pepsin, were named before an international ruling was made favoring the *-ase* ending. The root usually indicates the type of molecule on which the enzyme operates. Proteinases are enzymes that attack protein; peroxidases split peroxides; dehydrogenases remove hydrogen atoms; and so on.

Figure 3.3
A schematic version of a different type of enzyme catalysis. The overall reaction is the same as in Figure 3.2. However, in this case the phosphate is transferred first from A to the enzyme and then from the enzyme to B.

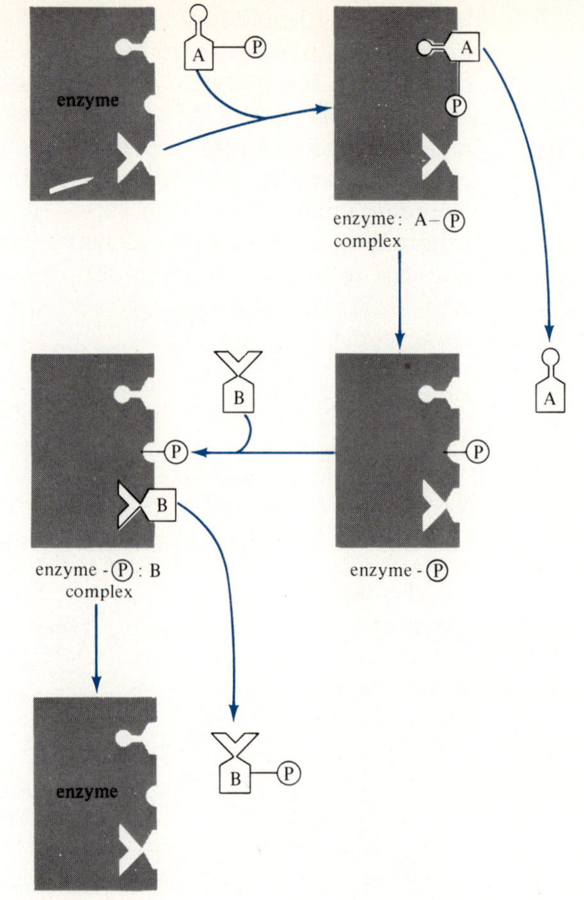

our current knowledge about enzymes. Who was right, Liebig or Pasteur? Pasteur was certainly wrong in his generalization that fermentation can occur only in living cells. However, he was correct in his fundamental thesis that fermentation is brought about in nature as a consequence of the living activities of microorganisms. Liebig's idea about the catalysis of fermentation by the decomposition of yeast was (and still is) inconsistent with the facts. Furthermore, his failure to recognize Pasteur's fundamental thesis was scientific blindness. In a more profound sense, however, Liebig was right in arguing that fermentation could be explained by chemistry. Today it is almost an article of faith that even the most complex biological process can be understood in chemical terms. Thus, there was some value in both points of of view. Fermentation is a biological process, resulting from a series of chemically intelligible reactions, each catalyzed by a specific enzyme. In the more general sense, it is the application of chemical principles to the solution of biological problems that best describes the discipline of **biochemistry.**

3.3 The Flow of Energy in the Biological World

With the principles of enzyme catalysis as a foundation, we are now ready to discuss energy transformations in biological systems. For purposes of discussion, cellular metabolism can be split into two groups, **energy-yielding** and **energy-requiring** reactions. In this section, the overall balance between the production and expenditure of energy is discussed in general terms. Subsequently, we examine in greater detail the three basic processes by which cells obtain energy: **fermentation, respiration,** and **photosynthesis**.

The most general energy-requiring process is termed **biosynthesis,** the series of reactions by which cells build large and complex molecules from small and relatively simple components. The biosynthesis of the bacterial cell wall, outlined in Chapter 2 (Figure 2.16), is an example of such a process. In Chapters 6 and 7 the biosynthesis of nucleic acids and protein is discussed in detail. Another energy-requiring process, **osmotic work,** is necessary for the transport and concentration of specific nutrients from the environment into the cell. Since molecules tend to randomize (second law of thermodynamics), it takes energy to concentrate the nutrients in the cell and pump the waste materials out. Many cells are able to concentrate chemicals in such a manner that the inside of the cell has a different electric charge from the outside fluid. The difference in charge between the cell and its surroundings makes possible **electric work,** the conduction of nerve impulses. A fascinating example of electric work is seen in the electric eel, which can deliver a shock of more than 300 volts. **Motion** of some sort is present in all cells, whether it is the crawling of amoeba, the swimming of bacteria, the contraction of muscle cells, or the movement of chromosomes during mitosis. In addition to these four general energy-requiring processes, some organisms need energy for specialized tasks, such as the production of light by fireflies and certain bacteria (bioluminescence).

For this general discussion, the two types of energy metabolism can be represented as follows:

$$\mathrm{A} \xrightarrow[\text{yielding}]{\text{Energy-}} \mathrm{B} \qquad\qquad \mathrm{C} \xrightarrow[\text{requiring}]{\text{Energy-}} \mathrm{D}$$

In the energy-yielding reaction, A is converted into a product of lower energy, B. Since the total energy of the system must be conserved during the process, the difference in energy between A and B has to be released in some other form. For example, when you digest a beefsteak, the complex protein molecules are broken down into molecules of lower energy content, carbon dioxide and water; during the process, part of the energy is released as heat and the remainder is utilized to promote energy-requiring processes. Conversely, the energy-requiring reaction $C \rightarrow D$, cannot proceed unless provided with additional energy. As it stands, the reaction is thermodynamically impossible because D is at a higher energy level than C. For D to be formed, energy must somehow by supplied. Since many energy-requiring processes are essential for life, a basic problem of cells is: How can the energy produced by energy-yielding reactions be used to "push" energy-requiring processes?

During the course of evolution, living organisms have developed an efficient method for linking reactions that consume energy with those that supply it. The method (Figure 3.4) utilizes as an intermediate "bridge" or linking system the interconversion of the all-important molecules **adenosine triphosphate** (ATP) and **adenosine diphosphate** (ADP).

The ATP:ADP system is similar to a battery. Forming ATP is equivalent to charging the battery, and requires energy. When ATP is broken down to ADP (discharging), the released energy can be used to do work. In living cells, the chemical linkage system works as shown in Figure 3.4. The energy released during the conversion of A to B is utilized to form ATP (charging); the energy needed for the conversion of C to D is supplied by the breakdown of ATP (discharging). In this way, energy-yielding and energy-requiring processes are intimately coupled.

In the economy of the cell, ATP has been compared to money in that it can be made and spent in a variety of ways. Adenosine triphosphate produced in one part of the cell can be stored and used anywhere in the cell whenever the need arises. Thus, ATP plays the primary role of intermediary between energy-producing and energy-requiring processes in all living organisms, whether they are microbes, animals, or plants.

The remaining sections of this chapter are devoted primarily to a more detailed discussion of the different ways in which cells produce ATP. In Chapters 6 and 7, we see how ATP is utilized as the immediate source of energy for cellular synthesis and growth.

3.4 Fermentation: "La vie sans air"

Fermentation is a relatively simple and primitive biochemical process by which certain living cells are able to obtain energy in the

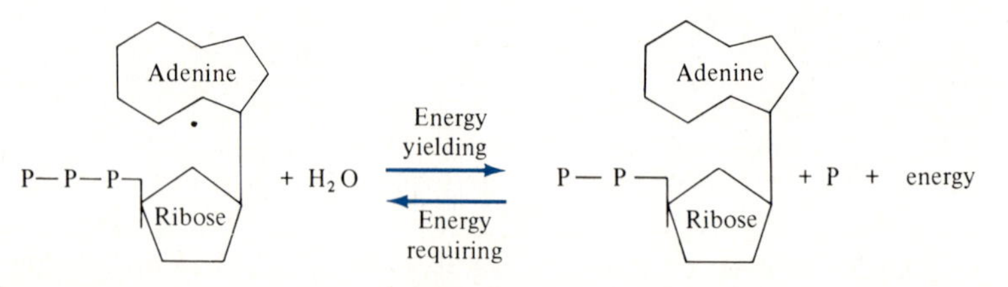

Figure 3.4
The interconversion of adenosine triphosphate (ATP) and adenosine diphosphate (ADP). The ATP molecule is composed of a base (adenine) joined to a sugar (ribose) that is also connected to three phosphate groups, abbreviated P. The ADP molecule has an identical structure except that it contains one less phosphate. Considerable energy is released when ATP and water react to form ADP and phosphate; conversely, energy is required for the reverse reaction, the synthesis of ATP.

absence of air. From a general thermodynamic point of view, the process can be stated as follows: Organic molecules are broken down into simpler substances of lower energy content; some of the free energy that is liberated during the process is captured in the formation of ATP. For example, when yeast transforms a molecule of glucose into ethyl alcohol and carbon dioxide, two ATP molecules are formed concurrently.

Following Buchner's discovery of a cell-free extract that could catalyze fermentation, it became possible to study the various intermediate steps of the process in great detail. By the late 1930s, owing to the experiments of a number of eminent biochemists, including the Englishman Arthur Harden and the German Otto F. Meyerhof, the exact fate of the glucose molecule during its degradation was established. For this discussion, however, we need to deliberate only on the essential features shown in Figure 3.5.

The fermentation of glucose can be considered to occur in two stages. In the first stage, one molecule of the sugar is broken down into two molecules of pyruvic acid and the equivalent of four hydrogen atoms.|| It should be pointed out that the formation of pyruvic acid from glucose actually entails at least ten distinct and sequential reactions. Each of these reactions is catalyzed by a specific enzyme. The enzymes work in tandem to produce two pyruvic acid and two ATP molecules for each glucose molecule consumed. Although there may be differences in detail, the overall reaction that takes place in the first stage of fermentation is a common theme of metabolism.

Fundamental differences in energy metabolism arise as a consequence of how the pyruvic acid and hydrogen are further metabolized. As we discuss in the next section, cells that are able to carry out the process of respiration burn pyruvic acid all the way to carbon dioxide

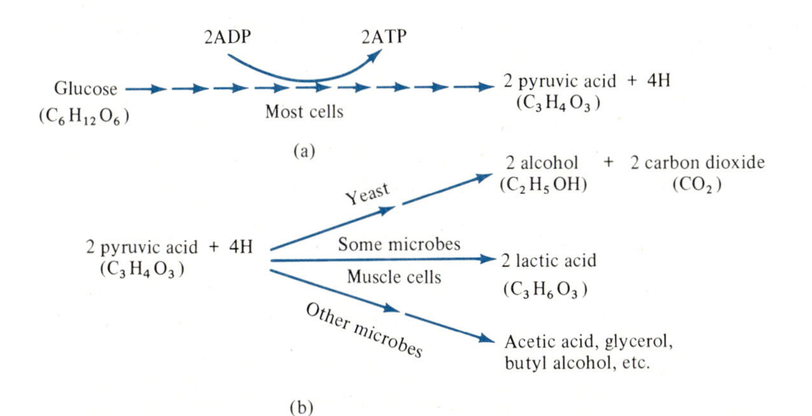

Figure 3.5
An outline of the sequence of reactions for the fermentation of glucose.

|| For simplicity, we will refer to hydrogen atoms in the remaining sections of this chapter. However, it should be realized that hydrogen atoms never actually exist free in the cell, but are transported by specific carrier molecules.

and water; the hydrogen atoms react indirectly with oxygen gas to form water. In the absence of oxygen, cells are faced with the serious problem of how to dispose of the hydrogen atoms. Essentially, the second stage of fermentation is the addition of hydrogen to pyruvic acid in one or more steps. The different fermentation products are a result of slight variations in this second stage. For example, in the simplest case, hydrogen is added directly to pyruvic acid to produce lactic acid. Yeasts, on the other hand, first split off carbon dioxide from the pyruvic acid and then add the hydrogen to form ethyl alcohol. In Chapter 10, microbial fermentation processes of economic importance are discussed from the viewpoint of the biochemical pathways that we have already discussed.

Whereas only a few types of microorganisms derive energy exclusively by fermentation, many types have the capacity to perform either fermentation or respiration. These **facultative** organisms carry out respiration as long as oxygen is available to them. When the oxygen supply is depleted, they switch to fermentation. This optional behavior is by no means limited to microorganisms. Certain tissues of plants and animals are also able to carry out fermentation when deprived of oxygen. A well-studied example is muscle tissue. Muscle contraction requires the expenditure of large quantities of ATP. During periods of vigorous exercise, muscle cells utilize oxygen much faster than it can be supplied by the heart and lungs. As the cells become starved of oxygen, they switch over to fermentation and produce ATP by the conversion of glucose to lactic acid. The fact that precisely the same steps are involved in lactic acid fermentation in a human muscle cell and in a lactic acid bacterium is another example of **unity in biology.**

3.5 Respiration

Respiration is the major means of ATP production in animals and most microorganisms. For our purposes, respiration can be defined as an energy-yielding metabolic process in which foodstuff reacts (indirectly) with oxygen, with the concurrent formation of ATP. The metabolic oxidation of glucose to yield carbon dioxide, water, and ATP is the most common example of respiration:

$$C_6H_{12}O_6 + 6\ O_2 \xrightarrow{38\ ADP \ \rightarrow\ 38\ ATP} 6\ CO_2 + 6\ H_2O$$

The greater efficiency of respiration (38 ATP per glucose) as compared with fermentation (2 ATP per glucose) is a result of the more complete breakdown of the sugar. When glucose is fermented to alcohol, for example, only a part of the energy stored in the sugar is released. This statement can be appreciated if we recall that considerable additional energy is liberated when alcohol is burned (oxidized) in the presence of air. The large number of steps involved in respiration makes it easier to control the rate of burning, so that energy can be efficiently captured as ATP, a form that can be used by the cell.

As previously discussed, the conversion of sugar to pyruvic acid is common to both fermentation and respiration. In respiration, however, pyruvic acid is further metabolized to carbon dioxide and water in a series of reactions called the **Krebs cycle** or **citric acid cycle** (Figure 3.6). This cycle was formulated in the early 1940s from biochemical data obtained primarily by two refugees of the Hitler regime, Hungarian-born Albert Szent-Gyorgyi[¶] and German-born Hans A. Krebs.

Before entering the cycle, pyruvic acid is converted into a two-carbon molecule with the release of carbon dioxide and a pair of hydrogen atoms. This two-carbon molecule combines with another cellular product that contains four carbon atoms, forming a six-carbon product, citric acid. Citric acid is then broken down in a series of reactions with the stepwise release of four pairs of hydrogen atoms and two carbon dioxide molecules. Finally, the original four-carbon compound is regenerated, which can then again combine with the two-carbon molecule and start the cycle all over.

The two principal benefits that the cell derives from the Krebs cycle are the creation of **building blocks** and production of **energy.** Some of the compounds formed during this cyclic system can be used to manufacture amino acids and other essential cellular components. For example, the amino acid aspartic acid ($C_4H_7O_4N$) is made by the enzyme-controlled addition of ammonia (NH_3) to the $C_4H_4O_4$ intermediate compound. In general, the Krebs cycle serves as the hub of the

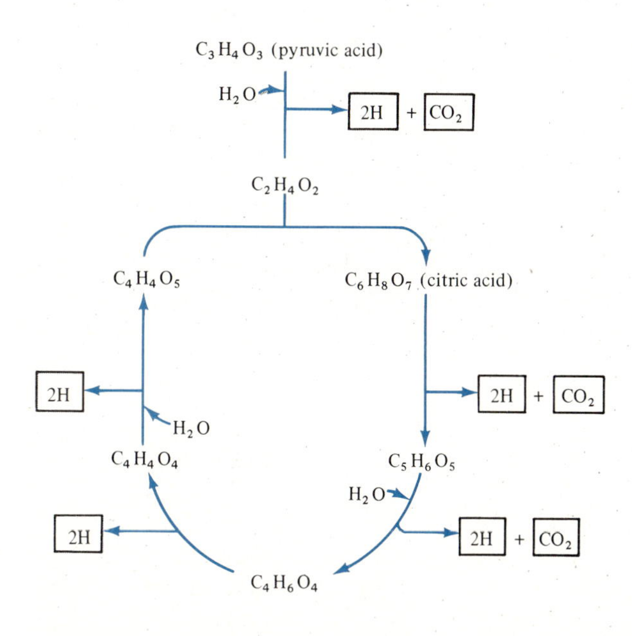

Figure 3.6
Schematic representation of the Krebs cycle. Each reaction is catalyzed by a separate enzyme.

[¶]Szent-Gyorgyi is one of many distinguished biochemists originally trained in medicine. His first research paper dealt with hemorrhoids. Later he went into physiology, then biochemistry, and finally physical chemistry. Reflecting on his scientific career, Szent-Gyorgyi recently stated he has only one regret: "I started science on the wrong end."

cell. Intermediate products can be siphoned off at different points in the cycle and utilized to form the variety of building blocks needed for biosynthesis; conversely, molecules that are not needed as building blocks can be fed into the cycle at different places, thus allowing the cell a greater diversity of usable foodstuffs.

The Krebs cycle explains how pyruvic acid is broken down to carbon dioxide, but it does not explain why oxygen is necessary for respiration or how ATP is produced. The specific aspect of respiration that is concerned with how oxygen utilization is coupled to ATP formation is called **oxidative phosphorylation.**

In eucaryotic cells, respiration, including both the Krebs cycle and oxidative phosphorylation, takes place in highly organized subcellular bodies, the **mitochondria** (see Figure 2.30). Almost all animal and plant cells contain numerous mitochondria in their cytoplasm. The detailed structure of the mitochondrion, as revealed by the electron microscope, is discussed in Chapter 2. It is now possible to separate mitochondria from other cell components. A suspension of these purified mitochondria under appropriate conditions is able to convert pyruvic acid to carbon dioxide and water with the formation of ATP. Since mitochondria can perform all the reactions involved in respiration, it follows that all of the enzymes and other components necessary for the Krebs cycle and oxidative phosphorylation must be located in mitochondria. **The mitochondria thus serve as self-contained powerhouses for the cell.**

Utilizing purified mitochondria, oxidative phosphorylation has been investigated during the last 30 years by a number of distinguished biochemists, including Britton Chance at the University of Pennsylvania, Albert L. Lehninger at Johns Hopkins University, Herman Kalckar in Denmark, and David E. Green at the University of Wisconsin. Although the exact details are unknown, the general scheme can be represented as shown in Figure 3.7.

First, a pair of hydrogen atoms combine with carrier A to produce AH_2. Next, AH_2 transfers the hydrogens to carrier B, producing BH_2 and regenerating A. This energy-yielding reaction is coupled to the formation of ATP from ADP and phosphate:

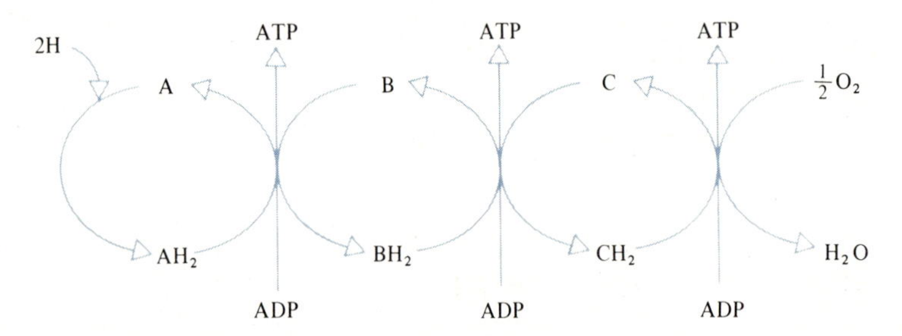

Figure 3.7
Schematic representation of oxidative phosphorylation. Hydrogen atoms produced during the breakdown of nutrient molecules are transmitted to oxygen by a series of carrier molecules (abbreviated A, B, and C). Three ATP molecules are formed for each pair of hydrogen atoms transferred to oxygen.

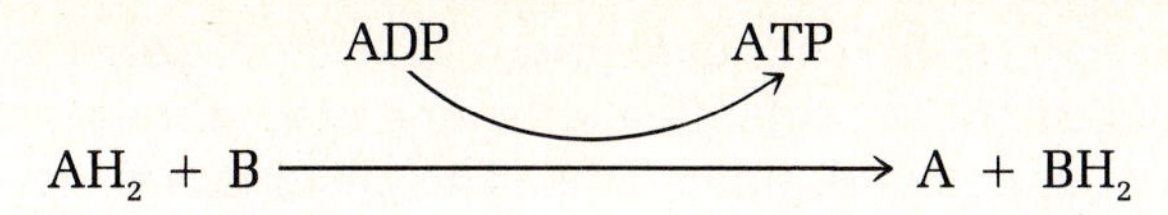

Carrier A again combines with a pair of hydrogen atoms to form AH_2; BH_2 transfers the hydrogens to carrier C, producing CH_2 and regenerating B. In this way, the hydrogens are transferred through the carrier** molecules, finally combining with oxygen gas to form water. **The net result of this chain of reactions in oxidative phosphorylation is the formation of three ATP molecules:**

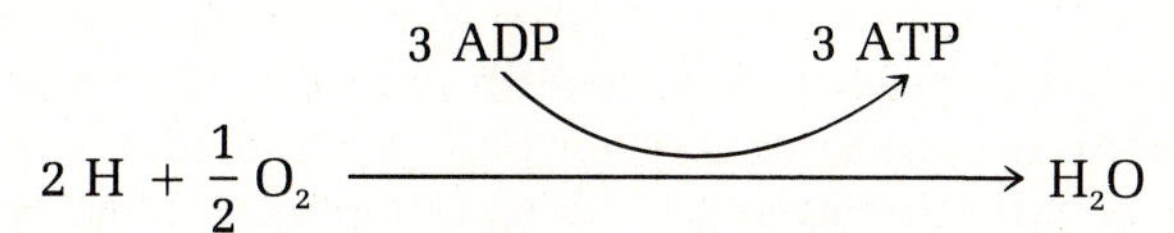

The importance of the oxidative phosphorylation chain to living organisms is demonstrated by the lethal effect of two well-known poisons, carbon monoxide and cyanide. Both toxic materials exert their effects by combining with carrier molecules used in the oxidative phosphorylation chain, thus preventing transfer of hydrogens and interrupting the chain. This results in an immediate halt in ATP production, and unless the poison is removed rapidly, death ensues.

We can now compute the total number of ATP molecules formed for each molecule of glucose consumed. In the first phase, the conversion of glucose to pyruvic acid, two ATP and two pairs of hydrogen atoms are formed. When the two pyruvic acid molecules are burned to carbon dioxide via the Krebs cycle (see Figure 3.5), ten pairs of hydrogens are produced, making a total of two ATP and twelve pairs of hydrogen atoms. Since each pair of hydrogens yields three ATP, a grand total of *38 ATP* molecules are manufactured for each glucose that is oxidized to carbon dioxide and water.

It is interesting to relate the ATP yield to the caloric value of food. For example, 7 grams of glucose (1 teaspoonful) contains approximately 2.5×10^{22} molecules. Since each glucose molecule gives rise to 38 ATP, $38 \times 2.5 \times 10^{22}$ (or 9.5×10^{23}) molecules of ATP are produced for each teaspoonful of glucose consumed. Biochemists have determined that the conversion of one ATP molecule to ADP releases 1.6×10^{-23} Calories†† of useful energy. Thus, a teaspoonful of sugar contains $(1.6 \times 10^{-23}) \times (9.5 \times 10^{23})$ or approximately 15 Calories.

3.6 Unique Energy-Generating Processes of Bacteria

Around 1885, Sergius Winogradsky, a Russian microbiologist, made a remarkable discovery. He found bacteria that use **inorganic**

**Actually, at least seven different carrier molecules are involved in transferring the hydrogens. However, in only three of these transfers is ATP produced.

††The Calorie used in nutrition is defined as the amount of heat energy necessary to raise one liter of ice-cold water 1°C.

materials as their sole food. One group of bacteria, for example, can grow in an open bottle containing only an aqueous suspension of sulfur and small amounts of other salts. The bacteria use oxygen in the air to oxidize the sulfur to sulfate. During this unique respiratory process, ATP is generated. The energy is used to convert carbon dioxide into organic material needed for cell growth. In a similar manner, different species of bacteria use other reduced inorganic compounds such as iron, ammonia, hydrogen gas, and carbon monoxide. Bacteria that gain energy by the oxidation of inorganic compounds and obtain their carbon primarily from carbon dioxide are called **chemoautotrophs.**

Another process peculiar to certain groups of bacteria is **anaerobic respiration.** In place of oxygen gas, specific species of bacteria can use other oxidizing agents, such as nitrate or sulfate. In a manner similar to the reduction of oxygen gas to water, nitrate goes to ammonia and sulfate to hydrogen sulfide. The processes of chemoautotrophy and anaerobic respiration can be viewed as modifications of the normal respiratory process. As shown in Table 3.1, for chemoautotrophy the organic nutrient is replaced by an inorganic compound as the energy source. In anaerobic respiration, an inorganic oxidizing agent substitutes for oxygen gas.

Studies on microbial modifications of respiration have been valuable for a number of reasons:

1. In microbiology, these peculiar bacteria have shown us a "way of life" that is not present in eucaryotic organisms. A scientist should not always have to justify research as being relevant to humans. Sometimes it is enough simply to satisfy one's natural curiosity.
2. For biology in general, a comparison of these curious bacteria with animal and plant cells has led to some important generalizations. One such example, discussed in the next section,

TABLE 3.1 Examples of Microbial Modifications of Respiration

Process	Nutrient	Oxidizing Agent	Products
Respiration	Organic	+ O_2	CO_2 + H_2O
Chemoautotrophy	Iron (Fe)	+ O_2	Rust (Fe^{3+}) + H_2O
	Ammonia (NH_3)	+ O_2	Nitrate (NO_3^-) + H_2O
	Sulfur (S)	+ O_2	Sulfate ($SO_4^=$) + H_2O
	Hydrogen (H_2)	+ O_2	Water (H_2O)
	Carbon monoxide (CO)	+ O_2	CO_2 + H_2O
Anaerobic respiration	Organic	+ nitrate (NO_3^-)	CO_2 + ammonia (NH_3)
	Organic	+ sulfate ($SO_4^=$)	CO_2 + hydrogen sulfide (H_2S)
	Organic	+ CO_2	CO_2 + methane (CH_4)

is van Niel's unifying hypothesis to explain the formation of organic matter from carbon dioxide.

3. For ecology, the importance of chemoautotrophy and anaerobic respiration in studying the environment cannot be overemphasized. As discussed in Chapter 9, these processes play a crucial role in the turnover of matter on earth. Many of the important geological changes on our planet are a result of the metabolism of these unique bacteria.

3.7 Photosynthesis: The Ultimate Source of Energy

As you probably already know, the ultimate source of energy for most living organisms on earth is the sun. A continuous series of nuclear reactions takes place on the sun, releasing vast quantities of energy. Part of this solar energy is in the form of visible light, and **photosynthesis** is the process by which plants (and certain microorganisms) utilize light as the source of energy for the production of carbohydrate.

Early History

Before discussing modern concepts of the mechanism of photosynthesis, let us sketch briefly some of the early history. Probably the first important study of photosynthesis was published in 1648 by the Dutch physician Jean-Baptiste van Helmont:

> I took an earthenware pot, placed it into 200 lb. of earth dried in an oven, soaked this with water, and planted in it a willow shot weighing 5 lb. After five years had passed, the tree grown therefrom weighed 169 lb. and about 3 oz. But the earthenware pot was constantly wet only with rain water; . . . . Finally, I again dried the earth of the pot, and it was found to be the same 200 lb. minus about 2 oz. Therefore, 164 lb. of wood, bark and root had arisen from the water alone.

In this classic experiment, van Helmont considered only soil and water as possible sources for the increased mass of the willow tree. Since the weight of the soil did not change appreciably, his conclusion was obvious—the tree grew at the expense of water alone. In implicating the importance of water as a raw material for plant growth, van Helmont was perfectly correct; we now know, however, that water is not the only essential ingredient.

In 1727 the English physiologist Stephen Hales published the results of an experiment that demonstrated for the first time that air as well as water was necessary for plant growth. Hales added ample quantities of water to two containers full of soil; in one of the containers he set a small peppermint plant; the other (the control) contained only soil and water. Both containers were then made air-tight by placing inverted glass jars over them. The mint grew for about nine months, then faded and died. Next, Hales exchanged the dead mint for a fresh plant utilizing a technique that prevented fresh air from entering the container during this transfer. This time, the plant died in four or five days. When a fresh plant was placed, using the same technique, in the control container whose air had also been confined for nine months, the plant lived several months. Hales concluded from these experiments that plants interact with the atmosphere, possibly removing a substance essential for growth. The combined data of van Helmont and Hales can then be summarized as

$$\text{water} + \text{air} \rightarrow \text{plant material}$$

The next important development was the discovery in 1772 by Joseph Priestley that green plants, "instead of affecting the air in the same manner as animal respiration, reverse the effects of breathing and tend to keep the atmosphere sweet and wholesome." One of Priestley's experiments is summarized in Figure 3.8.

It was already known that fresh air was necessary for both animal and plant life. Priestley's contribution was the discovery that plants and animals grown in the same atmosphere are mutually beneficial: Each restores something to the air that the other consumes. Plant growth could then be expressed as

$$\text{water} + \text{gas X} \rightarrow \text{plant material} + \text{gas Y}$$

with the added stipulation that X and Y are the same gases that are produced and consumed, respectively, during animal respiration.

Priestley's results were soon elaborated on by a Dutch physician named Jan Ingen-Houz. Ingen-Houz demonstrated that the production of gas Y occurred **only in the green parts of plants and only in the light.** In his own words,

> All plants possess a power of correcting, in a few hours, foul air unfit for respiration, but only in clear daylight, or in the sunshine . . . . This office is not performed by the whole plant, but only by the leaves and the green stalks that support them.

Since the production of plant material and gas Y required light,

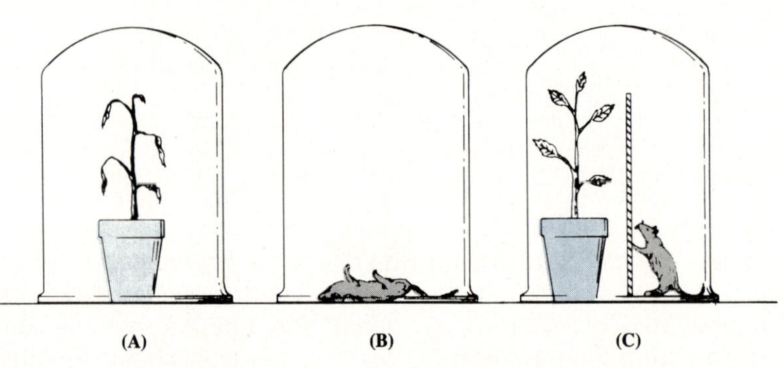

Figure 3.8
Priestley's experiment. A lone plant (A) or mouse (B) suffocated in a closed jar. When a plant and a mouse were enclosed in the same jar (C), both survived.

the process came to be called *photosynthesis,* which means "put together by light." By the end of the eighteenth century, photosynthesis could be expressed by the following equation:

$$\text{water} + \text{gas X} \xrightarrow[\text{green cells}]{\text{light}} \text{plant material} + \text{gas Y}$$

With subsequent developments in chemistry, the increased weight of plants during photosynthesis was shown to be due to the formation of glucose; gas X was carbon dioxide and gas Y was oxygen. Thus the overall photosynthetic process could be expressed in more precise chemical terms:

$$6\ H_2O + 6\ CO_2 \xrightarrow[\text{green cells}]{\text{light}} C_6H_{12}O_6 + 6O_2$$

Modern View

Early investigations of photosynthesis were primarily concerned with identifying each of the components involved in the process. As you probably have perceived from the previous equation, the net result of photosynthesis is the reverse of respiration. Since respiration is an energy-yielding process, it follows that the formation of sugar during photosynthesis is an energy-requiring process. In recent years, research into the photosynthetic process has focused on the problem of *how* light energy is trapped and utilized to "drive" the synthesis of sugar from carbon dioxide and water.

Important clues regarding the mechanism of photosynthesis came from studies of two specialized groups of bacteria. One group, the **chemoautotrophic bacteria,** obtains its energy (ATP) by burning (oxidizing) inorganic compounds. The other group obtains energy by performing an "abnormal" type of photosynthesis. In both cases, the energy is utilized to convert carbon dioxide from the atmosphere into cellular material.

In 1929, Cornelius B. van Niel of Stanford University proposed a hypothesis that applied not only to green plant photosynthesis but also to bacterial photosynthesis and chemosynthesis. It is truly enlightening to see how van Niel arrived at his unifying hypothesis from some rather simple analytical data.

The results of van Niel's quantitative experiments with the purple photosynthetic bacteria can be expressed by the following equation:

$$12\ H_2S + 6\ CO_2 \xrightarrow[\text{purple bacteria}]{\text{light}} C_6H_{12}O_6 + 6\ H_2O + 12\ S$$

This "abnormal" photosynthesis is similar to plant photosynthesis except that the bacteria use H_2S (hydrogen sulfide) instead of H_2O and produce sulfur instead of oxygen. The comparison is even more striking when six water molecules are added to each side of the equation for plant photosynthesis:

$$12\ H_2O + 6\ CO_2 \xrightarrow[\text{green plants}]{\text{light}} C_6H_{12}O_6 + 6\ H_2O + 6\ O_2$$

Combining the data from bacterial and plant photosynthesis and dividing by six, van Niel was able to express a general equation for photosynthesis:

$$2\ H_2A + CO_2 \xrightarrow[\text{all photosynthetic cells}]{\text{light}} (CH_2O) + H_2O + 2\ A$$

where A stands for either oxygen or sulfur and (CH_2O) symbolizes cell matter having a ratio of two hydrogens for each carbon and oxygen.

Since in bacterial photosynthesis the sulfur can come only from H_2S, van Niel argued by analogy that the **oxygen produced during photosynthesis arose by splitting H_2O.** (This part of van Niel's hypothesis was experimentally verified in the early 1940s by Samuel Ruben and his collaborators.) The generalized equation for photosynthesis can thus be considered as the sum of two processes:

$$(1)\ 2\ H_2A \xrightarrow{\text{light}} 4\ H + 2\ A$$

$$(2)\ 4\ H + CO_2 \xrightarrow[\text{or dark}]{\text{light}} (CH_2O) + H_2O$$

$$(1)\ \text{plus}\ (2)\ 2\ H_2A + CO_2 \longrightarrow (CH_2O) + H_2O + 2\ A$$

By writing photosynthesis as a two-step process, it is possible to derive an even more general concept of cell metabolism. Although reaction (1) occurs only in the light and is peculiar to photosynthetic organisms, reaction (2) is common to a large variety of cells and can proceed in the dark. In essence, the first reaction is the **conversion of light energy into chemical energy.** The energy is then able to "drive" the second reaction, the formation of cell material from carbon dioxide. The significance of separating the "photo" reaction from the "synthesis" reaction is that it now brings into the framework of this discussion, the chemoautotrophic bacteria mentioned in the previous section. This interesting group of bacteria obtain their energy by oxidizing inorganic substances, such as iron or sulfur. The hydrogen atoms and energy so derived are then utilized to convert carbon dioxide to cellular material exactly as in the second reaction of photosynthesis. In the most general sense, the formation of organic matter from carbon dioxide was expressed by van Niel as follows:

$$CO_2 \xrightarrow{4\ H} (CH_2O) + H_2O$$

where the hydrogens can be derived in a variety of ways, such as "photo" reactions or the burning of inorganic compounds.

It is interesting to compare the development of van Niel's conceptual scheme with the progressive abstraction that has taken place in modern art. The exquisite series of bas-reliefs of Henri Matisse serves to illustrate the point (Figure 3.9). Both Matisse and van Niel begin by examining in detail a specific case—for van Niel, the purple photosynthetic bacteria; for Matisse, the rear view of a beautiful woman. Both creators, scientist and artist, attempt to abstract from reality the essence of their subjects, slowly constructing a greater simplicity and universality. But as in art, so in science, the final form is forever tentative.

In the last 30 years, further progress has been made in elucidating the mechanisms of both the "photo" or **light reaction** and "synthetic" or **dark reaction.** From the experiments of Melvin Calvin and his associates at the Lawrence Radiation Laboratory of the University of California, the dark reactions of photosynthesis are now known in some detail. As shown in Figure 3.10, carbon dioxide does not react directly with hydrogen atoms to form organic molecules; instead, carbon dioxide is converted into an organic substance by combining with a five-carbon sugar diphosphate to produce a six-carbon sugar acid. It is not our intention to delve into the details of this important metabolic pathway. The crucial points are that ATP is necessary to form the five-carbon acceptor molecule and that *hydrogen* (reducing power) is necessary to convert the six-carbon acid to a six-carbon sugar. Adenosine triphosphate and hydrogen are thus necessary for the formation of carbohydrate from carbon dioxide even though they are not utilized in the specific reaction in which carbon dioxide is fixed.

Many different types of cells, including non-photosynthetic bacteria, are able to convert carbon dioxide to sugar **if provided reducing power (hydrogens) and ATP.** This implies that the function of the light reaction is to generate energy in the form of both hydrogen and ATP. As we have discussed, the hydrogens come from the splitting of water. But how is ATP formed?

At first it was thought that the ATP needed for fixing carbon dioxide in photosynthesis is produced in mitochondria by oxidative phosphorylation. According to this hypothesis, some of the hydrogens generated by the light reaction would function exactly like the hydrogens formed during the Krebs cycle.

It has been known for more than 100 years that photosynthetic eucaryotic cells contain in their cytoplasm specialized structures called **chloroplasts** (see Figures 2.29 and 2.33). In 1954, Daniel I. Arnon made the important discovery that **isolated chloroplasts can carry out the complete photosynthetic process.** It follows that all of the components necessary for producing ATP and hydrogen from light energy as well as the enzymes needed for the dark reactions must be present within the chloroplast. Furthermore, Arnon has provided direct experimental evidence for the generation of ATP in chloroplasts. This light-dependent, energy-yielding reaction is called **photosynthetic phosphorylation** to distinguish it from oxidative phosphorylation.

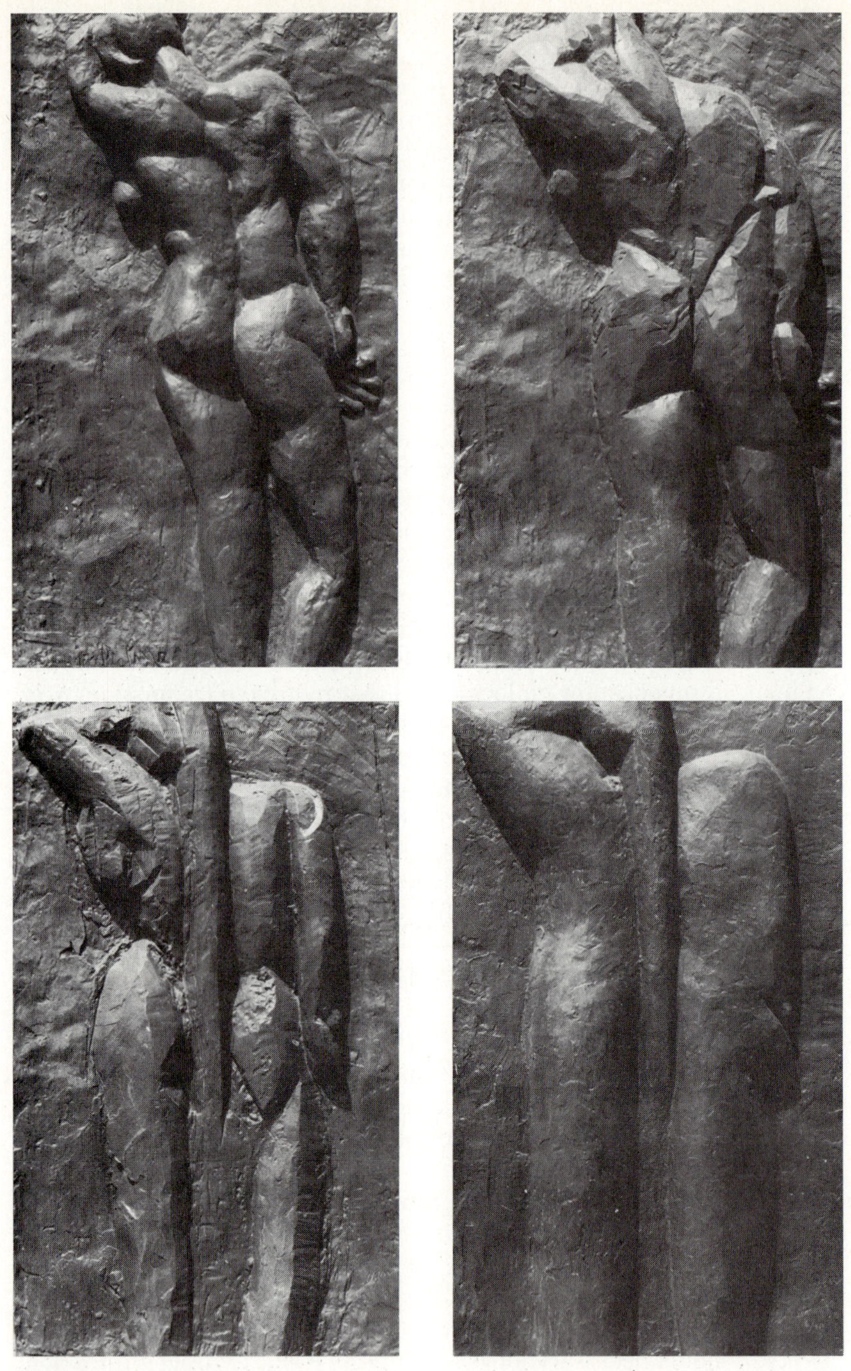

Figure 3.9
Series of bas-reliefs by Henri Matisse. (Franklin Murphy Sculpture Garden, University of California at Los Angeles.) (Courtesy of S. Laughner.)

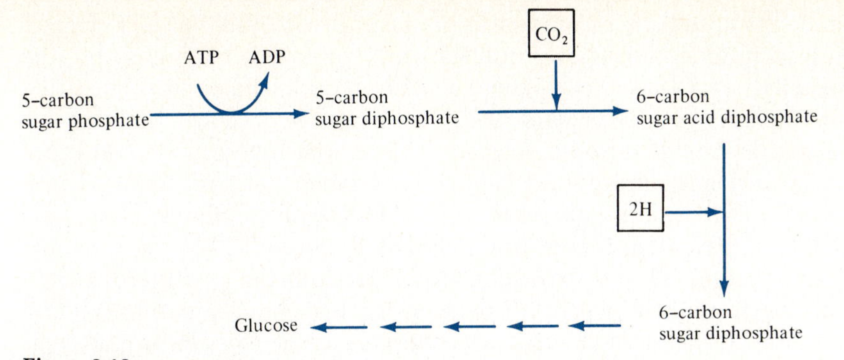

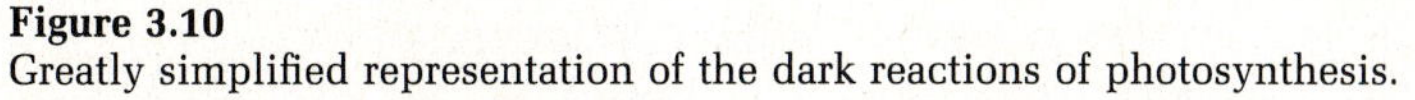

Figure 3.10
Greatly simplified representation of the dark reactions of photosynthesis.

Figure 3.11 is a schematic representation of one of the ways in which Arnon has suggested that ATP is formed in the chloroplast. The chloroplast contains the green pigment **chlorophyll,** which traps solar radiant energy. The light energy impels an electron into a higher energy state, and the chlorophyll molecule acquires, temporarily, a positive charge. The high-energy electron then passes through a series of carrier molecules, A,B, C, and D, finally relapsing to its original energy state and completing the cycle. The loss in energy as the electron drops from A back to chlorophyll is conserved in the formation of two ATP molecules. Although the physical details of this series of reactions are not completely understood, the importance of chlorophyll now becomes clearer: Through this molecule passes energy from the sun, which allows plants to grow, animals to move, and humans to think.

3.8 The Mitchell Hypothesis

Photosynthesis and respiration have one important feature in common. During both processes, electrons are passed from one carrier molecule to another, resulting in the formation of ATP. But precisely how is the transfer of electrons coupled to the generation of ATP?

In 1961, Peter Mitchell, working in a small laboratory in Eng-

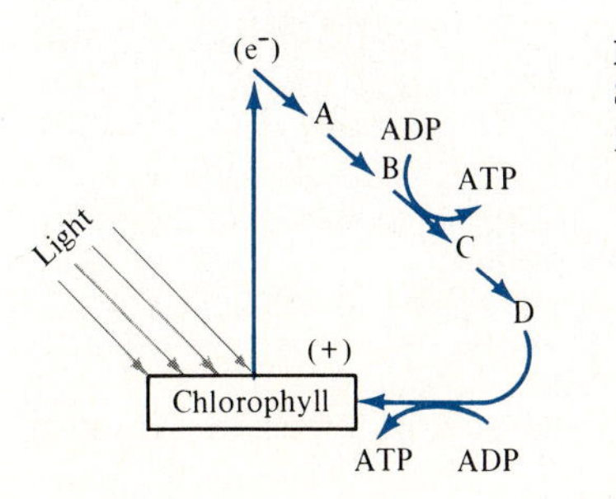

Figure 3.11
Schematic representation of photosynthetic phosphorylation.

land, proposed an ingenious hypothesis to account for ATP formation in mitochondria, chloroplasts, and bacteria. For many years, the Mitchell hypothesis was greeted with skepticism by the scientific community, both because nobody had ever heard of Mitchell and because his way of thinking was so original that biochemists could not (or would not) understand him. Unperturbed, he continued to perform careful experiments that supported his ideas. Slowly, scientists began to pay attention to the Mitchell hypothesis. His experiments were reproduced and expanded upon in a number of different laboratories around the world. Finally, in the last few years, the Mitchell hypothesis has gained wide acceptance and is recognized today as one of the major developments in modern biochemistry.‡‡

The three essential components of the Mitchell hypothesis are diagrammed in Figure 3.12.

1. **Membranes are not readily permeable to H^+ or OH^-.** It is a fact that biological membranes have low electric conductivities. Materials containing a positive or negative charge do not pass through these membranes without the assistance of specific transport systems.
2. **Transfer of electrons§§ from one carrier molecule to another results in the displacement of H^+ and the development of an electric potential.** In the case of mitochondria, each time a pair of electrons is transferred successively from one carrier

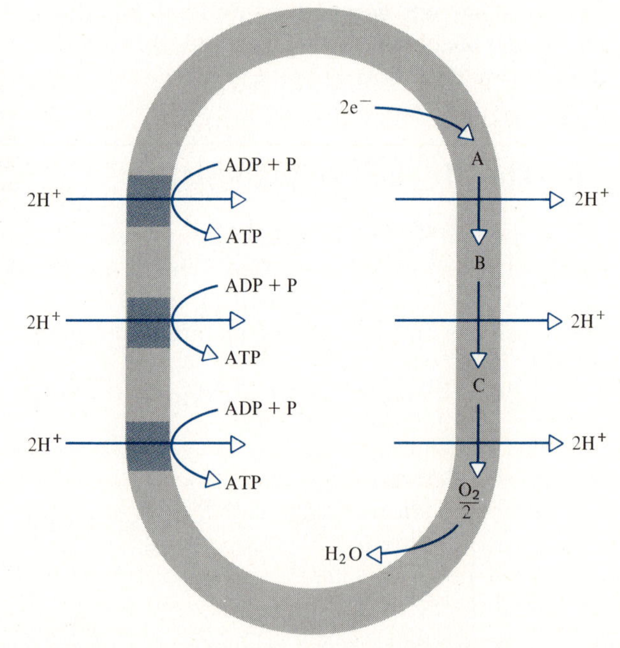

Figure 3.12
Schematic representation of the Mitchell hypothesis for ATP production in mitochondria. First, two electrons are transferred during respiration through carrier molecules A, B, C, and oxygen gas. During this process, six H^+ are expelled. Since the charge on the outside is more positive than on the inside, an electric potential develops. Second, the six H^+ flow back into the cell through special sites on the membrane, where the electric energy is converted into chemical energy in the form of ATP.

‡‡In 1978, Peter Mitchell received the Nobel Prize in chemistry.

§§The transfer of two hydrogen atoms as shown in Figure 3.7 is formally equivalent to the transfer of a pair of electrons.

to another, and finally to oxygen gas, six H^+ are expelled into the cytoplasm. As a result, the cytoplasm becomes more positively charged than the interior of the mitochondrion. Since the charge on the opposite sides of the mitochondrial membrane is different, an electric potential develops.

3. **Electric potential is the driving force for the formation of ATP from ADP.** The flow of H^+ back into the mitochondrion is analogous on a microscale to water falling through a dam. The water is channeled into a limited area so that when it falls, the energy can be trapped more efficiently. Similarly, the H^+ flow is channeled through a special site on the membrane where a molecular machine that includes the enzyme ATPase traps the energy in the form of ATP.

The coupling of electron flow and ATP generation is the same in chloroplasts as in mitochondria, except that H^+ is pumped into the chloroplast rather than expelled. The electric potential that develops is discharged when 2 H^+ flow out of the chloroplast through special sites, leading to the formation of a molecule of ATP. Non-photosynthetic bacteria behave like mitochondria with regard to H^+ flow; photosynthetic bacteria are similar to chloroplasts.

The Mitchell hypothesis leads to several predictions that are amenable to experimental testing:

1. The level of H^+ on opposite sides of the membrane should be different when the system is functioning.
2. It should be possible to generate ATP by artificially changing the H^+ concentration on the outside.
3. If the membrane is damaged so that H^+ can flow freely in and out of the organelle, then ATP generation should cease.

Each of these predictions has been experimentally verified. First, when isolated mitochondria were provided with oxygen gas in order to carry out respiration, the H^+ concentration outside the mitochondrion increased; in contrast, the H^+ concentration outside of chloroplasts decreased when illuminated. Second, in 1960, Jagendorf and Uribe at Johns Hopkins University demonstrated ATP production in chloroplasts *in the absence of light,* simply by suspending purified chloroplasts in medium containing a low concentration of H^+. Subsequently, Mitchell reported ATP production by isolated mitochondria in a medium containing a relatively high concentration of H^+. Third, a special class of chemicals was discovered that alters membranes in a specific manner, allowing H^+ to permeate the membrane freely. When these chemicals were added to mitochondria, chloroplasts, or bacteria, ATP production stopped immediately.

Although certain details of the Mitchell hypothesis are still controversial, the basic idea is now generally accepted. One of the major contributions of the hypothesis is that it focuses attention on the importance of cell structure in understanding biochemistry. The cell is not just a "bag of enzymes." Position is critical; the inside of the membrane is different from the outside. It is the challenge of the next generation of biochemists to begin to explain the highly organized internal structure of cells.

QUESTIONS

3.1 Which of the energy-yielding processes—fermentation, respiration, or photosynthesis—do you think evolved first on earth? Which was second?

3.2 Estimate how many different enzymes are required for the conversion of glucose to carbon dioxide with the formation of 38 ATP molecules.

3.3 Which of the following reactions are energy-yielding? Which require ATP?

a) Pyruvic acid + oxygen → carbon dioxide + water
b) Amino acids → protein + water
c) $C_4H_6O_4 \rightarrow C_4H_4O_4$ + 2 H atoms
d) Starch + water → glucose
e) Carbon dioxide + water → oxygen + cell material
f) Sulfate + water → sulfur + oxygen
g) Glucose + nitrate → carbon dioxide + ammonia

3.4 What is the fundamental similarity between photosynthetic plants and chemoautotrophic bacteria?

3.5 From the viewpoint of a chemoautotrophic bacterium, what is the importance of photosynthesis?

3.6 In 1861, Pasteur discovered that certain microbes can grow in the absence of air by a process called fermentation. Furthermore, Pasteur demonstrated experimentally that fermentation was an inefficient process; large amounts of food had to be consumed in order to obtain a little growth. As the distinguished biochemist Efraim Racker recently stated, "These experiments proved that, as in New York City today, life without air is possible, but expensive." Explain Racker's statement in terms of ATP and energy.

3.7 Distinguish between the "light reactions" and "dark reactions" of photosynthesis; between chemoautotrophy and anaerobic respiration.

3.8 Biology can be viewed in terms of unity and diversity. Which of the following statements apply to all living organisms and which are true of only certain organisms?

a) They produce their own ATP
b) They carry out respiration
c) They are motile
d) They have nuclear membranes
e) They are destroyed at 120°C for 20 minutes
f) They depend on other organisms for organic nutrients
g) They depend on other organisms for either organic or inorganic molecules
h) They have cytoplasmic membranes
i) They require oxygen for growth
j) They contain a large number of enzymes

3.9 Compare Mitchell's hypothesis with the manner in which a battery functions. What is the evidence supporting the Mitchell hypothesis?

3.10 Define entropy (see Appendix A). State the second law of thermodynamics in terms of entropy.

3.11 We know that living organisms are able to maintain and reproduce what appears to be a high degree of complexity and organization out of relatively random surroundings. Does this mean that living organisms do not obey the second law of thermodynamics? Explain. How about the construction of a building from bricks?

Suggested Readings

Asimov I: Life and Energy. New York: Bantam Books, Inc., 1962.

An introduction into the chemistry of energy processes by one of the world's foremost science writers.

Gabriel ML and S Fogel, Eds and Translators: Great Experiments in Biology. Englewood Cliffs, N J: Prentice-Hall, Inc., 1955.

An excellent collection of important papers, including the classical experiments of Buchner, Pasteur, van Helmont, Priestley, Ingen-Houz, and van Niel.

Hinkle PC and RE McCarty: How Cells Make ATP. Scientific American, March 1978.

A lucid account of the Mitchell hypothesis and the evidence supporting it.

Lehninger AL: Bioenergetics, 2nd Edition. Menlo Park, CA: W. A. Benjamin, 1971.

A more advanced treatment of many aspects of this chapter.

Levine RP: The Mechanism of Photosynthesis. Scientific American, December 1969.

Nash LK: Plants and the Atmosphere. In: Harvard Case Histories in Experimental Science, Vol. 11. Cambridge: Harvard University Press, 1967.

A historical account of photosynthesis.

CHAPTER 4

The Microbial World

This is the important step in every science: the construction of a first order which is reasonable in itself and which holds to the experimental facts that are known. The order is a selection of appearances. And any selection itself implies, and imposes, an interpretation.

N. Bronowski

The beauty and science of living organisms are best appreciated when viewed in terms of both unity and diversity. As many as ten million diverse kinds of living organisms inhabit our planet. Such a vast number of different organisms could never be comprehended without organizing or classifying them into groups. Each group consists of distinct members that share several important characteristics. The most significant features that are used to classify microorganisms are (1) cell morphology and (2) the manner in which they obtain energy. Since these topics were discussed in Chapters 2 and 3, we can now introduce the microbial world. The branch of biology that is concerned with the description, nomenclature, and classification of living organisms is known as **taxonomy.** As the distinguished Canadian-American-French microbiologist R. Y. Stanier has recently noted, "Taxonomists themselves can be broadly divided into two groups: 'lumpers,' who set wide limits to a species, and 'splitters,' who differentiate species on more slender grounds." In this chapter, we have chosen to be "lumpers" in order to help the student obtain an overview of the microbial world. Where possible, specific microorganisms are discussed as a means of illustrating characteristic features of the broader group.

4.1 Microbial Taxonomy: The Concept of Protists

Until the invention of microscopes and the discovery of microorganisms, the biological world was neatly divided into two kingdoms: **Plantae** and **Animalia.** At the macroscopic level, plants were generally not motile, were green, and grew thoughout their lives, whereas animals were motile, were not green, and had a more or less fixed size and shape as adults. Some properties that distinguish plants from animals are summarized in Table 4.1.

Scientific investigations on the structure and growth habits of microorganisms from the seventeenth century to the present have re-

TABLE 4.1 General Differences Between Plants and Animals

Characteristic	Plants	Animals
Active movement	Absent	Present
Photosynthesis	Present	Absent
Cell walls	Present	Absent
Primary energy source	Sunlight	Organic nutrients[b]
Primary carbon source	Carbon dioxide	Organic nutrients[b]
Principal reserve food	Starch	Glycogen, fat
Mode of growth[a]	Open	Closed

[a]Higher plants grow throughout their lives, reaching a size and a shape that are greatly influenced by the environment. The size and form of adult animals are more or less fixed.
[b]Animals are able to ingest solid food.

vealed that most microorganisms do not fit into either of the two traditional kingdoms. For example, there are microscopic algae that have the plant-like characteristics of carrying out photosynthesis and having a cell wall and the animal-like characteristics of swimming and having a closed mode of growth. Furthermore, certain algae can be grown indefinitely in the dark if provided with organic nutrients. In the case of bacteria, there are examples of both motile and non-motile forms, as well as those that perform photosynthesis and those that obtain their energy from organic nutrients. With minor exceptions, all bacteria contain cell walls. The structure of the bacterial cell wall, however, is fundamentally different from cell walls of plants. In addition, there are bacteria that obtain their energy from neither sunlight nor organic nutrients, but from inorganic compounds. With such diversity of characteristics, microorganisms cannot be properly classified as either animals or plants. Thus, a third kingdom, the **Protista,** was established in order to take care of the bacteria, algae, protozoa, and fungi.

If the characteristics used to distinguish plants and animals cannot be used to characterize protists, what then constitutes membership in the kingdom of Protista? The feature that separates protists from animals and plants is their **relatively simple biological organization.** Most protists are microscopic unicellular organisms. Size alone, however, does not define the protist kingdom. There are multicellular marine algae, for example, the seaweeds and kelp, that can reach lengths of more than 50 meters. In spite of their size, these algae are classified as protists because they lack the extensive tissue formation characteristic of higher plants and animals. The name Protista means literally "the very first." Although their detailed evolutionary relationships are not clearly understood, it is reasonable certain that most organisms that are included in the Protista appeared on earth before either animals or plants.

The kingdom of Protista is further divided into higher protists, consisting of algae, protozoa, and fungi, and lower protists, bacteria and blue-green algae. In recent years it has become clear that blue-green algae are very similar to bacteria; thus they are now called blue-green bacteria. The higher protists are eucaryotic organisms, while the lower protists are procaryotes.

4.2 The Algae: Grass of the Sea

Algae are eucaryotic protists that carry out photosynthesis and possess chloroplasts. They vary in size from microscopic unicellular diatoms with diameters less than 5 microns to the giant kelp of the Pacific Coast that grow to lengths of more than 50 meters in shallow water between Alaska and southern California. Growth of algae depends upon the availability of light and utilizable forms of nitrogen and phosphorus. Algae are most abundant in the upper layers of fresh water and seawater and on the surfaces of moist soils and rocks. Algae are the primary producers of organic matter in natural waters and, as such, make life possible for aquatic animals. The algae are classified into six major groups according to the chemical composition of their pigments, cell wall, and storage materials (Table 4.2). The cell walls contain varying amounts of pectin in addition to the characteristic compounds listed in Table 4.2. All algae contain the photosynthetic pigment chlorophyll a in their chloroplasts. The presence of additional pigments in the chloroplast imparts a characteristic color to each major group.

The **green algae** are grass-green because they contain primarily chlorophyll pigments. Green algae are similar to higher plants in that they contain cellulose-pectin cell walls, store starch granules, and have chlorophyll a and b. For these reasons, biologists have speculated that higher plants evolved from green algae. A simple example of a green alga is the unicellular organism *Chlamydomonas* (Figure 4.1).

Euglenids are green when grown in the light. Some strains, however, become bleached when kept in a rich medium at an appropriate temperature, because the cells multiply more rapidly than the chlo-

TABLE 4.2 Classification of Algae

	Characteristic Chemical Components of		
Group	*Chloroplast*	*Cell Wall*	*Reserve Food*
Green algae	Chlorophyll a + b	Cellulose	Starch
Euglenids[a]	Chlorophyll a + b	No wall	Polysaccharide
Diatoms	Chlorophyll a and carotenes	Silica	Polysaccharide, oils
Dinoflagellates	Chlorophyll a and carotenes	Cellulose	Starch, oils
Brown algae	Chlorophyll a and xanthophyll	Cellulose, algin	Polysaccharide, fats
Red algae	Chlorophyll a and phycobilins	Cellulose, agar	Polysaccharide

[a]Sometimes classified as protozoa because when placed in the dark, they lose their photosynthetic pigments and can grow on organic nutrients.

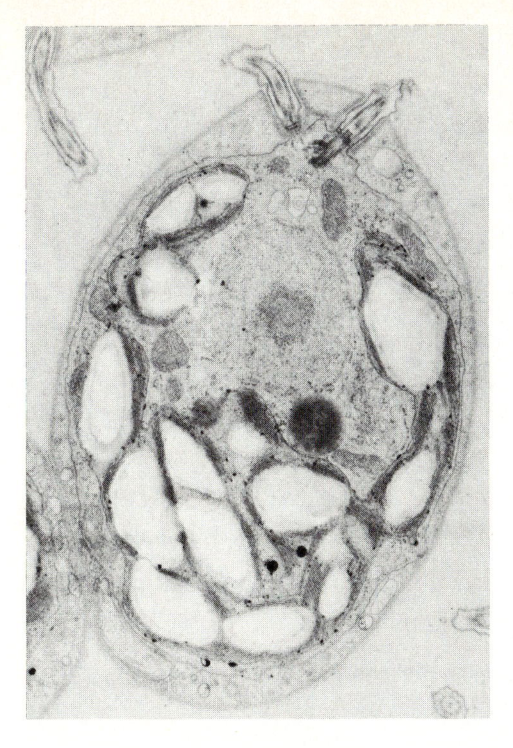

Figure 4.1
Electron micrograph of *Chlamydomonas reinhardi.* A common freshwater green alga, it is generally pear-shaped and moves rapidly with a characteristic darting motion caused by two flagella that protrude from one end of the cell. In preparing the thin section for microscopy, the flagella were cut, leaving only the proximal ends (16,000×). (Courtesy of Torill Torgersen.)

roplasts. The resulting colorless non-photosynthetic cells can survive indefinitely in the dark if provided with organic nutrients. These colorless forms closely resemble certain species of protozoa. Unlike other algae, euglenids lack a rigid cell wall, permitting the cells to change their shape. The most common and best studied member of the group is *Euglena* (Figure 4.2).

Diatoms are often called golden algae because they possess yellow-brown pigments in addition to chlorophyll. Diatoms are unicellular algae found abundantly in cold seawater. They are the principal component of phytoplankton, the photosynthetic cells that float near the surface of oceanic waters and that are the primary source of food for all water-dwelling animals. One of the unique features of diatoms is the extremely hard outer surface, which contains silicon imbedded in a pectin matrix. The glassy shells of diatoms are highly resistant to degradation and thus constitute some of the best preserved microfossils. Accumulated silicon shells of diatoms form the fine, crumbly sustance known as "diatomaceous earth," used as an abrasive in toothpaste and polishes, and as a filtering aid and insulating material. Some examples of diatoms are shown in Figure 4.3.

Dinoflagellates are primarily red unicellular algae that are most abundant in tropical waters. Like diatoms, they are important components of the phytoplankton. The infamous red tide that occurs periodically in warm coastal waters is the result of a bloom of red pigmented dinoflagellates. Thousands of fish die during a red tide as a result of the excretion of a potent nerve toxin by the dinoflagellates. Dinoflagellates are also toxic to humans who eat shellfish that have

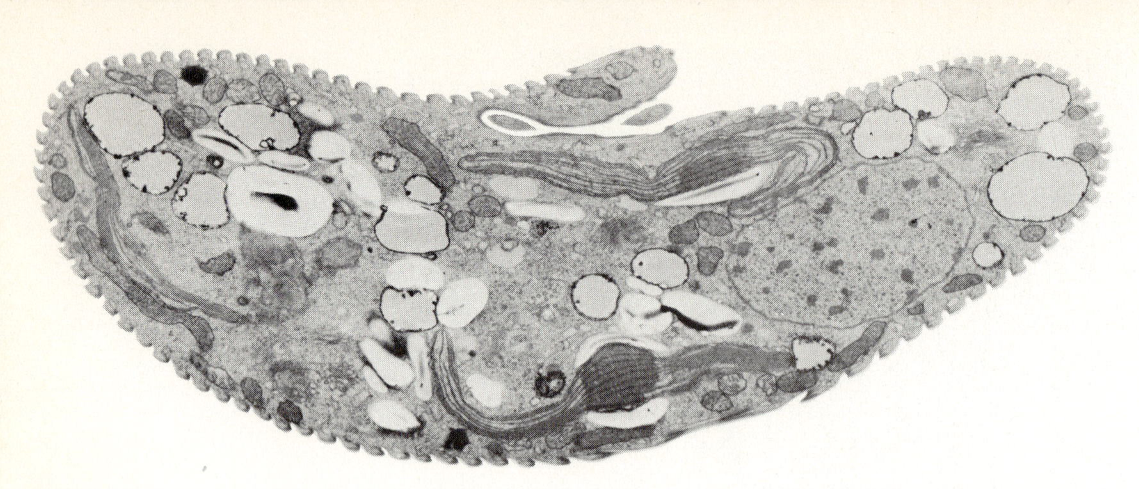

Figure 4.2
Thin section of *Euglena* in the electron microscope. The complexity of the eucaryotic cell is well illustrated (11,000×). (Courtesy of Y. Ben Shaul.)

Figure 4.3
Scanning electron micrograph of diatoms (2000×). (Courtesy of M. Kessel.

fed on the algae. A scanning electron micrograph of a dinoflagellate is presented in Figure 4.4.

Brown algae contain the pigment xanthophyll, which completely masks the green of the chlorophyll they also possess. The brown algae

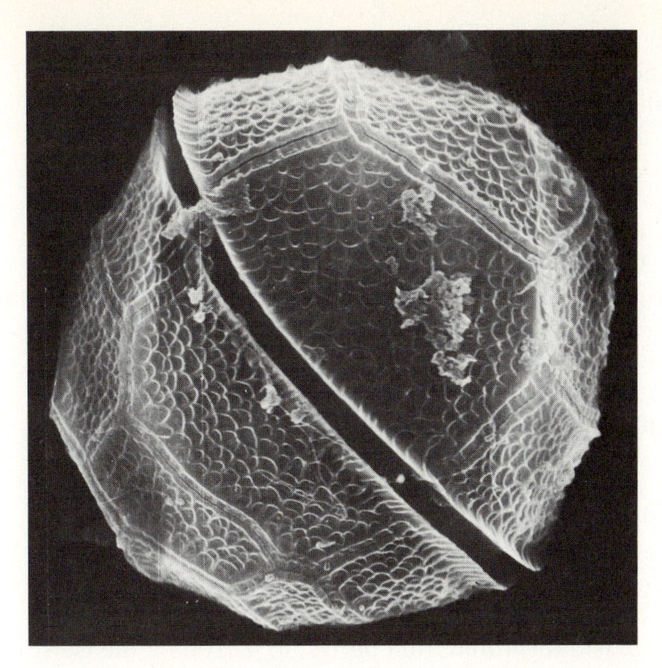

Figure 4.4
Scanning electron micrograph of a dinoflagellate from the Sea of Galilee (4000 ×). (Courtesy of Y. Ben Shaul.)

are a group of multicellular marine organisms that constitute the large seaweeds of the cold ocean waters. Many of them, such as the giant kelp, are attached to the substrate by primitive root-like appendages and extend through the water column to the surface; gas-filled spherical bulbs keep their leaf-like projections floating on the surface where they can more readily absorb the sunlight necessary for photosynthesis.

Most **red algae** are multicellular forms that are found in tropical and subtropical oceans. Their major pigments are phycobilins. The red algae are of practical importance to microbiologists because they are the commercial source of the polysaccharide agar, widely used in preparing solid media for microbiological research. In addition, certain species of red algae are used as a food in the Orient.

4.3 The Protozoa

You live in a world
Without a sound
Where you dart about
Or float around.
You show your face
Which you cannot hide,
Because you look the same
From every side.

Peter Daniel Nathan

Protozoa are non-photosynthetic single-cell eucaryotes that obtain their energy from metabolizing organic matter. Many of them feed on bacteria. The name protozoa, which means "first animals," is appropriate because all multicellular animals are believed to have evolved from protozoa. The protozoa are classified into the following four major groups according to their mode of locomotion: Sarcodina, Mastigophora, Ciliophora, and Sporozoa.

Sarcodina move by the flowing of their cytoplasm into temporary projections called pseudopodia. *Amoeba proteus* is the classic example of this group of protozoa; hence, this type of movement is referred to as amoeboid. Although the sarcodines contain no cell wall, some marine forms produce hard shells. The chalk cliffs of Dover, England, were formed from deposits of such shells. One parasitic form, *Entamoeba histolytica* (Figure 4.5), is the causative agent of amoebic dysentery, a serious disease.

Mastigophora (Figure 4.6) are motile by means of one or more flagella. It has been suggested that this group of protozoa has been derived from photosynthetic algae (the euglenoids), which irreversibly lost their chloroplasts. Most of these flagellates are free-living in lakes and ponds, but some are parasites. The latter includes *Trypanosoma gambiense,* the causative agent of African sleeping sickness. In humans, the parasite lives and grows primarily in the bloodstream. The parasite is transmitted from person to person by the tsetse fly, a blood-sucking insect found only in Africa. Another interesting flagellate is *Trichonympha collaris,* which lives in a symbiotic relationship with termites. The termite itself cannot digest the major component of wood, cellulose. *Trichonympha collaris,* which lives in the intestines of termites, has the enzymes necessary to break down the polymeric cellulose into sugars that can then be used by the in-

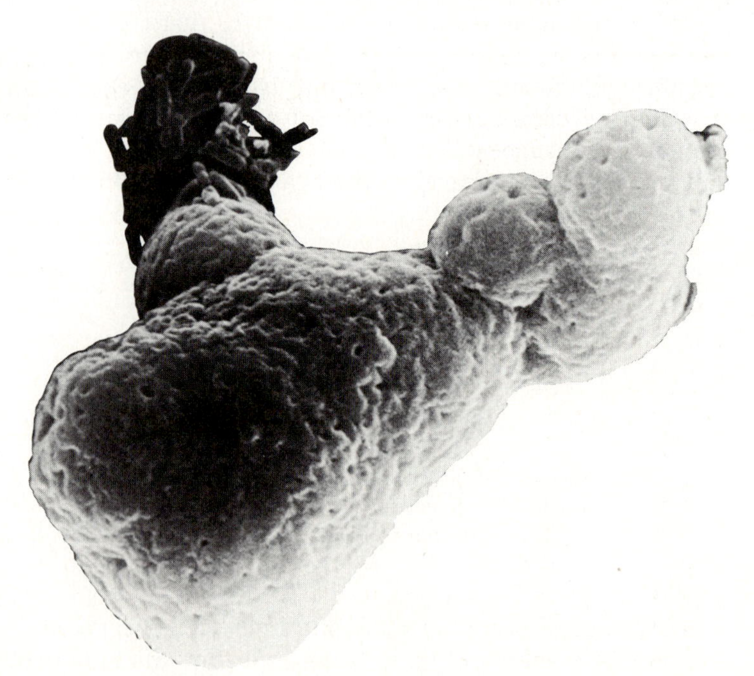

Figure 4.5
Scanning electron micrograph of *Entamoeba histolytica*. Note the much smaller bacteria above the amoeba. (Courtesy of D. Mirelman.)

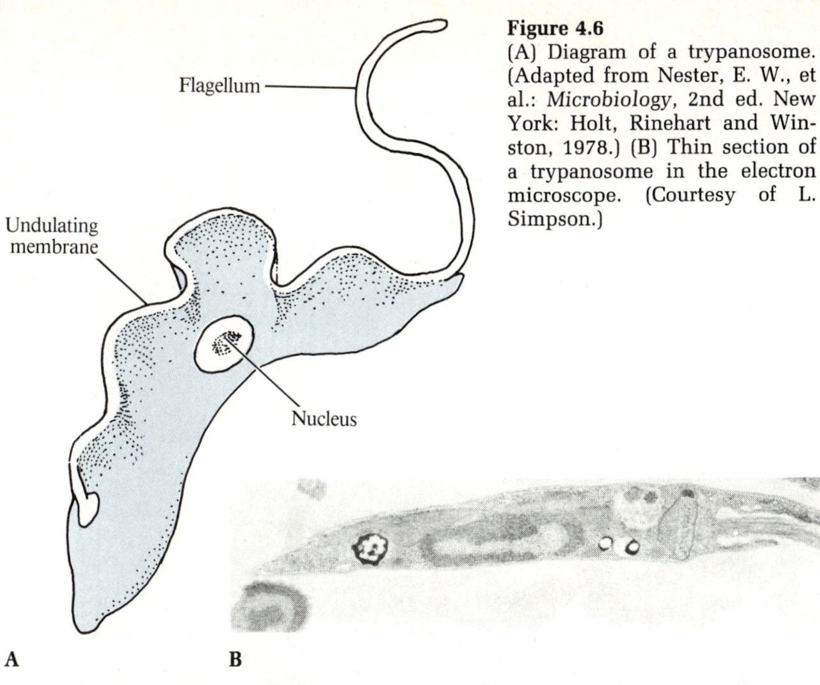

Figure 4.6
(A) Diagram of a trypanosome. (Adapted from Nester, E. W., et al.: *Microbiology*, 2nd ed. New York: Holt, Rinehart and Winston, 1978.) (B) Thin section of a trypanosome in the electron microscope. (Courtesy of L. Simpson.)

sect. In the absence of the protozoan, the insect starves to death although it continues to eat wood.

Ciliophora (or ciliates) move very rapidly by means of rhythmic beating of hundreds of short bush-like structures called cilia. The coordinated movement of these cilia also helps to propel food into the cell. Ciliates represent the most highly developed form of single-cell organisms. Figure 4.7 is a diagram of a paramecium exemplifying some of the complexity of ciliates.

The fourth class of protozoa is the **Sporozoa,** which are characterized by their complex life cycle and lack of motility. All sporozoans are obligate parasites. The best studied example is *Plasmodium vivax,* the protozoan responsible for malaria in humans. The disease is transmitted by the bite of a female Anopheles mosquito. The parasite must carry out part of its life cycle in humans and part in this specific mosquito. Since the Anopheles mosquito inhabits primarily the warmer parts of the world, malaria is predominantly a disease of the tropics. The World Health Organization (WHO) estimates that more than 200 million persons contract malaria each year, and about 1% die from it.

4.4 The Fungi: Yeasts, Molds, and Mushrooms

The fungi comprise a large heterogenous group of higher protists that are non-photosynthetic and generally immotile. They are characterized by a distinctive growth pattern, shown in Figure 4.8. The diversity of structures and growth habits of fungi can more readily be appreciated following a discussion of the different growth stages.

The life cycle of a typical fungus begins with spore germination.

Figure 4.7
Diagram of the ciliate protozoa, *Paramecium multinucleatum*. (Adapted from Nester, E. W., et al.: Microbiology, 2nd ed. New York: Holt, Rinehart and Winston, 1978.)

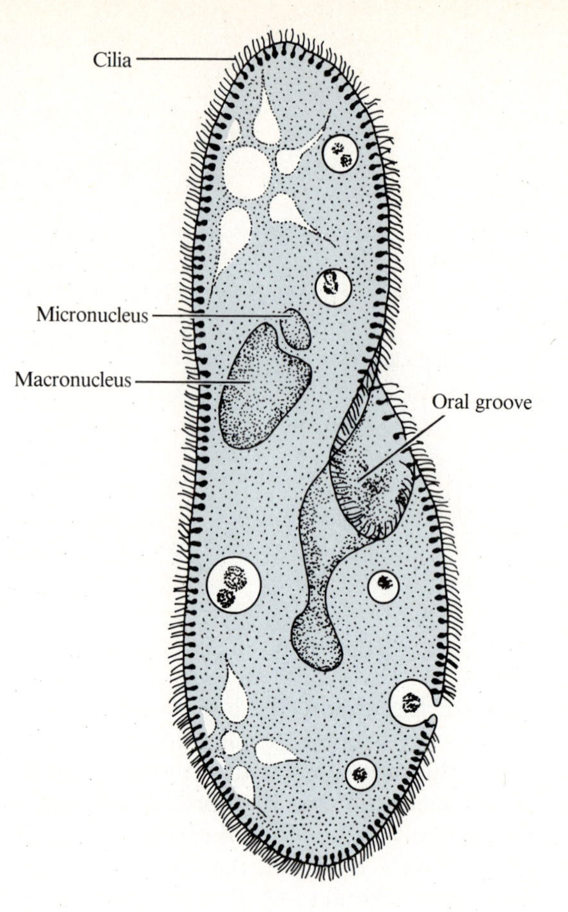

Figure 4.8
Life cycle of a fungus.

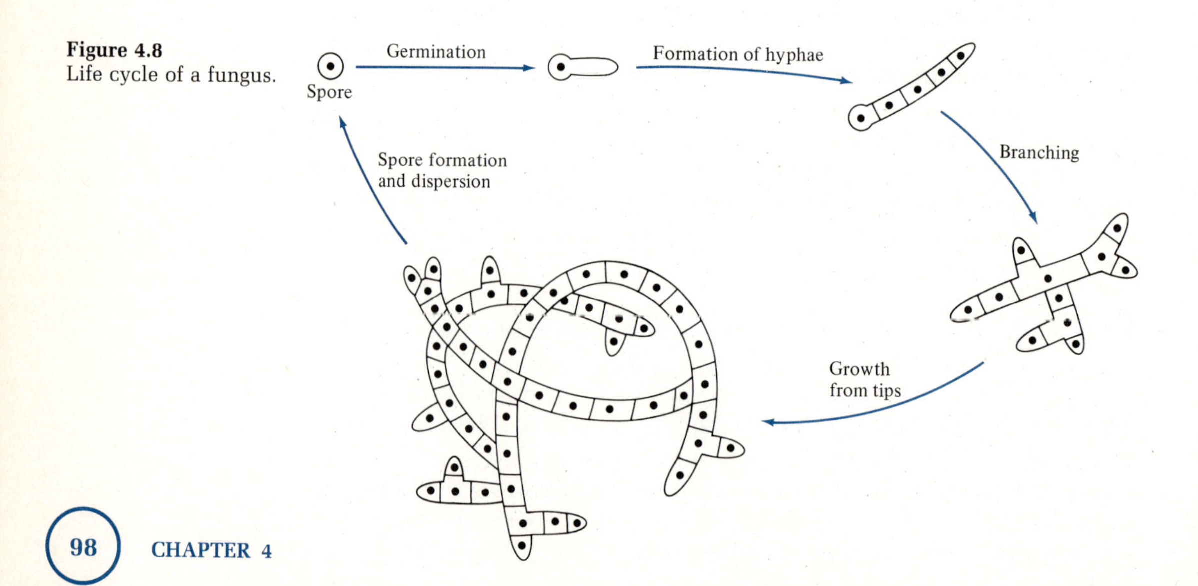

The spore is normally a resting cell. However, when the spore finds itself in an environment containing the appropriate concentration of nutrients, moisture content, and temperature, it germinates and begins to grow. Growth takes the form of walled cylinders, called **hyphae.** Some fungal hyphae contain cross-walls (as shown in Figure 4.8), while others lack entirely a dividing cross-wall. Where cross-walls exist, the partition is not completely closed, so that cytoplasm can flow freely from one compartment to another. During vegetative growth, the hyphae continually extend by apical growth and lateral branching to form a complex, tangled aggregate known as the **mycelium.** The mycelium is a multinucleate (containing many nuclei) mass of cytoplasm that is free to move within a highy branched system of tubes. The cottony growths of molds on decaying fruit and bread are commonplace examples of mycelia. The size of the mycelium is not fixed; in some cases, a single mycelium can extend in soil to a diameter of more than 15 meters. After a period of growth, reproduction occurs with the formation of spores that become detached from the parent and give rise to new mycelia.

Unlike protozoa, the cytoplasm of fungi is enclosed in **rigid cell walls** during all stages of growth. In most cases, the cell wall contains the polysaccharide chitin; the rigid cell wall prevents them from engulfing smaller microorganisms. Rather, fungi obtain nutrients necessary for growth by excreting enzymes into their immediate environment. These enzymes break down many of the complex organic materials in the soil into small molecules that are readily absorbed by the fungi. The large surface area of the mycelium makes this latter process very efficient.

Fungi reproduce both asexually and sexually. Asexual reproduction occurs either by the production of spores or by fragmentation of the mycelium (each fragment becoming a new individual). Sexual reproduction takes place by the union of hyphae from different mating types. The cells fuse, bringing together nuclei from the two parents. The nuclei combine either immediately or at some later time to form a zygote, which, by a series of nuclear divisions, gives rise to reproductive spores.

The fungi are divided into four groups according to the structure of their mycelium and the manner in which they produce reproductive spores (Table 4.3). **Basidiomycetes** contain a mycelium with

TABLE 4.3 Classification of Fungi

Group	Mycelia	Reproductive Spore	Example
Basidiomycetes	Cross-wall	Basidiospore	Mushrooms
Ascomycetes	Cross-wall	Ascospore	*Neurospora* (bread mold)
Phycomycetes	No cross-wall	Motile spore	Water molds
Imperfecti	Cross-wall	Unknown	Penicillium
Yeast	Absent	Basidiospore or ascospore	Bakers' yeast

cross-walls and produce four basidiospores at tips of the hyphae (as shown in Figure 4.8). The basidiospores are formed by a "pinching off" process at the upper end of a specialized club-shaped cell (see Figure 4.10A). Mushrooms and toadstools are familiar examples of basidiomycetes. Although there is no scientific distinction between mushrooms and toadstools, in common usage the word *toadstool* (from the German *tod-stuhl,* meaning "death-stool") refers to a poisonous mushroom. The main part of the mushroom is the underground mycelium. Only when conditions are suitable does the mycelium send up the familiar fruiting body structure (Figure 4.9). Vast numbers of basidiospores develop on the undersurface of the mushroom.

Although some mushrooms are good to eat, they cannot compensate for the food losses caused by two related groups of parasitic basidiomycetes, the smuts and rusts. The latter, especially, are responsible for serious losses of cereal crops such as wheat, oats, and rye. It has been estimated that in the United States alone, more than 100 million bushels of wheat are lost each year because of wheat rust diseases. In the face of increasing populations and food shortages in many parts of the world, no scientific problem is more important to mankind than finding a way to control these fungal diseases.

In **ascomycetes,** sexual reproduction always involves the formation of ascospores. A comparison of ascospore and basidiospore formation is shown in Figure 4.10. In both cases, a multinuclear specialized cell is formed at the tip of the aerial mycelium following a series of nuclear divisions. In basidiomycetes, four basidiospores are formed and subsequently ejected from the tip of the cell. In the case of ascomycetes, four, eight, or more ascospores are produced inside a heavily walled structure called an ascus (plural, asci). When the ascus ruptures, the ascospores are released. A common example of an ascomycete is the orange bread mold *Neurospora.* It is easy to grow this fungus in the laboratory and isolate the intact ascus. The fact that the four pairs of ascospores can be artificially removed in sequential

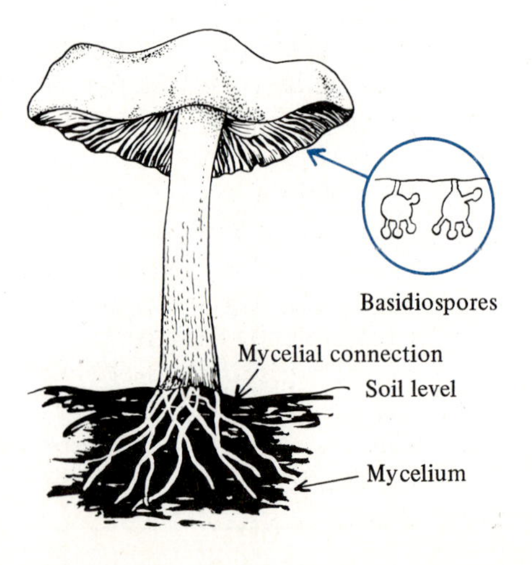

Figure 4.9
Diagram of a mature mushroom showing mycelium and fruiting body containing basidiospores.

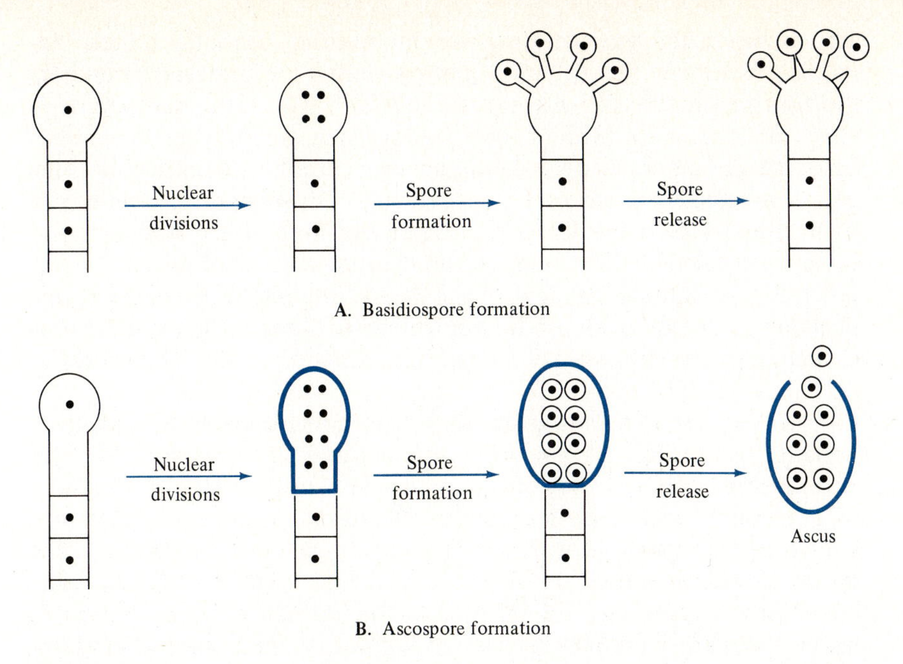

Figure 4.10
Comparison of (A) basidiospore and (B) ascospore formation.

order and then analyzed genetically has made *Neurospora* one of the most useful microorganisms for genetic studies (see Chapter 7). Many ascomycetes are parasitic on higher plants; an unfortunate example is the fungus responsible for Dutch elm disease.

Phycomycetes differ from other fungi in several respects: first, their hyphae contain no cross-walls. Second, they are the only group that produces flagellated, swimming spores. Third, they are also the only group that contains cellulose rather than chitin in their cell walls. Two examples of phycomycetes are the common black bread mold *Rhizopus* and the water mold *Chytrid*. Although most members of this group live on dead organic matter, some forms are parasitic and pathogenic, and, as one mycologist has stated, "at least two of them have had a hand—or should we say a hypha!—in shaping the economic history of an important portion of mankind." One of these pathogenic phycomycetes is the potato blight fungus, which caused the great Irish potato famines of the 1800s. As a result of this disaster, about one million Irish people died of starvation, and another million and a half emigrated to North America. Many settled in the rapidly growing cities of the United States, where they and their descendants have played an important role in all aspects of American life. The other economically important member of this group is *Plasmopara viticola*, the cause of downy mildew of grapes. This mildew threatened the entire French wine industry during the latter part of the nineteenth century.

The basidiomycetes and ascomycetes are distinguished by the type of sexual spore they produce, basidiospores in the first case, ascospores in the second. However, many fungi cannot produce sexual spores unless two different mating strains are present to serve as the two parents. There are some fungi of which only one strain is currently known or which, for other reasons, fail to reproduce sexually. In these situations, there is no way of telling whether the fungus is a basidiomycete or an ascomycete. Consequently, it is tentatively placed in a special group, the *Fungi Imperfecti*. Many of the human fungal parasites, such as the causative agent of athlete's foot, fall into this category. A few of the Fungi Imperfecti are of great importance to man in the production of certain cheeses (Roquefort and Camembert) and fermentation products, including citric acid and the antibiotic penicillin.

Yeasts are fungi that generally do not form hyphae and are thus unicellular. A typical oval yeast cell is shown in Figure 4.11. The familiar bakers' yeast that one can purchase in a market is simply a mass of such cells. One teaspoonful of bakers' yeast (approximately ten grams) contains more than ten billion cells. The predominant mode of multiplication in yeasts is budding (Figure 4.12). A small bulge develops on one side of the cell and gradually enlarges as cytoplasm flows from the parental cell to the bud. Nuclear division occurs in the parental cell, one of the nuclei moves into the bud, and then a double cross-wall is formed between the two cells. When the cells separate, a convex bud scar can be seen on the cell wall of the mother cell, while the daughter cell is left with a concave birth scar. In addition to the asexual budding process, yeasts can multiply sexually with the formation of either ascospores or basidiospores. Bakers' yeast and brewers' yeast (two varieties of *Saccharomyces cerevisiae*) form asci, each containing four ascopores, and are thus sometimes classified as ascomycetous yeasts.

Yeasts usually flourish in habitats where sugar is present in high

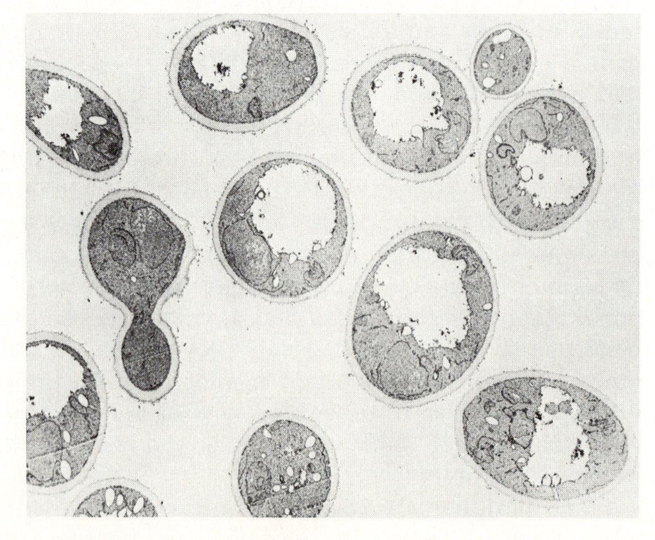

Figure 4.11
Thin sections of yeast cells in the electron microscope. One of the cells is in the process of budding division. (Courtesy of E. A. Bayer and E. Skutelsky.)

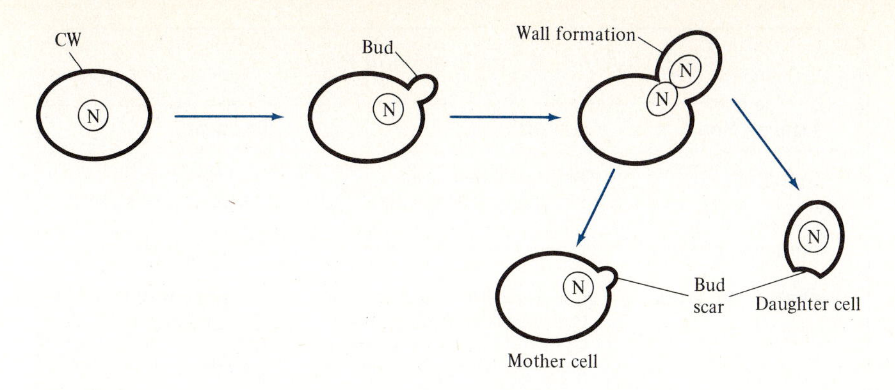

Figure 4.12
Budding division in yeast. After the bud develops on one end of the cell, the nucleus divides by mitosis, and a double cross-wall forms between the two cells. Following separation, the mother cell is left with a convex bud scar and the daughter cell with a concave birth scar.

concentration, such as the nectar of flowers and the surface of fruits. As discussed in the previous chapter, the yeasts can ferment the sugar into ethyl alcohol and carbon dioxide. This anaerobic process has been exploited by man for thousands of years in the production of alcoholic beverages and bread. More recently, the compressed yeasts cells, themselves, have become a valuable commercial product, serving as a rich source of protein, vitamins, and other nutrients. When the desired product is the yeast cells, it is more economical to provide oxygen and allow the cells to convert the sugar to carbon dioxide and water. The fact that respiration is a much more efficient process than fermentation results in a higher yield of yeast cells.

4.5 Bacteria: The Procaryotes

Bacteria are characterized by having the simpler procaryotic cell type. As discussed in Chapter 2, the more primitive and generally smaller procaryotic cell (1) harbors its genetic material in a single DNA molecule that is neither associated with histones nor bounded by a nuclear membrane; (2) lacks membrane-bound organelles in its cytoplasm; and (3) is generally enclosed by a rigid cell wall containing a unique peptidoglycan structure.

The smallness and relative simplicity of the procaryotic cell make it difficult to classify bacteria according to their appearances. Consequently, they must be organized into groups more by what they do (biochemistry) than by how they look (morphology). In the first instance, procaryotes are placed into three major divisions according to the manner in which they obtain energy (Table 4.4). The photosynthetic procaryotes are further divided into two groups, blue-green bacteria and photosynthetic bacteria, according to the nature of their pigments and the biochemical mechanism by which they perform

TABLE 4.4 Classification of Procaryotes

Energy Source	Group	Example
Sunlight	Blue-green bacteria	*Anabena*
	Photosynthetic bacteria	Purple bacteria
Inorganic chemical	Chemoautotrophs	Sulfur oxidizers
Organic chemical	Gram-negative aerobes	Pseudomonads
	Gram-negative anaerobes	*Salmonella*
	Gram-positive spore formers	Clostridia
	Gram-positive non-spore formers	Lactic acid bacteria
	Wall-less bacteria	Mycoplasmas

photosynthesis. Chemoautotrophs obtain energy by oxidation of inorganic chemicals. Bacteria that obtain energy from organic compounds can be divided into six groups according to the following criteria: (1) structure of cell wall—gram-positive, gram-negative, or absent; (2) type of metabolism—aerobic or anaerobic; (3) potential to produce spores; and (4) mycelial growth pattern.

Photosynthetic Procaryotes

Although blue-green bacteria contain all of the structural features of a procaryotic cell, they carry out photosynthesis in a manner very similar to eucaryotic algae and green plants:

$$6\ H_2O + 6\ CO_2 \xrightarrow[\text{blue-green bacteria}]{\text{light}} C_6H_{12}O_6 + 6\ O_2$$

The light-harvesting pigments of blue-green bacteria consist of chlorophyll a (identical in structure to that found in chloroplasts of higher organisms), carotenoids, and phycobiliproteins. Depending on the amount and distribution of these pigments, the cells may appear blue-green, red, violet, brown, or almost black. It is interesting that the Red Sea received its name because of the dense concentration of red-pigmented blue-green bacteria that float on its surface.

In general, the blue-green bacteria are **obligate photoautotrophs**—obligate because they cannot grow in the dark even if provided with organic nutrients, photo because they get energy from light, and autotrophs because they use carbon dioxide as their primary source of carbon for growth. In the process, oxygen gas is produced.

In addition to photosynthesis, most blue-green bacteria have the potential to carry out another process of utmost ecological importance—**nitrogen fixation**, the conversion of atmospheric nitrogen gas (N_2) into ammonia (NH_3). Nitrogen fixation occurs only in blue-green bacteria and a few species of bacteria. All other forms of life on this planet are dependent ultimately on these procaryotic nitrogen-fixers for their supply of usable nitrogen. The fact that blue-green bacteria can perform both photosynthesis and nitrogen fixation allows them to have the simplest nutritional requirements of all living organisms. They can multiply when supplied with only water, a few minerals, light, and air. Because of their nutritional independence, blue-green bacteria are found in a wide range of habitats, including soils and waters depleted of organic compounds and nitrogen salts.

Blue-green bacteria vary greatly in size and shape. Some are small unicellular organisms closely resembling rod-shaped or spherical bacteria; others form filamentous structures, consisting of rows of cells held together by a common outer cell wall (Figure 4.13; also see cover). Not all of the cells in a filament are structurally similar or metabolically equivalent. About 10 percent of the cells specialize in nitrogen fixation, whereas the remaining cells carry out photosynthesis. This is an excellent example of **cell differentiation** in a procaryote. Although completely lacking or not well developed in protists, cell differentiation is the general pattern of growth in plants and animals.

Blue-green bacteria lack cilia, flagella, or any other type of locomotive organelle. Nevertheless, some forms are capable of a gliding type of motility on solid surfaces, the mechanism of which is unknown.

The two types of photosynthetic bacteria, **purple bacteria** and **green bacteria,** perform photosynthesis anaerobically and without the production of oxygen gas. For example, some are able to use hydrogen sulfide (H_2S) in place of water (H_2O) and produce sulfur in place of oxygen:

$$12\ H_2S + 6\ CO_2 \xrightarrow[\text{photosynthetic bacteria}]{\text{light}} C_6H_{12}O_6 + 6\ H_2O + 12\ S$$

As discussed in Chapter 3, this unusual type of photosynthesis was the key that led van Niel to a generalized hypothesis for photosynthesis. In addition to hydrogen sulfide, photosynthetic bacteria can use hydrogen gas or certain organic compounds as reductants for photosynthesis.

Anaerobic photosynthesis imposes a severe restriction on where purple and green bacteria can grow in nature. They require an anaerobic environment in which there is adequate light intensity. One such

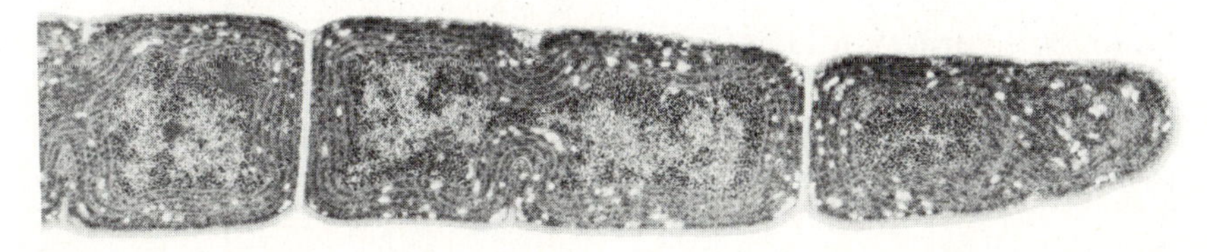

Figure 4.13
Thin section of a filamentous blue-green bacterium in the electron microscope (60,000×). (Courtesy of M. Kessel.)

ecological niche is shallow ponds, rich in organic matter. The oxygen in the pond is rapidly consumed by microorganisms growing on the organic nutrients. The photosynthetic bacteria then develop in the anaerobic water using either the organic matter or hydrogen sulfide as the reductant. Under these conditions, blue-green bacteria are able to multiply only in a very thin layer at the air–water interface. The fact that purple and green bacteria have different light-harvesting pigments than the algae allows them to absorb wavelengths of light not used by the algae.

Chemoautotrophs

By definition, a chemoautotroph is an organism that receives its energy from the oxidation of inorganic chemicals and its carbon from carbon dioxide. The chemoautotroph way of life is found only in a few groups of bacteria. The type of inorganic chemical that can be oxidized by a chemoautotroph is highly specific. For example, ammonia in the soil is oxidized to nitrate in a two-step process, each step carried out by a different bacterium:

$$\underset{\text{(ammonia)}}{NH_3} \xrightarrow[\text{bacteria}]{\text{ammonia-oxidizing}} \underset{\text{(nitrite)}}{NO_2^-} \xrightarrow[\text{bacteria}]{\text{nitrite-oxidizing}} \underset{\text{(nitrate)}}{NO_3^-}$$

Another important subgroup of chemoautotrophs is the **sulfur-oxidizing bacteria.** These microorganisms convert reduced sulfur compounds into sulfate:

$$\underset{\text{(hydrogen sulfide)}}{H_2S} \longrightarrow \underset{\text{(sulfur)}}{S} \longrightarrow \underset{\text{(sulfate)}}{SO_4^{=}}$$

In some cases, the bacteria accumulate sulfur granules inside the cell, and only later oxidize it to sulfate. Many of the sulfur-oxidizing bacteria are **facultative chemoautotrophs;** that is, they are also able to use organic compounds when available as the source of carbon and energy for growth.

A wide variety of aerobic bacteria are able to obtain their energy by the oxidation of hydrogen to water:

$$2\ H_2 + O_2 \xrightarrow[\text{bacteria}]{\text{hydrogen}} 2\ H_2O$$

All of the hydrogen bacteria are facultative chemoautotrophs.

The only other chemoautotrophs discovered so far are the **iron-oxidizing** bacteria. These microorganisms oxidize either reduced iron or manganese compounds.

Gram-Negative Aerobes

Hans Christian Gram, a Danish physician, discovered by trial and error almost one hundred years ago that bacteria could be divided into two classes according to the way in which they were stained by a certain dye complex. Strains of bacteria that appear purple following the staining procedure are referred to as gram-positive; strains that appear red are referred to as gram-negative. It is now known that the way in which a bacterium reacts to the Gram stain is strongly correlated with the detailed structure of its cell wall. Gram-positive bacteria have a single, thick cell wall; gram-negative bacteria have a thin, multilayered wall. (Compare the cell walls of the gram-positive *Bacillus subtilis* and the gram-negative *Escherichia coli*, shown in Figures 2.6 and 2.21, respectively.) In addition to its use in bacterial taxonomy, the Gram test has an important practical significance. Several important antibiotics are known to be effective only against one of the gram types. For example, penicillin is generally much more effective in killing gram-positive than gram-negative bacteria. Therefore, the Gram test is routinely performed in medical bacteriology laboratories once a disease-causing bacterium is isolated. The results of the test will influence the choice of drug to be used in treating the infection.

There is a very wide assortment of cell shapes, metabolic activities, and growth patterns found among gram-negative aerobic bacteria (Table 4.5). In fact, the only property they have in common, in addition to being gram-negative, is that they all obtain energy by the respiration of organic materials. Some of them are able to oxidize a wide variety of different organic compounds, while others tend to specialize in using only a few compounds as nutrients. The summation of the activities of all the gram-negative aerobes is largely responsible for the breakdown of organic matter in soil and water. There probably is not a single *naturally* produced organic compound that cannot be used as a source of carbon and energy by some gram-negative aerobe.

Pseudomonads are probably the most abundant microorganisms in soil and water. The short rods are motile by means of polar flagella (see Figure 2.20). **Acetic acid bacteria** oxidize nutrients to intermediate products that accumulate in the medium; for example, several species oxidize ethyl alcohol (wine) to acetic acid (vinegar). The ***Neisseria*** group are non-motile and generally spherical. The **spiral bacte-**

TABLE 4.5 Aerobic Gram-Negative Bacteria

Subgroup	Characteristic Features
Pseudomonads	Motile rods
Acetic acid bacteria	Motile rods; incomplete oxidation of nutrients
Neisseria	Non-motile cocci or short rods
Spiral bacteria	Helical-shaped cells; some are parasites
Azotobacter	Carry out nitrogen fixation
Intracellular parasites	Multiply inside of other cells
Rickettsias	Rods; cause of epidemic typhus
Bdellovibrios	Curved rods; predator on other bacteria
Chlamydias	Cocci; cause disease in birds and mammals
Caulobacteria	Attach to solids; unique life cycle
Myxobacteria	Gliding motility; complex life cycle

ria include two types of helical-shaped bacteria, spirilla and spirochetes, which differ in their cell walls and modes of locomotion. Spirilla have rigid cell walls and polar flagella. Spirochetes have flexible cell walls and are motile by means of internal structures called **axial filaments.** Studies with the electron microscope show that the axial filaments originate from each of the two ends of the cell, remain just under the outer cell wall, and overlap near the center of the cell. How the axial filaments propel spirochetes is not well understood. One example of a spirochete is the infamous *Treponema pallidum,* the causative agent of syphilis. The ***Azotobacter*** group is unique in its ability to use energy it obtains from oxidation of organic compounds to carry out nitrogen fixation. It has been estimated that azotobacteria are responsible for supplying soil with approximately five tons of nitrogen per square kilometer per year.

Caulobacteria, myxobacteria, and the intracellular parasitic bacteria, **rickettsias, bdellovibrios,** and **chlamydias,** have growth patterns that are distinctly different from other bacteria. The unique life cycle of a caulobacter is illustrated in Figure 4.14. Growth occurs when the caulobacter is attached to a solid surface by an extension of the cell that has an adhesive hold-fast at its terminus. Just prior to division, a flagellum develops on the opposite pole (Figure 4.15). Division results in the formation of two dissimilar sister cells, a stalk cell and a motile cell. The stalk cell can continue to grow and divide in the same manner. However, the motile cell must find a fresh solid surface upon which to adhere, develop a new hold-fast, and lose its flagellum before it can divide again.

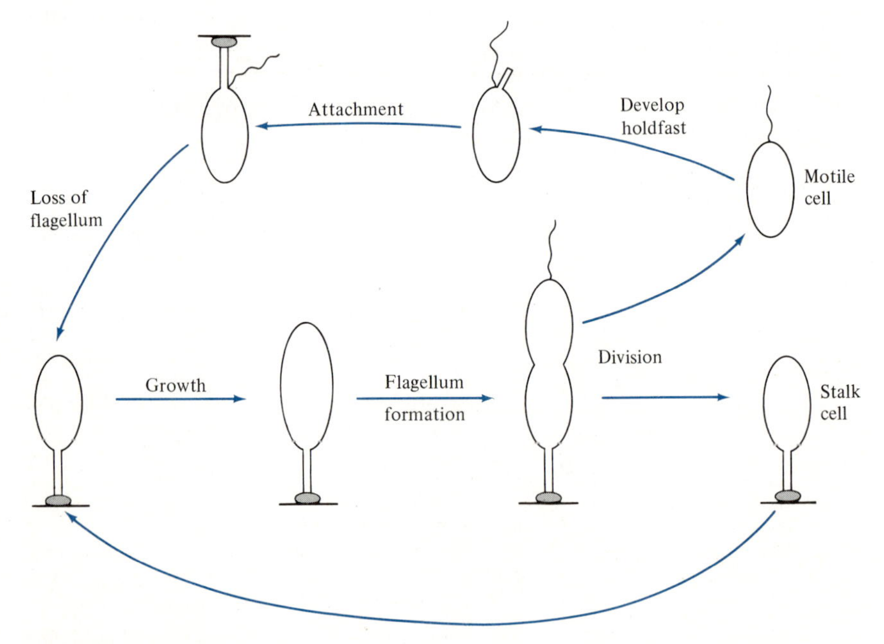

Figure 4.14
The life cycle of a caulobacter.

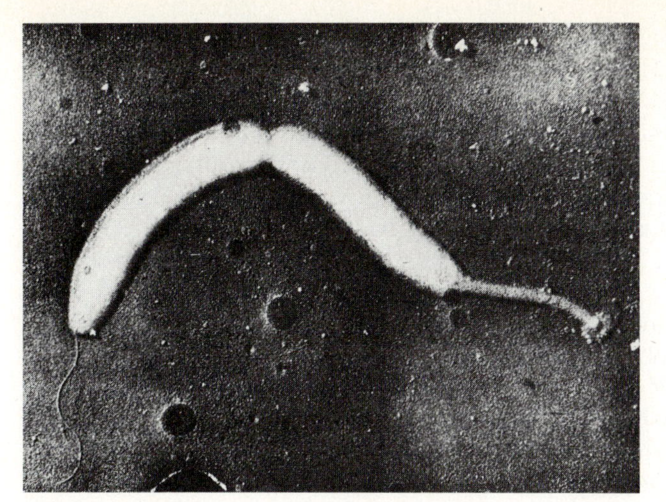

Figure 4.15
Electron micrograph of a caulobacter showing an adhesive hold-fast on one end and a flagellum on the other.

The myxobacteria have the most complex life cycle of any procaryote (Figure 4.16). When food is available, the rod-shaped cells grow and divide, forming a thin mass of cells that swarm over the soil in search of food. Although the cells have no known organelles of motility, they are able to glide over solid surfaces. Interestingly, myxobacteria appear to have developed a means of cell-to-cell communication; if a bacterium wanders too far from the swarm it receives a chemical signal to return. Since many types of myxobacteria feed on other bacteria, it has been suggested that cell-to-cell communication is necessary to ensure a high concentration of cells, a kind of "wolf-pack," needed for effective feeding. Once nutrients are depleted, the cells receive a different signal that causes them to aggregate. About a million cells pile one on top of another to form a visible mound called a

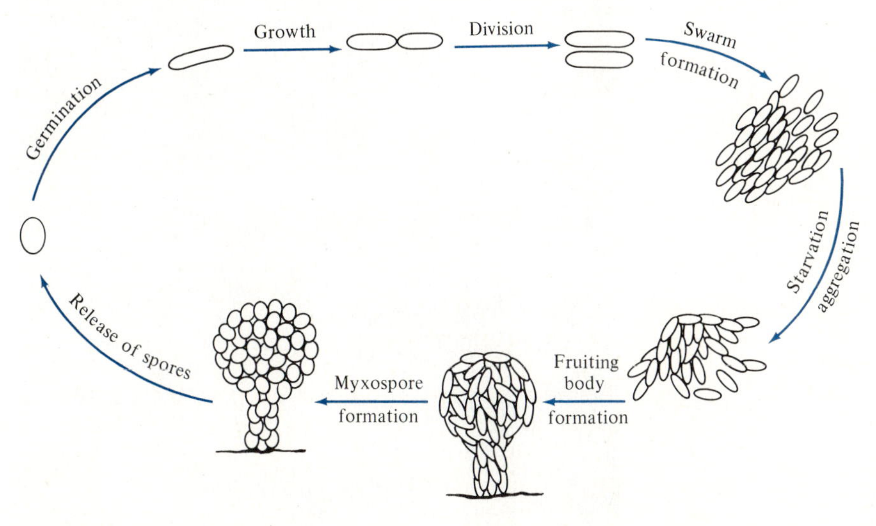

Figure 4.16
The life cycle of myxobacteria.

fruiting body (Figure 4.17). Each of the cells inside the fruiting body rounds up and forms a resting cell, called a **myxospore.** The myxospores are more resistant than the growing cells to a variety of chemical and physical conditions including heat, ultraviolet radiation, and desiccation. When conditions are appropriate, the myxospores are released, germinate, and begin to grow again. The myxobacterial life cycle indicates that communication and cooperation, traits generally associated with social organisms, exist even in the procaryotic world.

Certain bacteria are able to grow only inside other cells. Rickettsias are non-motile short rods that are normally parasites of fleas, lice, and ticks. If an arthropod carrying a rickettsia bites a vertebrate, the bacteria may be transmitted and cause severe and often fatal infections, such as epidemic typhus. Another intracellular parasite is chlamydia, small spherical bacteria that can be passed directly from one vertebrate host to another. One species of chlamydia is responsible for trachoma, the leading cause of blindness in the world. An interesting type of intracellular growth is exhibited by the bdellovibrio (Figure 4.18). This very small, slightly curved rod is highly motile by means of a single polar flagellum. When it collides with a large gram-negative

Figure 4.17
Scanning electron micrograph of a myxobacterial fruiting body. (Courtesy of P. Grillone.)

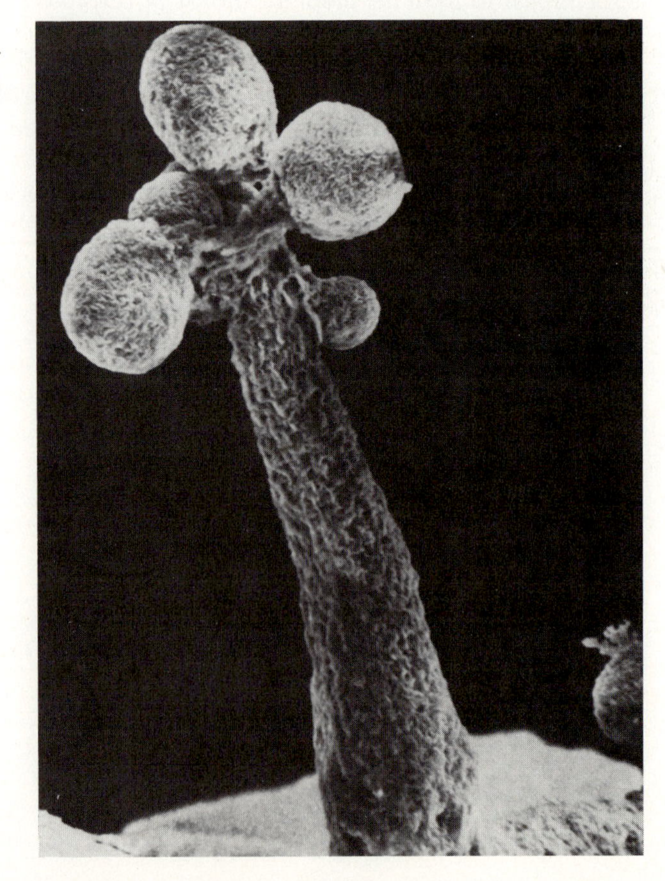

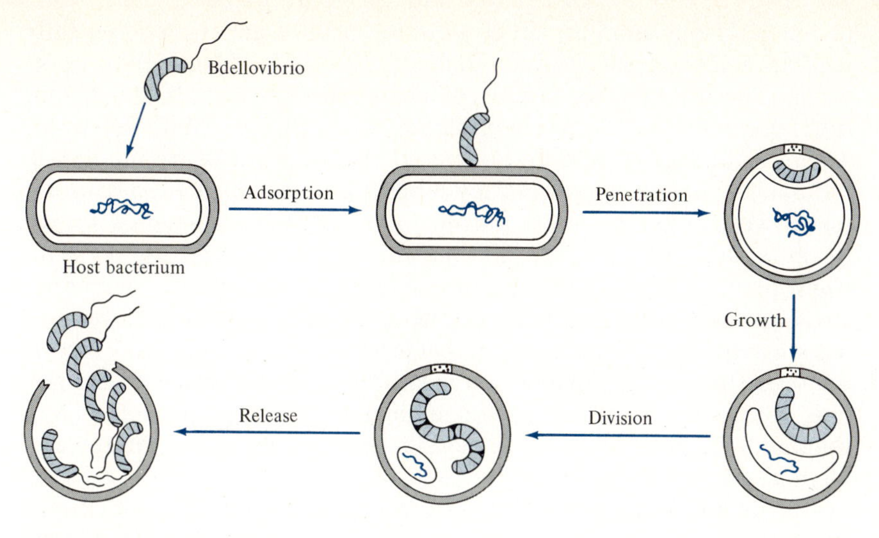

Figure 4.18
The life cycle of bdellovibrio.

bacterium, the bdellovibrio adheres to the host cell wall and slowly penetrates into the cell through a hole it drills in the wall. During the process of penetration, the bdellovibrio loses its flagellum and the host cell becomes spherical as a result of damage done to its wall. The bdellovibrio grows in the space between the repaired wall and cell membrane, deriving nutrients from the host bacterium. After the bdellovibrio has increased its length several times and the cytoplasm of the host cell has been largely consumed, cell division takes place. The four to 12 newly formed bdellovibrios develop flagella, produce enzymes that destroy the host cell wall, and then swim out into the medium searching for new victims.

Gram-Negative Anaerobes

Bacteria that utilize organic compounds in the absence of oxygen gas can be divided into two classes: **strict anaerobes** and **facultative anaerobes.** The strict anaerobes are rapidly killed by even small amounts of oxygen gas. The facultative anaerobes are able to grow either aerobically or anaerobically. In the absence of oxygen, they ferment carbohydrate materials; in the presence of oxygen, they carry out respiration using a wide variety of nutrients.

The gram-negative facultative anaerobes include a medically important group of bacteria, referred to as **enterics;** many of them can grow inside the intestine of human beings and other warm-blooded animals. The best studied example, *Escherichia coli,* is part of the normal flora of a healthy person; however, certain strains of *E. coli* are pathogenic and can cause severe gastroinestinal infections. In general, several types of enteric bacteria are known to cause diseases (discussed in Chapter 13), while other enterics are non-pathogenic. It is therefore of practical importance to be able to distinguish one type of enteric from another. Since enteric bacteria are morphologically similar, it is necessary to use biochemical tests for identification purposes. The use of three simple biochemical tests to distinguish four different enteric bacteria is summarized in Table 4.6. All of the strains except *Salmonella* are able to convert the amino acid tryptophan to indole; the indole is detected by adding a chemical that specifically reacts with indole to give a purple color. Of the four enterics shown in the table, only *E. coli* is able to utilize the sugar lactose as a nutrient for growth. The enzyme urease, which catalyzes the decomposition of urea into ammonia and carbon dioxide, is produced by *Proteus*. Therefore, the four enteric bacteria shown in Table 4.6 can be distinguished on the basis of three simple biochemical tests. In practice, a number of different biochemical tests are used routinely in microbiological laboratories to classify bacteria for both scientific and medical diagnostic purposes.

Strict anaerobes are difficult to study because of their sensitivity to oxygen gas. Special procedures must be used in handling these bacteria in order to avoid their direct contact with air. Many of these new techniques were developed by R. E. Hungate at the University of California at Davis. Using these techniques, Hungate and others found that certan bacteria have adapted to growth in anaerobic ecological niches inside body cavities of animals. For example, the upper digestive tract of cows and other herbivorous mammals is teeming with gram-negative strict anaerobes that are responsible for the breakdown of cellulose to (1) simple organic acids, such as acetic acid, and (2) gases, such as carbon dioxide and methane. The symbiotic relationship between these microorganisms and herbivores is discussed in Chapter 10.

Gram-Positive Bacteria That Form Endospores

> If one contemplates the infinite resourcefulness nature has adopted to perpetuate its biological representatives, . . . none is more marvellous than that device which confers relative indestructibility upon an organism, namely sporogenesis in bacteria.
>
> *Jackson W. Foster*

For most living organisms, the potential to survive "hard times" is as important as the ability to grow rapidly under favorable condi-

TABLE 4.6 Examples of Biochemical Tests Used in Identifying Bacteria

Enteric Bacterium	Indole Production	Lactose Fermentation	Urease Activity
E. coli	+	+	−
Shigella	+	−	−
Salmonella	−	−	−
Proteus	+	−	+

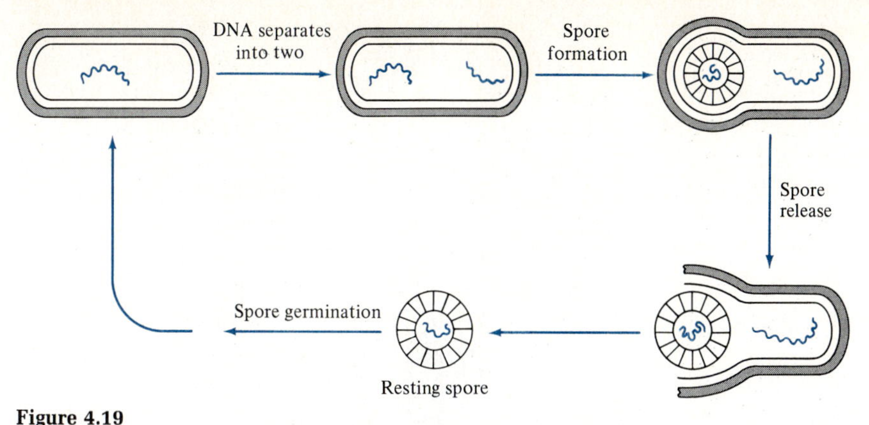

Figure 4.19
The formation and germination of a bacterial spore.

tions. In this regard, certain bacteria have developed a highly efficient mechanism for survival in unfavorable environments, the formation of resting cells called **endospores.** Although the exact mechanism for "triggering" spore formation is not known, it appears that cells begin to form spores when they become starved of certain nutrients.

As shown in Figure 4.19, the process begins with the separation of the nuclear region into two identical parts. One of the nuclear areas becomes successively enclosed in a cytoplasmic membrane, cell wall, and heavy outer coat. Since these changes take place inside the existing cell, the new cell is called an endospore. After the endospore matures, the mother cell is destroyed and the spore is released. At some later time, when the conditions are appropriate, the spore can be stimulated to germinate and commence growth and reproduction. An electron micrograph of a thin section of a bacterial endospore is shown in Figure 4.20. Endospore formation and germination in bacteria are not methods of reproduction because only one spore is formed and the parent is killed in the process.

From the point of view of survival, spores have two major advantages over growing cells. First, the spore is **dormant;** that is, its metabolic activity is greatly reduced. In this way, the spore conserves energy, allowing it to survive long periods of nutritional deprivation. Second, the spore is considerably more resistant than the vegetative cell to a variety of chemical and physical agents, such as high temperature, freezing, radiation, and dehydration. For example, bacteria are generally killed rapidly when they become dry, whereas spores can survive many years in a dry dormant state. As discussed in Chapter 1, the exceptional heat resistance of spores was one of the major reasons for the controversy over the spontaneous generation of bacteria. From the practical point of view, spores provide the major obstacle in the sterilization of foods, medical equipment, and other materials.

The majority of spore-forming bacteria are either *Bacillus* or *Clostridium* species. The bacilli are gram-positive **aerobic** organisms that

Figure 4.20
Thin section of a bacterial endospore. (Courtesy of R. G. E. Murray and coworkers.)

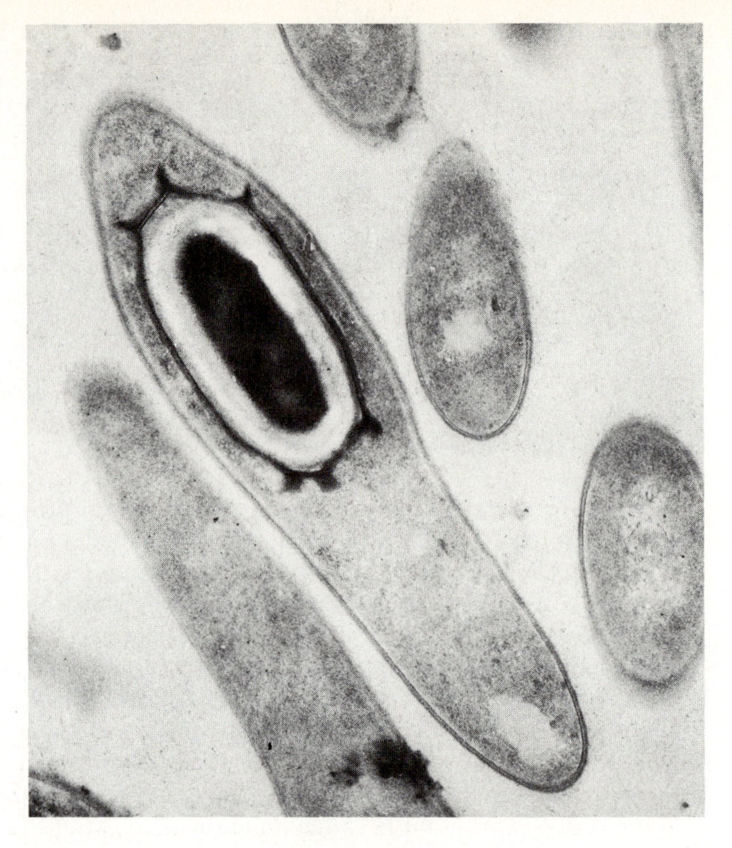

are widely distributed in soils, waters, and air. Most aerobic spore formers are chemoheterotrophs, capable of using a wide variety of organic materials. A few species are pathogenic for either insects or vertebrates. The clostridia are gram-positive strict **anaerobic** spore-formers. They obtain their energy either by the fermentation of carbohydrates or by the anaerobic decomposition of proteins ("putrefaction"). Some clostridia are pathogenic to humans; food poisoning (botulism), tetanus, and gas gangrene are examples of diseases caused by *Clostridium* species.

Gram-Positive Bacteria That Do Not Form Endospores

In addition to the spore-formers discussed in the previous section, there are three major groups of gram-positive bacteria: **lactic acid bacteria, coryneform bacteria,** and **actinomycetes** (Table 4.7). Each of these groups is important ecologically, medically, and industrially. The lactic acid bacteria receive their name as a result of their ability to ferment carbohydrates to lactic acid. The acidic conditions that develop around the lactic acid bacteria prevent most other microorganisms from growing. This is one example of how bacteria are able to change their microenvironment in order to enhance their growth

TABLE 4.7 Major Groups of Gram-Positive Bacteria

Group	Characteristics	Example
Bacillus	Aerobe; forms endospores	*Bacillus anthracis*
Clostridium	Anaerobe; forms endospores	*Clostridium tetani*
Lactic acid bacteria	Facultative anaerobes; produce lactic acid	*Lactobacillus bulgaricus*
Coryneform bacteria	Pleomorphic growth habit	*Mycobacterium tuberculosis*
Actinomycetes	Mycelial growth pattern	*Streptomyces griseus* (produces streptomycin)

and inhibit competition. Most lactic acid bacteria grow on decomposing plant materials. The preparation of sauerkraut, pickles, and fermented milk products, such as yogurt, buttermilk, and cheeses, depends on the conversion of sugar to lactic acid by specialized lactic acid bacteria. Included in the lactic acid bacteria are members of the genus *Streptococcus*, some of which are pathogenic to man and animals. *Streptococcus pneumoniae* is a common cause of bacterial pneumonia, ear infections, and meningitis. Other streptococci are involved in tooth decay, wound infections, urinary tract infections, and bacterial infection of the upper respiratory system, namely "strep throat."

Coryneform bacteria are an extremely diverse collection of gram-positive bacteria that often undergo shape changes during their growth cycle. For example, several coryneform bacteria of the genus *Arthrobacter* undergo sphere-to-rod and rod-to-sphere changes in cell shape, depending on the conditions of growth. *Mycobacterium tuberculosis*, the causative agent of pulmonary tuberculosis, begins to grow in the form of a long tube (mycelium), which subsequently fragments into thin rods. *Corynebacterium diphtheriae*, the causal agent of diphtheria, has an irregular club-shaped morphology.

Actinomycetes are a large group of filamentous bacteria that are common inhabitants of soils. In many respects, the actinomycetes resemble fungi; growth occurs at the tips of filaments that contain many nuclear bodies, usually not separated by cross-walls; branching leads to a cotton-like mycelium. Toward the end of the growth cycle, aerial filaments are produced that develop exospores. Although colonies of actinomycetes and fungi look similar macroscopically, it should be emphasized that the actinomycetes are procaryotic cells; they contain no nuclear membrane, have typical peptidoglycan bacterial cell walls, and are sensitive to antibacterial antibiotics.

The best studied genus of actinomycetes is the *Streptomyces*. These soil bacteria are strict aerobes that are nutritionally quite versatile. They produce extracellular enzymes that break down proteins, polysaccharides, and other complex materials in the soil. Interestingly, the characteristic "earthy" odor of certain soils is caused by *Streptomyces*. Most of the clinically important antibiotics are pro-

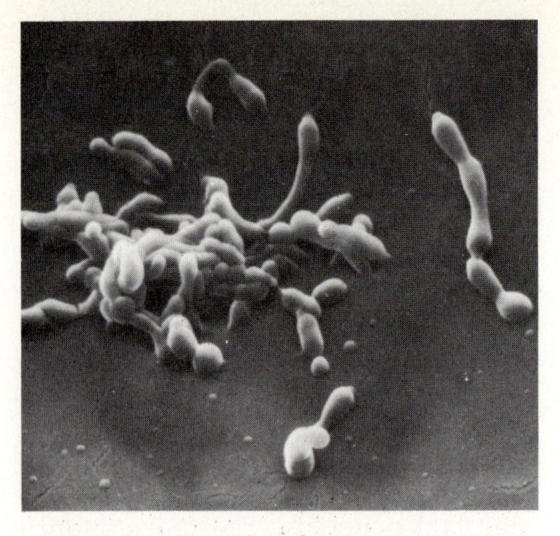

Figure 4.21
Scanning electron micrograph of mycoplasma gallisepticum. Note the pleomorphism and plasticity of the cells. (Courtesy of S. Razin.)

duced by strains of *Streptomyces*. The significance of antibiotics, from the point of view of both humans and microorganisms, is discussed in Chapter 11.

Wall-less Bacteria

The mycoplasmas are typically very small bacteria **without cell walls.** Absence of a rigid cell wall makes the cells more sensitive to osmotic lysis than other bacteria. The small size and plasticity of mycoplasmas allow them to pass through filters that retain other cells. Thus, for a long time, mycoplasmas were thought to be viruses. However, it is now clear that except for the lack of a cell wall and small size, the mycoplasmas are typical procaryotic cells. Because they lack cell walls, mycoplasmas have irregular and variable cell morphology (Figure 4.21).

Most mycoplasmas that have been studied are parasites of man or other vertebrates. One type of pneumonia is caused by *Mycoplasma pneumoniae*. Since these bacteria lack a cell wall, they are not sensitive to antibiotics that block cell wall synthesis, such as penicillin (see Figure 2.16). It is therefore important for a physician who diagnoses a case of pneumonia to ascertain if the causative agent is a pneumococcal bacterium, a mycoplasma, or a virus. In the first case, the disease can be treated with penicillin; in the second case, penicillin and other cell-wall antibiotics are not effective, but the disease can be successfully treated with certain other antibiotics, such as streptomycin; in the case of viral pneumonia, no antibiotics are known that are effective.

4.6 Viruses: The Ultramicrobes

In 1890, Dmitir Iwanowsky, a Russian biologist, traveled to Crimea where tobacco fields were infested with a disease that caused mottling and withering of the leaves. Since the mottling recalled a mosaic pattern, this pestilence was called the tobacco mosaic disease. Iwanowsky found that the tobacco mosaic disease was infectious; sap taken from a plant suffering from the disease, when spread on a healthy leaf, transmitted the mosaic disease.

By itself, this was not a surprising discovery, since Pasteur, Koch, and other microbiologists had already demonstrated that a number of diseases could be transmitted to healthy organisms by inoculating them with fluids containing disease-causing bacteria. What was remarkable was Iwanowsky's claim that the juice was still infectious after passing through filters that retained bacteria. Furthermore, nothing grew when the filtrate was inoculated into sterile broth. It is possible to conclude from these experiments that the material in the plant juice that caused the mosaic disease was (a) small enough to pass through the filter and (b) unable to multiply in broth.

The original findings of Iwanowsky were soon followed up by Martinus Willem Beijerinck (1851–1931), a Dutch microbiologist. First, Beijerinck confirmed the basic observation that the filtrate of the sap from an infected leaf could bring about the disease in a healthy tobacco plant. Next, he showed that the juice of a leaf that had been artificially infected by the filtrate could itself yield a filtrate that would bring about the mosaic disease in another plant. The process of passing the infection from one plant to another with filtrates could be repeated indefinitely. Soon afterward, other important diseases were shown to be transmitted by similar bacteria-free extracts. The agents responsible were called **viruses.**

Viral diseases of man include the common cold, mumps, poliomyelitis, rabies, measles, and viral hepatitis. In addition, there is strong evidence indicating that at least some forms of cancer can be induced by viruses. Viral diseases are known in most domestic mammals, fish, birds, and insects. Almost all plants are subject to viral diseases; symptoms induced vary from lesions on the leaves, as in the tobacco mosaic disease, to severe stunting of growth.

Until the fourth decade of this century, work in the area of plant and animal viruses centered on effective methods of prevention and therapy of diseases they produce. Knowledge concerning the basic nature of viruses was difficult to obtain for two technical reasons: It was impossible to see them, and they grew only inside living animal or plant cells. Viruses are usually not large enough to be seen under the light microscope, and thus escaped detection until special procedures were devised for this purpose. Especially significant was the introduction of the electron microscope into biological research in the 1940s. In Chapter 2, a number of electron micrographs of viruses are presented, and the student is urged to review these pictures while studying this section. Because viruses cannot multiply outside the cells of their host, investigators had to obtain and maintain large numbers of animals or plants for research purposes. The use of animals and plants as hosts for growing viruses made it difficult to carry out controlled experiments.

Despite the inherent difficulties of working with animal and plant viruses, the first advances were made with these materials. It was shown that viruses are relatively **host-specific**; that is, a given virus normally infects only one or a few closely related species. For example, the polio virus attacks humans and certain monkeys, but will not infect other animals.

In 1935, one man with one discovery changed completely the direction of research in virology. Wendell Stanley,* then a young biochemist working at the Rockefeller Institute in New York, succeeded in purifying and determining the chemical composition of a virus. Stanley chose to work with the tobacco mosaic virus because it was not dangerous to human beings and because characteristic infection could be demonstrated easily and rapidly. From the juice squeezed out of a ton of infected tobacco leaves, Stanley was able to isolate a few milligrams of a **crystalline** material that appeared to him, at first, to be a protein. The substance was a pure, homogenous, **non-cellular** material that could be stored, apparently indefinitely, in a dry state; however, if even one millionth of a milligram of the pure material was suspended in water and spread on a living tobacco leaf, the mosaic disease was produced. Extraordinary! The tobacco mosaic virus was at one moment a dry crystalline chemical and at another instant, inside the plant, a multiplying infectious agent. Careful chemical analysis of the pure virus showed that it was composed of 95 percent protein and 5 percent ribonucleic acid, and nothing else.

Stanley's discovery changed the course of biological thought and research. The virus was not a small bacterium; it was a large molecule. Thus, the problem of viruses fell within the domain of chemistry and physics as well as biology. The line of demarcation separating the living from the non-living became cloudy, and a new group of scientists† entered the field of virology. This group was more concerned with basic questions concerning the nature of virus multiplication than clinical applications of virus research. Unfortunately, detailed studies on the multiplication of viruses within the cells of the host they infect were not feasible in the complex environment of whole plants or animals. Thus, the new school of virologists turned their attentions to a group of viruses that attack bacteria, called **bacteriophages.**

Bacteriophages (or phages for short) were discovered independently in 1915 by F. W. Twort, a London bacteriologist, and in 1917 by Felix d'Herelle of the Pasteur Institute in Paris. A fictional but relatively accurate account of the discovery of bacteriophages and of efforts to use them to treat a plague epidemic is included in *Arrowsmith,* a novel by Sinclair Lewis. As in Lewis' story, phage therapy did not prove effective, but studies on the biological properties of

---

*For his fundamental research on the tobacco mosaic virus, Stanley was awarded a share of the 1946 Nobel Prize in chemistry.

†Included in this group of pioneers were three virologists who subsequently shared the 1969 Nobel Prize in physiology and medicine—Max Delbruck, Alfred Hershey, and Salvador Luria.

bacteriophages contributed greatly to our understanding of the chemical and biological interactions of viruses and living cells.

One of the most dramatic demonstrations of the effects of bacteriophages on their bacterial hosts occurs when a small amount of a bacteriophage suspension is added to a liquid medium, visibly teeming with bacterial growth. Within an hour, under proper conditions, the turbid bacterial culture starts to bubble and froth, and within a few minutes no visible trace of bacterial life remains. The bacteria have been destroyed, and the cleared culture liquid can now be shown to contain many more bacteriophages than were originally introduced into it. In some manner, the bacterial viruses have multiplied at the expense of their hosts.

The number of phage particles or infected bacterial cells in a suspension is easily determined by the plaque method. This consists of appropriately diluting the viral suspension and adding a small amount of it to a tube containing a few drops of a concentrated bacterial culture. A small amount of warm, molten nutrient agar is then added, and the entire mixture is poured onto the surface of a petri dish containing a nutrient agar medium. Each infected cell is surrounded by a large number of uninfected cells, which soon grow into a continuous sheet of cells called a **lawn.** During this time, the infected cells produce progeny phages that are released and start new cycles of infection in the surrounding cells. After several such cycles of infection, a clear area, or **plaque,** appears in the lawn, its position corresponding to that of the original infected cell. These plaques are then counted and their number related directly to the number of phages in the original suspension.

The sequence of events between the time bacteriophages are added to a bacterial culture and the time new bacteriophages appear is now fairly well known. This body of knowledge is the result of the efforts of a great many workers during the past 40 years. During the course of obtaining this information, important biological principles were established, as was an understanding of the niche that viruses occupy in the biological hierarchy.

The modern era of bacteriophage research was started by Max Delbruck at the California Institute of Technology about 1940 and was soon being carried on by many other researchers in the United States and elsewhere. Much of this research was confined to studies on a group of seven phages that infect *Escherichia coli* and are named T_1 (for Type 1), T_2, T_3, and so on. By coincidence, the even-numbered T phages (T_2, T_4, and T_6) happened to be closely related and quite similar to one another. These T-even phages have been most intensively studied and will serve as examples in our discussion of the virulent phage. **Virulent** phages are those that invariably lyse the cells they infect during the process of producing progeny phages.

The T-even phages are chemically and morphologically indistinguishable. They are composed almost exclusively of protein and DNA, which are present in approximately equal amounts. They have a polyhedral head to which a slender tail is attached, giving them somewhat the appearance of tadpoles (see Figure 2.35). The DNA of these phages, which corresponds to their chromosome, is one continuous strand approximately 50 microns long (this is about 25 times longer than an *E. coli* cell, but is only about $^1/_{30}$ as long as the bacterial chromosome) (see Figure 2.38). The phage DNA is tightly condensed in a protein-covered head. The tail is encased in a contractile sheath and has six tail fibers attached to its base.

Other phages have been discovered that are somewhat larger than the T phages, and a great many have been found that are smaller. Many of the smaller phages have no tail structure and are essentially spherical in shape; among these some have been shown to contain RNA as their genetic material instead of DNA.

Steps in the Replication of a Virulent Phage

Steps in the infection of a sensitive bacterium with a virulent phage are shown diagrammatically in Figure 4.22. The first step occurs when the non-motile phage collides with a bacterium as a result of random motion. The phage adsorbs to the cell by its tail fibers, which interact with specific chemical sites on the bacterial surface. When this occurs, the phage is irreversibly adsorbed to the cell. Figure 4.23 shows phage T_2 adsorption to *E. coli.* Figure 4.24 demonstrates that this phage can adsorb even to isolated cell walls. Following the adsorption process, a hole is made in the cell wall, probably by an enzyme in the phage tail. Then the tail core, a hollow needlelike tube that extends the length of the tail, is introduced into the hole by contraction of the sheath. Figure 4.25 shows a phage core extending from a contracted tail sheath. The DNA contained in the phage head is then injected through the core into the bacterium. After the DNA has penetrated into the cell, the proteinaceous phage components, which remain outside the cell wall, can be sheared off without affecting the outcome of the infectious process. Experiments that proved that only the nucleic acid enters the bacterium were performed by Alfred D. Hershey and Martha Chase and are presented in detail in Chapter 5.

After the phage DNA has been injected into the susceptible bacterium, it starts to multiply by a process called **vegetative growth.** Soon after infection, synthesis of bacterial products stops and the bacterial biosynthetic machinery is diverted exclusively to the manufacture of new phages. The genetic information of the bacterium is destroyed by the virulent phage. The phage's genetic information is substituted for that of its host cell and contains instructions for the synthesis of new phage DNA and protein components and for their assembly into infectious particles.

The sequence of events after infection is as follows. In approximately five minutes, new enzymes are made by the cell's synthetic machinery according to specifications from phage genes. These enzymes catalyze the synthesis of new copies of phage DNA. This is

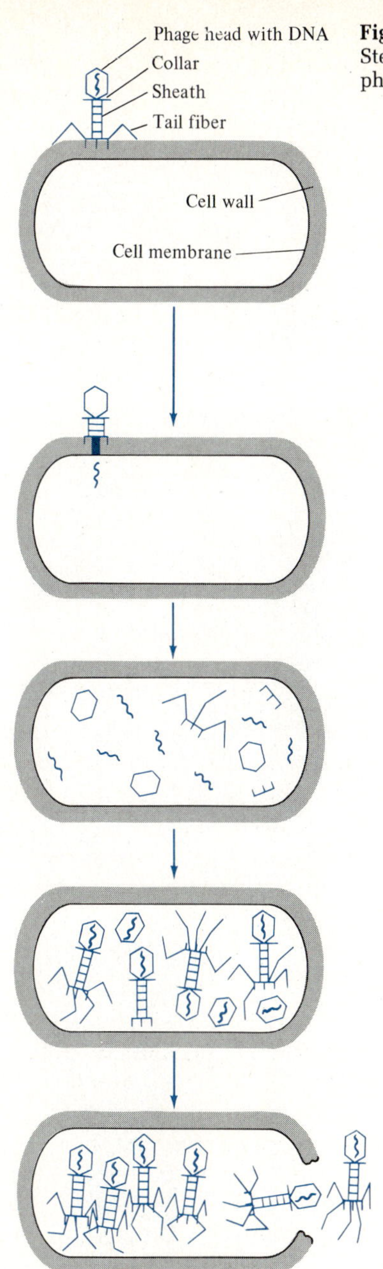

Figure 4.22
Steps in the replication of the lytic bacteriophage T_2.

followed shortly afterward by the synthesis of phage structural proteins such as those making up the head, tail, core, sheath, and fibers. In approximately 15 minutes, the completed phage chromosomes are condensed and encapsulated within a new protein covering to make the new phage heads. The newly synthesized tail components are then added to the preformed heads, thus completing the new infective particles. An electron micrograph showing intracellular phages

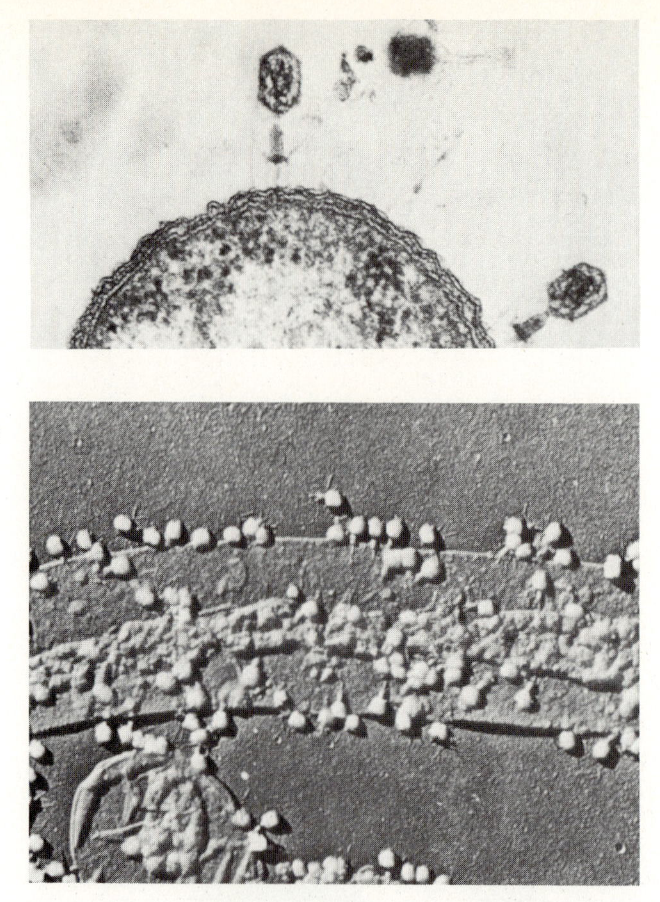

Figure 4.23
Adsorption of phage to a bacterium (144,000×). (Courtesy of L. Simon.)

Figure 4.24
Bacteriophage adsorbed to isolated cell walls (29,000×). (Courtesy of E. Kellenberger.)

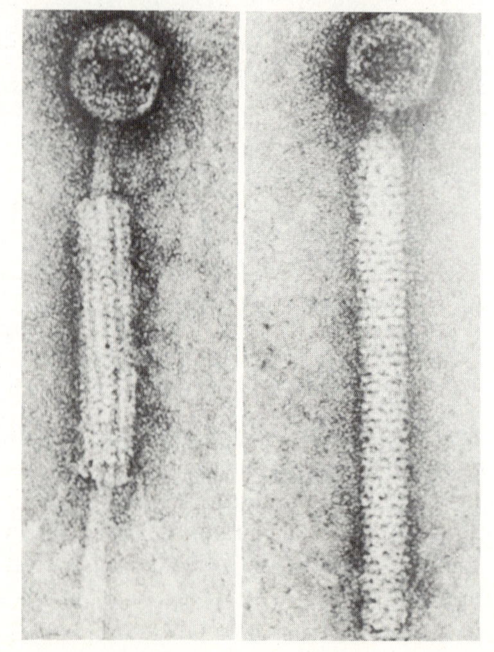

Figure 4.25
Electron micrograph of a *Bacillus subtilis* bacteriophage. Contraction of the tail sheath *(left)* is involved in the injection process (547,000×). (Courtesy of F. Eiserling.)

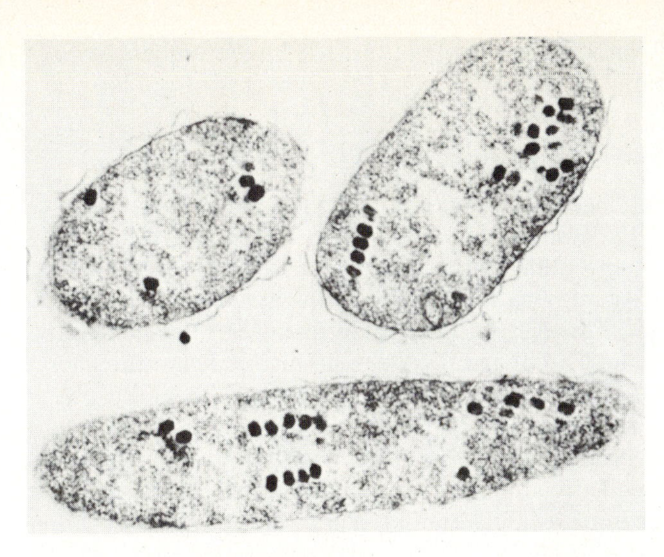

Figure 4.26
Intracellular condensation of phage DNA and phage head formation (45,000×). (Courtesy of E. Kellenberger.)

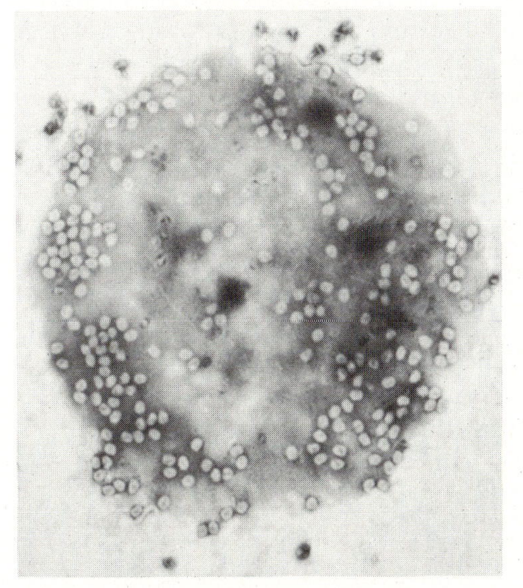

Figure 4.27
Lysis of a bacterial cell and liberation of the bacteriophages (24,000×). (Courtesy of F. Eiserling.)

is seen in Figure 4.26. When approximately 200 new phages are assembled in the infected bacterium, a phage-directed enzyme dissolves the cell wall and releases the mature progeny phages that are now able to initiate new infectious cycles (Figure 4.27).

Thus, **viruses multiply in a manner completely different than cells.** Under ideal conditions, an *E. coli* cell will grow in length for approximately 20 minutes and then divide into two identical bacteria. Hence, the number of cells will increase every 20 minutes in the progression 1–2–4–8. Multiplication of bacteriophages, on the other hand, resembles the production of automobiles; the parts are made initially and only subsequently assembled. Since the number of T_2 phages coming out of a single *E. coli* is about 200, the increase in phage number follows the progression 1–200–40,000–8,000,000.

The ease with which a few viruses can destroy large numbers of cells now becomes clear.

The Lysogenic Cycle of Phage Multiplication

Some bacteriophages have another extremely interesting mode of replication called the **lysogenic cycle** (Figure 4.28). The initial steps in the lysogenic cycle, phage adsorption and injection of its DNA, are similar to those occurring in the lytic cycle. Once the phage DNA is in the bacterium, however, a remarkable event can occur, leading to the lysogenic state. The phage DNA can be inserted into the bacterial chromosome. Once integrated into the bacterial DNA, the phage DNA is inactive except for a limited number of genes. It is replicated along with the bacterial chromosome and is distributed with it to each daughter prior to cell division. Bacteriophages existing in this latent form are termed **prophages.** Bacteria that contain prophages are called **lysogenic** bacteria.

How can we tell whether a bacterium is lysogenic? Lysogenic bacteria differ from non-lysogenic bacteria in two important properties. First, they are immune to further infection by the same kind of phage they carry; the prophage produces a protein that prevents similar phages from infecting the lysogenic bacterium. Second, they have the hereditary potential for phage production in the absence of external infection. When a culture of lysogenic bacteria is treated with various agents, such as ultraviolet light or ionizing radiation, they are induced to multiply like virulent phages. The phage DNA comes out of the chromosome and begins to replicate independently; phage parts are produced and assembled, the cell is lysed (hence the term lysogenic bacterium), and mature phages are released. The concept of lysogeny is fundamental to modern thinking about the relationship between genes and viruses, sex in bacteria (Chapter 6), and the induction of cancer (Chapter 14).

General Properties of Viruses

From the preceding discussion it is clear that viruses exist in two distinct forms: one extracellular and the other intracellular. Outside the cell, the virus has none of the properties associated with life. It is merely a biochemical substance composed of nucleic acid and protein molecules. (In the case of some animal viruses, the nucleoprotein is surrounded by a membrane.) Inside an appropriate host cell, the nucleic acid of the virus is capable of self-replication and gene expression, two properties characteristic of living organisms. Viruses differ, however, from cells in several basic properties:

1. Viruses contain only one type of nucleic acid, DNA or RNA, never both.
2. Viruses are produced by the assembly of their parts, whereas cells multiply by the orderly growth of all parts, followed by cell division.

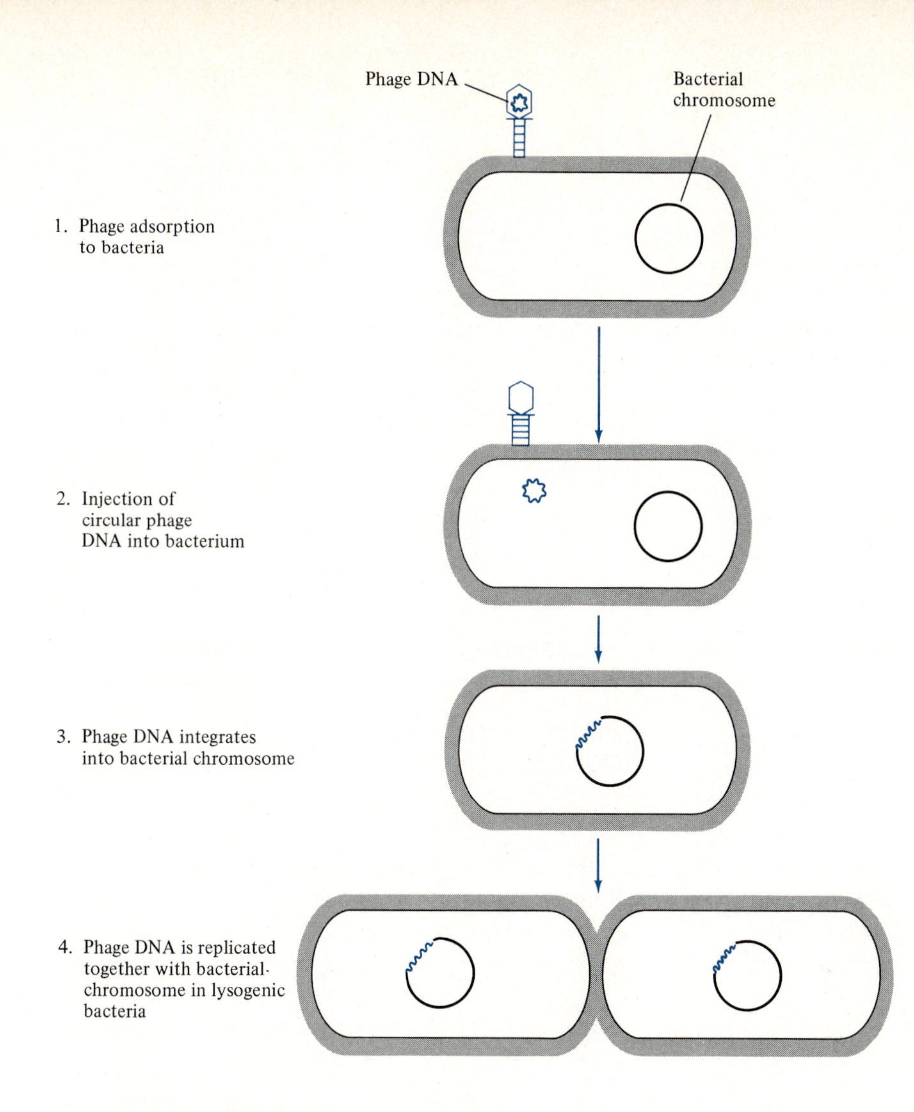

Figure 4.28
Lysogenic cycle of phage replication.

3. Viruses are genetic parasites. Although there are parasitic bacteria, fungi, and protozoa, the viruses demonstrate the ultimate in dependency. Lacking all of the machinery necessary for generating energy and synthesizing proteins, the virus is unable to carry out even the most elementary metabolic processes outside of its living host.

An interesting case of viral parasitism on a bacterial predator is shown in Figure 4.29. The outermost boundary is the cell wall of an *E. coli* cell that has been attacked by a bdellovibrio (see Figure 4.19)

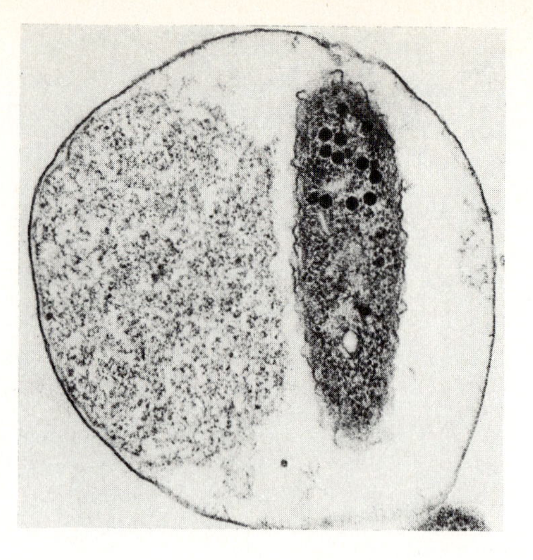

Figure 4.29
Electron micrograph of bdellovibrio growing inside an *E. coli* cell. The bdellovibrio itself is infected with phages (dark bodies). (Courtesy of M. Kessel and M. Varon.)

that has, itself, been infected with a phage (dark bodies inside the bdellovibrio). All this calls to mind a poem by Jonathan Swift:

> So, naturalists observe, a flea
> Hath smaller fleas that on him prey;
> And these have smaller fleas to bite 'em,
> And so proceed *ad infinitum*.

Swift could not have realized how infinitely small are viral "fleas"—a million of them would not cover the period at the end of this chapter.

QUESTIONS

4.1 The kingdom of Protista contains bacteria, blue-green bacteria, eucaryotic algae, fungi, and protozoa. Certain biochemical and morphological characteristics are used to distinguish each of these broad groups of microorganisms from the others, while different characteristics are used to sort out members of the same group. Explain how each of the following characteristics is used in classifying protists.

a) Contain chloroplasts
b) Amoeboid motility
c) Ability to fix nitrogen gas
d) Anaerobic photosynthesis
e) Cell wall composed of chitin
f) Gram-negative
g) Contain cilia
h) Ability to grow on lactose
i) Anaerobic spore formation

4.2 What characteristics do green algae possess that indicate they may have been the forerunners of all higher plants?

4.3 Give a fundamental similarity and difference between:

a) Animals and plants
b) Blue-green bacteria and eucaryotic algae
c) Blue-green bacteria and purple photosynthetic bacteria
d) Basidiomycetes and ascomycetes
e) Yeast and other fungi
f) Protozoa and fungi
g) Chemoautotrophs and photosynthetic bacteria
h) Streptomyces and fungi
i) Lytic and lysogenic cycles of phage multiplication
j) Rickettsias and viruses

4.4 What advantages does spore formation confer upon bacteria? Explain why endospore formation is not a method of reproduction in bacteria.

4.5 A large number of antibotics have been discovered that can be used to control diseases caused by bacteria. Why do you suppose none has been found that is useful in treating diseases caused by viruses?

4.6 What are the arguments, pro and con, for the proposition that *viruses are living?*

4.7 Figure 4.29 shows a phage multiplying inside a bdellovibrio bacterium that is itself growing inside an *E. coli*. Explain the dependence of the bdellovibrio phage multiplication on the presence of *E. coli*. At which step in the life cycle of bdellovibrio do you think it is susceptible to phage infection?

4.8 Recently, Salvador Luria and James Darnell, two of the pioneers of molecular virology, defined viruses as "entities whose genomes are elements of nucleic acid that replicate inside living cells using the cellular synthetic machinery and causing the synthesis of specialized elements that can transfer the viral genome to other cells." What are the most important elements of this definition? What are the "specialized elements" that transfer the viral genome?

Suggested Readings

Kellenberger E: The Genetic Control of the Shape of a Virus. Scientific American, December 1966.
A clear discussion with excellent electron micrographs of bacteriophages.

Pickett-Heaps JD: Green Algae. Sunderland, MA: Sinauer Associates, Inc., 1975.
A beautifully illustrated book on the structure, reproduction, and evolution of green algae.

Stanier R, EA Adelberg, and J Ingraham: The Microbial World, 4th edition. Englewood Cliffs, NJ: Prentice-Hall, 1976.
An excellent in-depth textbook of general microbiology for students with prior background in chemistry and biology.

Wood WB and RS Edgar: Building a Bacterial Virus. Scientific American, July 1967 (offprint 109).
A lucid account of assembly of viral components in the test tube.

The Genetic Material: Deoxyribonucleic Acid

Philosophy begins when someone asks a general question, and so does science.

Bertrand Russell

In the preceding chapters, we have been concerned with the morphology and metabolism of microorganisms. Another distinguishing characteristic of living things, perhaps the most basic of all, is their ability to reproduce themselves, that is, to transfer hereditary characteristics from cell-to-cell and generation-to-generation. In the last 30 years there have been astonishing developments concerning the molecular basis of inheritance. The significance of man's understanding of the very basis of heredity is incalculable—it will change all generations to come. Therefore, it is our obligation to gain some understanding of the work of the molecular geneticist.

The science of genetics can be traced to the carefully conducted studies of Gregor Mendel around the mid-nineteenth century. Using the common garden pea for his cross-fertilization experiments, Mendel was able to determine that traits are passed from parents to offspring with mathematical precision in small, separate information packets that subsequently become known as **genes.** Shortly after the rediscovery of Mendel's work in 1900, a group of scientists at Columbia University, led by Thomas Hunt Morgan,* began an extensive study of the genetics of a small fruit fly, *Drosophila melanogaster.* These investigators demonstrated that genes were located on chromosomes in a linear order.

The concept of a gene was invented in order to explain certain patterns of inheritance. The gene was characterized as an information packet that is located at a particular point on a particular chromosome. In effect, the genes were treated as though they were strings of beads of unknown composition that are duplicated and transmitted from parent to offspring by an unknown mechanism and that somehow interact with the environment to determine the characteristics of the organism. The fundamental questions were yet to be asked:

1. What is the chemical nature of the gene?
2. How are the genes reproduced so that each offspring gets an (almost) identical copy?
3. How is the genetic information converted into the observable physical trait?

In this chapter, we study some beautiful experiments that provide the answers to the first two questions. The third question is discussed in Chapter 7.

*For his pioneering genetic studies, Morgan received the Nobel Prize in 1934.

5.1 The Discovery of Transformation

In 1928, Fredrich Griffith, a medical officer in the British Ministry of Health, reported the results of his studies on what, at that time, was the number one killer of humans, pneumonia. It was known then that pneumonia was caused by a spherically shaped bacterium, the pneumococcus. Griffith began his studies by isolating the bacteria from the sputum of patients with pneumonia. When the bacteria were first obtained in pure culture, he found there was only one type of pneumococcus, later called type ***S*.** This type was characterized by the appearance of mucoid or smooth *(S)* colonies on agar medium; when type S bacteria were injected into mice, the mice contracted pneumonia and died. When type S bacteria were subcultured by transferring from one agar plate to another, there occasionally appeared (we now know, by mutation) a second type of pneumococcus, type ***R*.** This type gave irregularly shaped or rough *(R)* colonies and was unable to cause the disease.

Griffith realized that the basic difference between the *R* and S strains is that type S contains a polysaccharide capsule surrounding the cell. These capsules help produce the smooth appearance of the colony and, more important, protect the bacteria so that they can survive and grow in the animal body. Type *R*, lacking the capsule, gives rise to a rough colony and cannot survive inside the animal.

To understand how these capsules participated in the infectious process, Griffith performed a series of experiments on mice. Keep in mind that the motivation for these experiments was not a curiosity regarding the nature of the gene, but rather a quest for information about the nature of the infectious process. As so often happens in science, the correct interpretation for these experiments came much later. For the moment we shall consider only Griffith's experimental results; the interpretation and significance will emerge as subsequent experiments are discussed. Table 5.1 summarizes the results of Griffith's experiments.

Experiments 1 and 3 were controls. Neither type *R* nor heat-killed S was able to cause the disease when injected separately. However, when they were mixed together and then injected, the concoction was lethal. Furthermore, a postmortem examination of the mice from experiment 4 revealed *only live type S pneumococci*. These type S

TABLE 5.1 Griffith's Original Transformation Experiments

Experiment	Injection into Mice	Result
1	Type *R* only	No effect
2	Type *S* only	All mice dead
3	Heat-killed type *S*	No effect
4	Type *R* plus heat-killed type *S*	All mice dead

bacteria bred true; that is, the smooth characteristic was retained and reproduced in subsequent generations. The process by which type *R* bacteria were converted into type *S* came to be known as **transformation.** Although Griffith performed careful controls and his type transformation was confirmed by other workers, it did not attract due attention. One possible reason for the apparent lack of interest in Griffith's experiments was because he did not emphasize the genetic implications of his work. In 1942, Griffith was killed in London during a bombing raid. He did not live to see the significance of his experiments appreciated.

5.2 Transformation: The Chemical Nature of the Gene

In the 1930s, a group at the Rockefeller Institute led by Oswald T. Avery (1877–1955) became interested in the process of transformation. At first, Avery was skeptical of Griffith's work and asked one of his young assistants to repeat the experiments. Not only were Griffith's experiments reproducible, but it was soon discovered that mice were not needed to demonstrate transformation. In these later experiments, the bacteria were treated, mixed, and placed directly on the appropriate medium. The results were approximately as shown in Table 5.2.

Experiments 1 and 2 were controls. In experiment 1, only one of a million of the *R* bacteria gave rise to type *S* colonies. The probable cause of this infrequent event is spontaneous mutation, which is discussed in Chapter 6. Experiment 2 demonstrated that all of the bacteria were killed by the heat treatment. In experiment 3, 1 percent of the bacteria were converted from type *R* to type *S*. Since this frequency is much higher than the mutational rate, a process of transformation must have taken place.

It is reasonable to ask at this time: Why, when the mixture of heat-killed *S* and live *R* was injected into a mouse, were *all* of the bacteria found at the postmortem examination type *S*, whereas when placed directly on the agar medium, only 1 percent were type *S*? The answer is that in all probability, Griffith also obtained only 1 percent transformation, but that when injected into the mice, the type *R* was destroyed and did not show up in the postmortem examination.

Avery and his collaborators concluded from these experiments that heat-killed type *S* contains an active substance that enters into

TABLE 5.2 Avery's Transformation Experiments

Experiment	Number and Type of Bacteria Treated	Number and Type of Colonies Produced
1	10^6 (1,000,000) type *R*	1 type *S*, 10^6 type *R*
2	10^6 heat-killed type *S*	0
3	10^6 type *R* + 10^6 heat-killed type *S*	10^4 type *S*, 10^6 type *R*

the live type *R* and transforms it into type *S*. Since this trait is inherited in subsequent generations, it has the property that we uniquely associate with the gene. This active substance was called **transforming principle.** The Rockefeller group then began an extensive program to isolate and identify the transforming principle.

Biochemists have developed, mostly by trial and error, a series of techniques for isolating and identifying chemical substances from cellular matter. The first step in an isolation procedure is to prepare a cell-free extract by disrupting the cells. This can be accomplished in a number of different ways, such as freezing and thawing, grinding with powdered glass, or various chemical treatments. In the case of pneumococcus, the cells were broken by treatment with a chemical, sodium deoxycholate. Was the transforming principle still active? When the cell-free extract from type *S* was added to live type *R*, approximately 1 percent of the cells were transformed to type *S*. The next step was the separation of the components that are released when the cell is disrupted. By a variety of techniques, the chemicals can be separated according to their size, electrical charge, and other properties. In the case of the transforming principle, the key step in the isolation process was the use of ethyl alcohol. When two volumes of 95 percent alcohol were added to one volume of the cell extract, a stringy white precipitate formed in the solution. This precipitate was easily collected on a glass rod, leaving most of the other components behind either in the alcoholic solution or as granular precipitates. **All of the transforming principle was associated with the stringy precipitate.** The procedure was repeated a number of times until Avery was satisfied that he had the transforming principle in as pure a state as he could get it.

The next question: What is it? The observation that the material precipitates in alcohol indicates that it is a large molecule, a macromolecule. But is this stringy white precipitate a protein, deoxyribonucleic acid (DNA), ribonucleic acid (RNA), or polysaccharide?

With a few precious milligrams of the highly purified transforming principle, the Rockefeller group began their analyses. First, they determined the elementary composition of their unknown substance. It contained 36 percent carbon, 4 percent hydrogen, 16 percent nitrogen, 10 percent phosphorus, and 34 percent oxygen. This elementary chemical analysis, especially the relatively high phosphorus content, suggested that the substance was a nucleic acid, either DNA or RNA. As discussed in Chapter 1, proteins and polysaccharides do not contain appreciable amounts of phosphorus. The two types of nucleic acid can be distinguished by the type of sugar they possess. All DNA molecules contain the sugar deoxyribose, whereas RNA molecules contain the sugar ribose. Since the sugar analyses revealed that the unknown substance contained only deoxyribose, the transforming principle was evidently DNA.

It was still possible, however, that a minor component, not detected by the analyses, was responsible for the transforming activity. Therefore, as a final demonstration that the active substance was

DNA, a series of enzyme studies was conducted. The three enzymes were trypsin, ribonuclease (RNase), and deoxyribonuclease (DNase). Trypsin is a specific enzyme in that it destroys only protein; likewise, RNase destroys only RNA, and DNase destroys only DNA. The transforming principle was treated with each enzyme and then tested for activity. As shown in Table 5.3, only DNase inactivated the transforming principle.

Thus in 1944, after 15 years of persistent research, the Rockefeller group had clearly demonstrated that the transforming principle was DNA. Since the transforming principle has the property of a gene, and the transforming principle is DNA, **the gene must be made up of DNA** (at least in this case).

By a mechanism that we still do not understand, minute amounts of DNA enter the permissive cell and become permanently established. These recipient cells now act as if a new gene has been added to their genetic makeup. In Griffith's original experiments, the transformation process was made possible by the fact that the temperatures required to kill pneumococci (65°C) did not destroy the DNA. Thus DNA from the heat-killed type *S* entered the type *R* cells and transformed them into type S.

Avery's extraordinary results were soon enlarged on in two significant ways. First, it was shown that a large number of other traits could also be transformed. For example, DNA extracted from strains of pneumococci that were resistant to the drug streptomycin could transform streptomycin-sensitive pneumococci into the resistant form. Second, the process of transformation was demonstrated in a few other species of bacteria, including *Bacillus subtilis* and *Hemophilus influenzae*. Again it was shown that the active transforming principle was DNA. By 1950, the evidence from these and other experiments was convincing that in bacteria, at least, the genetic material is DNA.

5.3 DNA Is the Genetic Material in Viruses, Animals, and Plants

Although the transformation experiments clearly demonstrated that the gene is DNA in certain bacteria, there was at first some skepticism about the general applicability of Avery's discovery. Much of this skepticism vanished when it was shown by Alfred D. Hershey and Martha Chase of the Cold Spring Harbor Laboratories in New York that DNA was also the genetic material in viruses. These inves-

TABLE 5.3 Effect of Specific Enzymes on Transformation

Experiment	Treatment of Transforming Principle	Result of Transformation Experiment
1	Trypsin	Transformation achieved
2	RNase	Transformation achieved
3	DNase	No transformation

tigators used for their critical experiments a simple bacterial virus called phage T_2. In order to understand the Hershey-Chase experiment, you must remember (Section 4.6) that (1) phage T_2 is composed of only two kinds of macromolecules, protein and DNA; and (2) pictures taken with an electron microscope (review Figures 4.23 to 4.26) reveal that only part of phage T_2 enters a bacterium; once inside, the phage multiplies until a few hundred *complete* new phages are produced. Thus, that part of the phage that enters the bacteria must contain the genetic information for all of the phage parts. The question is, then, what part of the phage enters the bacterium?

Hershey and Chase prepared for their experiment by growing one batch of bacteria on a nutrient medium containing radioactive phosphorus atoms (^{32}P); another batch was grown on nutrient medium containing radioactive sulfur atoms (^{35}S). Each batch was then infected with phage T_2. As the phage grew inside the bacteria, some of the radioactivity was incorporated into its structures. The ^{35}S went exclusively into the protein part of the phage since DNA contains no sulfur. The ^{32}P went exclusively into DNA since phage proteins contain no phosphorus. In this way, the protein and DNA components of the phage were selectively labeled with radioactivity.

The ^{32}P and ^{35}S phages were then used to infect separate cultures of fresh bacteria. This time, however, there were no radioactive atoms in the medium. After allowing just enough time for the phages to infect the bacteria, the mixtures were shaken vigorously (a common kitchen blender was used) and the bacteria purified. The shaking was necessary to dislodge the parts of the phage that became attached to the bacteria but that did not actually penetrate it. Thus, only the part of the phage that entered the bacterium remained with it during purification. The purified bacteria were then examined for radioactivity. If radioactivity were found with the bacteria infected with the ^{32}P phages, it would mean that DNA penetrated the bacteria; if radioactivity were found with ^{35}S phage-infected bacteria, it would mean that the proteins penetrated. The result was that only the bacteria that were infected with ^{32}P phages were radioactive. **Therefore, the only part that enters the bacteria and must contain the genetic information is DNA.** As we discussed in Chapter 4, the phage protein serves primarily to aid the DNA in entering the cell and in protecting it against injury.

The Hershey-Chase experiment completed in 1952 was the first clear demonstration that DNA was the genetic material in a virus, phage T_2. Since that time, experiments have shown that DNA is also the genetic material in a number of other viruses. William R. Romig performed such an experiment at the University of California at Los Angeles. In this case, the experiments were performed with a virus called SP8 that infects the bacterium *Bacillus subtilis*. As mentioned previously, *B. subtilis* is one of the few kinds of bacteria that can carry out the process of transformation. Romig's technique was to mix purified DNA isolated from phage SP8 with its bacterial host, *B. subtilis*. The phage DNA entered the bacteria in the same manner as in the transforming principle. Once inside the cell, this purified phage DNA was able to multiply and produce several hundred new phages. Since the phages that are produced are complete (contain DNA and protein), the phage DNA must contain the genetic information for the entire phage. This interesting experiment is similar to the processes of both transformation and infection and has been termed **transfection.**

The statement that DNA is the genetic material is not universally true. It has been shown that a few viruses contain RNA instead of DNA as their nucleic acid component. In these cases, the genetic material has been shown to be RNA.

In higher animals and plants, a large body of circumstantial evidence indicates that here too the genetic material is DNA. Much of this evidence will be discussed in subsequent sections as it relates to specific biochemical processes. Some of the more significant data are as follows:

1. **Localization.** All cells contain DNA that is localized primarily† on chromosomes in the nucleus of the cell. This is consistent with the fact that the genes are also located specifically on chromosomes.
2. **Quantity.** Within the same species, the amount of DNA per (diploid) cell is the same in all organs. The (haploid) cells such as spermatozoa that contain one half the amount of genetic capability also contain one half the amount of DNA.
3. **Stability.** Most of the macromolecules of the cell are constantly being broken down and synthesized. If this were to happen to a gene, valuable hereditary information would invariably be lost. Of all the macromolecules in the cell, DNA is the most metabolically stable.
4. **Sensitivity to chemical and physical agents.** All agents that cause genes to be damaged (mutation) have also been shown to induce changes in the structure of DNA.
5. **Genetic engineering experiments.** Recently it has been possible to introduce pure animal DNA into bacteria. In certain cases, genetic information of the animal cell is expressed in the bacterium.

In summary, then, all evidence indicates that the genetic material is nucleic acid. In higher animals and plants, bacteria, and most viruses, the nucleic acid that carries the genetic information is DNA. In the few viruses that are devoid of DNA, the genetic material is RNA.

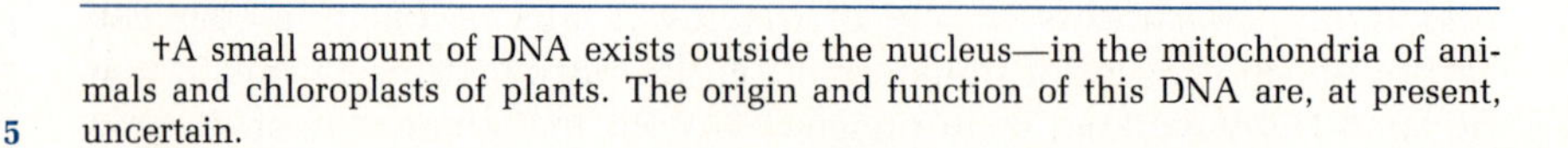

†A small amount of DNA exists outside the nucleus—in the mitochondria of animals and chloroplasts of plants. The origin and function of this DNA are, at present, uncertain.

5.4 DNA: Chemical Structure

Prerequisite to an understanding of how the genes function in the transmission and expression of genetic information is a knowledge of the chemical structure of DNA. Rather than simply presenting our current concept of the structure of DNA, the observations and experiments that led to this concept will first be offered. In this way, the student can obtain some insight into how a structure as complicated as a DNA molecule can be elucidated. Generally, the determination of the structure of a naturally occurring macromolecule involves the following steps:

1. Isolation and purification of the substance.
2. Determination of its component parts.
3. Clarification of how the parts are joined.
4. Determination of the three-dimensional structure of the molecule.

Isolation and Purification

In the same decade that Pasteur presented his swan-neck flask to the world, that Mendel reported his genetic studies on garden peas, and that Charles Darwin published *On the Origin of Species,* Friedrich Miescher (1844–1895), a student of the German chemist Felix Hoppe-Seyler, became interested in the chemistry of the nucleus. He chose to work with white blood cells because they contained a large and easily observable nucleus. A hospital at Tubingen supplied him with surgical bandages that had been peeled off purulent wounds. From the pus on the bandages he obtained the white blood cells, which he then treated with gastric juice (we now know that gastric juice contains an enzyme, pepsin, that digests protein). He observed microscopically that only the shrunken nuclei were left after the pepsin treatment; the remainder of the cell had dissolved. These nuclei were then analyzed and found to have a different composition from any cellular component then known. Since the material was isolated from nuclei, it was called **nuclein.**

Miescher continued his study of nuclein when he returned to his native city, Basel, in Switzerland. He soon found that a more convenient source of this material was salmon sperm. The sperm is exceedingly rich in nuclein. With this favorable material he was able to purify further the nuclein, and when all protein was finally removed, it became clear that the new material was an acid. It was then referred to as *nucleic acid.*

Determination of Component Parts

Once the nucleic acid molecule could be isolated in pure form, the next task was an analysis of its component parts. The nucleic acid was suspended in a solution of strong acid and then heated in order

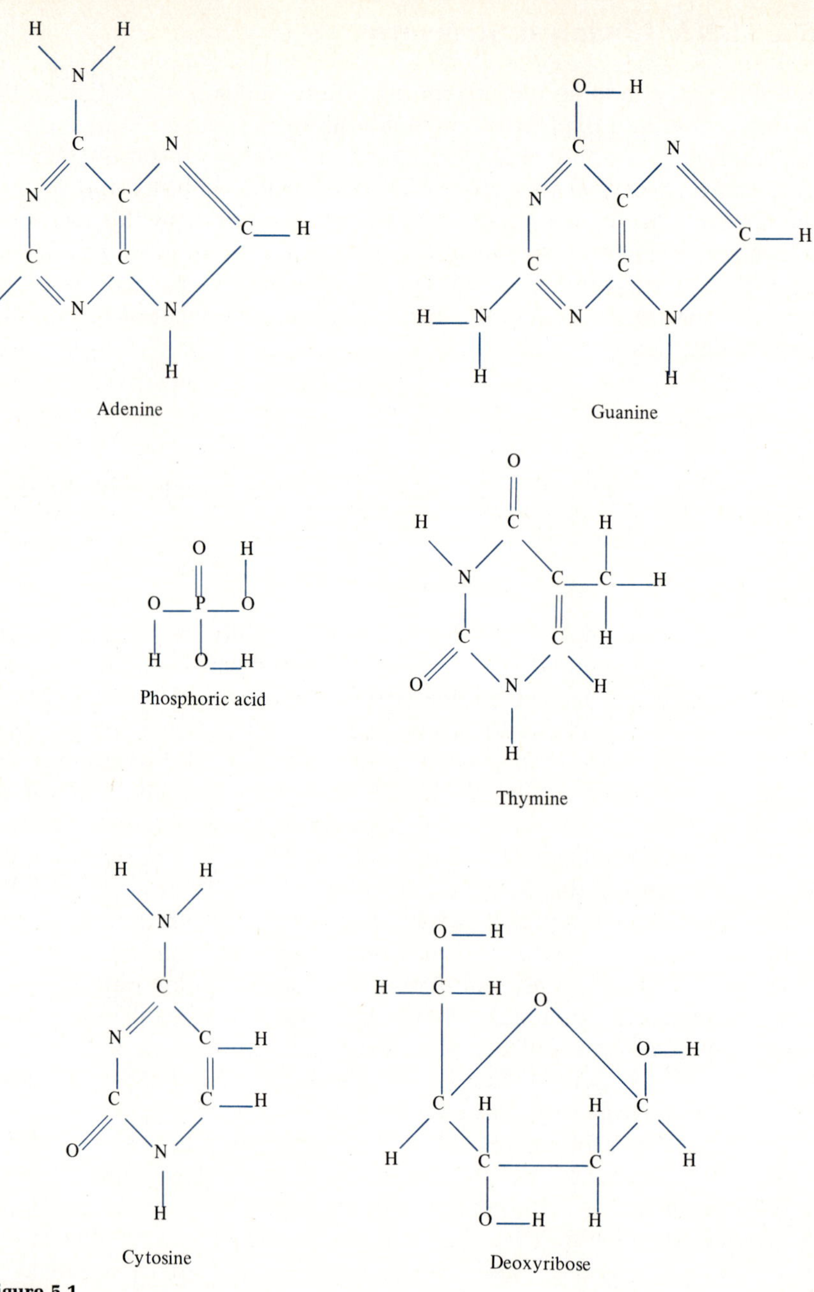

Figure 5.1
The component parts of DNA.

to break the large molecule into a mixture of many small pieces. The small pieces were then separated and identified. By 1930, each of the component parts had been characterized. Salmon sperm nucleic acid was composed of one part sugar, one part phosphoric acid, and approximately one fourth part each of the nitrogenous bases adenine, guanine, thymine, and cytosine (Figure 5.1). When the sugar was identified as deoxyribose, the nucleic acid came to be called **deoxyribonucleic acid,** or DNA.

Manner in Which the Component Parts are Linked

An intact DNA molecule is a giant structure of great complexity. Even the simplest phage DNA molecule contains more than 5000 submolecules each of deoxyribose and phosphoric acid, and more than 1250 each of adenine, guanine, cytosine, and thymine. How are they joined together? Two organic chemists, Phoebus A. Levene at the Rockefeller Institute and Lord Todd‡ in Cambridge, England, were most instrumental in demonstrating that the components of DNA were joined together to form a long chain of alternating deoxyribose and phosphoric acid units with side chains of the nitrogenous bases (Figure 5.2). An interesting feature of this structure is that although the phosphate-sugar chain is perfectly regular, the molecule as a whole need not be, because the order of bases on the sugar units can vary.

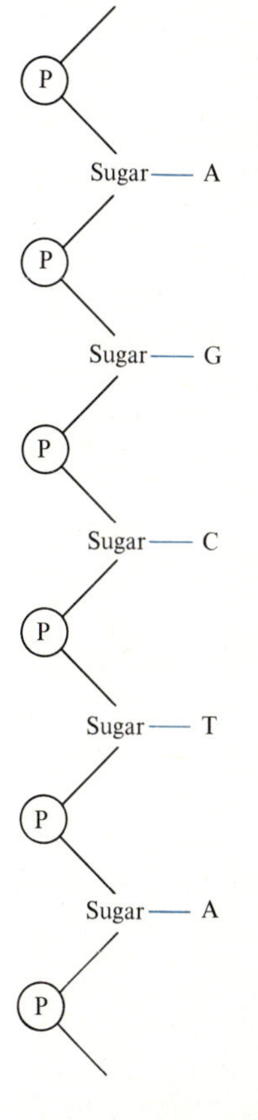

Figure 5.2
The chemical structure of DNA. The sugar is deoxyribose, P stands for the phosphate bridge connecting the sugars, A for adenine, G for guanine, C for cytosine, and T for thymine. The molecule must be imagined to extend a very great distance above and below this figure. For example, the smallest DNA molecules (found in viruses) contain approximately 5000 bases.

‡For his work on nucleic acids, Todd was awarded the 1957 Nobel Prize in chemistry.

5.5 The Three-Dimensional Structure of DNA

Although the elucidation of the chemical structure of DNA was one of the major achievements of the first half of this century, it was not until its three-dimensional structure became known that there was any clue as to how it might function as the genetic material. The determination of the three-dimensional structure of DNA required data that were obtained with two new techniques, paper chromatography and x-ray crystallography.

Paper chromatography is in practice an exceedingly simple procedure. A small drop of material is applied about two inches from the edge of a rectangular sheet of filter paper and allowed to dry there. The edge of the paper is then dipped in an appropriate liquid that slowly moves up the paper by capillary action. As the liquid passes over the point of application, the material is pulled along by the liquid at a speed that is characteristic of that substance and the liquid (solvent). By using this technique, a mixture of substances applied to the same spot can be separated from each other in a few hours. For example, if a mixture of the four nitrogenous bases found in DNA is applied on one spot, and butyl alcohol is allowed to pass over it, the four components will separate from each other in approximately 24 hours (Figure 5.3). The bases can be detected readily on the paper since they all absorb ultraviolet light. In fact, the precise amount of each base can be determined by how much ultraviolet light is absorbed.

Using this simple but sensitive technique of paper chromatography, Professor Erwin Chargaff and his collaborators at Columbia University analyzed the base composition of DNA from various sources. By the early 1950s, enough data had accumulated to allow some very interesting conclusions to be drawn. What would you deduce from the data in Table 5.4?

Conclusion One. The base composition of the DNA is characteristic of the species, differing in composition for different species, but not for the different tissues of any one species.

Conclusion Two. For all DNA samples the amount of adenine

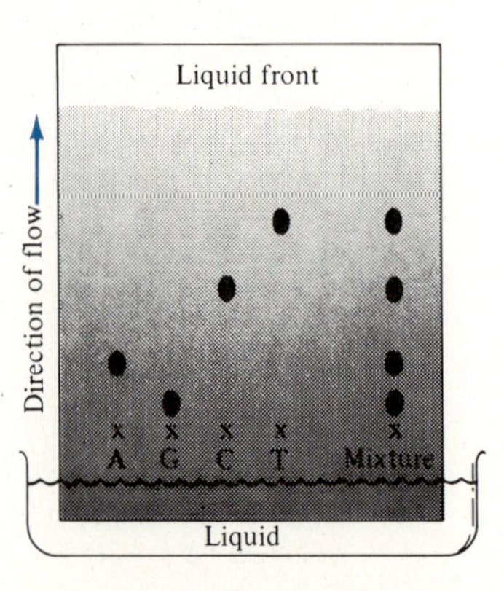

Figure 5.3
Paper chromatography of the nitrogenous bases of DNA. Each of the four bases in addition to a mixture was applied at the position marked with an X. As the liquid nears the top of the paper, the mixture becomes resolved, as viewed with an ultraviolet lamp.

TABLE 5.4 The Base Composition of DNA of Several Species[a]

Source of DNA	Adenine	Guanine	Cytosine	Thymine
Salmon sperm	29.7	20.8	20.4	29.1
Rickettsia (bacteria)	33.8	16.7	15.9	33.6
E. coli (bacteria)	25.2	25.0	25.5	24.3
Yeast	31.7	18.1	17.6	32.6
Micrococcus (bacteria)	14.5	36.1	35.9	13.5
Phage T_2	31.9	18.4	17.9	31.8
Calf thymus	29.8	20.4	20.7	29.1
Calf thyroid	29.6	20.8	20.7	29.1
Calf spleen	29.6	20.4	20.8	29.2

[a]These results are expressed as *mole percent*, which is the number of molecules of one base divided by the total number of molecules of all four bases, multiplied by 100. The data are accurate to ± 1 percent

equals thymine (A = T) and cytosine equals guanine (C = G). It follows that A + G (purines) = T + C (pyrimidines).

These conclusions are consistent with the concept of DNA as the genetic material. Since the different species contain different hereditary determinants, it is to be expected that their base composition would also be different. Different tissues of the same species, however, must have an identical base composition, since cells divide in such a way that each daughter cell gets an identical copy of genes.

The second conclusion, that A = T and C = G for all species of DNA, came as a complete surprise. You will remember that the purines, A and G, are double rings, and the pyrimidines, C and T, are single rings; so there is an equality of large and small rings. Although the significance of these regularities was not at first appreciated, they were soon to provide a major clue in unraveling the three-dimensional structure of DNA.

The next evidence came from x-ray diffraction images of DNA molecules taken by Rosalind Franklin in the laboratory of Maurice Wilkins at King's College, London. Because of their short wavelengths, x-rays can be used to examine the fine details of molecules. Although the theoretical aspects of x-ray diffraction need not concern us here, what is important is that the pictures were consistent with the following points:

1. DNA from different species give identical x-ray patterns, despite the fact that their base composition varies.
2. DNA molecules are shaped like spaghetti—long and thin (see Figures 2.38, 2.41, and 2.42); the actual length of the molecule is greater than 30,000 angstroms, whereas it is only about 20 angstroms thick (1 angstrom = one 100-millionth of a centimeter).
3. DNA has a repeating structure every 34 angstroms; that is, as you proceed along the length of the molecule, at regular intervals of 34 angstroms, there is a repetition of the structure.

These, then, were the pertinent experimental facts: the detailed chemical structure of DNA, Chargaff's base pairing rules, and the x-ray diffraction photographs of Franklin and Wilkins. There was only one more ingredient that was necessary for the solution of the three-dimensional structure. That ingredient was genius, and it was applied by James D. Watson and Francis H. Crick.

James Watson was a young American biologist at the time (1953), who, while still an undergraduate student, "became polarized toward finding out the secret of the gene." After some initial training with Professor Salvador Luria at Indiana University, Watson traveled to Europe, ending up at the Cavendish Laboratories in Cambridge, England. There he met Francis Crick, a physicist who had turned to molecular biology after spending the war years designing mines for the British Admiralty. Working together, they succeeded in a remarkably short time in building a scale model of DNA that fit all of the experimental data.§ How they arrived at this model is discussed in great detail in Watson's popular book, *The Double Helix*. Of great conceptual value was the discovery two years earlier by Linus Pauling at the California Institute of Technology that protein can exist in the form of a helix.

Figure 5.4 shows the Watson-Crick model for DNA. The most significant feature of the model is that it consists of two strands held together by the attractive forces of the nitrogenous bases. When Watson and Crick constructed scale models of the double-standard molecule, they soon discovered that the base-pairing must be specific; if there was adenine on one strand, there must be a corresponding thymine on the other strand; likewise, if guanine was on one, cytosine must be on the other. If they attempted to place two of the double-ring structures, adenine and guanine, opposite each other, it resulted in a bulge in the molecule and forced the strands apart. Two single-ring bases could not be brought close enough together to attract each other. Finally, adenine could not lie opposite cytosine, nor could guanine lie opposite thymine, because of the geometry of the molecules. Thus **A must always pair with T, and G with C.**

Furthermore, in order to maintain the pairing throughout the molecule, it was necessary to twist the two sugar-phosphate backbones in the form of a helical staircase (Figure 5.4B). The steps of the staircase would then be the DNA base pairs. When accurate measurements were made on scale models of this double helix, the diameter turned out to be exactly 20 angstroms and the length for a complete turn was 34 angstroms. There were ten bases on each chain for every complete turn of the helix. The distance between bases on the same chain was thus 3.4 angstroms.

As we have shown, the solution of the three-dimensional structure of DNA had a rather long history. Like almost every discovery, it cannot be attributed to any one person. Starting with Miescher, some-

§For this work on the three-dimensional structure of DNA, Wilkins, Watson, and Crick shared the 1962 Nobel Prize in medicine and physiology.

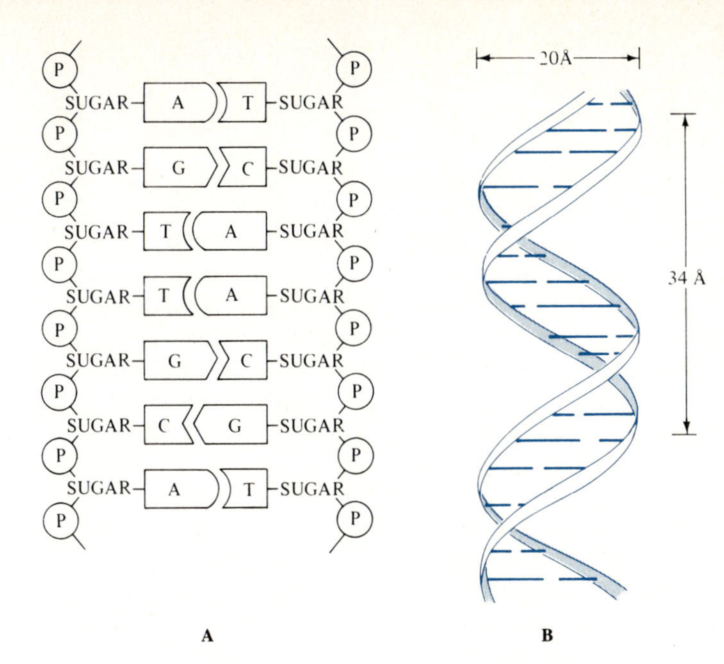

Figure 5.4
The Watson-Crick model for DNA. (A) A two-dimensional representation of a segment of double-stranded DNA. (B) The three-dimensional structure can be visualized as two strands wound around each other to form a double helix, the horizontal bars being the pairs of bases holding the chain together.

one found a bit here, another a bit there, and thus onward until a couple of theoreticians put the bits together and made the breakthrough.

> Science, like the Mississippi, begins in a tiny rivulet in the distant forest. Gradually other streams swell its volume. And the roaring river that bursts the dikes is formed from countless sources.
>
> *A. Flexner*

5.6 Replication of DNA: The Meselson-Stahl Experiment

The Watson-Crick model revolutionized biology and biochemistry, not so much because it explained a large body of data on the structure of DNA, but because it immediately suggested (to Watson and Crick) how DNA might create an identical copy of itself. Since the structure of DNA consists of two complementary strands, either chain thereby carries the information for making the entire molecule. The mechanism that Watson and Crick proposed for DNA replication is elegant in its simplicity: When ready to replicate, **the two strands separate and each strand serves as a template for the synthesis of its partner.**

Consider a segment of DNA consisting of six bases on each strand (Figure 5.5). When the strands separate, each of the original strands acts as a mold or template for the production of a new chain. The original ACGATT strand is copied to yield a fresh strand that must have the sequence TGCTAA because of the rigorous base-pairing requirements imposed by the DNA structure. Likewise, the original TGCTAA strand acts as a template for a new strand with the sequence

Figure 5.5
Watson-Crick model for DNA replication.

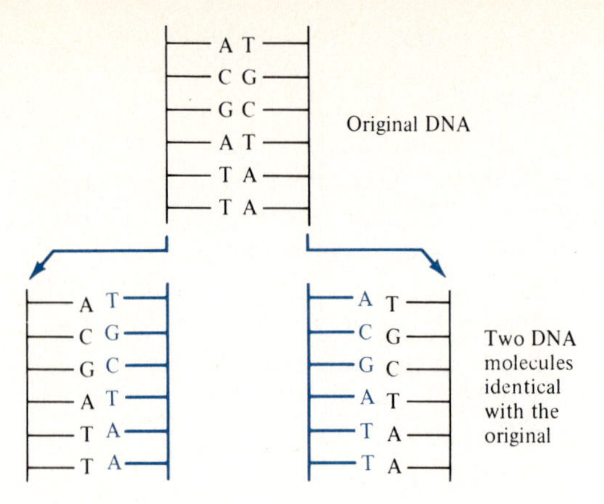

ACGATT. In this way, two complete sets of hereditary information can be produced, each an exact replica of the original.

The Watson-Crick hypothesis for the replication of DNA was put to the test in 1958 by an ingenious experiment designed and performed by two young scientists at the California Institute of Technology, Mathew Meselson and Frank Stahl. Meselson is a physical biochemist who, following a series of discussions with Stahl, developed a technique for separating DNA molecules according to their densities. He found that if certain salt solutions are centrifuged at very high speeds for approximately 48 hours, the salt molecules are forced toward the bottom of the tube because of the centrifugal force. Since salt is heavier than water, the higher concentrations of salt toward the bottom of the tube result in the formation of a gradient of densities, the highest density being at the bottom of the tube and the lowest at the top. Furthermore, if DNA is added to the salt solution and then centrifuged, the large DNA molecules collect in a thin band at that part of the gradient that has precisely the same density as the DNA. If a DNA molecule finds itself at too high a density, it floats until it reaches the correct density; if it is at too low a density (that is, too high in the tube), it will sink. Thus, DNAs that have different densities will concentrate at different positions in the tube and can thereby be separated.

Stahl, as a geneticist, provided the biological know-how for the crucial experiment. They began by growing the bacteria for several days in a medium that contained the heavy isotope of nitrogen, ^{15}N. As far as the bacteria are concerned, the presence of ^{15}N instead of the usual ^{14}N makes very little difference. The bacteria grow and multiply at a normal rate. However, when the DNA is extracted from these bacteria, the nitrogenous bases contain the heavier ^{15}N atoms; thus, the DNA is of greater density than normal ^{14}N DNA.

After the bacteria had become completely labeled with ^{15}N, they were transferred to a medium that contained ^{14}N and allowed to continue their growth. Henceforth all DNA would be made with the ^{14}N isotope only. At various times after the transfer to the ^{14}N medium, samples of the culture were removed, and the DNA was extracted and mixed with salt and centrifuged at high speeds. The results are shown in Figure 5.6. The bacteria that were used for these experiments divide every 30 minutes under the conditions used, so that 30, 60, and 90 minutes correspond to one, two, and three generations, respectively.

Before transfer to the ^{14}N medium (experiment 1), all of the DNA concentrated at a position near the bottom of the tube, corresponding to ^{15}N DNA. After one generation (experiment 2), all of the DNA had a density that was intermediate between ^{15}N and ^{14}N DNA, which will be referred to as hybrid DNA. After two generations (experiment 3), the DNA was 50 percent hybrid and 50 percent ^{14}N DNA. After three generations (experiment 4), the DNA was 25 percent hybrid and 75 percent ^{14}N DNA.

The Watson-Crick replication hypothesis predicts precisely the results of the Meselson-Stahl experiment (Figure 5.7). At the first rep-

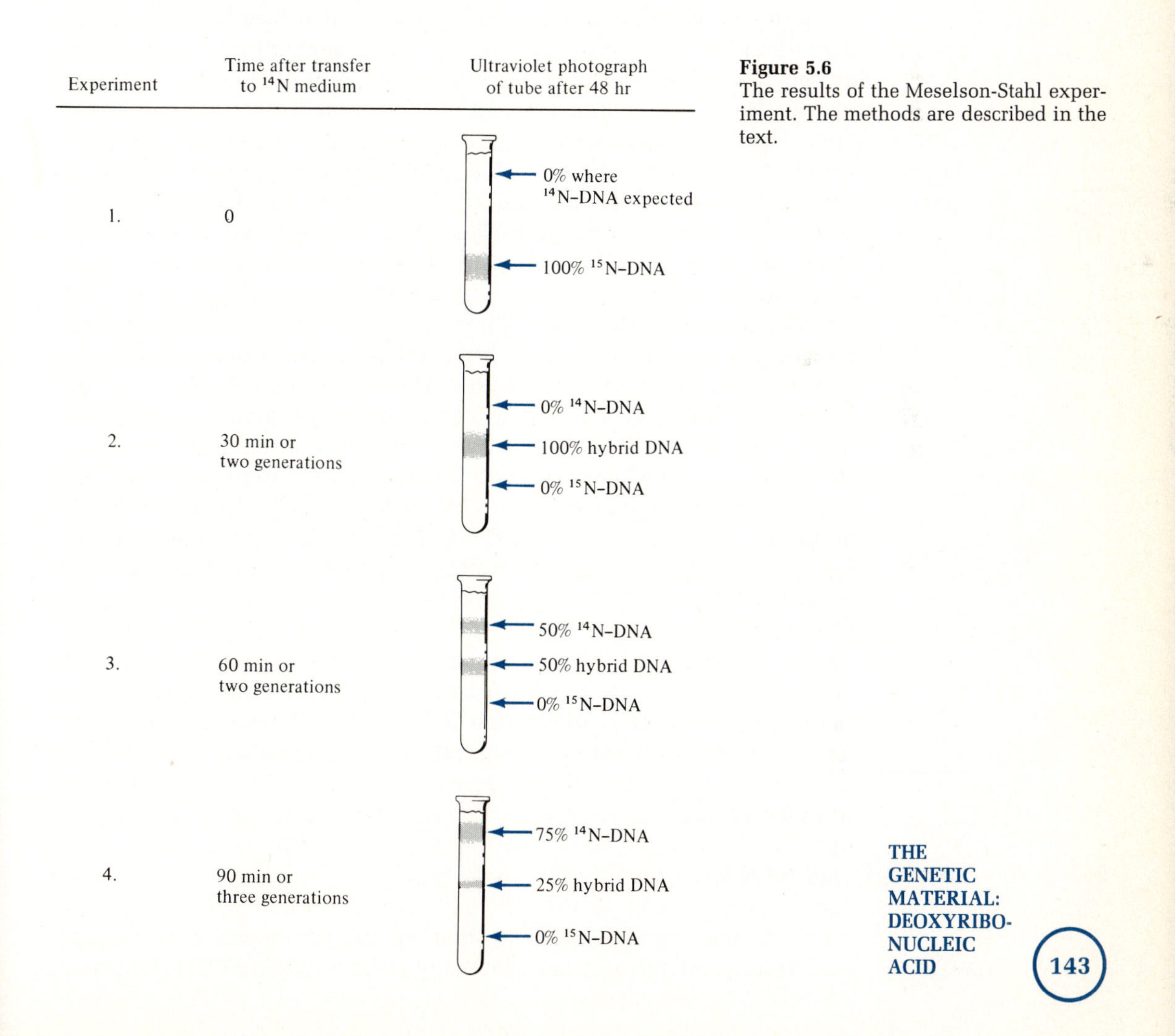

Figure 5.6
The results of the Meselson-Stahl experiment. The methods are described in the text.

Figure 5.7
Interpretation of the Meselson-Stahl experiment. Successive generations in the replication of DNA. (A) Parent DNA with two ^{15}N strands. (B) First generation, each with one ^{15}N and one ^{14}N strand. (C) Second generation, two hybrid molecules and two all ^{14}N DNAs.

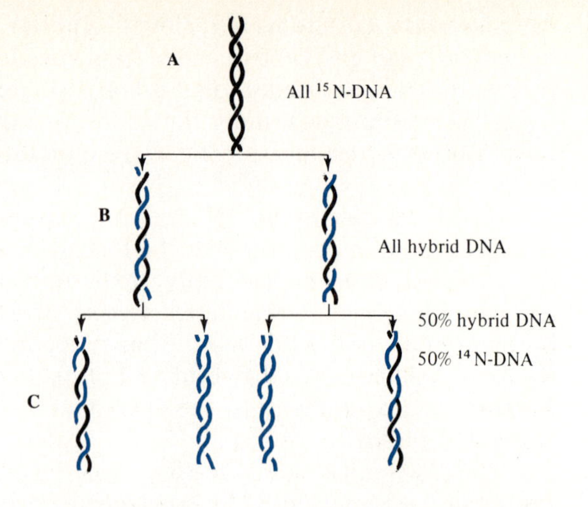

lication, the ^{15}N DNA would separate into two ^{15}N strands. With each of these acting as a template, two new strands would be constructed; they would contain only ^{14}N so that the first generation would consist of one ^{15}N and one ^{14}N strand; that is, all double strands would be hybrid. With the second replication, the strands would again separate. This time, however, half of the strands acting as templates would be ^{15}N and half would be ^{14}N. The new strands would, of course, be only ^{14}N. Thus the DNA would be of two classes, half containing two ^{14}N strands (^{14}N DNA) and half containing one ^{14}N strand and one ^{15}N strand (hybrid DNA). In each successive generation, the percentage of hybrid DNA would decrease by one half.

The Meselson-Stahl experiment does *not prove* the Watson-Crick replication hypothesis. It does make the hypothesis more acceptable, and more important, it eliminates several alternative hypotheses that are incompatible with the data. For example, prior to the Meselson-Stahl experiment, it was proposed that the entire double-stranded DNA molecule acts as a template for new DNA. This hypothesis predicts that after one generation in ^{14}N medium, half of the DNA should be the original conserved ^{15}N DNA and half should be composed of two new ^{14}N strands. Since all of the first-generation DNA was found to be hybrid, this alternative hypothesis is not tenable and has been discarded.

In addition to the Meselson-Stahl experiment, the Watson-Crick replication hypothesis has been tested in several different ways and in a number of diverse organisms. To date, their hypothesis is consistent with all the data and is generally applicable to the replication of DNA of all living organisms, from the simplest viruses to the most complex plants and animals.

5.7 The Biosynthesis of DNA

The experiments described thus far in this chapter indicate that the sequence of bases in new chains of DNA is specified by the parental DNA; the information comes from the "old" DNA. But precisely how are the new chains assembled and where does the energy come from?

While work was proceeding on the structure of DNA and the mode by which it replicates in the cell (in vivo), a group at Washington University in St. Louis, under the direction of Arthur Kornberg, was examining the possibility of replicating DNA in vitro (that is, outside the living cell). They reasoned that the minimum requirements for the biosynthesis of DNA molecules were (1) the component parts, that is, deoxyribose, phosphate, and the four bases; (2) energy to assemble the component parts; (3) DNA to act as a template; and (4) a catalyst to speed up the process.

The component parts could be synthesized in the laboratory or isolated from organic matter, but this would take months. Fortunately, there are several biochemical supply houses that store and sell a large number of research chemicals. To facilitate the experiments, Kornberg purchased the four bases, each of which was already connected to deoxyribose. These molecules that contain one base and one deoxyribose are called **deoxyribonucleosides.** The four deoxyribonucleosides they used in their experiments can be abbreviated dT, dC, dG, and dA. As discussed in Chapter 3, the source of energy for almost all biological reactions is ATP. Thus, as a source of energy, Kornberg used commercially available ATP. Items 3 and 4 were supplied by extracts of bacteria. The extracts were prepared by grinding the bacteria with powdered glass in a mortar with a pestle. This process disrupts the bacteria and releases the DNA and enzymes of the cell. This, then, was how they obtained the four ingredients for their initial experiments.

The next question was how to measure the newly synthesized DNA. Initially they expected to make very little DNA, if any at all; thus, they needed an extremely sensitive test for DNA synthesis. The method they chose made use of radioactive tracers. In order to understand this method, it is essential to realize that deoxyribonucleosides are soluble in cold acid, whereas DNA is insoluble. One of the deoxyribonucleosides used in the incubation mixture was made radioactive. The test for DNA synthesis was the conversion of acid-soluble radioactivity (for example, dT) to an acid-insoluble form (DNA).

The initial Kornberg experiment consisted of incubating dG, dC, dA, dT (this one being radioactive), and ATP with bacterial extracts (Figure 5.8). At timed intervals, cold acid was added and the acid-insoluble radioactivity was measured in a Geiger counter. Since acid-insoluble radioactivity was found at the later incubation times, the experiments indicated for the first time that **DNA could be synthesized in a test tube.**

For the next ten years, Kornberg and his associates (now at Stanford University) performed a series of brilliant experiments that demonstrated in detail how the DNA was synthesized.|| The sequence of

||For his work on DNA synthesis, Kornberg shared the 1959 Nobel Prize in medicine and physiology.

Figure 5.8
The first experiment showing the synthesis of DNA in vitro. The four deoxyribonucleosides (dG, dC, dA, dT) were mixed with ATP and a bacterial extract. The newly synthesized DNA was measured by the formation of acid-insoluble radioactivity. At the beginning of the experiment, all of the radioactivity was acid soluble (dT); as the dT became part of the new DNA, the radioactivity became insoluble in cold acid.

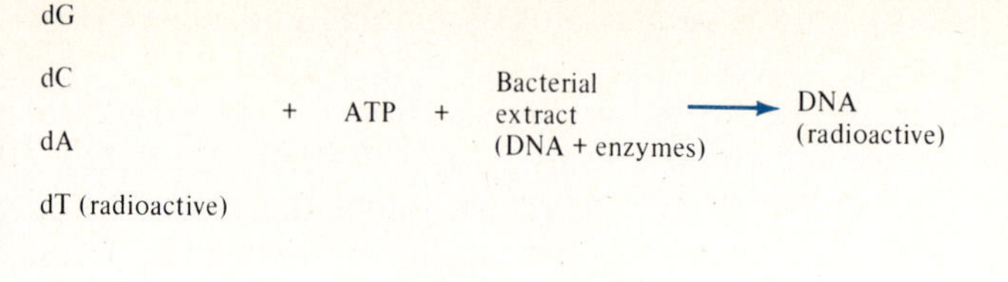

reactions shown in Figure 5.9 not only demonstrates how DNA is made, but, more significantly, illustrates the following general principles in the biosynthesis of all macromolecules.

1. **The synthesis of a macromolecule is a multistep process.** Starting with the deoxyribonucleosides, three steps are required to form each of the deoxyribonucleoside triphosphates and an additional step to polymerize them, a total of 13 different reactions.
2. **A different specific enzyme is needed for each step.** The 13 enzymes indicated in Figure 5.8 have been separated and purified. Enzyme 13, which unites the four triphosphates, is called DNA polymerase.
3. **No ATP is required for the final step.** The four deoxyribonucleoside triphosphates are similar to ATP in that they are "high-energy" compounds. As they are linked together, the terminal two phosphates are released. The energy released by this breakdown is conserved in the formation of new DNA.
4. **The information for the sequence of assembly of the building blocks in the macromolecule is inherent in the template.** Since the sequence of the units that comprise the macromolecule determine its properties, and since the properties of the macromolecule are genetically determined, the sequence of

Figure 5.9
The sequence of events in the biosynthesis of DNA. To deoxythymidine (dT) is added first one phosphate (dTp), then a second (dTpp), and finally a third (dTppp). In the same manner, the other three deoxyribonucleosides are converted into triphosphates. Each reaction is catalyzed by a different enzyme. The source of the added phosphates is ATP, which is converted into ADP. The four triphosphates are then joined together by enzyme 13 (DNA polymerase) to form a new chain of DNA, whose order is specified by the template DNA.

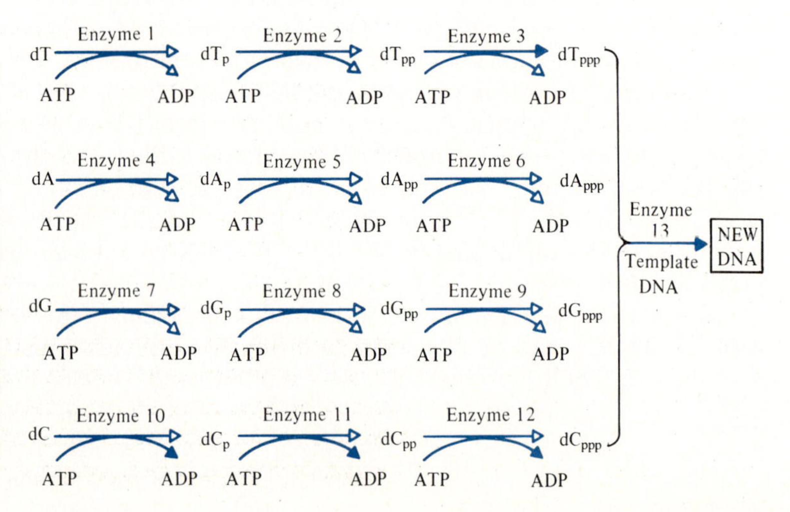

the assembled bases must be determined (either directly or indirectly) by the genes (DNA).

In the case of DNA biosynthesis, Kornberg clearly demonstrated that the base composition of the newly synthesized DNA is specified by the template DNA and not by the enzyme DNA polymerase or the deoxyribonucleoside triphosphates, which are the building blocks of DNA. Evidence for this conclusion is summarized in Table 5.5. In the first experiment, both the DNA polymerase and template DNA came from the bacterium *Escherichia coli*. The newly synthesized DNA had, of course, the base composition of *E. coli*. Experiment 2 demonstrates that all four triphosphates, DNA polymerase, and the template DNA are absolutely essential for the synthesis of DNA. If any one of them is omitted, no DNA is made. If the enzyme is obtained from *E. coli*, but the template DNA from a different source, the new DNA has a base composition indistinguishable from the template DNA (experiments 3 and 4). However, if the template is *E. coli* DNA, but the enzyme is obtained from a different source, the new DNA has the base composition of *E. coli*. The conclusion is clear: The base sequence of the DNA product is determined by the template DNA.

5.8 DNA Ligase Is Also Needed for DNA Replication

In the last few years, it has become clear that the details of the final steps in DNA synthesis inside the cell are considerably more

TABLE 5.5 The Enzymatic Synthesis of DNA In Vitro

Experiment	Components in the Incubation Mixture	Base Composition of Newly Synthesized DNA[a]
1	Complete system: dATP, dGTP, dCTP, dTTP, DNA polymerase, and template DNA from *E. coli*[b]	51% A + T
2	Omit any one or more ingredients of experiment 1	None formed
3	Like experiment 1 except *E. coli* DNA replaced by micrococcus DNA[b]	29% A + T
4	Like experiment 1 except *E. coli* replaced by phage T_2 DNA[b]	63% A + T
5	Like experiment 1 except *E. coli* DNA polymerase replaced by micrococcus DNA polymerase	51% A + T

[a]The composition of newly synthesized DNA is expressed as mole percent adenine plus thymine; the remainder was guanine and cytosine. In all cases, A = T and C = G.
[b]The A + T composition of the template DNAs used was as follows (see Table 5.4): *E. coli* 50 percent, micrococcus 28 percent, and phage T_2 63 percent. These values are accurate to ±2 percent.

complicated than originally supposed. There are several different DNA polymerases in *E. coli*. The enzyme that Kornberg studied extensively is not solely responsible for DNA synthesis in vivo. In fact, at least seven different protein catalysts are necessary for DNA replication in growing bacteria. One of these enzymes, **DNA ligase,** has particular significance because it plays such a key role in modern genetic engineering. A diagram showing the role of DNA ligase in DNA replication is presented in Figure 5.10. In the first step of the elongation process, DNA polymerase replicates one of the parental DNA strands in the same direction as the growth of newly forming bacterial chromosomes. The second strand, however, is synthesized in small fragments in an antiparallel manner. The reason for this complication is that the polymerase can catalyze DNA synthesis in only one direction, whereas the two strands of DNA are antiparallel. In the next step, DNA ligase brings about the union of the DNA fragments. The DNA polymerase and DNA ligase work in harmony so that the actual number of small DNA pieces at any time is very low under ordinary circumstances.

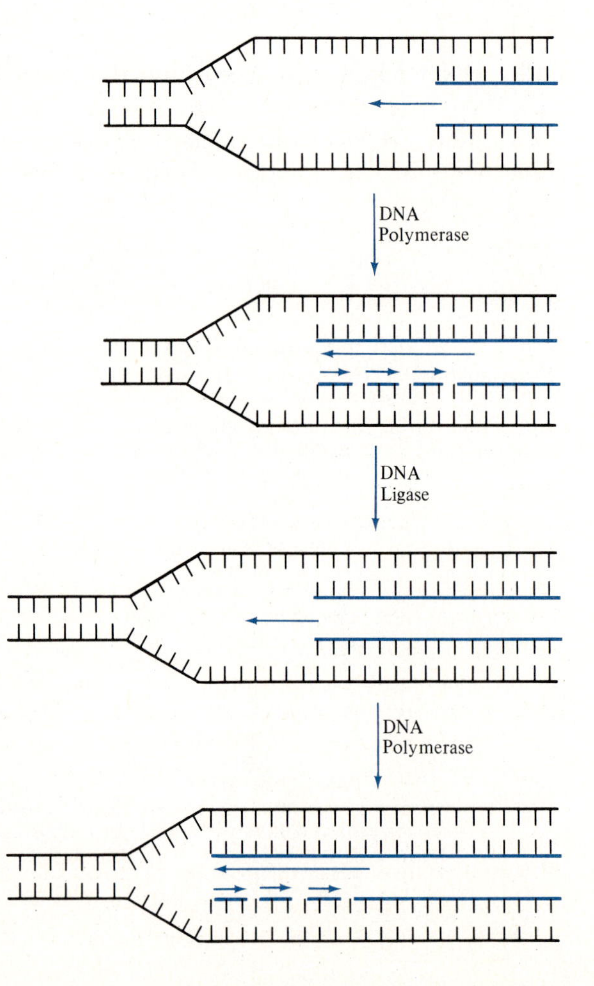

Figure 5.10
Replication of DNA at the growing point of the bacterial chromosome. One of the new strands is synthesized in small fragments and only subsequently joined by DNA ligase.

DNA ligase is one of the most important implements used in genetic surgery. Any operation in which a new piece of DNA (gene) is introduced into an organism requires that the transplanted DNA become linked to the existing genetic material. DNA ligase is used both in vivo and in vitro to carry out this crucial reaction. An example of the use of DNA ligase to construct DNA molecules in vitro is presented in the next section.

5.9 A Sequel: Synthesis of Biologically Active DNA in the Test Tube

As in many other fields, biology is rapidly catching up with science fiction. What follows is a brief report of one such episode.

Kornberg now asked the crucial question: Was the DNA synthesized in vitro an exact replica of the template DNA? In the January 1968 issue of the *Proceedings of the National Academy of Science (USA)* came the first report of the in-vitro synthesis of biologically active DNA. For this experiment, Kornberg collaborated with Robert L. Sinsheimer of the California Institute of Technology and Mehran Goulian of the University of Chicago.

For more than ten years, Sinsheimer had been investigating several interesting properties of the small spherical virus ϕX174 (see Figure 2.35). The DNA of this bacteriophage is unusual in several ways. First of all, ϕX174 is the smallest known naturally occurring DNA molecule, containing only 5500 bases. Second, rather than being double-stranded, the DNA isolated from phage ϕX174 contains only a single strand of DNA. Furthermore, the DNA is in the form of a closed ring (Figure 2.41). Most important of all, Sinsheimer had discovered that purified ϕX174 DNA could infect *E. coli* if the cells were pretreated with a specific enzyme that destroyed the rigidity of the cell wall. With the destruction of the cell wall, the bacteria assumed a spherical shape, called **spheroplasts.** A single-strand ring of ϕX174 DNA can enter *E. coli* spheroplasts, multiply, and then lyse the cell with the release of several hundred new, complete ϕX174 bacteriophages.

The techniques used to reproduce ϕX174 DNA in vitro are shown in Figure 5.11. The synthesis can be considered in three steps. First, the ϕX174 template was replicated in the presence of DNA polymerase and the four deoxyribonucleoside triphosphates as previously described. In this case, however, one of the triphosphates contained a heavy atom so that the newly synthesized DNA was more dense than the original template. In the second step, DNA ligase was used to join the ends of the new DNA, forming a double-strand hybrid molecule. Separation of the synthetic DNA from the original template was accomplished by (1) introducing a single break into the hybrid with DNase; (2) heating to force the broken strands off the remaining closed rings; and (3) purifying the closed rings by centrifugation. The introduction of a single break into the hybrid molecule was necessary because the two closed rings were interwoven in such a way that they

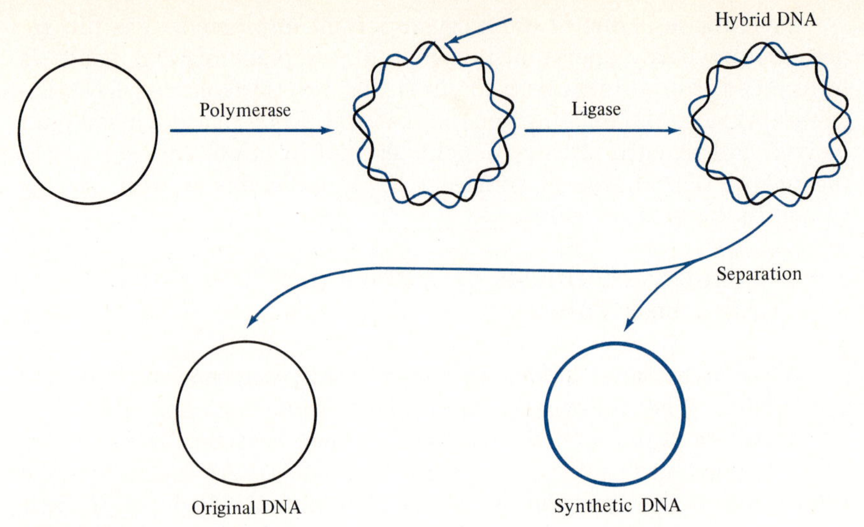

Figure 5.11
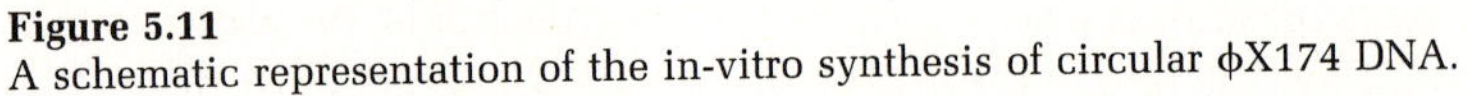
A schematic representation of the in-vitro synthesis of circular φX174 DNA.

would not come apart even when heated. Since the breakage by DNase was random, the same proportion of original and synthetic rings was opened. Thc rcmaining closed rings were separated by centrifugation in a salt gradient (see Section 5.6) because the synthetic DNA had a greater density. In this way, single-strand closed rings of φX174 DNA were synthesized and separated from the starting template.

The synthetic φX174 DNA was then tested for infectivity with *E. coli* spheroplasts. The DNA made in the test tube was infectious! It penetrated the spheroplasts, multiplied, and then lysed with the release of φX174 particles that were indistinguishable from the naturally occurring φX174 phages. This exciting experiment demonstrates that all of the essential genetic information of a φX174 is faithfully reproduced in the synthetic DNA. But does it mean that *life* has been produced (or copied) in a test tube?

QUESTIONS

5.1 What are the similarities and differences between the following:

Chemical structure of DNA and RNA?
Chemical structure of yeast DNA and human DNA?
Transformation and mutation?
Transformation and transfection?

5.2 Outline an experiment that demonstrates that the genetic information of viruses is carried by the nucleic acids.

5.3 What do the following terms signify?

Radioactive isotope
DNA polymerase
Paper chromatography
Transforming principle

DNase
Template DNA
DNA ligase

In vitro
Purines and pyrimidines
Angstrom

5.4 List and explain the function of each of the ingredients necessary for the in-vitro synthesis of DNA.

5.5 The bacterium *Escherichia coli* has a generation time of 30 minutes. Starting with a single radioactive cell and then allowing it to multiply in a non-radioactive environment, how many cells would you have in two hours? What fraction of those cells would contain some radioactive DNA? Of those cells that do contain radioactive DNA, what fraction of their DNA is radioactive?

5.6 Would you expect Chargaff's base-pairing rules to be applicable to ϕX174 DNA?

5.7 How many twists are there in a phage T_2 DNA molecule (approximately 200,000 bases per molecule)?

Suggested Readings

Allfrey VG and AE Mirsky: How Cells Make Molecules. Scientific American, September 1961 (offprint 92).

Crick FHC: Nucleic Acids. Scientific American, September 1967 (offprint 54).
A lucid description of the Watson-Crick model for the structure of DNA.

Hotchkiss RD and E Weiss: Transformed Bacteria. Scientific American, November 1956 (offprint 18).
A popular account of the discovery and early development of transformation.

Judson H: The Eighth Day of Creation. Antioch, OR: S & S Publishers, 1979.
A popular account of the development of molecular biology.

Kornberg A: The Synthesis of DNA. Scientific American, October 1968.
A popular account of self-replication of DNA in vitro.

Watson JD: Molecular Biology of the Gene, 3rd Edition. New York: W. A. Benjamin, Inc., 1975.
A more advanced treatment of many aspects of this book.

Watson JD: The Double Helix. New York: Atheneum, 1968.
A personal account of the discovery of the structure of DNA. The book was on the best seller list for a number of weeks, attesting to the great interest the public has in modern biology. Many reviews of this controversial book are as interesting as the work itself. (See, for example, PB Medawar: The New York Review of Books, March 28, 1968.)

CHAPTER 6 Microbial Genetics

Well, I have also joined the ranks of microbiologists. It used to take us ten years before we could obtain results in the breeding of higher animal forms; it took us up to ten months when lower forms of life were used; then only ten weeks or less when we used insects (Drosophila); now it takes only a few days to obtain results by using bacterial cultures.

M. Demerec

The science of genetics is the study of **heredity** and variation. By heredity we mean the process whereby animals, plants, protists, and viruses reproduce themselves, or at least something unmistakably like themselves. Early developments in genetics focused on patterns of inheritance in higher organisms of a given trait (e.g., flower color, blue eyes, and short tail) from parent to offspring. It was shown by hybridization experiments that these traits were dictated by genes that were aligned in a linear fashion along the chromosomes. In the previous chapter, experiments were presented demonstrating that the chemical composition of the gene is deoxyribonucleic acid (DNA) and that precise replication of DNA occurs by a base-pairing mechanism with the aid of DNA polymerase. Thus, one of the fundamental aspects of heredity, duplication of the gene, was explained at the molecular level.

Variations are exceptions to the general rules of heredity; occasionally there appears in a population an individual that exhibits inheritable traits not previously observed in its ancestry. For example, T. H. Morgan found one white-eyed male fruit fly that arose in a culture containing several thousand wild-type flies. Since the appearance of the white-eyed fly could not be explained by the laws of heredity, it was termed a genetic variant. The formation of variants in a population is as fundamental to life as heredity; without variants, change would be impossible. As Lewis Thomas has stated so vividly:

> The capacity to blunder slightly is the real marvel of DNA. Without this special attribute, we would still be anaerobic bacteria and there would be no music.

Because of their small size and rapid rate of multiplication, enormous populations of bacteria can be produced in small culture volumes. These conditions make possible the study of rare events, such as variation. From quantitative studies of variation in bacteria was born bacterial genetics and its daughter science, phage genetics. The use of these simple organisms as tools for genetic studies has been largely responsible for the dramatic increase in our knowledge, in recent years, of the molecular basis of genetic variation and has thrown open the door to the much-debated discipline of genetic engineering. The following three general ways in which variants arise in a population are the subjects of this chapter:

1. Mutation
2. Sexual recombination
3. Genetic engineering (artificial technique)

Before discussing mutations and how they are brought about, it is first necessary to distinguish between the terms **genotype** and **phenotype.** The genotype of an organism is its genetic makeup; the phenotype is the way it appears, each particular feature being known as a trait. The genotype of an individual is relatively stable, whereas the phenotype can vary greatly, depending upon age, environmental conditions, and so forth. For example, most male Caucasians have genes for facial hair and the potential to develop brown skin pigmentation if exposed to sunlight. However, very few ten-year-old boys in Minnesota have beards or suntans in the winter. The experimental science of genetics depends on the examination of the phenotype traits; since genes cannot be visualized directly, the genotype can only be inferred from the phenotype.

The problem of gaining detailed information about genes from observing their phenotypic effects is further complicated by the diploid nature of eucaryotic organisms. Inheritance of a gene is not tantamount to inheriting a trait. In hybrids possessing dominant and recessive genes, only the phenotype of the dominant gene is expressed. Thus, eucaryotic cells possess many silent genes.

Unlike plants and animals, bacteria are normally haploid and reproduce by simple binary fission. In this form of asexual reproduction, the cell grows in size about two-fold, then divides to form two new unicellular bacteria. Normally, both of the new cells have exactly the same properties as the cell from which they were derived. Furthermore, all the bacterial genes can be expressed immediately since no problems of dominance and recessiveness arise in haploid cells.

6.1 Mutation

Mutations can be defined as "sudden heritable changes," and **mutagenic agents** or mutagens as substances that increase the frequency of mutation.

A simple experiment illustrating mutation and selection in bacteria is shown in Figure 6.1. The experiment starts with a petri dish containing several *Escherichia coli* colonies on nutrient agar medium. Since each of these colonies arose from a single haploid cell, all of the bacteria in a colony have the same genotype. The ability to start with such a pure culture is basic to all experiments in microbial genetics. As soon as *E. coli* cells from one of the colonies are inoculated into a tube containing broth, the bacteria begin to grow and divide. In about six hours, the culture becomes turbid with bacteria. Then penicillin is added to the tube and the culture allowed to continue incubation. Within an hour, a rather dramatic event occurs; suddenly, the culture becomes clear; the antibiotic has caused the bacteria to lyse. If, however, the culture is allowed to continue incubation overnight, the tube again becomes turbid with bacteria. What is the explanation for this growth-lysis-growth phenomenon?

The parent *E. coli* used in the experiment was sensitive to the antibiotic penicillin. This can be readily seen from the fact that the culture lysed upon addition of the antibiotic to the bacteria growing in the tube. Nevertheless, when one milliliter of the six-hour culture was spread on a nutrient agar plate **containing penicillin,** ten colonies developed. Since the six-hour culture contained a total of one

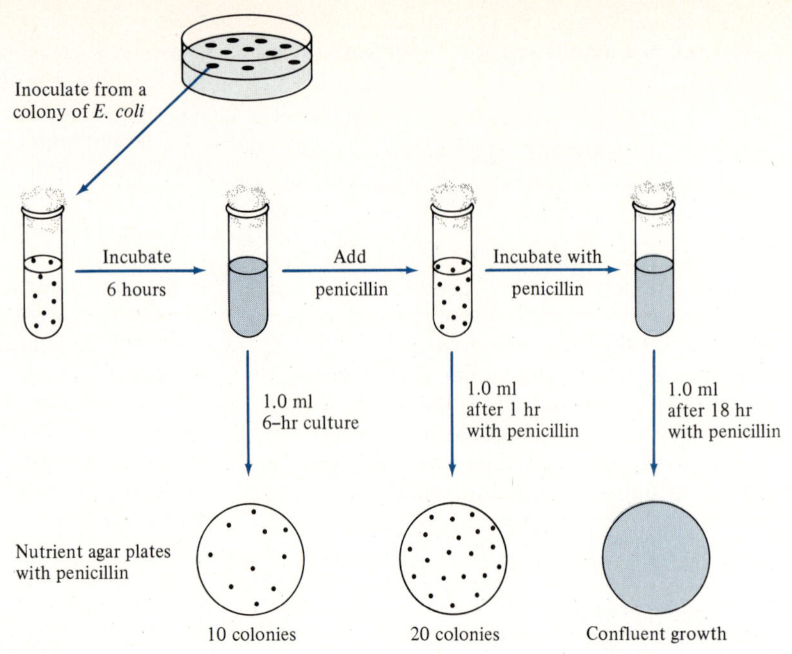

Figure 6.1
An experimental demonstration of mutation and selection to penicillin resistance in *Escherichia coli.*

hundred million (10^8) cells per milliliter, the frequency of penicillin-resistant *E. coli* in the culture must have been $10/10^8$ or one per ten million (10^{-7}). It follows that sometime during the six-hour incubation period, prior to the addition of penicillin, a genetic error was made that gave rise to a penicillin-resistant bacterium. Such an abrupt change in a particular hereditary trait that affects only a small fraction of the individuals present in a given population, fits precisely the definition of a mutation.

When penicillin was added to the tube, 99.99999 percent of the cells lysed, causing a clearing of the culture. The penicillin-resistant mutants, however, were not killed by the drug and continued to multiply, so that one hour after penicillin addition they had actually doubled to 20 mutants per milliliter. On further incubation overnight in the presence of the antibiotic, the culture became turbid with penicillin-resistant mutants. Thus, the results of the experiment outlined in Figure 6.1 are explainable in terms of a spontaneous mutation to penicillin resistance, at a frequency of 10^{-7} followed by a selection for the mutant bacteria.

Spontaneous Mutation or Directed Mutation?

During the early part of the nineteenth century, the naturalist Jean Baptiste Lamarck proposed a theory of evolution based on the concept of **inheritance of acquired characteristics.** According to his theory, organs of animals and plants become stronger or weaker as a result of use or disuse, and these changes are subsequently transmitted from parent to progeny. For example, the neck of a giraffe becomes longer by constantly stretching in order to reach leaves on high branches; the characteristic of a longer neck is then passed on to its offspring, which again stretches his neck, and so on.

In 1859, Charles Darwin published his classic study, *On the Origin of Species,* which presented a fundamentally different mode of evolution. Darwin's theory of evolution depends on the natural selection of variants within a population. Environmental conditions are the principal driving force, selecting or "favoring" some variants and eliminating or "discouraging" others. But where do variations come from? According to Darwin's theory of evolution, variations occur absolutely at random. They are produced neither by the environment nor by the conscious or unconscious striving of the organism. A variation that gives an animal even a slight advantage makes that animal more likely to produce offspring.

The theories of Lamarck and Darwin represent two opposing mechanisms for explaining how change occurs in biological systems. Moreover, the implication of these ideas extends far beyond biology—into philosophy, sociology, and politics. Consequently, the decision as to which of these two general ideas is more correct has been based not always upon valid experimentation, but sometimes on political expediency. After World War I, Trofim Lysenko, a Russian agronomist, won a gruesome struggle to become the most powerful Soviet biologist because he supported the idea of inheritance of acquired characteristics. Although this is not the place to discuss in detail Lysenko's experiments with winter wheat, it should be pointed out that many of his experiments are not reproducible and the others cannot be interpreted in an unambiguous manner. Stalin embraced Lysenko's ideas because they fit with his concept of Marxist theory: Political and social institutions can free man from his antisocial animal instincts, thereby producing a better society. By strongly supporting Lysenko and cruelly suppressing other Russian geneticists, Stalin made the tragic mistake of not distinguishing between biological evolution and social change. Although man can transmit acquired knowledge and wisdom from generation to generation in the form of great works of art and science, biological evolution depends on heritable changes in genes that take place randomly and without direction.

Using bacteria, it is possible to demonstrate experimentally that variation is a result of spontaneous mutation. The experiment shown in Figure 6.1 can be interpreted in two different ways: (1) The Lamarckian explanation is that the antibiotic provoked the mutation, that is, a small fraction of the *E. coli* was induced to become penicillin-resistant following direct contact with the drug. (2) The alternative explanation, spontaneous mutation and selection, assumes that a few penicillin-resistant variants already existed in the population prior to addition of the drug; addition of the antibiotic simply selected for these pre-existing mutants. The crucial question then is: Were the penicillin-resistant mutants in the population **before exposure to the drug?** The question might, at first, appear paradoxical since it is impossible to detect the mutants without using the antibiotic. In fact, the question was answered directly by the application of an ingenious technique called **replica plating.**

In 1952, Joshua and Esther Lederberg devised the replica plating procedure in order to examine a large number of clones for variants. The principle of the technique is shown in Figure 6.2. The initial plate contains seven colonies. Instead of testing each of the colonies individually by transferring a portion to the test media, the Leder-

Figure 6.2
Replica plating. Colonies on an agar medium are pressed onto the velveteen. A few of the adhering bacteria from each colony are then transferred to test media. After allowing time for growth, it can be seen that two of the colonies are penicillin-resistant and one colony is streptomycin-resistant.

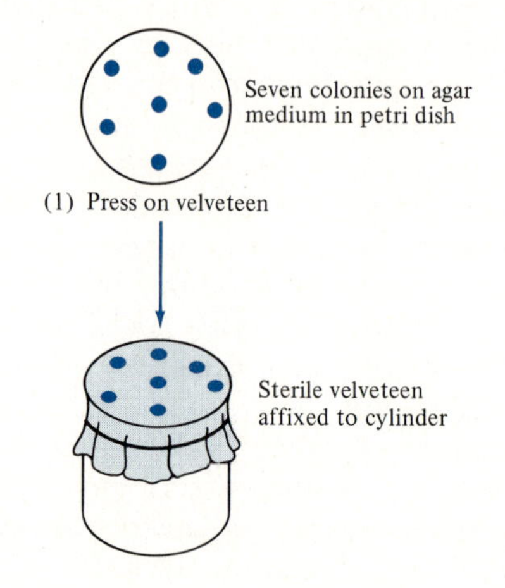

(2) Transfer imprint from velveteen to each of the three dishes shown below

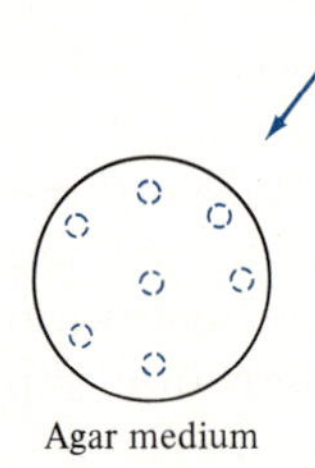

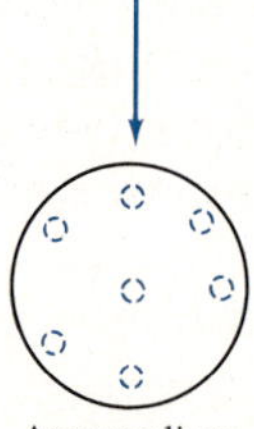

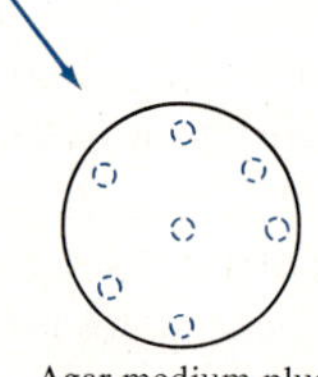

(3) Incubate at 37°C

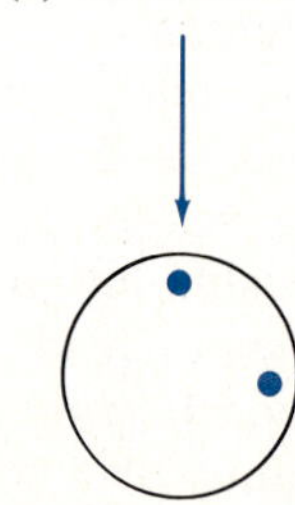

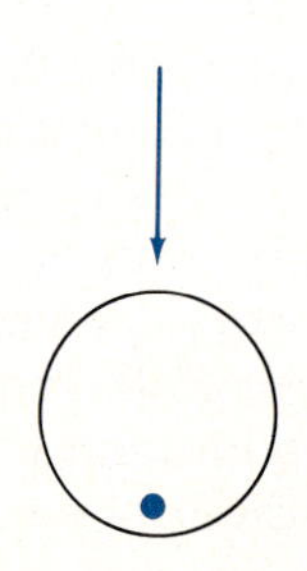

bergs pressed the surface of the agar, on which the individual bacterial colonies were growing, onto a piece of sterile cloth. Large numbers of individual cells in each clone were retained in the fibers of the velveteen, forming a velveteen "stamp pad" that was a replica of the initial plate. The "stamp pad" was then used to inoculate three different agar media. In the medium without antibiotic, seven clones developed in precisely the same relative positions as in the original plate; in the media containing penicillin and streptomycin, two and one clones developed, respectively. By comparing the positions of the colonies on the original plate to the positions of the clones that developed on the second set of plates, it was an easy matter to determine which of the colonies on the original plate were antibiotic-resistant. For example, the colony at a position corresponding to six o'clock was penicillin-resistant. This simple replica plating technique has found many valuable applications in microbial genetics. In practice, about 100 colonies per dish can be replica plated onto six different media, thereby performing what would have been 600 manual operations in less than one minute.

Now, let us see how the replica plate technique can be used to demonstrate that spontaneous penicillin-resistant mutants arise in a population **prior to addition of the drug.** The experimental outline for such an experiment is illustrated in Figure 6.3. Initially, about ten million bacteria from a pure culture are spread on a nutrient agar plate. Each cell multiplies for a few hours, giving rise to a microcolony; these microcolonies merge to give confluent growth, or what is called in the bacteriologist's jargon a lawn of bacteria. Using the velveteen "stamp pad," the lawn is replica plated onto an agar medium containing penicillin; only colonies of resistant bacteria appear on the replica plates, sensitive cells having been destroyed by the antibiotic. The position of these resistant colonies indicates where on the original plate the mutant cells were concentrated. Approximately one hundred thousand cells from the area enriched in the mutant are then transferred to a second nutrient agar plate. Again, a lawn develops consisting of many microcolonies. The penicillin-resistant mutants are again located by replica plating and transferred to a third nutrient agar plate. Since only approximately one hundred cells are transferred in this step, discrete colonies develop. The precise position of the mutant clones can now be determined by replica plating. In this manner, the Lederbergs were able to isolate pure culture of drug-resistant *E. coli* without directly exposing the bacteria at any time to the antibiotic. In addition to the Lederberg experiment, many similar studies have provided a formidable array of evidence that bacterial variation is due exclusively to spontaneous mutation.

If mutation to penicillin resistance occurs as a random mistake in the replication of the bacterial genetic apparatus, it may be expected that other mistakes will also occur during the many gene duplications required to produce a large population of bacteria. This prediction has been amply verified. Spontaneous mutations affecting almost every detectable trait occur in all organisms at frequencies at least as high

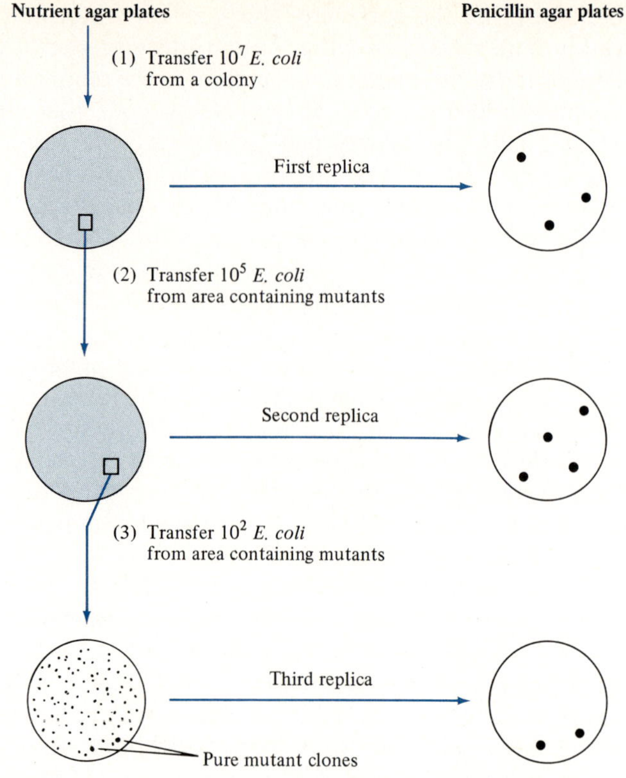

Figure 6.3
Use of replica plating technique for the isolation of mutant clones. The replica plates are used to locate the areas on the nutrient agar plates that contain microcolonies of mutant bacteria. On each successive transfer, the frequency of mutants is increased until finally pure mutant clones that have never been in contact with the antibiotic can be isolated.

as the penicillin-resistant mutation of *E. coli*. Some examples are shown in Table 6.1.

Mutagenic Agents and the Mechanism of Mutation

From experiments described in the previous chapter, we know that the hereditary information of cells is stored in DNA in the form of a coded sequence of the four bases, adenine, guanine, cytosine, and thymine. Since mutation is an alteration in the heredity information, it follows that mutation must result from a change in the base sequence of the DNA. What kind of changes can we envisage? If we think of the genetic information as a particular sequence of bases in DNA, it can be compared to a message written in a four-letter alphabet, and a mutation would be equivalent to a misprint in a line of type. Here are some examples of possible misprints in popular songs:

SIMGING IN THE RAIN
BLOWING IN THE WHIND
WE ALL LIVE IN A YELLOW SUBMRINE
ROCK AROUND THE BLOCK

In the first case, an incorrect letter has been **substituted** for the correct one; in the next two examples a letter has been added or left out, giving rise to errors that are referred to as **insertions** and **dele-**

TABLE 6.1 Examples of Spontaneous Mutation Rates

Organism	Gene	Mutation Rate
Virus (phage T_2)	Host–range	3×10^{-9}
Bacteria (*E. coli*)	Streptomycin resistance	10^{-9}
	Inability to use the sugar lactose	2×10^{-7}
	Inability to produce histidine	2×10^{-6}
	Penicillin resistance	10^{-7}
Corn (*Z. mays*)	Shrunken seeds	10^{-5}
	Purple seed color	10^{-6}
Fruit fly (*D. melanogaster*)	White eye	4×10^{-5}
	Brown eye	3×10^{-5}
	Coat color	3×10^{-5}
Humans (*H. sapiens*)	Hemophilia	3×10^{-5}
	Albinism	3×10^{-5}
	Total color blindness	2×10^{-5}

tions, respectively. In an analogous manner, mutations in the base sequence of DNA can occur by substitution, insertion, and deletion. Occasionally, a misprint (or a mutation) can give rise to a message that still makes sense, but has an altered meaning; in the last example, a simple substitution has changed the meaning of a song of the "Fifties." Such "sense mutations" are the raw material upon which natural selection operates.

A large number of agents and treatments speed up the rate at which mutations appear. As one would predict, these mutagenic agents act directly on the DNA. Figure 6.4 illustrates how the mutagen hydroxylamine causes **base-pair substitutions** in DNA. Hydroxylamine reacts with amino groups (NH_2) in DNA and converts them into keto groups (C═O). In this way, cytosine, for example, is transformed into uracil. Because the molecular geometry of uracil resembles cytosine more closely than thymine, it forms a base-pair with adenine, rather than guanine, during DNA replication. The result is that one of the daughter cells receives a DNA molecule with a UA base-pair in place of a CG pair. On subsequent replications, the mutant DNA containing the UA base-pair gives rise to progeny DNA molecules containing TA base-pair substitutions.

In addition to hydroxylamine, a large number of other chemicals increase the frequency of mutation. These **chemical mutagens** include caffeine, hydrogen peroxide, sodium nitrite, base analogs, and the highly toxic mustard gases used in World War I. Base analogs are compounds whose chemical structures closely resemble the nucleic acid bases found in DNA. An example of a base analog is bromouracil (Figure 6.5). Because these analogs are so similar to the natural constituents of DNA, the cell takes them up along with its foodstuff and uses them, instead of the normal bases, to make DNA. Their presence in the genetic material induces high mutation rates as a result of errors in replication.

Figure 6.4
Action of the mutagenic agent hydroxylamine on DNA. (A) Hydroxylamine converts amino groups (NH_2) to keto groups (C=O), thereby transforming cytosine to uracil. (B) If the reaction takes place on DNA, the uracil forms a base pair with adenine instead of guanine during DNA replication, causing a substitution mutation.

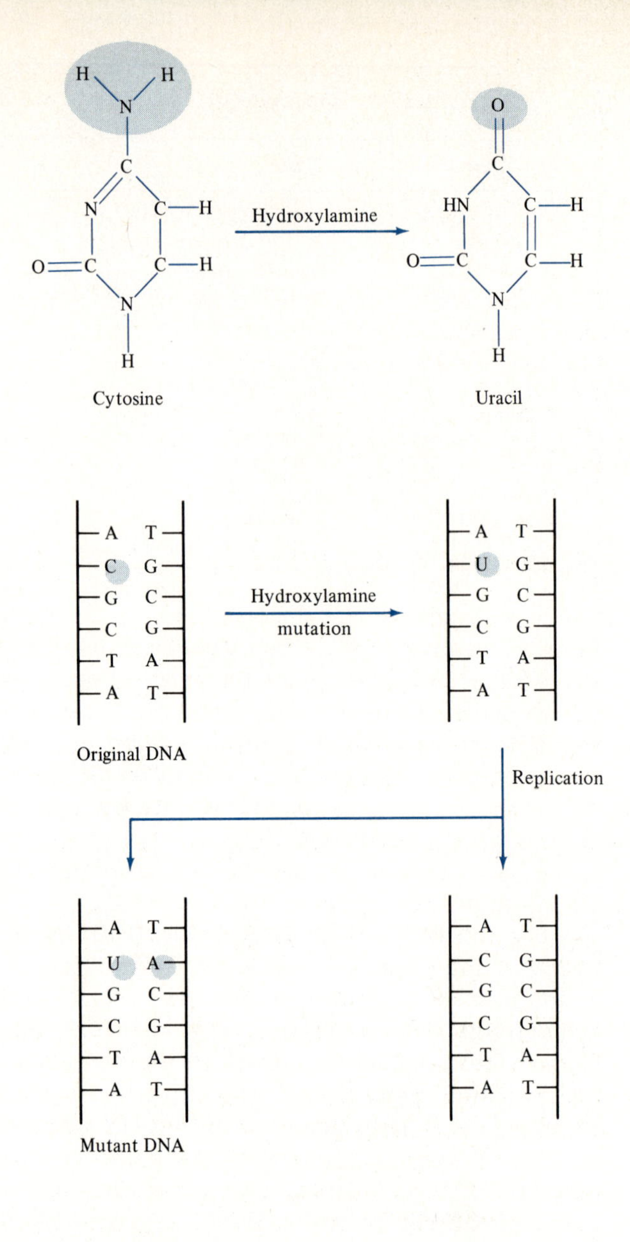

Figure 6.5
Bromouracil is an analog of thymine. Because the chemical structure of bromouracil is so close to that of thymine, the cell incorporates the analog into its DNA by mistake. The bromine atom (Br) is similar in size to the methyl group (CH_3).

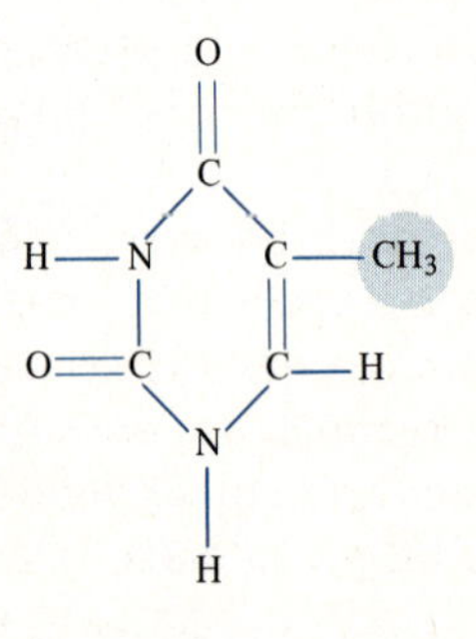

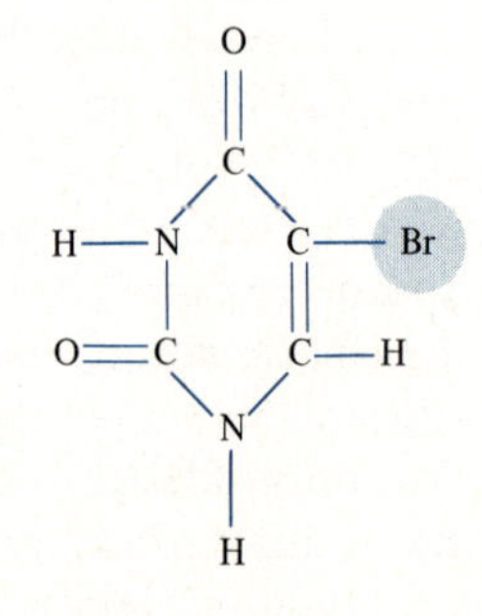

Mutations are also induced by certain types of **radiation.** Especially powerful mutagenic activity is exhibited by ultraviolet light, x-rays, and ionizing radiation produced by the disintegration of radioactive elements. Since all of these categories of radiation are encountered frequently in our environment, they deserve special attention. X-rays used for medical and dental purposes should be kept to a minimum. Sunlight contains considerable amounts of ultraviolet radiation, and individuals, light-skinned people in particular, should avoid overexposure. Radioactive contamination produced by nuclear reactors or nuclear explosions has proved to be mutagenically active; there does not seem to be a threshold below which they are ineffective. Furthermore, some of the radioactive compounds that are formed are preferentially incorporated into biological materials where they can continue to emit radiation for many years.

Increasing the mutation rate in our environment presents a serious danger to mankind for several reasons: First, mutagenic agents are generally so strongly reactive that they kill a large proportion of the organisms exposed to them. Second, among the survivors, the frequency of mutation is ten to ten thousand times higher than the normal spontaneous frequency. Since all mutations are random mistakes, the vast majority of them are deleterious and cause grave suffering to the surviving mutant individual; the deficiency is transmitted from generation to generation since mutations are inheritable. Third, most mutagenic agents are **carcinogenic;** that is, they cause cancer.

6.2 Environmentally Induced Mutation and Cancer: The Ames Test

A recent development of potentially enormous value is the use of bacteria to assess chemicals for possible carcinogenicity. For at least 30 years, biologists have realized that most mutagenic agents could also induce cancer in laboratory animals. The reciprocal was also true: Most carcinogens were also mutagenic. However, there were some notable exceptions. Hydrocarbons, for example, were potent carcinogens even though they were not mutagenic in bacteria. In spite of these well-documented exceptions, Bruce Ames, then at the National Institutes of Health in Bethesda, Maryland, was sufficiently impressed with the correlation between mutagenesis and carcinogenesis to investigate the relationship more directly. One of the first things that Ames and his associates realized was that many of these chemical agents were not inert; once inside the cell they could be transformed by enzymes into more active or less active compounds. In fact, it was already known from the research of several biochemists that aromatic hydrocarbons are oxidized by enzymes in mammalian cells to toxic compounds that react strongly with DNA, and thus are potentially mutagenic. On the other hand, David Gibson at the University of Texas showed that bacteria oxidize aromatic hydrocarbons in a completely different manner, so that no toxic chemicals are formed as intermediate compounds. It is therefore possible that hydrocarbons are mutagenic in mammals, but not in bacteria. If this were the case, then the correlation between mutagenesis and carcinogenesis would be much stronger. So much for scientific speculation; where is the experimental evidence?

Ames demonstrated that certain carcinogenic chemicals, including the aromatic hydrocarbons, that are not mutagenic to bacteria become potent mutagens when an extract of rat-liver is added to the mixture. The enzymes in the extract convert the chemicals into active compounds that rapidly enter the bacteria and cause mutations. These experiments became the basis of what has come to be called the **Ames test.** The test contains three main components:

1. **Tester bacteria.** In principle, any microorganism in which mutants can be detected readily may be used. Ames used a gram-negative enteric bacterium, *Salmonella typhimurium*. Before using the bacterium for the test, a mutant was selected that could not grow unless supplied with the amino acid histidine in the medium. This makes it very easy to detect subsequent mutations; if *S. typhimurium* his^- is spread on a glucose-salts-agar medium, only mutants that have regained the ability to synthesize histidine will form colonies. Thus, the number of colonies that develop on the glucose-salts-agar is a direct measure of cells that have mutated from $his^- \rightarrow his^+$. These types of mutants are referred to as revertants. Actually, the strain of *S. typhimurium* that Ames developed for the test has some additional mutations in it that make the bacterium more sensitive to mutagens; one of these is an alteration in the cell wall so that chemicals can penetrate more easily.
2. **Test chemical.** Any chemical or mixture of chemicals that can be suspended in water can be tested for its ability to increase the frequency of mutation.
3. **Rat-liver extract.** Since the liver is the primary site in the body where chemicals are metabolized, the enzyme from a rat-liver extract should convert the test chemical into metabolites that are mutagenic—if there are any.

Table 6.2 illustrates the use of the Ames test to demonstrate that the food preservative AF-2 is mutagenic. The first tube shows that the spontaneous mutation frequency of $his^- \rightarrow his^+$ is about one per million bacteria. Neither the food preservative (tube 2) nor the rat-liver extract (tube 4) by itself is mutagenic. However, a mixture that contains both the preservative and the extract (tube 3) causes a greater than hundred-fold increase in the frequency of revertants. Therefore, the enzymes convert AF-2 into a potent mutagen.

The simplicity, sensitivity, and accuracy of the *Salmonella* test has resulted in its being adopted in hundreds of laboratories around the world, in order to test for environmental mutagens. **More than 90 percent of the chemicals that give positive Ames tests also induce**

TABLE 6.2 Ames Test for Detecting Chemical Mutagens

Tube	Components in the Mixture[a]	Colonies that Develop on Glucose-Salts-Agar Medium[b]
1	10^6 *S. typhimurium* his$^-$	1
2	10^6 *S. typhimurium* his$^-$ plus food preservative AF-2	1
3	10^6 *S. typhimurium* his$^-$ plus food preservative AF-2 plus rat-liver extract	150
4	10^6 *S. typhimurium* his$^-$ plus rat-liver extract	1

[a]The bacterium used is a histidine-requiring mutant of *Salmonella typhimurium*.
[b]Since the agar contains no histidine, only bacteria that have mutated from histidine-requiring (his$^-$) to histidine-independence (his$^+$) can give rise to colonies; this type of mutation is called a *reversion*.

cancer in mice. This is of enormous practical significance since (1) it has been shown that a large proportion of cancers in humans are caused by chemicals in our environment and (2) it is inordinately expensive and time-consuming to screen a large number of materials for cancer-producing potentials in animals. To test a single compound for its ability to induce cancer in animals requires two to three years, costs about $100,000, and necessitates sacrificing several hundred laboratory animals. With the development of the Ames test, thousands of environmental materials can be screened for potential carcinogens rapidly and relatively inexpensively.

In 1975, undergraduates at the University of California at Berkeley, under the supervision of Ames, examined several hundred commercial products for potential carcinogens. They found that almost all the hair dyes sold on the U.S. market at that time contained highly mutagenic substances. As a result of that laboratory student exercise, the cosmetics industry removed the dyes from the market and replaced them with new formulations. Today, an increasing number of industries are utilizing the Ames test to test their products.

6.3 Genetic Recombination in Bacteria: Transformation

From the evolutionary point of view, the essence of sex is that offspring are formed with **new mixtures of genes,** resulting from the recombination of genes from two parents. Sexual or genetic recombination must be an important natural process since it is found throughout the entire biological world—from the simplest viruses, through the protists, to the higher animals and plants. The processes leading to sexual recombination, however, are quite different in eucaryotes, procaryotes, and viruses.

In all eucaryotic organisms, including animals, plants, and the higher protists, sexual recombination takes place by the fusion of two specialized cells called **gametes.** The resulting fusion cell, or **zygote,** contains two complete sets of genetic determinants, one derived from each parent. If the gamete contains n number of chromosomes (haploid), the zygote contains $2n$ chromosomes (diploid). Some time following zygote formation, the number of chromosomes must be halved to form the haploid gametes that are necessary for sexual reproduction. The process by which this reduction of chromosome number occurs is called **meiosis.**

In the case of humans, the zygote or fertilized egg develops into a multicellular diploid organism containing 46 chromosomes per cell ($2n$); just prior to sexual reproduction, meiosis takes place to produce haploid eggs or sperm cells, each containing 23 chromosomes (n). Alternatively, in many eucaryotic protists, meiosis occurs immediately after zygote formation; the resulting haploid organism has n chromosomes throughout most of its life cycle. Although there is considerable variation with regard to the timing of the two processes, the common theme of sexual reproduction in eucaryotic organisms is (1) the formation of a diploid zygote by the fusion of two haploid gametes and (2) meiosis leading to the haploid gametes.

In bacteria, a true fission cell has never been observed. Rather, sexual recombination takes place by the transfer of DNA from a donor cell to a recipient cell by one of the following three processes:

1. **Transformation**—DNA released from one kind of bacterium is taken up from the medium and genetically incorporated into a recipient cell.
2. **Conjugation**—DNA is transferred from a male to a female bacterium by direct contact.
3. **Transduction**—DNA is transferred from one related bacterium to another by bacterial viruses.

In Chapter 5 we discussed the discovery of genetic transformation in bacteria by Griffith and the subsequent proof of Avery and co-workers that the transforming principle was DNA. Bacterial transformation with pure samples of DNA provided the first and still most convincing demonstration that genes are made up of DNA.

The experimental procedure for transferring genes from one strain to another by bacterial transformation is quite simple. Consider, for example, two strains of *Bacillus subtilis,* a mutant that is resistant to the antibiotic streptomycin and a wild-type strain that is sensitive to the drug. DNA isolated from the streptomycin-resistant *B. subtilis* (the donor) is added to a culture of the wild-type strain (the recipient), and after enough time has been allowed for the DNA to penetrate the cell, the mixture is diluted and spread on an agar medium containing streptomycin. Under appropriate conditions, approximately 1 percent of the recipient cells are transformed genetically into streptomycin-resistant cells. Control experiments in which the DNA has been omitted show that the spontaneous frequency of streptomycin-resistant mutants is less than one per million *B. subtilis.*

Bacterial transformation can be used to locate the position of different genes on the bacterial chromosome. The following example illustrates the logic of the approach: DNA is isolated from a mutant of *B. subtilis* that is resistant to three different antibiotics, streptomycin, neomycin, and kanamycin; the recipient strain is sensitive to the three antibiotics and requires the amino acid histidine for growth. The transformation experiment can be abbreviated as follows:

$$\underset{\text{(donor DNA)}}{\textit{B. subtilis}\ \text{str}^R\ \text{neo}^R\ \text{kan}^R\ \text{his}^+} \times \underset{\text{(recipient cell)}}{\textit{B. subtilis}\ \text{str}^S\ \text{neo}^S\ \text{kan}^S\ \text{his}^-}$$

After allowing enough time for the DNA to enter the cells, the mixture is diluted and spread on four different agar media, three that contain histidine plus one of the antibiotics, and one that contains neither histidine nor an antibiotic. Only cells that have been genetically transformed will grow into colonies on these agar plates. The results of such an experiment are shown in the first two columns of Table 6.3. The frequency of transformation for each of the four genetic markers varies slightly from 0.1 to 0.3 percent.

More interestingly, the transformants appearing on the initial agar medium can be tested to see if they have also picked up the other three genes. Results shown in columns 3 and 4 of Table 6.3 show that 12 percent of the streptomycin-resistant transformants are also neomycin-resistant, whereas only 0.1 percent and 0.3 percent of them are kanamycin-resistant and histidine-independent, respectively. Thus, the genes for streptomycin and neomycin resistance are **linked;** that is, the genes are close enough together so that there is a good chance that the same DNA molecule that is picked up by the recipient cell will contain both genes. On the other hand, the kan gene and the his gene are far enough away from the str gene that it is unlikely that they

TABLE 6.3 Mapping of Genes by Transformation[a]

Agar 1 Containing	% Recipients Forming Colonies	Agar 2 Containing	% Colonies from Agar 1 that Grow on Agar 2
Streptomycin	0.1	Neomycin	12
		Kanamycin	0.1
		No histidine	0.3
Neomycin	0.2	Streptomycin	12
		Kanamycin	20
		No histidine	0.3
Kanamycin	0.1	Streptomycin	0.1
		Neomycin	20
		No histidine	0.3
No histidine	0.3	Streptomycin	0.1
		Neomycin	0.2
		Kanamycin	0.1

[a]The donor DNA was from *B. subtilis* $\text{str}^R\ \text{neo}^R\ \text{kan}^R\ \text{his}^+$; the recipient was $\text{str}^S\ \text{neo}^S\ \text{kan}^S\ \text{his}^-$.

will be on the same piece of DNA. (The bacterial chromosome is broken into about 100 pieces of DNA during the isolation procedure.) Similarly, one can see that the genes for neomycin resistance and kanamycin resistance are linked. If the str gene is linked to the neo gene, and the neo gene is linked to the kan gene, it follows that the neo gene must be between the kan and str genes. The order on the *B. subtilis* chromosome must then be str-neo-kan.

Using this basic technique with hundreds of different mutations has allowed microbial geneticists to draw a genetic map of the chromosome of *B. subtilis* (Figure 6.6). Somewhat surprisingly, all of the genes are located on a single circular DNA molecule.

It might seem a bit ludicrous to consider bacterial transformation a sexual process. Nevertheless, offspring are produced with new mixtures of genes, resulting from the recombination of genes from two parents. Although it is not certain that transformation is an important means of genetic exchange in nature, it does provide the molecular geneticist with a valuable tool for studying the effects of various chemical, physical, and biological agents on DNA molecules. In effect, genes can be isolated from certain bacteria, manipulated in test tubes, and then introduced into another bacterium. As discussed later in this chapter, DNA transformation is a prerequisite to the modern techniques of genetic engineering.

6.4 Sexuality in Bacteria: Conjugation

An important mode of sexual recombination in bacteria, called **conjugation,** was discovered by J. Lederberg and E. L. Tatum at Yale

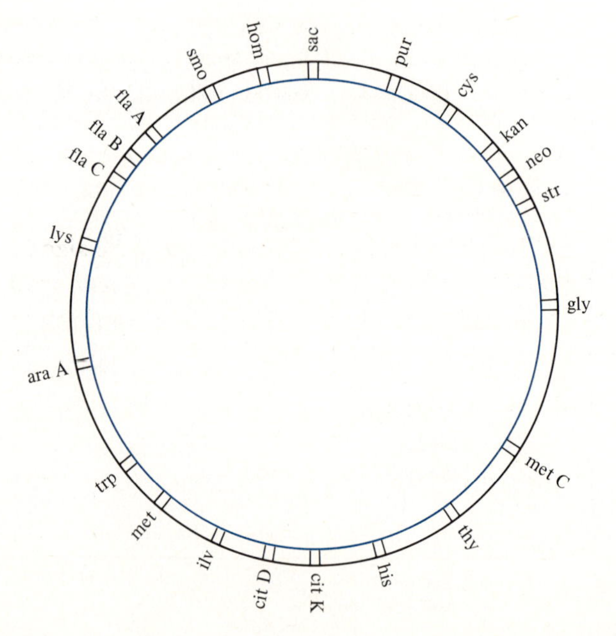

Figure 6.6
Genetic map of the *Bacillus subtilis* chromosome.

University.* Lederberg, a graduate student at the time, had dropped out of medical school when he became turned-on to research in microbial genetics. His mentor, Tatum, provided Lederberg with two mutant strains of *E. coli* that he had isolated at Stanford University. One of the mutants, designated 58-161, required the amino acid methionine for growth, whereas mutant W-1177 required the amino acid leucine. Their strategy for examining the sex life of *E. coli* was simple: If the two mutant strains could genetically recombine, then offspring should be produced that could grow in the absence of both amino acids.

To begin the experiment, each mutant was grown separately in a "complete" medium to which the required amino acids, methionine in the case of 58-161 and leucine in the case of W-1177, were added. They were then washed free of the growth factors, intimately mixed together, and a mixture containing about one hundred million (10^8) of each type was placed on an agar medium free of any amino acids. **Several hundred colonies developed.** Control experiments in which the same number of bacteria were placed on the agar medium separately gave only ten colonies. Thus, their results cannot be explained by the formation of revertants by spontaneous mutation.

Lederberg and Tatum argued that sexual recombination must have taken place between 58-161 and W-1177, resulting in the formation of progeny that have the genetic know-how of both mutants. Since the presence of DNase did not inhibit genetic recombination in *E. coli,* the process is fundamentally different from bacterial transformation. If you recall, transformation is sensitive to DNase because free DNA molecules are taken up by the recipient bacteria. Lederberg and Tatum suggested that gene transfer took place between strains 58-161 and W-1177 by a mechanism involving direct contact between the two parents. Although bacteriologists had occasionally observed under the microscope pairs of bacteria in suggestively close proximity, it was impossible to tell whether they were participating in sexual union or if it was only a random artifact of sticky bacteria. To test whether direct contact between parents was necessary for recombination in *E. coli,* B. Davis at Harvard University developed and applied the "U-tube" test.

The U-tube (Figure 6.7) is a device consisting of two arms separated by a sintered-glass filter whose pore size is just small enough to hold back bacteria. After placing the two parental strains of *E. coli* in the different arms of the tube, the medium was flushed through the filter from one side to the other. Since the number of recombinant offspring was no larger than the spontaneous mutation frequency, Davis concluded that cell-to-cell contact was necessary for genetic exchange in *E. coli.*

Once bacterial conjugation was demonstrated, scientists all over

---

*For their pioneer studies in bacterial genetics, Lederberg and Tatum received the Nobel Prize in medicine and physiology in 1952.

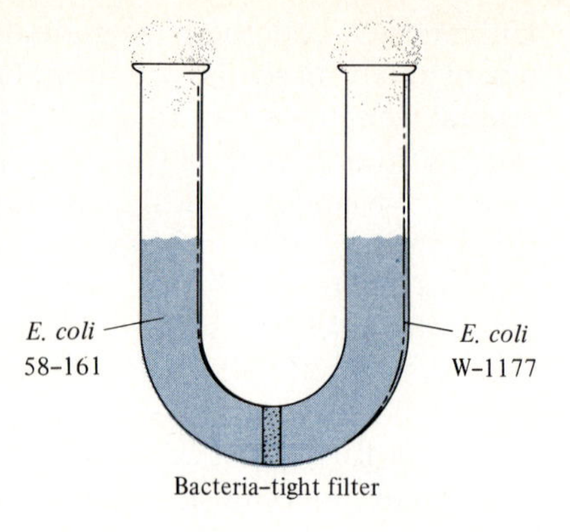

Figure 6.7
The U-tube is used to test whether cell-to-cell contact is necessary for sexual recombination in bacteria. The two arms of the U are separated by a sintered-glass filter, which is impervious to bacterial cells but allows free passage of the growth medium. The two parents are inoculated into opposite arms.

the world began to study the process in order to determine the precise manner in which *E. coli* exchanges genetic information. As a result of this effort, today *E. coli* may well be the best understood organism on earth with regard to its biochemistry and genetics. The intense interest in *E. coli* resulted from the anticipation that a detailed understanding of the genetics of this "simple" bacterium would help gain new insights into genetic mechanisms of higher forms. In fact, that hope has already been partly realized.

We now know that cells of *E. coli* can exist in either of two sexual states, called F^+ (for fertility) and F^-. F^+ cells are genetic donors, or males; F^- cells are genetic recipients, or females. Male bacteria possess, in addition to their normal chromosomal DNA, a small circular piece of DNA called the F(ertility) factor. The F factor contains genes for the production of extracellular appendages, called **sex pili.** When a female comes into contact with the sex pilus of a male, it sticks and a conjugation bridge is formed between the two cells. Within a few minutes, the two cells move into a position of direct contact, most likely brought about by the retraction of the pilus into the male cell. The F^+ donor will transfer a copy of the F factor to the F^- recipient, while keeping at least one copy for itself. The female receiving the factor is converted into an F^+ male. The process of "sex change" in *E. coli* is illustrated in Figure 6.8.

Normally F^+ males transfer only the F factor; however, mutants of F^+ cultures were found that efficiently transferred their chromosomal genes to the female recipients. These "supermales" were called Hfr for **high frequency of recombination.** When Hfr males were mixed with female recipients, genetic recombinants were found at a rate 20,000 times higher than that reported in the original experiments of Lederberg and Tatum. The availability of these Hfr mutants made possible more detailed studies on the mating process in *E. coli* than would otherwise have been possible.

Many of the important experiments on the mating process in *E. coli* were performed by Elie Wollman and François Jacob of the Pasteur Institute in Paris and by William Hayes in London. They found that the basic difference between Hfr males and F^+ males is

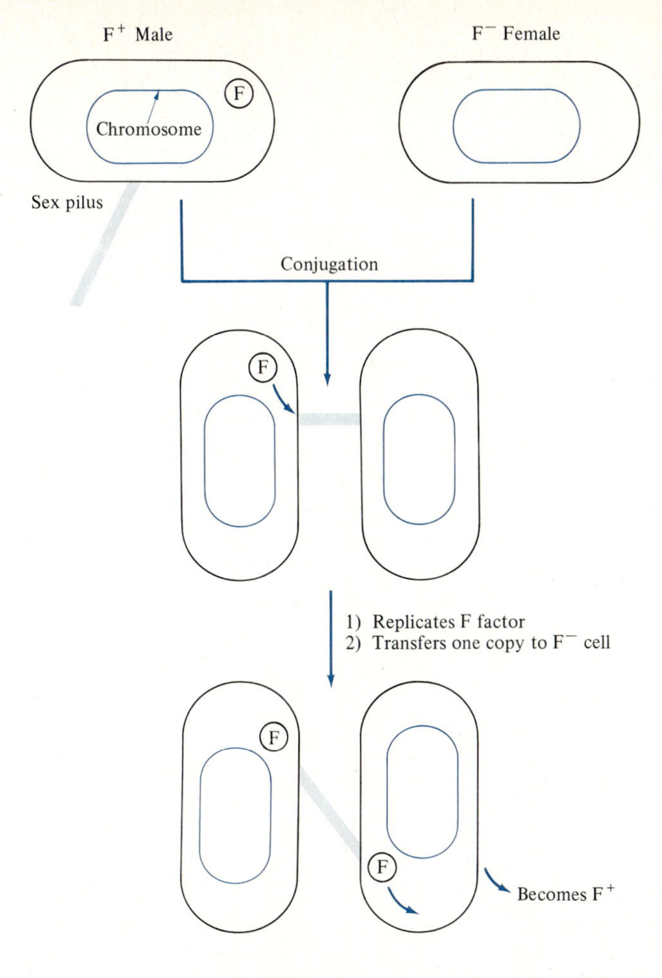

Figure 6.8
Sex change in *E. coli*. When an F$^-$ female adheres to the sex pilus of an F$^+$, a conjugation bridge is formed; the pilus of the male then retracts, bringing the two cells into direct contact. After replicating the F$^-$ factor, one copy is transferred to the F$^-$ female, converting it into a male.

that in Hfr males the F factor is integrated into the bacterial chromosome. In this regard, it is similar to the lysogenic bacteriophage (discussed in Chapter 4). When conjugation takes place between an Hfr male and an F$^+$ female, the bacterial chromosome breaks at the point within the fertility factor. The part of the chromosome adjacent to the attached fertility factor then enters the conjugation tube and slowly threads its way into the female. Genes are transferred to the female in precisely the same order that they exist on the chromosome. Figure 6.9 illustrates the conversion of an F$^+$ male into an Hfr male and subsequent conjugation with an F$^-$ female.

Usually the Hfr does not transfer its entire chromosome to the recipient because the conjugation bridge through which the Hfr chromosome moves is very fragile and ruptures spontaneously before transfer is complete. Wollman and Jacob took advantage of the weakness in the mating tubes to map the genes on the *E. coli* chromosome. These scientists mixed together Hfr males and F$^-$ females. At closely spaced intervals after mixing, they removed samples and interrupted the mating by subjecting the samples to high-speed mixing in a

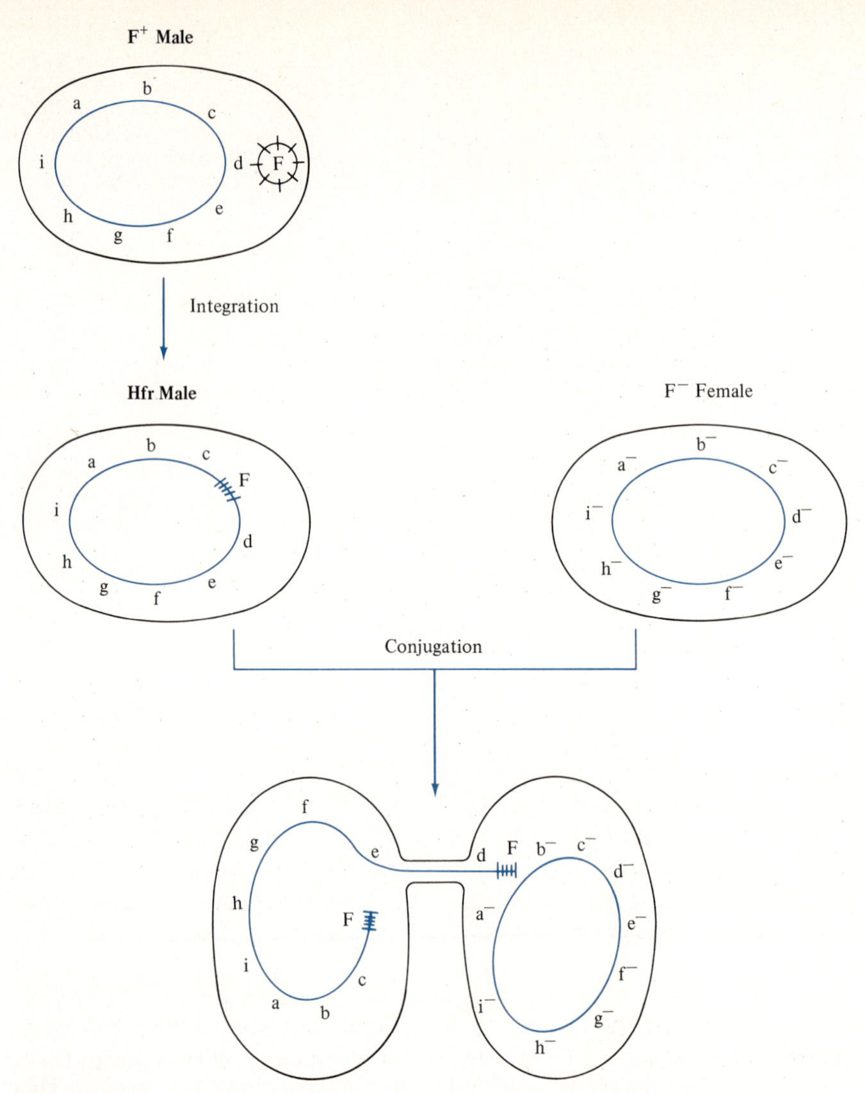

Figure 6.9
Formation of an Hfr male and conjugation with an F^- female. First, the F factor of an F^+ male becomes integrated between genes c and d of the chromosome, producing an Hfr male. During conjugation, the chromosome is broken within the F element and genes are transferred to the female in linear order, starting with a part of the F factor and then gene d.

kitchen blender. The shearing forces of the swirling waters break the conjugation tube and bring the nuptial activities to a sudden halt. In this manner, they were able to determine which genes had been transferred to the F^- by the Hfr at successive intervals after mixing them together.

Typical results from an **interrupted mating** experiment are shown in Figure 6.10. The Hfr was formed by the integration of the F factor between genes c and d. During the first 20 minutes of mating, only genes d and e were transferred to the F^-. If conjugation is allowed to

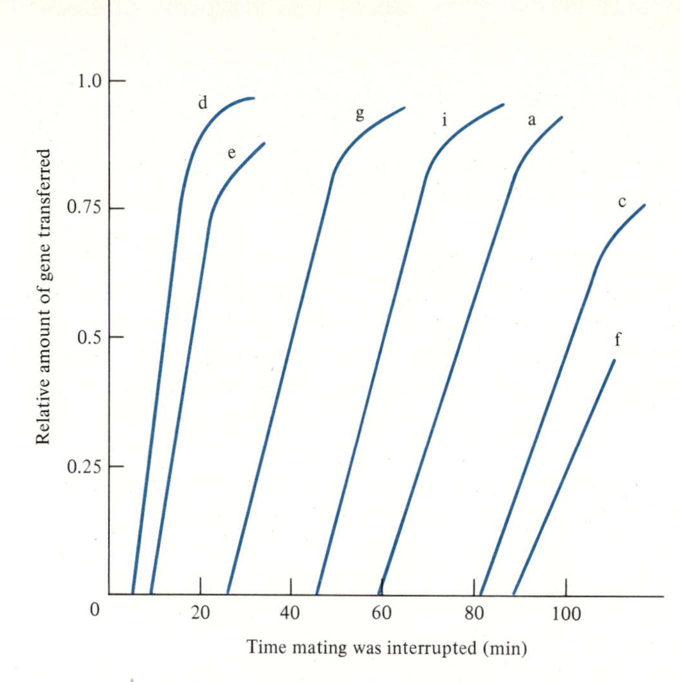

Figure 6.10
Timing of gene transfer from Hfr males to F^- females by interrupted mating. The hypothetical mating pair shown in Figure 6.10 is used for illustrative purposes.

proceed for 40 minutes, genes d, e, f, and g will be transferred; by 60 minutes, genes h and i will also be transferred. The time it takes to transfer the entire *E. coli* chromosome is approximately 90 minutes. The last genetic element to be transferred is the remaining part of the F factor. Since chromosomal transfer is usually incomplete in laboratory experiments, most F^- recipients do not receive that part of the Hfr chromosome with the attached sex factor. Thus, F^- recipients are rarely converted to males after mating with an Hfr.

Reversion of Hfrs to the F^+ state occurs by reversal of the insertion process. When this happens, the fertility factor seems to replicate once again as a small circular DNA and is again freely transmissible to the recipients. Occasionally, however, when the fertility factor is released from the bacterial chromosome, it incorporates adjacent bacterial genes into its structure. When these strains, called F′ (F prime), are mated with recipient F^- cultures, they transfer efficiently their fertility factor to them. The recipients thus acquire the fertility factor and also whichever bacterial genes were incorporated into it. This process whereby bacterial genes are transmitted from donor to recipient along with the fertility factor is called **sexduction.** It was used extensively to investigate the properties and control of enzymes involved in the utilization of the sugar lactose by *E. coli.* These studies are discussed in Chapter 8.

The F factor is only one example of an extrachromosomal DNA molecule found in bacteria. In 1952, Lederberg coined the term **plasmid** as a general name for all extrachromosomal hereditary determinants. In Section 6.7, the different types of plasmids are discussed with special reference to their significance in public health problems associated with the indiscriminate use of antibiotics.

6.5 Bacteriophage-mediated Recombination: Transduction

After the discovery of genetic exchange in *E. coli*, sexual recombination was demonstrated in a number of other bacterial species. One of the species exhibiting genetic recombination, *Salmonella typhimurium*, was studied in detail by N. D. Zinder and J. Lederberg. (This time Lederberg was the mentor and Zinder the graduate student.) As with *E. coli*, they found that by mixing together two different mutants, offspring were obtained that contained genes from both parents. To test whether direct contact between parents was necessary, they applied the same U-tube test that was used to demonstrate the need for cell-to-cell contact in conjugation (review Figure 6.7).

When this test was applied to *Salmonella* by Zinder and Lederberg, a completely unpredicted result was obtained. Recombinants *were* formed, but only in one arm of the tube. Furthermore, genetic recombination was not inhibited by DNase. Thus, the process was radically different from conjugation and transformation. They reasoned that some product was released from the donor strain, passed through the glass filter, and changed the recipient parent to a new genotype. After a series of ingenious experiments, Zinder and Lederberg discovered that the filterable product was a temperate bacteriophage carrying some bacterial genes.

The process whereby temperate phages transfer bacterial DNA from one bacterium to another is called **transduction,** and the phages that carry the bacterial genes are called **transducing phages.** Transducing phages are found at a frequency of about one per million normal phages. They are formed when the temperate phages infect the bacterial cells, enter into the phase of vegetative growth, and start encapsulating their newly synthesized DNA into phage particles. Occasionally, a piece of the bacterial DNA, which in these cases is broken down into small pieces, is encapsulated into the head of an otherwise normal phage particle. The adventitious inclusion of bacterial DNA in phage particles occurs very infrequently, but seems to occur equally well for any section of the bacterial chromosome. As a result, when recipient bacteria are infected with these transducing particles, bacterial DNA is injected into them. Different particles contain different sections of the bacterial chromosome, any of which can be integrated into the recipient's chromosome, bringing about genetic recombination. Some of the details of transduction are presented in Figure 6.11.

Note that the net results of transduction, transformation, and conjugation are very similar. In each case, a piece of bacterial DNA from a donor is inserted into a recipient that possesses its own complete chromosome. A section of the recipient's chromosome can then be replaced with the corresponding section of the inserted donor DNA. The main difference in these three forms of genetic exchange is

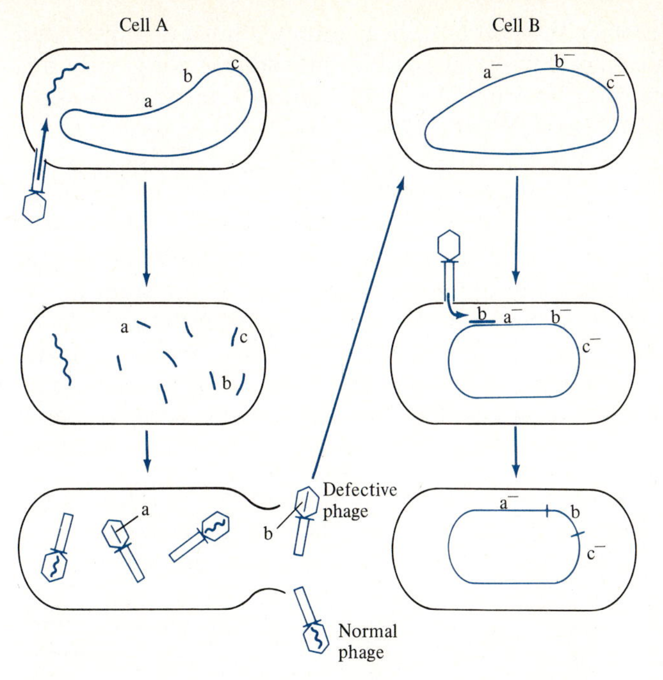

Figure 6.11
In transduction, bacterial genes are transferred from one cell to another via viruses. After the bacteriophage adsorbs to cell A and injects its DNA (wavy line), the bacterial DNA is broken into small pieces. Occasionally a piece of bacterial DNA is packaged into the virus coat in place of viral genes. When such a defective phage infects cell B, bacterial DNA (in this case, gene b) enters the cell. Instead of destroying the cell, genetic recombination occurs.

the means of DNA delivery: In transformation, free DNA enters the recipient cell; in transduction, the DNA is carried by a transducing phage; and in conjugation, DNA is delivered to the recipient through a conjugation tube connecting the two parents. In the latter case, sexually differentiated bacteria are required and a much larger piece of DNA can be transferred.

6.6 Viruses Also Can Recombine Genetically

Bacteriophages not only carry genes from one bacterium to another, but also have an efficient mechanism of genetic recombination themselves. When two genetically distinguishable phage mutants simultaneously infect the same bacterium, both kinds replicate within it and approximately equal numbers of both kinds of progeny are produced. During vegetative replication following mixed infection, both phage chromosomes, prior to being encapsulated in the protein head, come into intimate contact and can exchange segments of DNA. To demonstrate genetic recombination in bacteriophages, two phage mutants can be used. One mutant forms large round plaques that are clear; the other mutant forms small turbid plaques. When these two mutants simultaneously infect the same cell, four kinds of progeny phage are produced: the two parental types and two recombinant types, large turbid and small clear plaques. Each of these four types breeds true; that is, they continue to produce progeny with the same characteristics in subsequent single infections. Under standard conditions of mixed infection, the frequencies at which the four kinds of phage are produced are quite reproducible. Similar kinds of crosses between large numbers of different phage mutants have demonstrated that viral chromosomes can be mapped by procedures not greatly different from those used by Mendel and Morgan in their studies of inheritance in garden peas and *Drosophila*.

6.7 Plasmids

Extrachromosomal genetic elements, called **plasmids,** are found in many gram-negative and gram-positive bacteria. In addition to sex factors, such as the F factor, other bacterial plasmids contain genes that code for a wide variety of functions, including antibiotic resistance, induction of tumors in plants, production of poisons, and degradation of certain chemicals. Plasmids have the following general properties:

1. **Plasmids consist of circular, double-stranded DNA molecules,** about $^{1}/_{50}$ the length of the bacterial chromosome; this is enough DNA to code for about 100 genes. An electron micrograph of a purified plasmid is shown in Figure 6.12.
2. **Plasmids can replicate independently** of the bacterial chromosomes. Although the plasmid and the chromosome usually duplicate at the same rate, ensuring that the number of plasmids per cell remains constant, it is possible to uncouple the two processes; during starvation, for example, plasmids continue to multiply whereas synthesis of chromosomal DNA ceases, resulting in a cell that contains many copies of the plasmid. Conversely, certain drugs, such as acridine orange, selectively inhibit plasmid replication; when the acridine-treated cell divides, one of the daughter cells does not receive the plasmid and is considered "cured."
3. **Plasmids add optional genes.** While the chromosome ordinarily includes all the genes required for bacterial growth and multiplication, plasmids add genes that are not absolutely es-

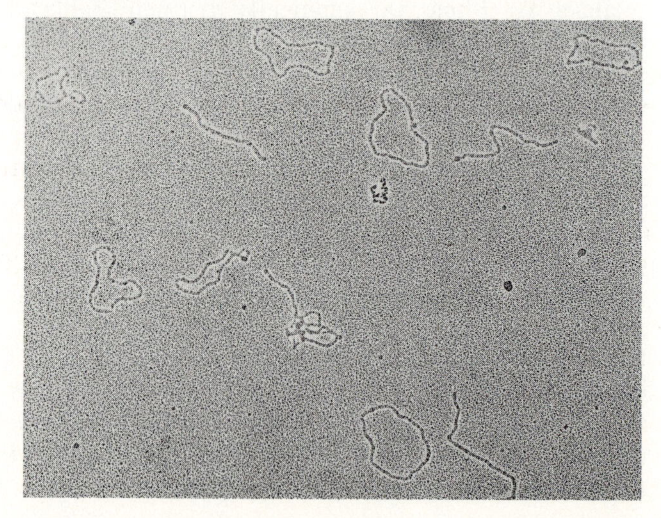

Figure 6.12
Electron micrograph of purified hydrocarbon degradation plasmids.

sential; plasmid genes broaden the range of ecological niches that the microorganism can inhabit. For example, plasmids that contain genes for hydrocarbon degradation allow bacteria, which house these plasmids, to use petroleum as a nutrient. If cells are "cured" of these degradative plasmids, the bacteria can no longer grow on petroleum, but they can still grow on other nutrients, such as glucose.

4. **Some plasmids are episomes.** Plasmids that alternate between a state in which they are integrated into the chromosome and a state in which they multiply autonomously are called **episomes.** The F factor of *E. coli* is an example of an episome; when integrated into the chromosome, the cell is converted to an Hfr male. When an episome detaches from the chromosome, it sometimes carries with it a few chromosomal genes. By entering and leaving the chromosome, an episome can move genes from one part of the chromosome to another.
5. **Some plasmids are promiscuous.** Plasmids tend to be mobile and are readily transferred between strains of the same species by direct cell-to-cell contact. Moreover, some plasmids will induce mating between widely different bacteria. One such promiscuous plasmid can be transferred among several different gram-negative bacteria.

One of the most important groups of plasmids from the public health point of view is the **resistance** or **R factors.** These plasmids carry genes for resistance to antibiotics and other chemicals. Different R factors have different combinations of resistance genes, ranging from one to more than eight. When a pathogenic bacterium receives an R factor containing multiple antibiotic resistance, the pathogen immediately becomes resistant to all of these drugs simultaneously. The difficulty in treating infections caused by such pathogens is clear. The problem is not just theoretical. In 1974, 12,000 people died in Guatemala from dysentery caused by a resistant bacillus harboring an R factor. More than 90 percent of the *E. coli* isolated recently from a major hospital in the United States were resistant to six or more antibiotics. In considering how this serious problem arose and how it might best be alleviated, let us review the general properties of plasmids outlined with specific regard to the R factor.

The double-stranded circular DNA molecule that constitutes the R factor is made up of two linked parts: One DNA segment is similar to the F factor in that it contains genes for pilus production and controls replication and transmission of the plasmid (RTF factor); the other segment contains a variable number of resistance genes (r determinants).

Bacteria in nature that have not come into contact with antibiotics contain the RTF segment but lack the r determinants. When a population of bacteria is exposed to an antibiotic, most of the cells die, but a few resistant mutants survive and multiply. For reasons that are not completely understood, the gene conferring resistance is frequently transferred to the RTF segment. If the population containing the RTF segment and linked r determinant is treated with a different antibiotic, the same events can occur—mutation, selection, and transfer to the R factor. In this stepwise manner, R factors can be constructed containing many r determinants, which can replicate independently and be transferred as a single unit.

R factors are particularly common among bacteria that live in the intestinal tract of humans and animals, including not only agents of disease, such as the dysentery and typhoid fever bacilli, but also *E. coli*. Since R factors are promiscuous plasmids, multiple antibiotic resistance can be transferred between pathogens and non-pathogens. Thus, harmless bacteria can provide a "pool" of R factors that can be passed on to virulent ones.

Recent surveys show a large rise in the number of bacteria containing R factors for multiple drug resistance. Many scientists have argued that the indiscriminate use of antibiotics is the major cause of this serious problem. Many doctors prescribe antibiotics when they are quite unnecessary. The vast majority of sore throats, upset stomachs, and other minor ailments clear up without any treatment. It is certainly better that they should be dealt with by the body's natural defenses, rather than be flooded with antibiotics.

Antibiotics should be reserved for when they are actually needed. The massive application of antibiotics as growth promoters for fowl, cattle, and other farm animals has provided ideal conditions for R factors to flourish and spread. Resistant strains develop in the animals, leave their host in excreted fecal matter, and find their way into soils, sewers, and polluted rivers; there they can freely transfer their plasmids to human pathogens. The public health hazard caused by the use of antibiotics in animal feed has led several countries to pass laws limiting the type of antibiotic that can be used in agriculture. However, bacteria are blind to national borders, and laws restricting the agricultural use of antibiotics will be much more effective when adopted on a world-wide basis.

6.8 Cloning DNA: Genetic Engineering

Up to now we have discussed three basic mechanisms by which bacteria sexually recombine: transformation, conjugation, and transduction. These are **natural** processes of genetic exchange between closely related bacteria. In the last few years, a combination of biochemistry and microbial genetics has led to the development of an **artificial** procedure for transferring genes into bacteria. The remarkable feature of this man-made genetic recombination technique is that DNA from any source can be successfully introduced into bacteria. Before exploring the scientific and social implications of this important invention, let us briefly describe the general procedure.

Genetic engineering was made possible by three separate discoveries in bacteriology: DNA-mediated transformation, plasmids, and "restriction" enzymes. The first two of these have already been discussed in this chapter. "Restriction" enzymes were discovered in bacteria about 15 years ago. These enzymes protect bacteria against infection by foreign DNA. "Restriction" enzymes were subsequently isolated by biochemists and found to be sequence-specific DNases. Each enzyme cleaved DNA at a different site dictated by a short sequence of four to eight base-pairs. The restriction sequences were further found to have palindromic symmetry; the base sequence is such that the complementary strand has the same sequence if read from the opposite end. For example, the complementary sequence to CATATG is GTATAC, which when read backward generates the original sequence.

To illustrate the technique of genetic engineering, let us consider the steps in cloning the yeast histidine gene in *E. coli* (Figure 6.13). Before beginning the experiment, four main components have to be

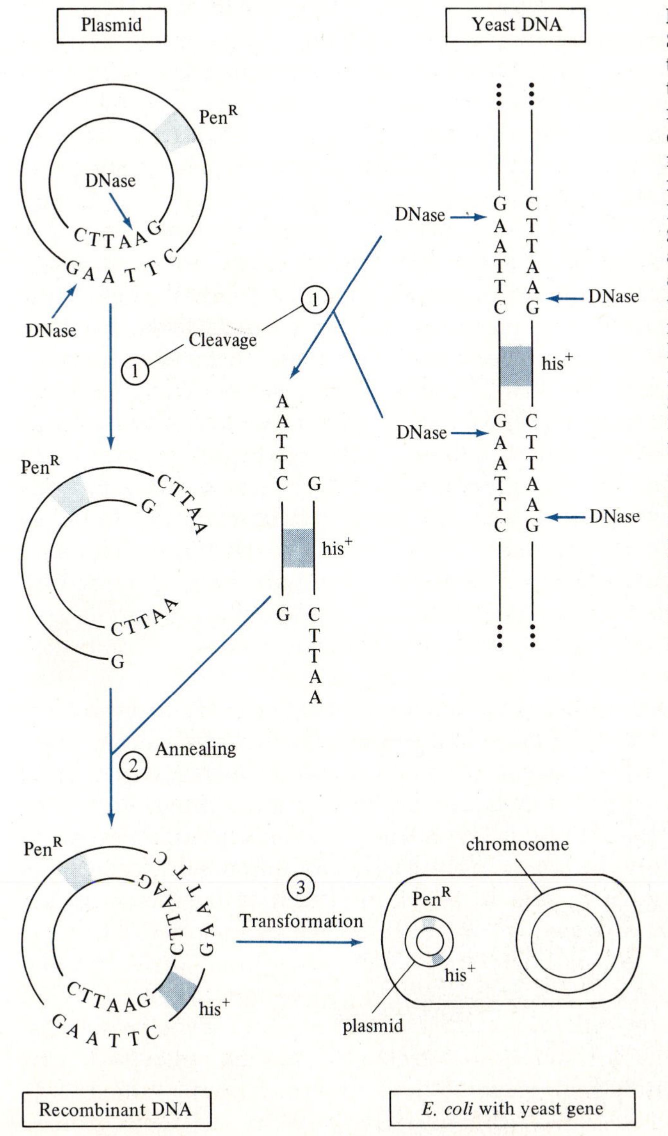

Figure 6.13
Simplified diagram illustrating the construction and cloning of a recombinant plasmid. In the first step, a specific DNase cuts the plasmid DNA and yeast DNA between the G and A of the sequence GAATTC. Only one cut is made on each strand of the plasmid DNA. The much larger yeast DNA is cleaved into many pieces, one of which contains the histidine gene. In the second step, the broken plasmid and yeast DNA molecules are heated slightly and then slowly cooled (annealed). Because the ends of the DNA are complementary, they match up to form a variety of recombinant DNA molecules. In the example shown, a piece of yeast DNA containing the histidine gene is inserted into a plasmid DNA containing the gene for penicillin resistance. In the third step, the treated DNA is mixed with *E. coli* his$^-$ and spread on agar plates containing penicillin but lacking histidine. Only those cells that have taken up the recombinant plasmid and thereby have been transformed into both drug resistance and histidine independence will be capable of forming colonies on the agar. By this technique, the yeast histidine gene was cloned in *E. coli*.

obtained (they can now be purchased) or prepared by standard procedures:

1. A **plasmid** that can multiply in *E. coli* and that contains a drug-resistance marker.
2. **Yeast DNA.**
3. A **"restriction enzyme"** that introduces only two breaks in the plasmid DNA, one in each strand. In this example, a DNase isolated from *E. coli* called EcoRI can be used.
4. ***E. coli* his$^-$** (a mutant strain that requires histidine for growth).

First, the plasmid DNA and yeast DNA are cleaved with the EcoRI; this special DNase splits the DNA molecules only between G and A in the sequence GAATTC. Next, the DNA is warmed gently, causing the DNA to break apart at the cleavage sites. Notice that the short single-strand extensions at the ends are all complementary to each other. When the DNA is slowly cooled (annealed), some of the yeast DNA pieces are inserted into the plasmid DNA before the circle is closed. The result is a heterogenous mixture of DNA molecules: original plasmid DNA, fragments of yeast DNA, and a variety of recombinant plasmids containing yeast genes incorporated into plasmid DNA.

Since the yeast chromosome contains several thousand genes, only a very small fraction of the recombinant plasmids will carry the histidine gene. Therefore, powerful techniques have to be used for selecting and cloning the specific gene of interest. In this case, the treated DNA is used as donor DNA in a transformation experiment with *E. coli* requiring histidine as the recipient. After allowing enough time for the DNA to penetrate the cell, the mixture is spread on an agar plate containing penicillin but lacking histidine. All cells that pick up the plasmid DNA will become penicillin-resistant, but only those very rare cells that become a host for the specific recombinant plasmid containing the yeast histidine gene will be both **penicillin-resistant and histidine-independent.** Such cells will multiply on the penicillin agar (lacking histidine), giving rise to clones of *E. coli* containing the yeast histidine gene.

Using genetic engineering techniques similar to those described, genes from a variety of plants and animals, including humans, have been introduced into bacteria. The fact that eucaryotic genes can function inside procaryotic cells is one of the strongest arguments for the unity of life at the molecular level. The seemingly only limitation of the cloning technique is the ingenuity of the microbial geneticist in choosing appropriate media for selecting the hybrid plasmid of interest.

Applications of Genetic Engineering

Cloning genes in bacteria has tremendous potential with regard to both basic biological research and practical applications. It is now possible to study isolated genes of higher organisms inside the simplified background of a bacterium. Unlimited quantities of any gene can be obtained in pure form for biochemical studies, including the determination of the exact sequence of bases in the specific DNA.

One of the major problems in biology today is the regulation of gene function. How are genes turned on and off? It is obvious, for example, that hair cells and toe cells of an individual are very different, yet both cell types have an identical set of genes. It follows that the expression of genes is under careful control. Cloning of eucaryotic genes in bacteria opens the possibility of understanding how gene expression is regulated in higher organisms. Furthermore, an aberration in the regulation of gene function is thought to be an important aspect of the cancer problem. Regardless of what causes cancer, the disease(s) is(are) characterized by an abnormality in the regulation of gene expression leading to a loss over the control of cell growth.

The potential practical exploitation of genetic engineering staggers the imagination. Already the human genes for producing insulin and the pituitary growth hormone have been cloned in bacteria. Microbial cells containing these genes manufacture products identical to those produced in humans. For thousands of diabetics who are allergic to insulin isolated from the pancreas of cows or pigs, human insulin, produced in bacteria, promises an immediate relief. Since treating a child for dwarfism requires the pituitary glands of about 50 cadavers a year, the microbial-produced hormone is an important medical advance. In general, the opportunity of synthesizing a variety of human hormones, antiviral substances, antibodies, and other natural products of man by cloning the specific genes in bacteria is one of the most exciting challenges facing the pharmaceutical industry today.

The ability to fix atmospheric nitrogen is restricted to a few groups of procaryotes. Recently, the genes responsible for nitrogen fixation (*nif* genes) have been cloned in other bacteria; under appropriate conditions, the transformed bacteria are able to fix nitrogen. The possibility of incorporating the *nif* genes into higher plants, in order to eliminate the need for nitrogen fertilizers, is currently under intensive investigation.

The technique of cloning DNA can also be applied to eucaryotic cells. Yeasts have already been transformed with plasmid DNA; there is every reason to expect that in the not too distant future, specific genes will be introduced into higher organisms, including humans. The most obvious potential application of the technique in medicine is that of using a plasmid, containing the necessary genetic information, to treat patients with an enzyme deficiency disease. In order to obtain enough of the specific plasmid in pure form for performing the genetic surgery, the recombinant plasmid can initially be cloned in bacteria. Gene therapy—the replacement of defective genes—opens the door to "preventive medicine" in the most permanent sense.

Potential Dangers of Genetic Engineering: The Recombinant DNA Debate

If the benefits of genetic engineering are clear, so too are the unpredictable dangers. With this in mind, in 1975 a group of concerned scientists publicly called for a temporary moratorium on certain types of experiments with recombinant DNA and immediately arranged an international meeting to discuss various technical and moral problems arising as a result of the new DNA cloning technology. Several questions were raised at the gathering, which took place at the Asilomar Conference Center in Pacific Grove, California:

1. What are the risks of accidentally producing novel, dangerous microorganisms that would be hazardous to humans or other living things?
2. What is the risk that a laboratory worker will be infected with such a man-made pathogen?
3. What is the hazard that such a pathogen will spread outside the laboratory, initiating an epidemic—the "Andromeda strain" scenario?
4. What precautions can be taken to minimize the potential dangers? How can these precautions be enforced?
5. What are the moral implications of genetic engineering?

The first three questions deal with the potential biohazards of recombinant DNA research. The scientists agree that certain types of experiments are clearly dangerous, others have a low but significant probability of producing dangerous strains, and finally, there are many types of DNA cloning experiments in which there is no hazard whatsoever to any human being. As an example of a dangerous experiment, consider the introduction of a gene for toxin production into a bacterium that is highly infectious and antibiotic-resistant. Because such experiments are dangerous and of little scientific value, the scientific community recommended that these types of experiments be prohibited.

The introduction of mammalian genes into bacteria (by procedures similar to those described for cloning the yeast histidine gene in *E. coli*) is the prototype of an experiment that has considerable scientific merit but may inadvertently produce dangerous strains. For example, mammalian cells may contain tumor virus genes that will be incorporated into the plasmids; following transformation with the recombinant DNA, there is a low but significant chance that a bacterium carrying the plasmid will infect a human and spread the tumor virus. The recommendation of the scientists was that these experiments should be performed, but restricted to (1) laboratories that are specially equipped for working with pathogenic microbes, so-called "containment laboratories," and (2) bacteria that are poor transmitters of disease and cannot survive outside the laboratory. A mutant *E. coli*, sensitive to heat and requiring several growth factors, was developed specifically for these experiments. These bacteria fail to survive passage through the intestinal tracts of experimental animals, even when injected in large numbers.

The discussions at the Asilomar Conference, as well as subsequent meetings, called attention to some of the potential risks of recombinant DNA experiments and led, in 1976, to the formulation of a set of guidelines by the National Institutes of Health (NIH) for the conduct of recombinant DNA research; similar guidelines have been adopted in Canada and England and are under consideration in several other countries. The guidelines indicate, in considerable detail, which DNA cloning experiments are permissible and under what conditions they have to be carried out. The guidelines are under constant review; during the past few years, a growing body of knowledge has indicated that the original guidelines were excessively conservative; accordingly, restrictions on several types of experiments have been relaxed. It should be emphasized that the NIH guidelines deal with the "conjectural biohazards" of recombinant DNA research and as such are concerned primarily with safety considerations rather than ethical, social, and legal implications of genetic engineering.

The extraordinary advances made in molecular genetics during the last 35 years, many of which have been described in the last two chapters, have brought biology to a point where genetic engineering in microorganisms is an accomplished fact and genetic engineering in higher organisms is clearly on the horizon. Certainly the issues go far beyond the question of safety of laboratory experiments—in many respects, those are the easiest to answer. The real problem is to comprehend the social implications of what science can enable us to do and to promote scientific and technological research in that direction which maximizes the benefit to mankind. To give some idea of the scope of the problem, Robert Sinsheimer stated;

> For eon after eon, creature has given rise to creature upon this earth—blindly, each generation usually like the former, occasionally—by accident—a little different. Of all the creatures that have lived upon this earth we are the first to understand this process. . . . The ultimate significance of this understanding of the very basis of heredity is incalculable. It will change all the eons to come.

Who will decide what these changes will be? James Watson wrote in a letter to the *Washington Post*, "The matter is much too important to be left in the hands of scientists whose careers might be made by achieving a given experiment." Furthermore, today's scientists are usually too highly specialized for broad policy-making (there are, of course, a few notable exceptions). Even in the specific case of safety in recombinant DNA research, the early practitioners of this art, who initially called attention to the potential risks, were *not* the most competent scientists to evaluate biohazards. Medical microbiologists, especially epidemiologists, had considerably more experience in working with dangerous microorganisms than had molecular geneticists. More importantly, scientists are not specially trained to deal with moral, legal, and social problems. Probably most scientists

would support the statement made by Arthur Kornberg to the United States Congress Subcommittee on Government Research in 1968:

> One way to prepare ourselves for the developments in genetic engineering is to educate all our citizens so that biologists and chemists are not the only people called upon to make crucial judgements . . . what you do in congressional legislation has an enormous impact on our future. . . . For my part, I see no alternative but to learn more and share my knowledge.

QUESTIONS

6.1 Evolution depends on diversity in the biological world. What is the physical basis of organic diversity?

6.2 What are the similarities between the following:

Mutation and selection?
Genotype and phenotype?
Spontaneous and induced mutation?
Plasmid DNA and chromosomal DNA?
Conjugation and mutation?
Transduction and conjugation?
Genetic engineering and conjugation?
Plasmids and temperate bacteriophages?

6.3 You have two strains of *E. coli:*

Type A: Resistant to streptomycin, sensitive to penicillin, and requires sugar for growth;
Type B: Resistant to penicillin, sensitive to streptomycin, and requires sugar for growth.

The following experimental results were obtained:

10^8 A cells placed on sugar and penicillin yields 100 colonies.
10^8 A cells placed on streptomycin with no sugar yields 10 colonies.

What is the mutant frequency of penicillin sensitive to the penicillin resistant?
How many colonies would you expect if you placed 10^9 B cells on penicillin with no sugar? If you placed 10^9 A cells on penicillin and no sugar?

6.4 How can the "U-tube" be used to distinguish transformation, conjugation, and transduction?

6.5 What is the basis of the Ames test? Although currently used to detect potential carcinogens, many scientists claim its real purpose is to protect future generations. Explain.

6.6 Explain how bacterial genes can be mapped by

a) interrupted mating,
b) co-transformation, and
c) co-transduction.

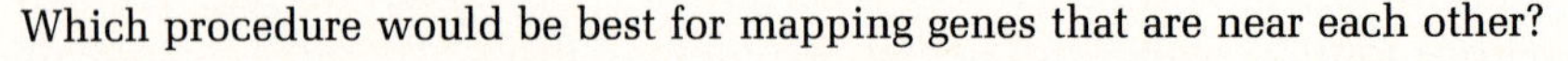

Which procedure would be best for mapping genes that are near each other?

6.7 Do *you* think the molecular geneticists at the Asilomar Conference had the right (authority or responsibility) to call a halt to certain types of DNA cloning experiments? In general, who should decide which genetic experiments should be permitted?

Suggested Readings

Ames BN, HO Kammen, and E Yamasaki: Hair dyes are mutagenic: Identification of a variety of mutagenic ingredients. Proceedings of the National Academy of Sciences, U.S.A. 75: 2423–2427, 1978.

This original research paper is readily comprehensible and of great significance.

Cohen S: The Manipulation of Genes. Scientific American, July 1975.

A lucid description of the experiments that led to DNA cloning in bacteria by one of its early practitioners.

Davis BD: The Recombinant DNA Scenarios: Andromeda Strain, Chimera and Golem. American Scientist, 65: 547–555, 1977.

An analysis of the techniques and risks of working with recombinant DNA in E. coli; *contains a list of most of the important references on the subject.*

Devoret R: Bacterial Tests for Potential Carcinogens. Scientific American, August 1979.

An up-to-date discussion of the Ames test and modification of it.

Goodenough U: Genetics, 2nd Edition. New York: Holt, Rinehart and Winston, 1978.

An excellent modern textbook in general genetics. Chapters 6, 10, and 11 deal with microbial genetics.

Grobstein C: The Recombinant DNA Debate. Scientific American, July 1977.

Hayes W: The Genetics of Bacteria and Their Viruses, 2nd Edition. New York: John Wiley & Sons, 1968.

A well-written reference textbook that is used by professional bacterial geneticists.

Jacob F and EL Wollman: Viruses and Genes. Scientific American, June 1961 (offprint 89).

An illuminating discussion of both heredity and infection.

Peters JA, ed: Classic Papers in Genetics. Englewood Cliffs, NJ: Prentice-Hall, Inc., 1959.

Especially pertinent are the translation of the original classic paper of Mendel, a short paper by Muller on induced mutation, and the original reports of sexual recombination in bacteria by Lederberg and Tatum and Zinder and Lederberg.

Wills C: Genetic Load. Scientific American, March 1970.

A stimulating discussion of mutation and the viability of the species.

Wollman EL and F Jacob: Sexuality in Bacteria. Scientific American, July 1956 (offprint 50).

A lucid discussion of conjugation by two of the men who have made great contributions to the field.

CHAPTER 7 Protein Synthesis and the Genetic Code

The frontiers are not east or west, north or south, but wherever a man confronts a fact.

Henry David Thoreau

In Chapter 5, evidence was presented that indicated that the genetic material is deoxyribonucleic acid (DNA). Furthermore, we described the double helical model of DNA and showed how the model led to a relatively simple scheme for the duplication of DNA. We now turn our attention to the crucial question of how the genetic information is translated by living organisms into observable physiological characteristics. For example, how does a "gene for blue eyes" actually cause an individual to exhibit that characteristic pigmentation? In the last two decades, there has been a major breakthrough in our understanding of the mechanism by which the genetic information stored in the sequence of bases in DNA exerts its control over the cell. This chapter is an attempt to explain the background for the breakthrough and to discuss our current understanding of gene action.

7.1 The Pioneer Work of Sir Archibald E. Garrod

In 1902, the English physician Archibald E. Garrod published in *Lancet* the first of a series of papers that dealt with the physiological defect called **alkaptonuria.** This disease, which has been known to the medical community for more than 300 years, is characteristically diagnosed by the blackening of a patient's urine on exposure to air. Usually the condition is noticeable in infants from the discoloration of soiled diapers. Garrod began his study by isolating from the urine of patients with alkaptonuria the chemical that turns black on prolonged contact with air. The agent was purified and identified as **homogentisic acid.**

Homogentisic acid is present in the urine of patients with alkaptonuria but absent from the urine of normal individuals. Garrod traced the origin of homogentisic acid to certain foods in the diet, namely the amino acids, tyrosine and phenylalanine. These amino acids are broken down to produce homogentisic acid in all humans. Moreover, normal individuals continue to metabolize homogentisic acid until only carbon dioxide and water are formed. Those afflicted with alkaptonuria, however, cannot metabolize homogentisic acid and therefore excrete it in their urine. The larger the quantity of tyrosine or phenylalanine in their diet, the more homogentisic acid that is excreted.

The diagnosis of alkaptonuria in infants at birth suggested to Garrod that the disease might be inherited. This was supported by examining the marriage records of families having alkaptonuric children. Garrod found that a much larger than expected number were first-cousin marriages. Furthermore, examination of the genealogy of some of his patients revealed that alkaptonuria was passed down from parent to child as a simple recessive Mendelian character.

From the data at hand, Garrod conjectured as early as 1908: "We may further conceive that the splitting of the benzene ring in normal metabolism is the work of a special enzyme, that in congenital alkaptonuria this enzyme is wanting." In other words, Garrod was stating for the first time a **relationship between genes and enzymes.** The normal gene results in the production of a specific enzyme, whereas with a defective gene this enzyme is missing.

Garrod reported his results in research papers, books, and even gave a series of lectures to the Royal Society. In spite of these attempts to popularize his theory of gene action, he failed to arouse the interest of biochemists and geneticists sufficiently to follow through in these studies. It may have been that biochemistry and genetics, both young disciplines at that time, did not have the necessary tools to probe more deeply into this problem. The human organism is a difficult subject to study genetically and biochemically—a life cycle that is too long, offspring too few, and severe limitations on subjecting humans to chemical analysis.

It may also have been that Garrod, like Mendel, was so far ahead of his time that his contemporaries were not ready to consider seriously his far-reaching gene–enzyme concept.

7.2 The One Gene–One Enzyme Hypothesis

As genetics developed during the first half of this century, certain organisms came to the forefront that were well suited for careful genetic and biochemical analyses. One of these was the common bread mold, *Neurospora crassa*. This fungus possesses many advantages for a study on the mechanism of gene action. It can be maintained easily in the laboratory in pure culture. Starting with a single cell, a genetically homogeneous population can be obtained in a few days. Furthermore, genetic analysis is simplified by the fact that during the greatest part of its life cycle, the bread mold has only a single set of genes (haploid) instead of the two sets (diploid) found in higher organisms. Thus, complication of dominance and recessiveness is avoided, since genes cannot be hidden by their dominant counterparts. Most important, *Neurospora* can be grown in a minimal medium consisting of water, various inorganic salts, sugar, and one vitamin of the B group, biotin. From these few substances the mold can synthesize all the other components of the cell.

The ease with which both genetic and biochemical studies could be conducted on *Neurospora* induced two scientists at Stanford University, George W. Beadle and Edward L. Tatum, to choose that organism for a careful investigation of the gene–enzyme relationship. As Beadle recollects,

> In 1940 we decided to switch from *Drosophila* to *Neurospora*. It came about in the following way: Tatum was giving a course in biochemical genetics, and I attended the lectures. In listening to one of these—or perhaps not listening as I should have been—it suddenly occurred to me that it ought to be possible to reverse the procedure we had been following and instead of attempting to work out the chemistry of known genetic differences, we should be able to select mutants in which known chemical reactions were blocked. *Neurospora* was an obvious organism on which to try this approach, for its life cycle and genetics had been worked out by Dodge and by Lindegren, and it probably could be grown in a culture medium of known composition. The idea was to select mutants unable to synthesize known metabolites, such as vitamins and amino acids which could be supplied in the medium. In this way a mutant unable to make a given vitamin could be grown in the presence of that vitamin and classified on the basis of its differential growth response in media lacking or containing it.

The experimental design for the isolation of mutants of *Neurospora* is outlined in Figure 7.1. The same general principles are applicable to many other microorganisms. First, the mold culture is irradiated by x-rays or ultraviolet light. This increases the frequency of mutation several thousand fold. Next, a single irradiated cell is placed in a complete medium. If the cell is viable, it multiplies in the complete medium, giving rise to a genetically homogeneous (pure) culture of *Neurospora*. The complete medium contains a wide assortment of nutrients, all of the amino acids, vitamins, purines, pyrimidines, and so on. A sample of the pure culture is then transferred to a minimal medium that contains only sugar, biotin, salts, and water. **Failure of the mold to grow on minimal medium is taken as evidence that a mutation has occurred.** Since the mutant *Neurospora* grows on the complete medium but not on the minimal medium, there must be one or more components in the complete medium that the experimentally produced mutant can no longer synthesize.

The final step is to classify the mutant by identifying the component in the complete medium that is necessary for its growth. This is done by testing the mutant for growth in minimal medium supplemented, in separate cultures, with vitamins, amino acids, or other nutrients. In the example illustrated, the mutant failed to grow when supplemented with vitamins, purines, or amino acids, but grew normally when supplemented with a mixture of the three pyrimidines. By adding the three pyrimidines to the minimal medium one by one, it was determined that the mutant mold had lost the ability to synthesize thymine. The mutant is thus classified as a thymine-requirer.

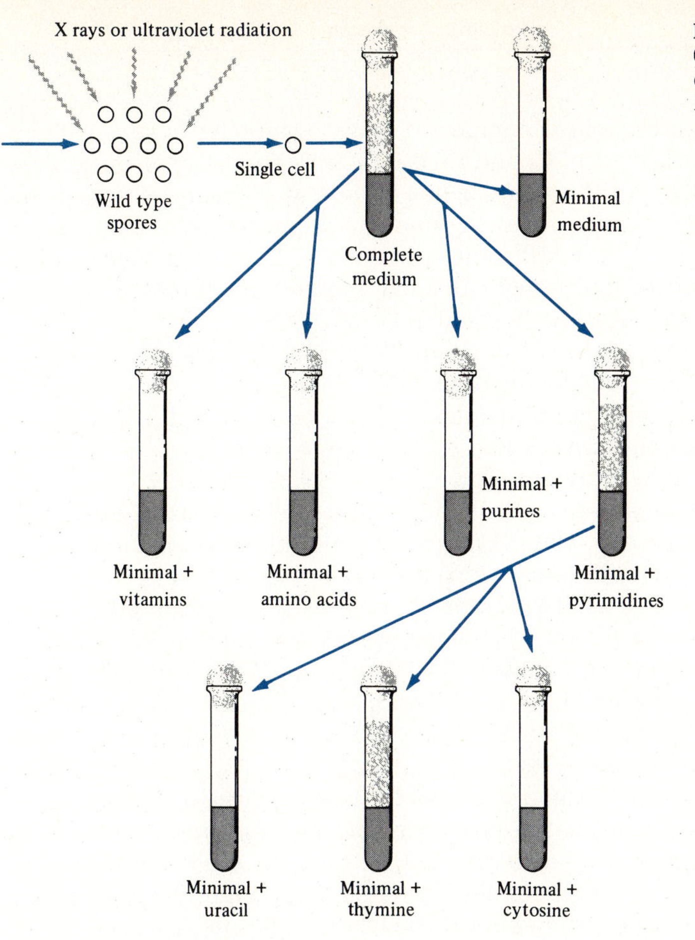

Figure 7.1
Outline of the procedure for producing, detecting, and classifying mutation in *Neurospora crassa*.

Using this systematic technique, Beadle and Tatum isolated a large number of different mutant strains of *Neurospora*, each of which required some nutrient for growth. Each of these mutants was characterized by the loss of the capacity to synthesize a specific nutrient; since all chemical syntheses within the cell are mediated by enzymes, they concluded that the mutational change resulted in a loss of some enzyme(s) involved in the biosynthesis of the nutrient in question. Genes must somehow give rise to specific enzyme(s). This was, of course, what Garrod had suggested (with less evidence) some 20 years earlier.

The second conclusion of Beadle and Tatum was that a *single* genetic factor was responsible for the production of a *single* enzyme, the so-called **one gene–one enzyme hypothesis.** This conclusion was derived from a careful examination of the data from one class of mutants—those that require the amino acid arginine for growth.

The wild-type *Neurospora* normally synthesizes arginine by a sequence of enzymatically catalyzed reactions that can be abbreviated as follows:

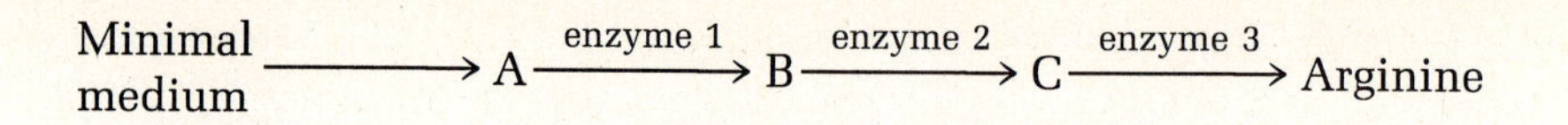

In the final three steps, a compound that we can refer to as A is transformed into B, then to C, and finally to arginine. Each of these steps is catalyzed by a different enzyme. Beadle and Tatum reasoned that some of their arginine-requiring mutants might be lacking enzyme 1, others enzyme 2, and still others enzyme 3. To test this possibility, they investigated the growth of their mutants on minimal medium plus compound A, B, C, or arginine (Table 7.1).

The arginine-requiring mutants fell into four classes (I–IV) according to their growth requirements. Class I mutants grew only if supplemented with arginine; class II grew on C or arginine; class III grew on B, C, or arginine; and class IV grew on A, B, C, or arginine. In no case could a mutant grow on A or B but not on C. These data are logically consistent with the following interpretation: Class I mutants lack enzyme 3, class II lack enzyme 2, class III lack enzyme 1, and class IV lack a necessary enzyme for the production of compound A. Consider, for example, class III mutants. Since they are able to grow on minimal plus B, they must contain enzymes 2 and 3, which are necessary for the conversion of B into arginine. However, since they cannot grow on A, class III mutants must lack enzyme 1, which catalyzes the conversion of A into B. The same type of reasoning also applies to the other classes.

Two follow-up experiments were then performed that provided additional support for the one gene–one enzyme hypothesis. First, genetic recombination experiments (similar to those described in Chapter 6) with *Neurospora* demonstrated that each of the four classes of arginine-requiring mutants was deficient in a different gene, localized on a different part of the chromosome. Second, direct biochemical tests revealed that in each case, the appropriate enzyme was indeed missing.

The Beadle and Tatum experiments* thus provided conclusive

TABLE 7.1 Growth Requirements of Four Classes of Arginine-Requiring Mutants of *Neurospora*

Arginine-requiring Mutant	Growth[a] on Minimal Plus				
	Nothing	*A*	*B*	*C*	*Arginine*
I	−	−	−	−	+
II	−	−	−	+	+
III	−	−	+	+	+
IV	−	+	+	+	+

[a] + indicates growth; − indicates no growth.

*For their experiments on the gene–enzyme relationship in *Neurospora*, Beadle and Tatum received a share of the 1958 Nobel Prize for medicine and physiology.

support for the one gene–one enzyme hypothesis. Since that time, similar conclusions have emerged from biochemical genetic studies on various other organisms, from bacteria to humans. The gene–enzyme relationship is now known to be a general biological phenomenon.

7.3 The Mechanism of Protein Synthesis

During the 1950s, molecular biologists realized that genes (DNA) exert their control by the production of specific enzymes (protein). However, it was not at all obvious at the biochemical level *how* DNA could dictate the structure of proteins. In the first place, there was no apparent complementation in chemical structure between DNA and protein as had been demonstrated previously for the two strands of DNA. Furthermore, careful cytochemical analyses had revealed that DNA is located in the nucleus of eucaryotic cells, whereas most of the protein is synthesized in the cytoplasm of these cells. How, then, can DNA direct protein synthesis? This paradoxical situation only recently has been resolved. As the details of the mechanism of protein synthesis have been unraveled, the indirect but controlling role by DNA on protein synthesis has emerged.

In 1953, Paul C. Zamecnik and co-workers at Harvard University reported for the first time the synthesis of protein in vitro. Their experiment consisted of mixing the following four ingredients in a test tube: (1) ATP; (2) 20 amino acids, one of them radioactive; (3) cellular particles called ribosomes (see Figures 2.28 and 2.30); and (4) soluble extract of cells. After 30 minutes of incubation, the reaction mixture was chilled, and cold acid was added to it. The measurement of protein synthesis was based on the fact that the amino acids are soluble in acid, whereas proteins are insoluble. The reaction mixture was then centrifuged to separate the soluble and insoluble fractions. Zamecnik and his associates observed radioactivity in the acid-insoluble fraction and thus concluded that protein was synthesized during the experiment. From careful studies on the role of each of the four components in the synthetic process came our current understanding of protein synthesis.

The first step in protein synthesis is the "activation" of the amino acids by ATP:

$$(1) \qquad \text{ATP} + \text{amino acid} \xrightarrow{\text{enzyme}} \text{Amino acid–AMP} + \text{P-P}$$

In this reaction, the ATP and amino acids are combined with the release of pyrophosphate (P-P) from ATP. The energy lost in splitting off the phosphates is conserved in the formation of the amino acid–AMP complex. For each of the 20 amino acids there is a different enzyme that catalyzes the activation. The source of these enzymes in Zamecnik's concoction was the soluble extract.

The second step in protein synthesis was discovered by a combination of an accident and subsequent careful experimentation. One of Zamecnik's associates, Mahlon B. Hoagland, considered the possibility that RNA, as well as protein, was being synthesized in their reaction mixture. To test this hypothesis he added to the protein-synthesizing system one additional ingredient, radioactive uracil. As mentioned previously, uracil is a component of RNA, but not DNA. If RNA were synthesized, the radioactive uracil would also be found in the acid-insoluble fraction. Afterward, protein and RNA could then be distinguished, since RNA is soluble in *hot* acid, whereas protein remains insoluble.

The experiment was performed. The first result Hoagland obtained was that the cold acid-insoluble fraction was radioactive. This could be due to amino acids entering into proteins and uracil entering RNA or both. Next, the acid-insoluble fraction was suspended in acid and heated for one hour. As expected, radioactivity was found in the hot acid-insoluble fraction, indicating that protein was synthesized. The exciting result was that the hot acid-soluble fraction was also radioactive. The tentative conclusion was that RNA had also been synthesized. In one of Hoagland's several controls, he left out the radioactive uracil although, of course, the radioactive amino acid was still present. Nevertheless, he still found radioactivity in the hot acid-soluble fraction containing the RNA. The logical conclusion was that some of the radioactive amino acid had become attached to RNA and was thus rendered soluble in hot acid. These experiments are summarized in Figure 7.2.

The demonstration of an amino acid–RNA complex led to an understanding of the second step in protein synthesis:

$$\text{(2)} \quad \text{Amino acid–AMP} + \text{RNA} \xrightarrow{\text{enzyme}} \text{Amino acid–RNA} + \text{AMP}$$

In this reaction, an amino acid is transferred from AMP onto an RNA molecule. This type of RNA is now referred to as transfer RNA (tRNA).

The tRNAs, containing approximately 80 nucleotides, are the smallest naturally occurring nucleic acid molecules. There is at least one specific tRNA for each amino acid; in the case of some amino

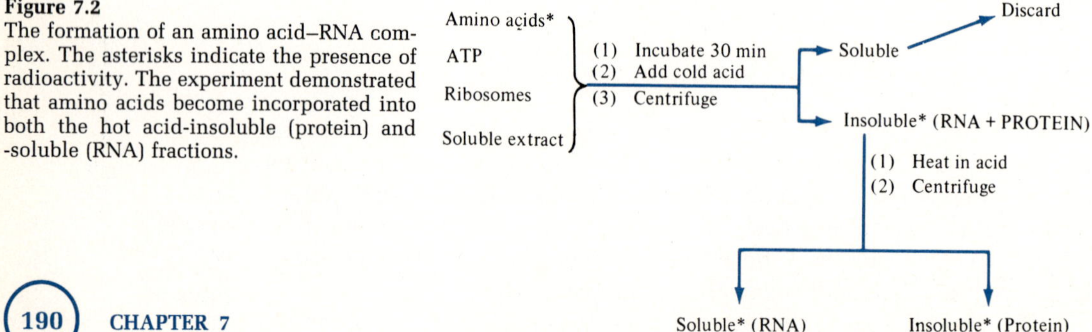

Figure 7.2
The formation of an amino acid–RNA complex. The asterisks indicate the presence of radioactivity. The experiment demonstrated that amino acids become incorporated into both the hot acid-insoluble (protein) and -soluble (RNA) fractions.

acids, there may be as many as five different, specific tRNA molecules.

In one of the great achievements of modern chemistry, Robert W. Holley and his co-workers in 1965 determined the entire sequence of bases in a tRNA for the amino acid alanine. Since then, about 75 different tRNAs have been sequenced. Figure 7.3 shows the two-dimensional structure of alanine tRNA. Several general features of tRNA are exemplified by alanine tRNA: (1) A large percentage of the nucleotides contain unusual purine and pyrimidine bases; in the case of alanine tRNA, 10 of the 77 nucleotides are neither uracil, adenine,

Figure 7.3
The two-dimensional structure of alanine tRNA. The 77 nucleotides fold in the form of a "cloverleaf." In addition to 8 adenine (A), 25 guanine (G), 23 cytosine (C), and 11 uracil (U) residues, there are 10 unusual bases (blue boxes). The dots between C . . . G and U . . . A indicate the base-pairing.

guanine, nor cytosine. (2) Although tRNAs are single-stranded, they spontaneously fold into "cloverleaf" configurations as a result of intramolecular base-pairing; alanine tRNA contains 19 base-pairs. (3) The site of amino acid attachment to tRNA is at the end containing the single-strand sequence ACC. (4) One of the loops in the "cloverleaf" structure contains a sequence of three bases, called the **anticodon,** which plays a crucial role in protein synthesis.

Reactions (1) and (2) can occur when a soluble extract of cells is mixed with ATP and the amino acids. All the enzymes and tRNAs necessary for the formation of amino acid tRNA complexes are contained in the extract. However, the reaction ceases, and no protein is formed unless **ribosomes** are also added. We can thus express the next reaction of protein synthesis as follows:

$$(3) \qquad \text{Amino acid–tRNA} \xrightarrow[\text{ribosomes}]{\text{soluble extract}} \text{Protein} + \text{tRNAs}$$

Investigators in many laboratories throughout the world have studied the ribosome and its reactions. Ribosomes have been found in the cytoplasm of all types of cells, from bacteria to mammals. The ribosome is an almost spherical granule, consisting of approximately equal weights of protein and RNA. Ribosomal RNA is much larger than tRNA, containing more than 1000 bases per molecule. Each ribosome contains more than 50 different proteins; some of these proteins are enzymes that control the linking together of the amino acids and the release of tRNA.

Several independent experiments have demonstrated that the ribosomes are the site of protein synthesis in living cells. In one such experiment, growing bacteria were exposed to radioactive amino acids for only 5 seconds. The bacteria were then chilled with ice to prevent further reactions, and the cells were rapidly harvested and disrupted. After separation of the broken cell extract into a soluble fraction and a ribosomal fraction by centrifugation, each was analyzed for acid-insoluble (protein) radioactivity. With this very short labeling period, the radioactivity was associated with the ribosomal fraction, indicating that it was the site at which the protein was manufactured. In a parallel experiment, the bacteria were labeled for 5 seconds, but then allowed to incubate with excess non-radioactive amino acids for 2 minutes prior to chilling. Under these conditions, no radioactivity was found in the ribosomal fraction. It would appear, therefore, that protein is first synthesized on the ribosomes, then liberated into the soluble portion of the cell.

Although the discovery of the role of ribosomes in protein synthesis was highly important, it did not solve the fundamental problem of how DNA controls the formation of protein. One approach to this problem was to ask the question: Do the ribosomes determine the *kind* of protein being synthesized? From the pioneer (1948) experiments of Seymour S. Cohen at the University of Pennsylvania, it was known that immediately after phage T_2 infects *Escherichia coli,* synthesis of ribosomes stops. Since the phage coat and other phage-specific proteins are produced in large quantities after infection, it follows that *E. coli* ribosomes are non-specific; prior to infection the ribosomes synthesize *E. coli* proteins, and after infection they synthesize phage proteins. Thus the ribosomes are non-specific protein assembly factories. The information for specifying the type of protein to be synthesized must reside elsewhere.

In 1960, François Jacob and Jacques Monod at the Pasteur Institute in Paris put forth a suggestion that explained the source of the information-containing component. They proposed that the information for determining the protein structure was transmitted from the genes to the ribosomes by a third type of RNA molecule, which they called messenger RNA (mRNA). According to their model the mRNA is **transcribed** from the DNA template, using base-pairing rules, and would thus have embodied in it the genetic information. Once synthesized, the mRNAs migrate to the cytoplasm, where they become attached to the ribosomes. The message would be translated at the ribosomes when the various tRNAs, each bearing its specific amino acid, became aligned in a sequence dictated by the mRNA. Finally, when the amino acid–tRNA complex is fixed in position, an enzyme unites the amino acid to its neighbor, with the release of the tRNA. A schematic representation of the processes of transcription and translation is shown in Figure 7.4.

Several predictions can be made from the Jacob–Monod model: (1) The size of the mRNA should vary considerably, depending on the size of the protein it codes for; (2) the mRNA should have a base composition similar to DNA (except uracil should replace thymine); (3) the mRNA should be complementary to a portion of one DNA strand; and (4) the mRNA should stimulate protein synthesis in vitro and also determine the type of protein made. Using modern techniques of centrifugation, mRNA has been isolated from a variety of cells. In all cases, the purified mRNA has properties that coincide with those predicted. In addition, it has been shown that in many cells, mRNA is metabolically unstable. In bacteria, for example, the mRNA seems to break down as soon as it has done its job of programming a few hundred protein molecules.

The discovery of the role of mRNA in protein synthesis brought together the various bits of genetic and biochemical data, thus forming a unified theory for gene action. The scheme, which is summarized in Figure 7.5, may seem complicated, but it is certainly ingenious.

The following points should be emphasized: (1) All three species of RNA are made from the DNA template; (2) the ribosome is formed from ribosomal RNA and protein, and becomes programmed for protein synthesis only when an mRNA is attached to it; (3) the amino acids enter the programmed ribosome after they have been activated and linked to a specific tRNA. The tRNA acts as a two-handed molecule, one hand holding onto the specific amino acid, the other binding to a specific sequence of bases on the mRNA. In this manner, the

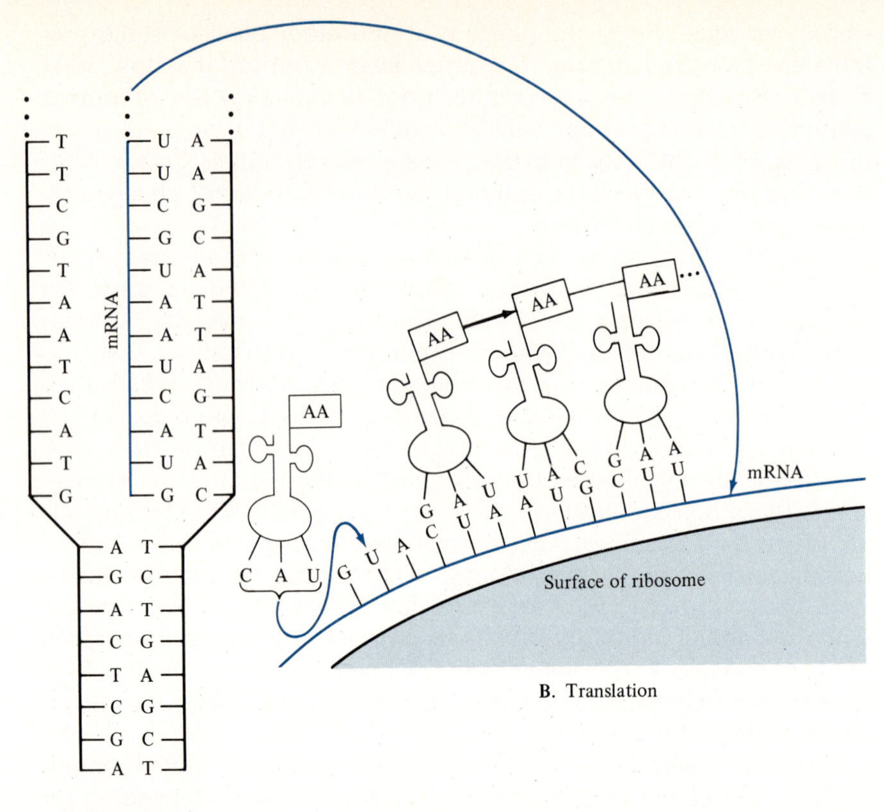

Figure 7.4
Schematic view of protein synthesis. (A) Transcription. The messenger RNA is copied from a segment of DNA, according to the base-pairing rules, except that uracil takes the place of thymine. (B) Translation. The mRNA then migrates to the cytoplasm where it becomes attached to a ribosome. The tRNA–amino acid complexes then wait their turn to line up on the mRNA template. The precise order in which the tRNA and amino acids line up is determined by base-pairing between the anticodon and the mRNA. When the amino acid (AA) has been joined to the growing protein chain, the "empty" tRNA is released.

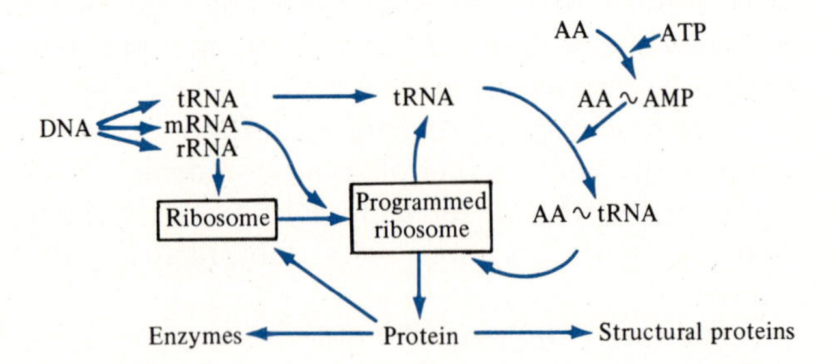

Figure 7.5
Summary of reactions involved in protein synthesis.

amino acids are incorporated in the order coded for by the mRNA, which in turn is coded for by the DNA.

7.4 The Genetic Code

In the light of what is currently known about the mechanism of protein synthesis, the one gene–one enzyme concept can be restated as follows: One segment of DNA is responsible for the production of an mRNA which codes for a specific sequence of amino acids. But precisely which sequence of bases on the mRNA "spells out" which amino acid? What is the code?

From a purely theoretical argument, we can conclude that it requires more than two bases to code for a single amino acid. Suppose only one base were to be used as a code for an amino acid. Then the four bases could specify only four amino acids. Since there are 20 different amino acids found in proteins, the single base code is excluded.

Using two bases to code for one amino acid yields 4 × 4, or 16, possible combinations:

AC	CC	GC	UC
AA	CA	GA	UA
AG	CG	GG	UG
AU	CU	GU	UU

Since this is still not enough to account for the 20 amino acids, we must conclude that more than two bases code for one amino acid. If three bases are used in each code word, there are 4 × 4 × 4, or 64, possible triplets, which is more than adequate.

Although these simple theoretical considerations were discussed as far back as 1952, it was not until 1961 that biochemistry progressed to the point where it was possible to "crack the code." Then, at the Fifth International Congress of Biochemistry in Moscow, a young biochemist from the National Institutes of Health, Marshall W. Nirenberg, reported his experiments on protein synthesis in vitro. His results amazed the scientific world. He had made the breakthrough that led directly to deciphering the code.

Nirenberg and his associates performed a series of experiments similar to those of Zamecnik, except for one important step: Nirenberg added deoxyribonuclease (DNase) in addition to the ATP, ribosomes, amino acids, and soluble extract. Under these conditions, protein synthesis stops after about 20 minutes (Figure 7.6).

Why does the reaction stop? To an empiricist, such a question suggests a simple experiment: After 30 minutes with DNase, add back

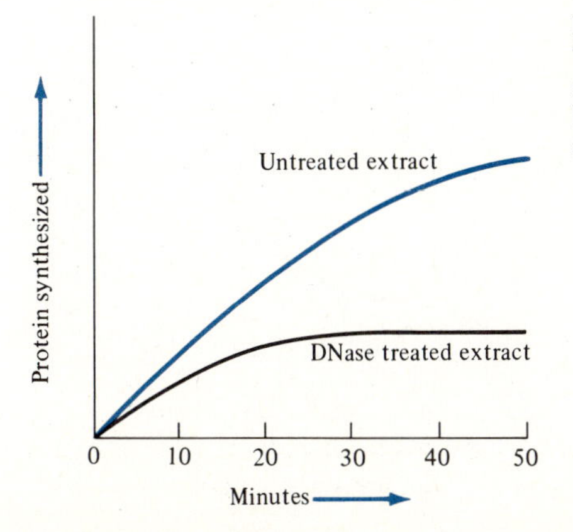

Figure 7.6
The effect of DNase treatment on protein synthesis in vitro. Protein was measured by the incorporation of radioactive amino acids into the acid-insoluble fraction.

to the incubation mixture in separate tubes, ATP, amino acids, ribosomes, tRNA, and mRNA. The results of such an experiment are shown in Figure 7.7. The answer is clear: Only the mRNA stimulates. They reasoned that during the initial 30 minutes, the mRNA was broken down. Since no template DNA was present (DNase destroyed it), the mRNA could not be replenished and protein synthesis ceased.

These experiments paved the way for a comparative study of the effect of various mRNAs on protein synthesis. Once the incubation mixture becomes depleted of the mRNA originally present in the soluble extract, the extent and type of protein synthesized depend entirely on added mRNA. Nirenberg and his associates found, for example, that RNA from virus-infected cells greatly stimulated protein synthesis, whereas RNA isolated from ribosomes stimulated only slightly. Then, either by design or by chance (only Nirenberg knows for sure), they decided to test an artificial RNA-like polymer that is called "poly-U." Poly-U is a synthetic macromolecule that has a structure identical to that of RNA, except that it contains only one type of base, uracil.

Their experiments revealed that poly-U was a surprisingly efficient messenger for protein synthesis. But if the RNA message was UUUUUUUUUU . . . , what was the protein product? Chemical analysis revealed that the reaction product was a polymer containing only one amino acid, phenylalanine.

Thus, Nirenberg concluded, "Addition of poly-U resulted in the incorporation of phenylalanine alone into a protein resembling polyphenylalanine. Poly-U appears to function as a synthetic template, or messenger RNA, in this system. One or more uridylic acid residues appear to be the code for phenylalanine." Assuming a triplet code, the first word (codon) in the genetic dictionary then becomes UUU = phenylalanine.

With the path clearly prepared by Nirenberg's initial discovery, progress was rapid. Artificial RNA polymers were synthesized containing every possible combination of the four bases. They were then tested for their ability to stimulate protein synthesis in vitro. By ana-

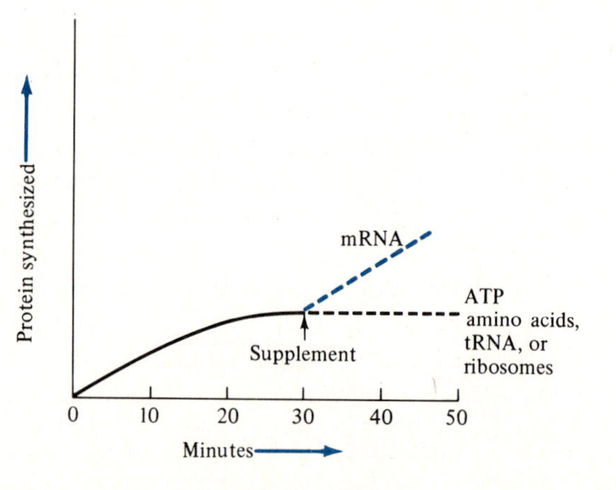

Figure 7.7
The stimulation of protein synthesis in vitro by ATP, amino acids, ribosomes, tRNA, and mRNA. The initial conditions were the same as in Figure 7.5 (plus DNase). After 30 minutes, the incubation mixture was divided into five portions; each was then supplemented with one of the substances.

TABLE 7.2 Assignment of Codons Using Synthetic mRNAs

Messenger RNA	Amino Acid Incorporated	Codon
ACACACAC . . .	Threonine	ACA
	Histidine	CAC
CAACAACAA. . .	Threonine	ACA
	Glutamic acid	CAA or AAC
	Asparagine	CAA or AAC

lyzing the relative amount and distribution of each amino acid incorporated into protein under the guidance of these synthetic RNAs, it was possible to assign codons to every amino acid. Especially useful were the nucleic acids chemically synthesized by H. G. Khorana and his associates at the University of Wisconsin. The specific example (Table 7.2) illustrates how some of these assignments were made.

An artificial messenger containing an alternating sequence of adenylic and cytidylic acids, ACACACACACAC . . . , led to the incorporation of two amino acids, threonine and histidine, also in alternating sequence. This is consistent with a triplet code in which one of the amino acids is coded for by ACA, the other by CAC. It was also found that the synthetic mRNA containing the sequence CAACAA-CAACAACAA . . . led to the incorporation of three amino acids, glutamate, asparagine, and threonine. Depending on where the reading begins, the possible codons are CAA, AAC, and ACA. Since threonine is the only amino acid incorporated in both these examples and since ACA is the only triplet present in both messengers, then ACA must be the codon for threonine. It follows that CAC codes for histidine.

By the end of 1968, the total assignment of the 64 triplet codons was completed (Table 7.3).† From an analysis of the data, several generalizations emerge.

1. **The code is degenerate.** Degeneracy means that several different triplets can specify the same amino acid. For example, both UUU and UUC code for phenylalanine.
2. **There are codons for start and stop.** The mRNA is not simply read from one end to the other. Instead, there are specific codons that determine where reading begins and ends. Chain growth is initiated by the codons GUG and AUG and is terminated by UAA, UAG, or UGA.
3. **The code is largely, if not entirely, universal.** A given mRNA will be read in the same manner by species at opposite ends of the evolutionary scale. For example, poly-U stimulates phenylalanine incorporation in cell extracts from bacteria to mammals.

†The "triplet," R. W. Holley, H. G. Khorana, and M. W. Nirenberg, shared the 1968 Nobel Prize in medicine and physiology for their independent but related experiments on the involvement of RNA in protein synthesis.

TABLE 7.3 The Genetic Code[a]

First Base	Second Base				Third Base
	U	*C*	*A*	*G*	
U	Phe	Ser	Tyr	Cys	U
	Phe	Ser	Tyr	Cys	C
	Leu	Ser	"Stop"	"Stop"	A
	Leu	Ser	"Stop"	Trp	G
C	Leu	Pro	His	Arg	U
	Leu	Pro	His	Arg	C
	Leu	Pro	Gln	Arg	A
	Leu	Pro	Gln	Arg	G
A	Ileu	Thr	Asn	Ser	U
	Ileu	Thr	Asn	Ser	C
	Ileu	Thr	Lys	Arg	A
	Met[b]	Thr	Lys	Arg	G
G	Val	Ala	Asp	Gly	U
	Val	Ala	Asp	Gly	C
	Val	Ala	Glu	Gly	A
	Val[b]	Ala	Glu	Gly	G

[a]This table shows the codon assignments of all 64 triplets. Standard abbreviations are used for the 20 amino acids. The left-hand column lists the four bases representing the first letter of the codon; the second letter is one of the bases listed across the top, and the third letter is listed vertically in the last column. The codons indicated by "stop" cause chain termination.

[b]Two codons have been shown to be the signals for chain initiation, AUG and GUG, when they occur either at the beginning of the mRNA or following a "stop." The codon AUG specifies methionine, both at the beginning of the protein and internally. The codon GUG normally specifies valine; however, when the GUG occurs at the beginning of an mRNA or following a "stop," it codes for methionine.

Chapters 5, 6, and 7 have concentrated on a molecular approach to genetics, an approach which in the last 30 years has yielded the most extraordinary results in the history of the life sciences. The gene has been identified and chemically characterized; gene duplication, mutation, and sexual recombination can now be appreciated at the molecular level; and giant strides have been taken toward an understanding of the mode of gene action. However, it would be wrong to leave the impression that the genetic processes are now understood completely at the molecular level.

Unraveling the genetic code is equivalent to learning the alphabet. We must still learn to read—and to comprehend what we read. Buried in the sequence of bases in DNA are the genetic histories of species, for only through changes in the order of bases in DNA are mutation and subsequent evolution possible. Very recently, procedures have been developed for determining the sequence of bases in small DNA molecules. With further refinements of these techniques, we will begin to read the "book of life."

7.5 Information Flow in the Opposite Direction: Reverse Transcriptase

The flow of information in the biological world is normally from DNA to RNA to protein. This simplified concept is so deeply rooted in modern biology that it has come to be known as the "central dogma" of molecular biology. But what happens with viruses whose genetic information is encoded in RNA rather than DNA?

The three different modes of messenger RNA formation that occur in RNA viruses are illustrated in Figure 7.8. In one group of viruses, the infecting viral RNA serves directly as the messenger RNA for protein synthesis; once the RNA penetrates the cell, it becomes attached to ribosomes where it is immediately translated into viral proteins. In a second group of RNA viruses, the mRNA is a complement of the viral RNA; a special RNA-directed RNA polymerase is needed for the viral RNA⟶mRNA transcription reaction. The final group of RNA viruses, the so-called RNA tumor viruses, have an entirely different mode of messenger RNA formation. Genetic information flows "in reverse"—from RNA to DNA. This RNA-directed DNA synthesis is catalyzed by a unique enzyme appropriately called **reverse transcriptase.** Messenger RNA is then transcribed from the complementary DNA in the normal manner.

The discovery of reverse transcriptase in 1970 by Howard M. Temin at the University of Wisconsin and David Baltimore of the Massachusetts Institute of Technology was important for a number of reasons. First, the enzyme can be used as an extremely sensitive detector

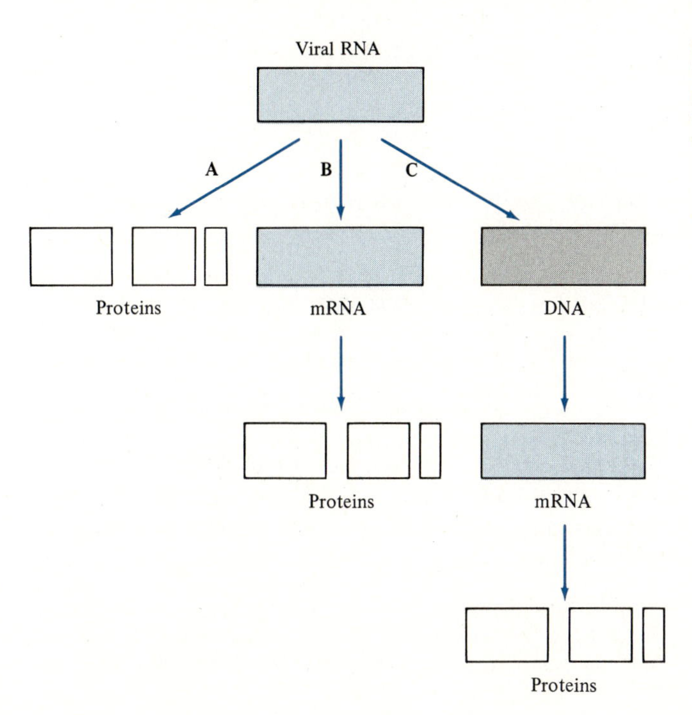

Figure 7.8
Three modes of messenger RNA formation by RNA viruses: viral RNA is translated directly to viral proteins (A); viral RNA is transcribed to complementary mRNA prior to protein synthesis (B); and viral RNA is transcribed to complementary DNA, which acts as the template for mRNA production (C).

of RNA tumor viruses; such early detection methods are of great potential medical significance because a number of animal cancers have been shown to be caused by RNA tumor viruses. Second, reverse transcriptase provides an explanation (at least, for some RNA viruses) for the phenomenon of viral latency. Certain viruses seem to disappear shortly after infecting an organism only to reappear several months or years later. Once the infectious RNA has transferred its genetic information to DNA, it could remain latent in a cell and be replicated by the normal cellular enzyme system. As a result of a later activation, the virus could reappear as an infectious agent. Third, reverse transcriptase has stimulated an enormous amount of research on the relationship of RNA viruses to cancer. The virus theory of cancer is discussed in Chapter 16.

QUESTIONS

7.1 The wild-type *E. coli* can grow on sugar as its sole source of carbon. Five mutants were obtained that required vitamin B_1 in addition to the sugar for growth. When these mutants were tested for growth with precursor molecules (G, H, I, R, and T), the following results were obtained:

Organism	Vitamin B_1	Nothing	G	H	I	R	T
Wild-type	+	+	+	+	+	+	+
Mutant 1	+	−	+	+	−	−	+
Mutant 2	+	−	−	+	−	−	+
Mutant 3	+	−	+	+	+	−	+
Mutant 4	+	−	−	−	−	−	+
Mutant 5	+	−	−	−	−	−	−

a) In the formation of vitamin B_1, what is the order of the five precursors?
b) Where is each mutant blocked?
c) How does mutant 1 differ from the wild-type with respect to the detailed structure of its DNA, RNA and protein?

7.2 Assume that a particular gene segment of DNA consisted of the tetranucleotide TTTC repeated 30 times; that is, TTTCTTTCTTTCTTTCTTTC$(TTTC)_{25}$

a) How large a protein (number of amino acids?) would you predict would result from this strand of DNA?
b) Using Table 7.3, what would be the amino acid sequence in the resultant protein? (Use standard abbreviations).
c) Draw the structure of the relevant tRNAs, indicating the anticodons.
d) Draw schematically the complex that is formed between the mRNA derived from this gene and the tRNA amino acids. Show the base sequences wherever possible.

7.3 A non-motile *E. coli* mutant was isolated. Analysis of its flagella protein indicated that it differed from the wild-type in having an extra lysine and one less isoleucine. What can you infer about the change in its DNA base sequence?

7.4 Define transcription and translation as they are used in molecular genetics. What two modifications of the transcription process occur with RNA viruses?

7.5 Compare the different types of RNA with respect to their size, base composition, and function in protein synthesis.

7.6 Explain the poem in Section 1.9 with respect to DNA, RNAs, and protein.

7.7 Although DNA is a double helix, we know that only one of the two strands is used to produce protein. Suggest a mechanism that allows one strand to be transcribed, but not the other.

7.8 Can you determine the exact base sequence of a gene if you know the amino acid sequence of its protein product? Explain.

Suggested Readings

Beadle GW. Genes and Chemical Reactions. Science, 129: 1715, 1959.
Nobel Prize lecture given in Stockholm on the one gene–one enzyme hypothesis.

Borek E. The Code of Life. New York: Columbia University Press, 1965.
A non-technical account of biochemical genetics.

Crick FHC. The Genetic Code: III. Scientific American, October 1966.

Nomura M. Ribosomes. Scientific American, October 1969.

Rich A and SH Kim. The Three-Dimensional Structure of Transfer RNA. Scientific American, January 1978.
The relationship between the structure of tRNA and its function in protein synthesis is discussed in some detail.

Sinsheimer RL. The Book of Life. Reading, MA: Addison-Wesley Publishing Co., 1967.
An easily read discussion of the nature of the genetic code. Conceptual relations are clearly developed, independent of the specific biochemistry.

Temin HM. RNA-Directed DNA Synthesis. Scientific American, January 1972.
A well-illustrated, clear discussion of the discovery of reverse transcriptase and its possible role in virus-induced cancer.

Microbial Nutrition and Growth

As the area of light increases so does the circumference of darkness.

Albert Einstein

Previous chapters have concentrated on the subcellular and biochemical aspects of life. To a biologist, these subjects take on added significance when they lead to a better understanding of the organism as a whole. The most dramatic property of whole living microorganisms is growth. Growth may be defined as the orderly increase of all cellular components leading to an accurate duplication of the existing pattern. The study of growth is an all-encompassing subject, including such fundamental processes as energy generation, biosynthesis, DNA replication, and protein synthesis. The amazing feature of growth lies in the integration of all these biochemical processes.

In this chapter, we discuss nutritional requirements for growth, the growth cycle, and finally the regulation of gene expression. The discussion focuses on the growth of bacteria. The relative ease of growing large numbers of single-cell organisms under well-defined laboratory conditions makes them ideal subjects for elucidating the basic principles of nutrition and growth.

8.1 Nutritional Requirements for Growth

The concept of the unity of biochemistry proposes that the basic chemistry of all earthly organisms, whether protists, plants, or animals, including humans, is essentially the same. All cells transform energy from their environment into ATP, and then use the energy stored in ATP to perform biological work. All utilize DNA as their genetic material, and are replicated by a similar mechanism. All express their inheritance through the production of enzymes (proteins) that are synthesized on ribosomes, utilizing messenger RNAs as templates and tRNAs as adapter molecules. Although cells are composed of similar macromolecules, constructed from an identical set of building blocks, they differ and express their uniqueness in the precise order with which the monomeric units are placed in the polymers. Since the biosynthesis of polymers requires ATP, the first nutritional requirement for growth of any cell is a **source of energy.**

The great variety of energy sources used by different bacteria emphasizes the remarkable diversity of the microbial world. As discussed in Chapter 3, in addition to obtaining energy from sunlight (photosynthesis) and the oxidation of organic compounds (respiration), certain classes of bacteria derive their ATP from oxidizing inorganic compounds, such as iron, sulfur, hydrogen, and ammonia; other groups of bacteria can utilize organic compounds in the absence of oxygen (fermentation) or in the presence of nitrate or sulfate (anaerobic respiration). Certain individual bacteria are highly specific in the type of organic compound they can use for energy, while others are extremely versatile. For example, a certain bacillus can grow only on methyl alcohol, whereas *Pseudomonas cepacia* can use any one of more than 100 different organic compounds as its sole source of energy.

In addition to an appropriate energy source, all cells require **raw materials** to synthesize the cellular constituents. The elementary composition of a typical microbial cell is shown in Table 8.1. After drying (water makes up about 80 percent of the weight of most cells), carbon is the most abundant cellular element. Together with oxygen and hydrogen, carbon is found in essentially all biological compounds. Nitrogen is a major component of both proteins and nucleic acids; phosphorus is found primarily in nucleic acids, while most cellular sulfur is in proteins. The remaining 5 percent of the dry weight of a cell is composed of a large variety of inorganic materials, including relatively high amounts of magnesium, calcium, potassium, sodium, and iron. In order for a cell to grow, all of these elements must be supplied in a usable form.

Living organisms can be divided into two broad groups on the basis of how they obtain carbon: **Autotrophs** can utilize carbon dioxide as their principal source of carbon. These organisms have the simplest of all nutritional needs; they require only inorganic salts, water, and carbon dioxide, deriving their energy from light or by the oxidation of inorganic materials. Organisms dependent on an organic carbon source, such as glucose, are termed **heterotrophs.** In most heterotrophic organisms, part of the carbon compound that is used as an

TABLE 8.1 Elementary Composition of a Typical Cell

Element	Percentage of Cell *(Dry Weight)*	Major Constituent of
Carbon	50	All organic compounds
Oxygen	18	Water and all organic compounds
Nitrogen	13	Proteins and nucleic acids
Hydrogen	10	Water and all organic compounds
Phosphorus	3	Nucleic acids
Sulfur	1	Proteins
Other inorganics	5	

energy source can also be used as the raw material for synthesizing cellular material.

Many organisms require one or more **growth factors** in addition to a general source of carbon and energy. These growth factors include amino acids, purines, pyrimidines, and vitamins. Again it is useful to consider the question of growth factors in terms of unity and diversity of the biological world. The unity principle is that *all* organisms need *all* of the factors; nutritional diversity arises as a result of the relative ability of different organisms to synthesize these factors for themselves. For example, all organisms require the vitamin biotin for carrying out essential life processes: The bacterium *E. coli* makes biotin itself and therefore does not require the vitamin as a nutrient; the mold *Neurospora* lacks the necessary enzymes for synthesizing biotin and must receive the vitamin in its diet in order to grow.

The elements **hydrogen** and **oxygen** are readily available to cells in the form of water and most organic compounds. **Nitrogen, phosphorus, sulfur,** and other inorganics are generally obtained by microorganisms as salts dissolved in water. As far as we know, all organisms can utilize ammonia (NH_3) for synthesis of nitrogen-containing compounds of the cell. Many but not all microorganisms can use nitrates (NO_3^-) as a nitrogen source. Finally, a few specialized bacteria can satisfy their nitrogen requirements by converting atmospheric nitrogen gas (N_2) to ammonia by a process called **nitrogen fixation.** The elements phosphorus and sulfur are normally supplied by phosphate (PO_4^{3-}) and sulfate (SO_4^{2-}) ions, respectively. In addition to these elements, the most important mineral requirements for microorganisms are potassium, sodium, magnesium, calcium, and iron.

The basic principles of nutrition are exemplified in the widely varying compositions of the growth media of four different bacteria shown in Table 8.2. The autotrophic blue-green bacteria can grow in the light in a medium containing a few salts in water in contact with air. If the medium also contains glucose as a carbon and energy source and ammonia as a nitrogen source, then simple heterotrophs like *E. coli* can multiply. The myxobacterium *M. xanthus* is an example of a heterotroph that cannot use glucose; it derives its energy, carbon,

TABLE 8.2 Growth Media for Different Bacteria

Requirement[a]	**Ingredient Necessary to Satisfy Requirement for**			
	Blue-Green Bacteria	*E. coli*	*M. xanthus*	*Lactobacilli*
Energy source	Light	Glucose	Protein	Glucose
Carbon source	Carbon dioxide	Glucose	Protein	Glucose
Nitrogen source	Nitrogen gas	Ammonia	Protein	Ammonia
Growth factors	None	None	5 amino acids	Yeast extract

[a]In addition, all organisms require water, phosphate, sulfate, magnesium, calcium, potassium, iron, and trace amounts of other elements, including cobalt, copper, zinc, manganese, sodium, and molybdenum.

and nitrogen from the metabolism of proteins. Since *M. xanthus* cannot synthesize five different amino acids, these specific amino acids must be provided as growth factors in the medium. Lactic acid bacteria have extremely limited synthetic abilities; they require ten different vitamins, three purines, uracil, and almost all of the amino acids. Rather than providing each of these requirements separately, it is technically easier to add yeast extract, which is a rich source of most growth factors.

In essence, diversity in nutritional requirements is a manifestation of evolution. As discussed in Chapter 1, primordial cells presumably arose in "organic soups," rich in a multitude of complex nutrients. We may speculate further that as these requisite compounds were exhausted from the "soup" as a result of their consumption by the cell population, the process of biosynthetic evolution was initiated; that is, by mutation and natural selection, organisms arose that contained enzymes whose function was to convert chemicals in rich supply to those that were in demand by the cells. In this way, primordial cells gradually developed their biosynthetic machinery and evolved from highly nutritionally demanding creatures with a paucity of biosynthetic enzymes to nutritionally independent organisms rich in biosynthetic properties. Eventually autotrophic organisms evolved with the capacity to obtain their cellular carbon by fixation of carbon dioxide gas from the atmosphere. With the advent of highly evolved autotrophic forms, other cells underwent a gradual loss of biosynthetic powers—a type of retrograde evolution. They obtained their needed nutrients pre-formed from other cell types. Humans, for example, require ten amino acids and several vitamins for growth. These are obtained by eating animal or plant materials. Certain disease-causing bacteria are so nutritionally demanding that they have not yet been grown outside their living host animals.

8.2 Influence of Temperature on Microbial Growth

The human body has a comfortable temperature range of approximately two degrees Celsius; if the body temperature rises above 41°C or falls below 34°C, we require urgent medical treatment. Microorganisms are much more versatile; most microbes grow over a temperature range of approximately 30°C, with each type exhibiting well-defined upper and lower limits. Microorganisms that grow best at low temperatures (below 20°C) are called **psychrophiles;** those that grow fastest between 25° and 45°C are referred to as **mesophiles;** organisms that prefer even higher temperatures are called **thermophiles.** The rates of growth of a typical psychrophile, a mesophile, and a thermophile are shown in Figure 8.1.

For each organism there is a **minimum temperature** below which growth no longer occurs, an **optimum temperature** at which the rate of growth is fastest, and a **maximum temperature** above which growth does not take place. In general, the growth rate increases slowly as the temperature is raised, until the optimum temperature is

Figure 8.1
Growth rate as a function of temperature of a psychrophilic, a mesophilic, and a thermophilic microorganism.

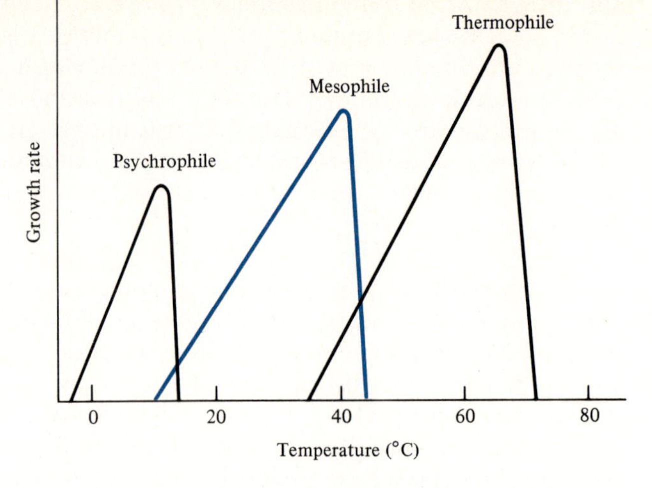

reached; then the growth rate falls sharply, so that at about 5°C higher than the optimum temperature, growth ceases completely.

The maximum growth temperature is not an absolute value. It depends somewhat on other factors, such as nutrients in the medium. In an interesting series of experiments, E. Ron and B. Davis of Harvard University showed that *E. coli* could grow at temperatures of up to 45°C in a nutrient broth medium (containing a wide assortment of amino acids, vitamins, purines, and other components), but only to 42°C in a defined medium containing only glucose and salts. They then asked the question: What is there in the nutrient broth that allows the cells to grow at 45°C? By adding compounds one at a time to the glucose-salts medium and testing for growth at 45°C, they eventually found that the amino acid methionine is the critical component. *E. coli* will grow at up to 45°C in a glucose-salts medium supplemented with methionine. Further investigation showed that one of the enzymes necessary for the synthesis of methionine is **temperature sensitive.** The methionine-synthesizing enzyme functions at a maximum of 42°C. Above this temperature, the three-dimensional structure of the enzyme is destroyed (i.e., the protein becomes denatured) and activity is lost.

The experiments of Ron and Davis provide direct evidence that the effect of temperature on growth of an organism is largely dependent upon its enzymatic makeup. In general, as the temperature increases, the enzymes work faster and faster, resulting in a more rapid rate of growth, until a critical temperature is reached where the enzymes become unstable and growth ceases. This critical temperature is much higher for thermophiles than for mesophiles and psychrophiles. For example, the DNA polymerase of a particular thermophilic bacillus was recently shown to be most active in the temperature range of from 55° to 70°C; at these temperatures, the DNA polymerase of the mesophilic *E. coli* is completely denatured and fails to function.

Somehow the sequence of amino acids in a protein (coded in the DNA) determines its temperature sensitivity. Thermophiles have evolved proteins that can withstand relatively high temperatures, whereas psychrophiles contain proteins adapted to function best at low temperatures. In the next chapter, we discuss some specific ecological niches where these microorganisms can be found.

8.3 The Growth Curve

The growth of the bacterium *E. coli* has been studied more extensively than that of any other living form. A simple medium for the growth of *E. coli* is shown in Table 8.3. Glucose serves as both the carbon and the energy source, ammonium chloride as the nitrogen source, potassium phosphate as the source of phosphorus, and magnesium sulfate for both sulfur and magnesium. The trace elements necessary for growth are needed in such minute quantities that they are usually present as contaminants in water or in the other ingredients.

Let us now consider what happens when a single *E. coli* cell is placed in a drop of the growth medium at 25°C and observed under a microscope. After a certain period of time, the bacterium elongates and then divides approximately in half to form two "sister" cells. Both "sister" cells immediately commence growth and, when they reach a critical size, again divide to produce four cells. The process of growth and division is repeated, yielding an exponentially increasing number of bacteria—8, 16, 32, 64, 128, and so on. These data are displayed graphically in Figure 8.2.

The two curves shown in Figure 8.2 present the same information in slightly different forms. In Curve A, the ordinate (vertical axis) is simply the number of bacteria on a linear scale. In Curve B, the ordinate is arranged on the exponential scale, 2^n, 2^{n+1}, 2^{n+2} . . . . In both cases, the abscissa (horizontal axis) shows time in hours on a linear scale. There are several advantages in plotting the logarithm of cell number as a function of time in growth experiments. First, a large number of cell doublings can be shown conveniently on the same graph. (Note that Curve A goes off scale after six hours.) Second, plotting the logarithm of cell number versus time yields a straight line for exponentially growing cells. This makes it much easier to estimate the

TABLE 8.3 Simple Medium for the Growth of *E. coli*

Ingredient	Amount
Distilled water	1 liter
Glucose	5 grams
Ammonium chloride	1 gram
Phosphates	1 gram
Magnesium sulfate	0.5 gram
Trace elements	0.2 milligram of each

Figure 8.2
Growth of bacteria as a function of time. In Curve A, the number of bacteria is plotted directly, whereas in Curve B, the logarithm of bacteria to the base 2 is plotted.

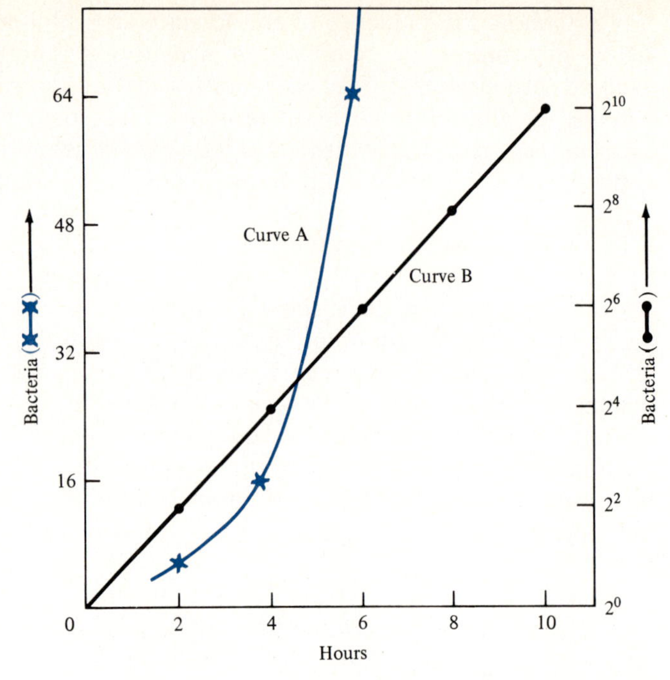

generation time. In the illustrated example, *E. coli* has a generation time of one hour in glucose medium at 25°C.

Let us now consider the events that take place when a flask containing glucose-salts medium is inoculated with *E. coli* and incubated on a shaking machine to provide adequate aeration. (Oxygen is required by cells for production of ATP via respiration.) The shape of the growth curve that is obtained (Figure 8.3) is characteristic of the growth of virtually all populations, whether viruses, bacteria, rabbits, or giraffes. Only the rates vary.

The growth curve may be divided conveniently into four phases or periods: (1) **lag phase,** (2) **exponential phase,** (3) **stationary phase,** and (4) **death phase.** The lag phase, which occurs immediately after the inoculum is introduced into a new medium, is a period of no increase in cell number. Following a short period of increasing growth rate, the culture enters the exponential phase, sometimes called the logarithmic growth phase. During this period, all viable cells are dividing at the maximum rate. This phase is followed by a short period of decreasing multiplication rate, and the culture then enters into the stationary phase. The last phase of the growth curve is called the death phase.

The Lag Phase

The lag phase is a period of adjustment. It is observed when a fresh medium is inoculated with cells from an old culture, or when the environmental conditions, such as the composition of the me-

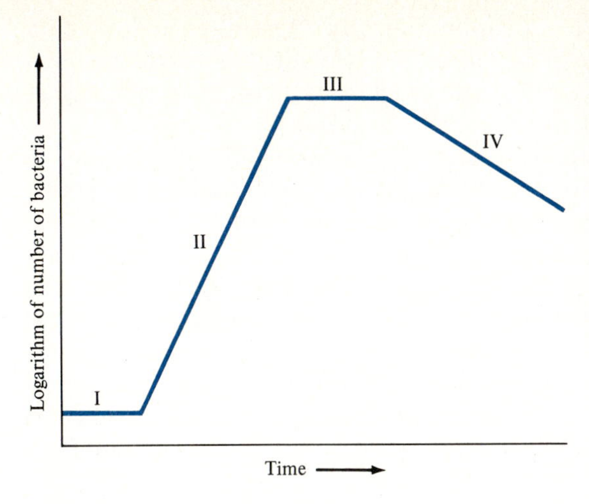

Figure 8.3
The growth curve. A flask containing sterile medium is inoculated with bacteria. Samples of the culture are withdrawn at intervals and the logarithm of the number of cells plotted against time. Four phases are usually observed: (I) lag phase, (II) exponential phase, (III) stationary phase, and (IV) death phase.

dium or the temperature of growth, are changed. In the early 1920s, microbiologist A. T. Henrici noted that bacterial cells increased their size two or three times when introduced into a fresh medium. Later it was shown that cells in lag phase synthesize nucleic acids and proteins. It appears, then, that there is a greater delay in cell division than in actual growth as represented by the synthesis of cell material. Why, then, if cells are undergoing net synthesis of nucleic acids and proteins, is there a lag in cell division? Old cells that are no longer dividing may be deficient in certain substances that are critical for the process of cell division. These substances must be synthesized and properly balanced with the other cellular constituents before division takes place. As an illustration, Moselio Schaechter at the University of Florida recently demonstrated that old bacteria, that is, cells in the stationary phase of growth, have fewer ribosomes than cells in the exponential phase. These cells must therefore synthesize ribosomes before entering into exponential growth. **The lag phase constitutes a period of adjustment for initiation of exponential growth.**

It follows that cells that are growing exponentially, and are therefore fully adjusted for rapid cell division, should not show a lag when transferred to fresh medium of the same composition. The lag phase can be completely eliminated by using as an inoculum a culture from the exponential phase. The lag phase may last anywhere from minutes to many hours to even days, depending on the size and age of the inoculum.

The Exponential Phase

The exponential growth phase is that period during which the rate of increase of cells is maximum and constant. Every organism has a rate of growth during this phase that is characteristic for that creature under the particular conditions of cultivation. This represents the maximal reproductive potential of the organism. What are the factors that affect the generation time? Of course, the primary factor is the organism itself. It takes *E. coli* approximately 20 minutes to double its mass under the best growth conditions. However, a rabbit at birth requires approximately six days to double its mass; a guinea pig at birth, 18 days; a human child at birth, 180 days.

Another factor of major significance is the medium. If *E. coli* is grown in a very rich medium containing all the natural amino acids, purines, pyrimidines, vitamins, and other nutrients, it will have a generation time of approximately 20 minutes at 37°C. The same organism, however, when placed in minimal medium with glucose as the sole carbon and energy source, will have a generation time of 50 or 60 minutes. The reason for this is that in a rich medium, the organism does not have to synthesize the myriad of enzymes required for the biosynthesis of all of the amino acids, purines, vitamins, and so on. It uses preferentially those nutrients from the medium and thus it conserves energy and carbon. The conserved energy and carbon are used for growth and division rather than for the biosynthesis of building blocks for macromolecules. This phenomenon is discussed later in the chapter when we consider regulation of enzyme synthesis.

Exponential growth cannot proceed indefinitely, either in nature or in the laboratory. A single *E. coli* cell with a generation time of 20 minutes would produce in 48 hours of exponential growth 2.2×10^{43} cells! The total weight would be about 2.5×10^{25} tons, or roughly 4000 times the entire mass of earth.

What are the factors that cause exponential growth to terminate and bring on the maximum stationary phase? The two most common factors are the exhaustion of nutrients and the production of toxic metabolic wastes by the culture. During the process of growth and division, nutrients are constantly being assimilated by the cells; hence the concentration of these nutrients in the medium is being depleted. When the concentration of the limiting nutrient is exhausted, growth ceases. Under these circumstances, then, the extent of growth will be directly proportional to the concentration of the limiting nutrient. In the laboratory, exponential growth may terminate as a result of oxygen deprivation. In cultures of high population densities, the rate of oxygen utilization may become greater than the rate of oxygen diffusion into the medium. As a result, the culture may become starved for oxygen and bring the exponential growth phase to an end.

Also during growth, organisms produce and excrete metabolic wastes which may reach concentrations that are toxic. It is an interesting general point that as a consequence of growth, the environment becomes less and less favorable for further growth to occur; the growth rate decreases, and eventually cell division ceases and the maximum stationary phase is established.

The Stationary Phase

In this phase of the growth curve, the population has reached the maximal level that the closed environment permits. This period may last for hours or even days. Once the stationary phase has been attained, organisms are no longer involved with the problems of growth and division, but are now faced with the more immediate problem of survival. Energy is required for this survival process—for the maintenance of proper osmotic pressures, for motility, for the resynthesis of proteins and repair of DNA that are constantly being broken down during the stationary period, and so on. This is termed the **energy of maintenance.**

If growth has ceased as a result of the complete exhaustion of the carbon and energy source in the medium, where does the cell obtain energy for maintenance? Most cells accumulate storage products during exponential growth. In liver cells, for example, the storage product is glycogen; in plant cells, it is starch; and in microbial cells, it may be glycogen, starch, or other organic storage materials. Storage products are of large molecular size and are usually stored in the form of insoluble granules. Once cells reach the stationary phase, these storage materials are degraded to obtain the essential energy for maintenance. When these storage products have been consumed, the cell is faced with the necessity of degrading other cellular components to obtain energy for survival. It is apparent that this destruction of cellular constituents cannot go on for long without having deleterious effects. Death will soon ensue.

The Death Phase

In this period of the growth cycle, the number of viable cells, that is, the cells capable of giving rise to progeny, decreases sharply. Populations contain individuals (presumably mutants) having varying degrees of resistance. Those creatures most susceptible to the adverse environment will, of course, die first; the more resistant ones will have a prolonged viability. It is often observed that after the majority of the population has died, the rate of death decreases, and a small number of survivors may persist for many months or even years. This may be due to cannibalism. Those cells that have survived grow and divide at the expense of the nutrients released from the decomposition of the dead cells.

As discussed in Chapter 4, some microbial cells produce resistant forms known as **spores.** One of the outstanding features of spores is their tremendous resistance to adverse environmental conditions. Viable spores, for example, have been isolated from mummies buried in Egypt more than 2000 years ago.

8.4 Diauxic Growth and the Regulation of Enzyme Synthesis

In 1941, Jacques Monod of the Pasteur Institute in Paris published a monograph dealing with studies on bacterial growth (Figure 8.4). The last sections of this work describe experiments in which growth was measured in media containing mixtures of two sugars. Monod observed that in the presence of certain sugar mixtures (e.g.,

Figure 8.4
A portrait of Jacques Monod (1910–1976), painted by his father. (Courtesy of I. Monod.)

glucose and lactose), two distinct exponential growth phases were obtained, separated by a short lag phase (Figure 8.5). He termed this phenomenon **diauxic growth** (double growth).

Monod performed a few simple experiments in order to understand the nature of diauxic growth. In the first series of experiments, he measured the extent of growth when the relative amounts of glucose and lactose in the media were varied. The results were clear: The number of bacteria at the end of the first growth period was directly proportional to the quantitiy of glucose, whereas the increase in *E. coli* during the second growth period depended on the amount of lactose. If, for example, the amount of glucose was tripled, then essentially three times as much growth was observed in the first phase. These experiments provided circumstantial evidence that during the initial exponential growth phase, only glucose was being used as a nutrient.

Figure 8.5
The diauxic growth of *E. coli* on a defined medium containing a mixture of the two sugars glucose and lactose.

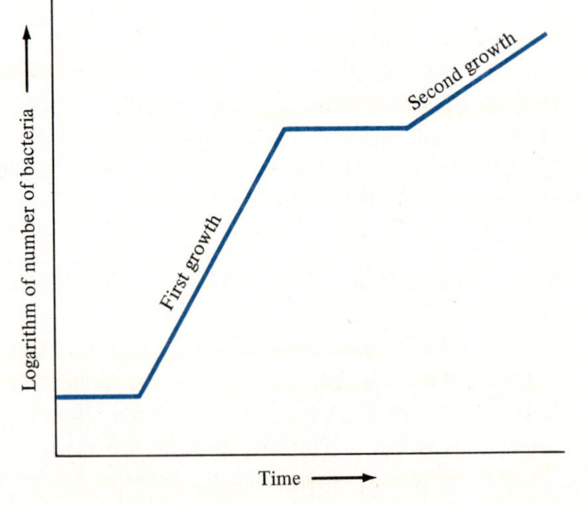

Next, Monod provided direct proof for his hypothesis by measuring chemically the amounts of glucose and lactose in the medium throughout the diauxic growth experiment (Figure 8.6). Glucose was used exclusively during the first growth phase; lactose was metabolized only after all the glucose was consumed.

Before Monod's discoveries, biochemists had shown that lactose utilization by *E. coli* required a specific enzyme called **β-galactosidase.** This enzyme functions by splitting the disaccharide lactose into its constituent monosaccharides, glucose and galactose. When Monod measured β-galactosidase in *E. coli* during diauxic growth, he found that this enzyme activity was very low during the first growth period. However, just prior to the second growth phase, the β-galactosidase activity increased several hundred times. Thus, *E. coli* is able to **regulate the synthesis of enzymes.** Somehow it is able to "turn on" and "turn off" the production of enzymes.

It should be emphasized that the appearance of β-galactosidase was not a result of mutation and selection of a new cell type; rather it was a temporary change in the expression of existing genetic potential. That is, no new genetic information had been introduced into the population; a change in the environment induced the expression of pre-existing genes. This implies that cells are able to regulate, within limits, their enzymatic constitution in response to changes in the environment. The number of enzymes that a cell can produce, of course, is dictated by its genetic potential; that is, a cell can elicit an enzyme only if it has the genetic information requisite for the structure of the specific enzyme protein. The capacity to alter its enzymatic constitution in response to environmental changes is of profound significance to the survival of cells. Microbial cells are in direct contact with their environment, which may undergo drastic fluctuation in composition. Let us consider, for example, an *E. coli* cell growing on glucose in the soil. Suddenly the cell is swept into a new environment

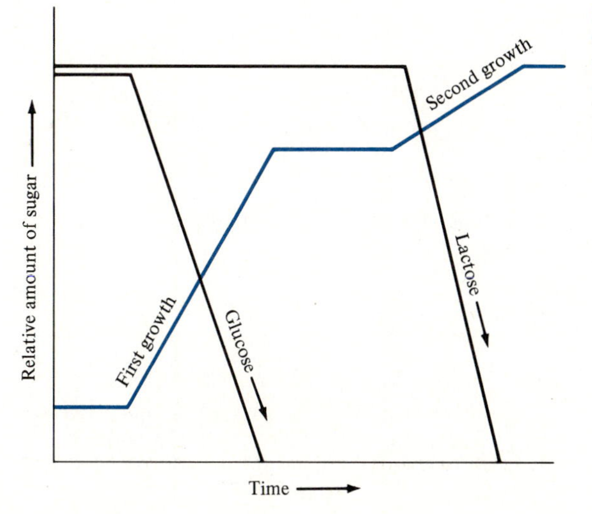

Figure 8.6
Changes in glucose and lactose quantities during diauxic growth of *E. coli*. The diauxic growth is shown as a blue line.

in which the major source of carbon and energy is lactose. The cell responds to this change in nutrients by inducing the synthesis of β-galactosidase. This economical arrangement means that bacteria do not squander their materials or energy in synthesizing enzymes when they are not needed.

These pioneering experiments by Monod provided the impetus for research on the regulation of enzyme synthesis for which he later (1965) shared the Nobel Prize with two of his colleagues at the Pasteur Institute, François Jacob and Andre Lwoff.

The two basic mechanisms by which cells regulate the production of enzymes, **induction** and **repression,** are discussed in the next two sections.

8.5 Regulation of Enzyme Synthesis: Induction

Enzyme induction may be defined as the increased production of an enzyme by cells in response to the presence of specific chemicals in the environment. Such enzymes are referred to as **inducible enzymes,** and the chemicals that induce the enzyme are called **inducers.** The best studied example of an inducible enzyme is β-galactosidase in *E. coli*. A simple experiment demonstrating the induction of β-galactosidase is presented in Figure 8.7. Inducer is added to a culture of *E. coli* growing in minimal medium with glycerol* as the carbon source. At a later time, the inducer is removed. The β-galactosidase activity is measured at different times during the experiment.

Several important conclusions may be deduced from this experiment: (1) Before the addition of inducer, the level of enzyme in the culture is very low (basal level); (2) shortly after addition of the in-

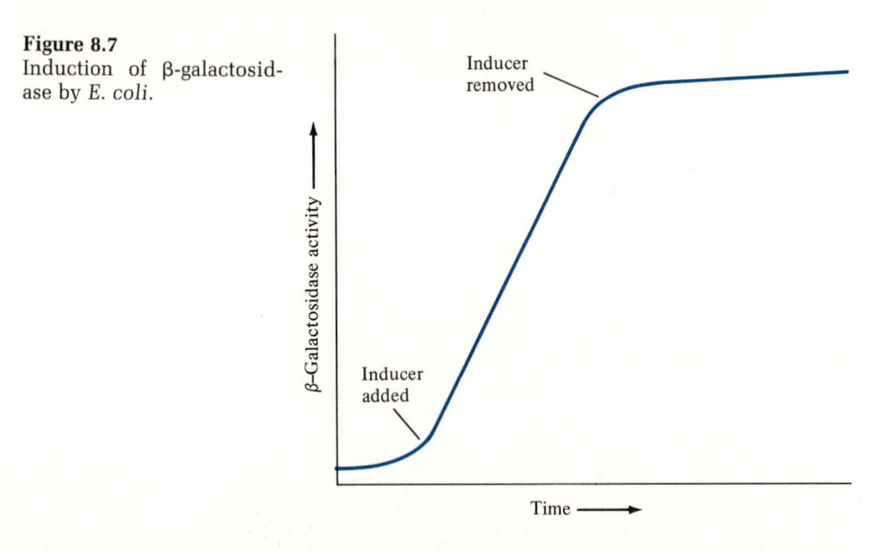

Figure 8.7
Induction of β-galactosidase by *E. coli*.

*Glucose cannot be used as the carbon source because it prevents β-galactosidase production as seen in the diauxic growth experiment. This special "glucose effect" is discussed in Section 8.6.

ducer, the amount of enzyme activity increases dramatically and continues to rise at a constant rate of as much as 1000 times the uninduced value as long as the inducer is present; (3) after removal of the inducer, no further increase in enzyme activity is observed—the enzyme level remains constant. This type of experiment describes in general terms the induction process, but provides little information about the mechanism, that is, the precise molecular basis of induction.

How is it possible to regulate the production of an enzyme protein? From our current knowledge of protein biosynthesis, it is possible to suggest three different steps where enzyme synthesis might be regulated: at the level of transcription (synthesis of mRNA from the DNA template); at the translational level (synthesis of polypeptide chains on the ribosomes); or at the post-translational level (conversion of an inactive protein into an active enzyme).

The Precursor for β-Galactosidase

Many attempts were made to resolve the question of the site of regulation of β-galactosidase synthesis. The following data provided circumstantial evidence that the regulation was not at the activation of an inactive protein.

1. Energy was required after addition of inducer before β-galactosidase activity appeared. For example, in the presence of certain compounds such as sodium azide, cyanide, and dinitrophenol, which inhibit ATP generation, β-galactosidase was not formed.
2. Cells starved of amino acids failed to be induced for β-galactosidase.
3. The drug chloramphenicol, which specifically inhibits protein synthesis, also inhibits the induction of the enzyme.

Since energy, as well as amino acids, was required for enzyme induction, and since chloramphenicol also prevented β-galactosidase synthesis, these experiments suggested but did not prove that enzyme synthesis was a result of de novo protein synthesis (synthesis of protein from their component amino acids). In 1955 this question was answered conclusively by decisive experiments carried out independently at the Pasteur Institute and at the University of Illinois. *Escherichia coli* was grown in a minimal medium in the presence of radioactive sulfur (^{35}S) but in the absence of inducer. As a result of growth in the presence of ^{35}S, cellular proteins became radioactively labeled. After removal of all the ^{35}S that was not incorporated into proteins, the culture was transferred to a medium containing inducer and nonradioactive sulfur (^{32}S), and growth was permitted to continue. The β-galactosidase was isolated and purified, and the level of radioactivity in the protein was determined. The β-galactosidase contained insignificant levels of radioactivity. If the enzyme had been synthesized from pre-existing protein precursors, the purified β-galactosidase would have been radioactively labeled, because the precursors were labeled with ^{35}S during the initial part of the experiment. Since the enzyme was not found to contain radioactive ^{35}S, this implied that the β-galactosidase was synthesized exclusively from materials assimilated after the addition of inducer. Hence, proteins present in the non-induced cells did not serve as precursors of the active enzyme. In summary, induced enzymes are synthesized completely de novo from their respective amino acids; control of induced enzyme synthesis must therefore reside at the level of either transcription or translation.

Galactose Permease and Galactose Transacetylase

An important clue to the site of regulation of β-galactosidase came from a comparative study of inducer molecules. Certain compounds, such as lactose, acted as both inducers and substrates for the enzyme; initially lactose induced the β-galactosidase, later it was broken down by the enzyme. Other compounds, called **gratuitous inducers,** behaved strictly as inducers and were not metabolized by the enzyme. This latter point was of crucial significance because it led to the discovery of a second protein involved in lactose utilization by *E. coli.* When radioactively labeled gratuitous inducers were added to a fully induced culture, there was a rapid accumulation of the labeled inducer in the cells. Bacteria not previously induced did not accumulate the compound. These, and many genetic experiments, led to the conclusion that the accumulation of inducer was caused by the action of an enzyme protein, distinct from β-galactosidase, which they called **galactoside permease.** Permeases, in general, function in the transport and concentration of specific compounds from the medium into the cell. Galactoside permease was found to be inducible. Furthermore, its synthesis was controlled by the same inducers as β-galactosidase. It should be stressed that the discovery of the permease was greatly facilitated by the use of gratuitous inducers. It is difficult to measure the accumulation in cells of inducers such as lactose that are rapidly broken down by the induced enzyme.

A third protein, **galactoside transacetylase,** was later discovered in Paris. This protein was also found to be inducible, and its activity appeared almost simultaneously with the activities for β-galactosidase and permease. Thus, three independent and distinct proteins are synthesized when a single inducer is added to a culture of *E. coli*.

Genes to Control Genes

Investigations using the refined tools of modern biochemical genetics furnished another major clue to the mechanism of induction. The Paris group under the leadership of François Jacob and Jacques Monod isolated a large number of mutants of *E. coli* that were deficient in some aspect of lactose metabolism. Characterization of these mutants revealed the existence of **three distinct genes** involved in lactose utilization: The z gene determines the structure of the β-galactosidase protein; the *y* gene determines the structure of β-galactoside permease; and the *a* gene determines the structure of transacetylase. Mutants with all possible combinations of genotypes were obtained. For example, $z^+y^+a^+$ cells possess the genetic information for the synthesis of all three enzyme proteins upon induction (β-galactosidase, permease, and transacetylase); $z^-y^+a^+$ cells have the genetic information for making permease and transacetylase, but not β-galactosidase.

In addition to mutants lacking the potential for producing one or more of the enzymes, they isolated another class of mutants called **constitutives** for β-galactosidase; that is, the enzyme was synthesized in the *absence* of inducer. Surprisingly, these constitutive mutants synthesized not only β-galactosidase in inducer-free medium but also permease and acetylase, even though the structure of these three different proteins was specified by three distinct genes. This observation led to the suggestion that the mutation to constitutivity had taken place in still another gene (*i*) distinct from the other three (*z*, *y*, and *a*). Mutation of the *i* gene from i^+ to i^- allowed the **expression** of z^+, y^+, and a^+ genes in the absence of inducer. This was a striking phenomenon without precedence in biology—one genetic element regulating the expression of three other genes.

All bacterial genes described prior to this time functioned in determining the amino acid sequence of a protein (structural genes). Since proteins are relatively easy to isolate and characterize, structural genes are therefore amenable to study. By investigating the alteration in the protein, one could study the end result of a mutation. But how does one study a regulatory gene such as the *i* gene? The mechanism of bacterial conjugation, which had just been clarified by Jacob and Wollman in Paris and Hayes in London (see Section 6.4), provided an experimental tool for these studies. It was known that in bacterial conjugation, the DNA of the male (Hfr) bacterium was injected into the female (F^-) recipient without any cytoplasmic exchange. The zygote that is formed thus contains genes from both parents, but the cytoplasm is entirely female. In 1959 Arthur Pardee, along with Jacob and Monod, performed an ingenious set of experiments designed to investigate the expression of the i^+ and z^+ genes when injected into a female bearing z^-i^- genes. Basically, the question was whether the i^+ gene product could exert some controlling effect on the expression of the *z* gene. This famous experiment is affectionately called the PaJaMo experiment in reference to the authors. The formation of β-galactosidase by zygotes was measured.

Two bacterial mating experiments were performed:

Experiment A: Hfr $(z^+i^+) \times F^-(z^-i^-)$
Experiment B: Hfr $(z^-i^-) \times F^-(z^+i^+)$

None of the four parental types could synthesize the enzyme in the **absence of inducer** because the potential to produce an active β-galactosidase (z^+) was always accompanied with inducibility (i^+). However, the zygotes (z^+i^+/z^-i^-) formed in experiment A did synthesize the enzyme for about two hours after DNA injection in the absence of inducer (constitutive synthesis). In experiment B, no significant amount of β-galactosidase was detected at any time after mating in the absence of inducer. These results are summarized in Figure 8.8.

The zygotes formed in mating experiments A and B were genetically identical (z^+i^+/z^-i^-), yet zygotes from mating A produced enzymes constitutively, whereas zygotes from mating B did not! The difference must therefore reside in the cytoplasm of the female recipient. Somehow the i^+ gene produced a cytoplasmic substance that inhibited the expression of the z^+ gene (experiment B), whereas in experiment A, the female cytoplasm lacked this hypothetical substance because it was i^-. This interpretation suggested that the i^+ inducible gene was dominant and that the i^- was recessive. This predicts that an i^- cytoplasm, lacking the cytoplasmic substance, should become inducible after receiving the i^+ gene from a male, but only after the i^+ gene has had an opportunity to be expressed (that is, transcribed and translated in the zygote). Under the conditions in which the PaJaMo experiment was performed, it took approximately two hours for the i^+ gene product to turn off β-galactosidase synthesis (Figure 8.8).

In summary, the i^+ gene directs the synthesis of a cytoplasmic substance that prevents the production of β-galactosidase (permease and transacetylase) unless inducer is added to the medium. In i^- mutants, no such substance is produced (or if produced, it is not functional), and the β-galactosidase (permease and transacetylase) is formed even in the absence of inducer. Pardee, Jacob, and Monod proposed that (1) the i^+ gene product was a specific **repressor** that inhibited the synthesis of β-galactosidase; and (2) inducers counteract the action of the repressor, permitting enzyme formation. Since constitutive mutants did not make the repressor, inducer was not necessary for enzyme formation in i^- mutants.

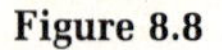

Figure 8.8
Production of β-galactosidase in the absence of inducers immediately following mating of z^+i^+ and z^-i^- parents.

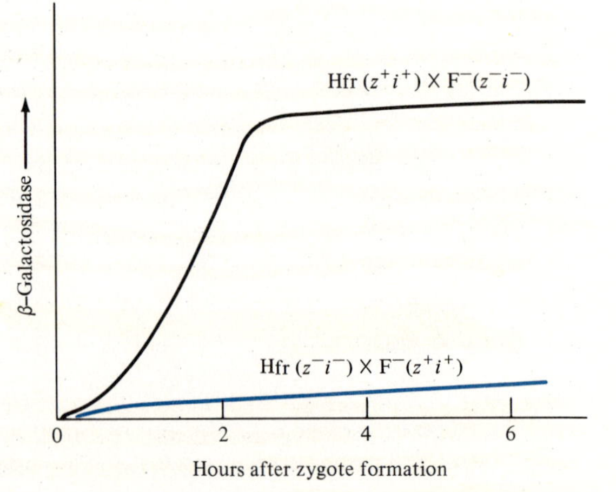

The Operon Model

As a consequence of these experiments, Jacob and Monod in 1961 proposed the **operon model** for the regulation of gene expression (Figure 8.9). According to this model and subsequent modifications, the DNA information for the three structural genes (z^+, y^+, a^+) is transcribed into a single mRNA, starting at a site on the DNA called the **promoter**†. Translation of this mRNA then takes place on the ribosome, resulting in the three enzyme protein products. The i^+ gene exerts its effect on this system by directing the synthesis of a repressor which acts on the DNA at a site called the **lactose operator** (*o*) region. The reaction between repressor and lactose operator region prevents transcription by blocking the RNA polymerase from reaching the promoter site. Inducers somehow prevent the repressor from combining with the operator region, thus permitting mRNA synthesis. In essence, reaction of the repressor with the operator prevents transcription. Inducers function by interfering with the reaction between repressor and operator, thus allowing transcription to occur. The three structural genes whose expression is regulated by an operator plus the *o* region were called the lac operon; thus the name, the operon model.

The operon model makes several predictions. It predicts that (1) the repressor is the product of the i^+ gene; (2) the repressor can combine with the operator region (*o*) of the lac operon; (3) the repressor must also react with the inducer (for mRNA synthesis to take place the inducer must be able to displace that repressor which is in combination with the operator region); (4) when the repressor is in combination with the operator, no mRNA synthesis for enzymes of the lactose operon will take place; (5) when in combination with the inducer, the repressor is inactive in interfering with the synthesis of lac operon mRNA and thus enzyme formation can take place.

Before the operon model could be experimentally verified, the repressor substance had to be isolated. In December 1966, Walter Gilbert and Benno Muller-Hill of Harvard University reported the isolation of the lac repressor. The repressor was found to be a protein. Verification of these predictions soon followed.

Evidence that the i^+ gene product was the repressor was obtained using i^- mutants that either manufactured faulty repressor or failed to produce repressor completely. Repressor could be isolated from i^+ strains but not from the i^- mutants. The first prediction was thus verified—the i^+ gene product is the repressor.

DNA from *E. coli* having the normal lac operon was isolated and purified. The lac repressor protein was shown to combine with this DNA. On the other hand, DNA from cells that did not possess the lac operon did not bind the lac repressor. Furthermore, DNA isolated

†The transcribing enzyme, RNA polymerase, binds to a site on the DNA called the **promoter,** and transcribes sequentially all the genes of the operon. If the promoter is blocked, RNA polymerase cannot bind to the DNA and no transcription will occur.

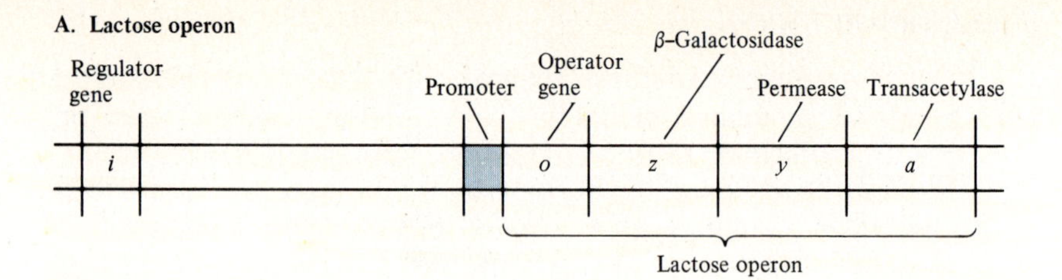

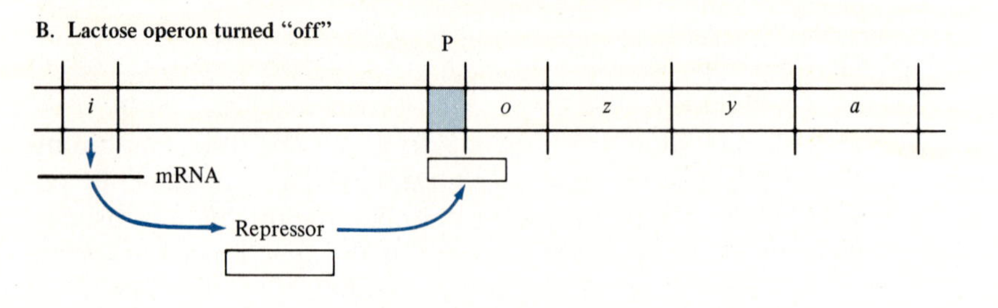

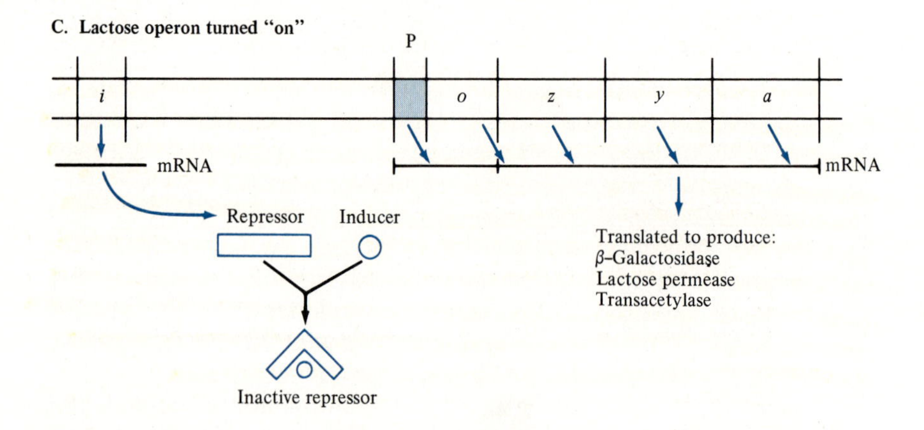

Figure 8.9
The operon model. (A) The lactose operon contains the operator gene, the promoter, and the three structural genes for β-galactosidase, permease, and transacetylase. A regulator gene, *i*, controls the operon. (B) The operon is turned "off" when the repressor (produced by transcription and translation of the *i* gene) binds to the operator gene and prevents transcription at the promoter site. (C) The operon is turned "on" when the inducer binds to the repressor, causing it to be inactive.

from cells having a mutation at the operator region of the lac operon failed to bind the repressor. Thus it was shown that the repressor not only combined with DNA but also appeared to combine specifically with the operator region of the lac operon.

The initial isolation of the lac repressor was based on the prediction that it had to react with inducers of the lac operon. Gilbert and Muller-Hill isolated a protein from an inducible strain of *E. coli* that

combined with radioactive inducer. Moreover, they showed that adding inducer to the repressor–DNA complex released the repressor from the DNA.

In 1968 a research group at the University of Wisconsin reported that isolated repressor inhibited mRNA synthesis in vitro. Presumably, inhibition of mRNA synthesis by the isolated repressor took place by blocking the RNA polymerase from transcribing the DNA at the promoter site.

Several independent lines of evidence have demonstrated that addition of inducer to an i^+ cell resulted in the synthesis of mRNA for the lac operon. This mRNA was not made unless the inducer was added.

It is apparent that all the predictions derived from the operon model have been experimentally verified. Although this discussion has been restricted specifically to β-galactosidase and the lac operon, investigations into the mechanism of induction of other enzymes reveal that the operon model may well be a general mechanism for enzyme regulation. The concept that regulatory genes, such as the *i* gene, function by directing the synthesis of repressors that regulate DNA transcription in response to environmental changes is probably the underlying mechanism of enzyme regulation.

8.6 Regulation of Enzyme Synthesis: Repression

A regulatory mechanism of even greater economic significance to the cell than induction is that of **repression**—the effective **inhibition of formation of enzymes** of a biosynthetic pathway by the end product of that pathway. Repression is the reverse of induction; enzyme synthesis is inhibited rather than induced. The reactions affected by repression are usually those that cells use to build up cell materials rather than those that digest foodstuffs. For example, *E. coli,* when growing in a minimal medium with glucose as the carbon and energy source, must produce all of the enzymes involved in the biosynthesis of the amino acid isoleucine. However, if isoleucine is added to the growth medium, *none* of the enzymes responsible for the synthesis of this amino acid is produced. The formation of enzymes of other biosynthetic pathways is not affected. In other words, the presence of isoleucine in the medium specifically represses the formation of its own biosynthetic enzymes.

The phenomenon of repression has been observed in numerous biosynthetic pathways in bacteria. Regulation of the enzymes of the histidine biosynthetic pathway has been extensively investigated. Histidine is an amino acid found in many proteins; it is therefore essential to the cell. The biosynthesis of histidine requires 11 enzymatically catalyzed reactions. Because two of the enzymes carry out two steps in the pathway, nine structural genes are involved in the formation of these enzymes. These biosynthetic enzymes function in harmony; the product of one enzyme serves as the substrate for the next enzyme in the pathway, and so on, until histidine is formed. Histidine is then used for the biosynthesis of proteins.

Structural genes for enzymes of several biosynthetic pathways have been found in clusters on the chromosome; each cluster has its own operator region. The genes determining the structure of the enzymes in the histidine pathway are also clustered on the chromosome of some bacteria; that is, they lie adjacent to each other on the DNA. All nine histidine structural genes are under the control of one operator region. Thus we have a *his*tidine operon similar to the lac operon. The various genes in the operon are transcribed in a coordinated fashion. For example, bacteria make slightly more histidine than is required for protein synthesis. If, however, the intracellular concentration of histidine becomes limiting, all of the enzymes in the histidine biosynthetic pathway are produced in increased and equal quantities **(coordinate depression).** The ultimate effect is a harmonious and coordinated increase in all of the enzymes of the biosynthetic pathway, resulting in an increased rate of histidine production to meet the increased demand. When the synthesis of histidine is no longer required, either because of reduced demands or because it is supplied to the culture, the formation of *all* of the enzymes in the pathway is repressed coordinately **(coordinate repression).** The cell makes use preferentially of nutrients supplied to it rather than expend the energy and the carbon required to make them. It is unnecessary for the cell to manufacture the numerous enzymes in a biosynthetic pathway if it already contains enough of the final product; it can channel this conserved energy for other uses. Thus, regulation of enzyme synthesis by repressing the formation of biosynthetic enzymes ensures the cell of balanced amounts of precursor molecules for macromolecular synthesis.

The molecular mechanism of repression has been conceived as a modification of the mechanism described for induction. As an example, consider the biosynthesis of a hypothetical amino acid X (Figure 8.10). A repressor protein is produced by the regulatory gene of the biosynthetic operon (similar to the *i* gene of the lac operon). In its native state, this repressor substance is inactive. The inactive repressor has no effect on the transcription of the structural genes of the operon since it cannot combine with the operator region of the operon. However, when the end product of the biosynthetic pathway (amino acid X) is present in excess, it combines with the inactive repressor and changes it to an active repressor. The active repressor can then combine with the operator region and prevent transcription of the structural genes of the operon.

Repressors, then, may exist in either active or inactive form. In inducible systems, such as the lac operon, combination of the repressor with the inducer renders the repressor inactive and non-functional; in repressible systems, such as the histidine operon, combination of the inactive repressor with histidine results in an active, functional repressor.

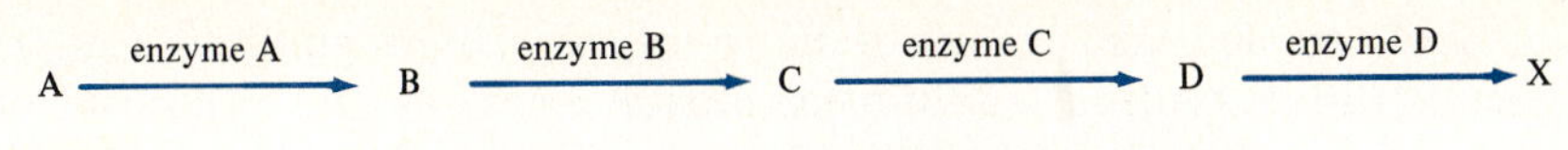

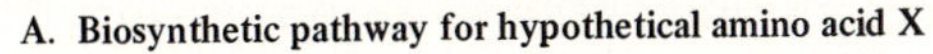

A. Biosynthetic pathway for hypothetical amino acid X

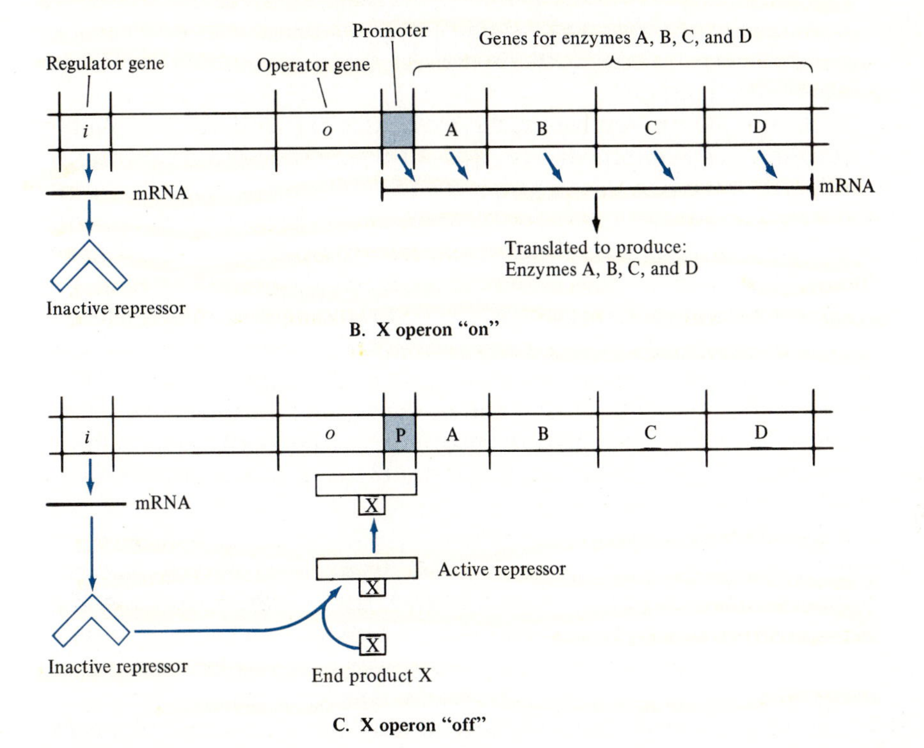

Figure 8.10
Repression of enzyme formation by the end product of a biosynthetic pathway. (A) A hypothetical end product, amino acid X, is made from compound A by a sequence of reactions, each one catalyzed by a specific enzyme. (B) In the absence of amino acid X, the operon is turned on because the repressor protein is not, by itself, active. (C) When amino acid X is in excess it binds to the repressor, resulting in the formation of an active repressor; the active repressor binds to the operator, preventing transcription at the promoter and thereby blocking production of enzymes A, B, C, and D.

Catabolite Repression and Cyclic AMP

In addition to repression of specific enzyme synthesis by the end product of a biosynthetic pathway, an even more general type of repression exists in virtually all cells, called **catabolite repression.** First described by Monod to explain his diauxic growth experiments (review Figure 8.5), catabolite repression overrides induction. During growth of *E. coli* on glucose plus lactose, β-galactosidase is *not* induced until the glucose is consumed. This inhibition of induction of the lac operon was initially called the "glucose effect." Subsequently, it was shown that the effect was not restricted to glucose; it occurred with different bacteria when they were growing on energy sources that they could rapidly metabolize. For example, *Pseudomonas putida* metabolizes sodium succinate more rapidly than glucose. Consequently, in *P. putida*, sodium succinate represses the synthesis of several inducible enzymes *even in the presence of the inducer*. Rather than call the phenomenon the "succinate effect" in *P. putida*, the "glucose effect" in *E. coli*, and so on, this type of general inhibition was termed catabolite repression because it was assumed that the repression was due to the accumulation of some common breakdown product (catabolite).

During the last few years, the molecular mechanism of catabolite repression has been uncovered. First, it was shown that the extent of catabolite repression is directly proportional to the concentration of ATP in the cell; the higher the intracellular concentration of ATP, the stronger the repression. What determines the ATP concentration in the cell? Clearly the level of ATP is determined by the relative rates at which it is produced (by respiration or fermentation) and consumed (for the growth and maintenance of the cell). In the presence of rapidly metabolizing energy sources (e.g., glucose in *E. coli*), ATP is generated more rapidly than it can be consumed, thus reaching a high intracellular level. At the other extreme, during starvation, little or no ATP is produced, and consequently the ATP concentration drops precipitously.

But how do high concentrations of ATP cause catabolite repression? For reasons that are still unclear, when the ATP concentration goes up, the concentration of a special regulatory molecule, called **cyclic AMP,** goes down. Conversely, when the level of ATP drops, the level of cyclic AMP goes up. What makes cyclic AMP so special is that its only function in the cell is to regulate enzymes. In the case of inducible operons, such as the lac operon, cyclic AMP is necessary for induction. The cyclic AMP binds the catabolite activator protein (CAP, for short) to form a cyclic AMP–CAP complex. Only when this complex is bound near promoter sites can the RNA polymerase efficiently transcribe the adjacent operons. Figure 8.11 summarizes our current understanding of catabolite repression.

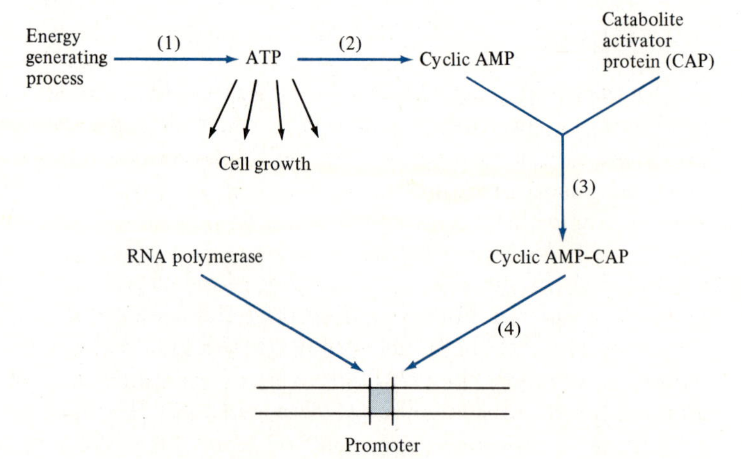

Figure 8.11
The mechanism of catabolite repression. (1) The cell generates ATP from a metabolizable energy source; (2) cyclic AMP is produced from ATP but *only when the ATP level is low;* (3) cyclic AMP binds to the catabolite activator protein (CAP), forming a cyclic AMP–CAP complex, which (4) binds to DNA near promoter sites and stimulates (5) RNA polymerase to transcribe the adjacent operons.

Now let us return to the diauxic growth experiment and see if we can explain the phenomenon in terms of both catabolite repression and induction of enzyme synthesis. During the early stages of growth, when glucose is being metabolized, the cells generate ATP rapidly, causing the level of cyclic AMP to fall. Without cyclic AMP, the CAP cannot bind to the lac promoter, so transcription of the lac operon does not occur, even though the inducer is present. Without β-galactosidase to break down the lactose, *E. coli* grows exclusively on glucose during the first growth phase. When the glucose is finally consumed, there is a short lag period during which time the cells induce the formation of β-galactosidase, lactose permease, and the transacetylase. These enzymes can now be induced because (1) the inducer is present and (2) the cell is free of catabolite repression. The latter occurs because the glucose level drops, causing the ATP level to fall and the cyclic AMP concentration to rise, thereby allowing for the formation of cyclic AMP–CAP, which in turn allows for efficient transcription of the lac operon. The presence of the inducer, lactose, causes inactivation of the lac repressor, so that all systems are "go" for production of enzymes of the lac operon and the subsequent second growth phase.

8.7 Regulation of Enzyme Activity: Feedback Inhibition

We have discussed cellular mechanisms of regulation that function by turning on or turning off the transcription process. These mechanisms permit the formation of enzymes when they are required (induction) or prevent enzyme formation when they are no longer necessary (repression). A third regulatory mechanism that has been extensively studied is called **feedback inhibition.** This control device functions by an **inhibition of the activity** (not the formation) of enzymes. Usually the first enzyme in a biosynthetic pathway is feedback inhibited by the end product of that pathway. Thus it is complementary to the repression of enzyme formation previously discussed.

One of the earliest examples of feedback inhibition was reported in 1953 by H. Edwin Umbarger working at the Long Island Biological Laboratories in New York. Umbarger found that when *E. coli* was grown on a minimal medium with glucose as the carbon and energy source, the organism produced the amino acid, isoleucine, by a series of three reactions schematically represented in Figure 8.12. If, however, excess isoleucine was fed to the culture, the bacteria preferentially used the supplied amino acid and *immediately* ceased to produce their own.

Feedback inhibition is a remarkably efficient and economical control device for the cell. The *instant* that the cellular concentration of isoleucine reaches an adequate level, the cell *immediately* stops making more of it. The specificity of the system became apparent when Umbarger demonstrated that the first enzyme in the biosynthetic pathway (enzyme A), and *only* the first, was inhibited by the end product. Since, as pointed out previously, biosynthetic enzymes function

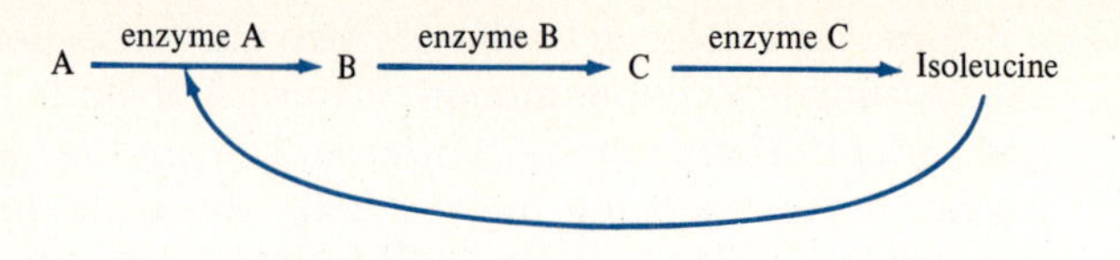

Figure 8.12
Feedback inhibition of isoleucine biosynthesis. The end product of the pathway, the amino acid isoleucine, inhibits the activity of the first enzyme of the pathway.

in tandem (one enzyme forming a product that becomes the substrate of the next enzyme in the pathway, and so on) inhibition of the first enzyme effectively shuts off the activities of all the enzymes in the sequence. Moreover, the inhibition of the first enzyme in the pathway by the end product need not be absolute. Rather, it may be a partial inhibition depending on the relative level of end product present (Figure 8.13). This fine control of end product formation is crucial for the harmonious functioning of the cellular machinery, since the rate of isoleucine utilization (or other end products) for protein synthesis may vary from time to time. Only the end product of the pathway exerts the inhibitory effect on the first enzyme of the pathway. Even compounds very closely related to isoleucine have no effect.

Feedback inhibition has been demonstrated for many different biosynthetic pathways; several amino acids, vitamins, and the purine and pyrimidine biosynthetic pathways have been shown to be regulated by feedback inhibition. It appears to be a general mechanism for regulating the concentration of intracellular building blocks.

8.8 Allosteric Interactions: A Unifying Concept

The molecular mechanism of regulation of enzyme activities, by either negative control (inhibition) or positive control (activation), has been investigated in several laboratories in recent years. It has long been known that enzymes participate in biological reactions by combining with their specific substrates. A specific site on the enzyme, called the **active site,** is responsible for this union with sub-

Figure 8.13
Inhibition of the first enzyme in the isoleucine biosynthetic pathway by varying concentrations of isoleucine. As the concentration of isoleucine increases, progressively greater inhibition is observed until finally enzyme A is completely turned off.

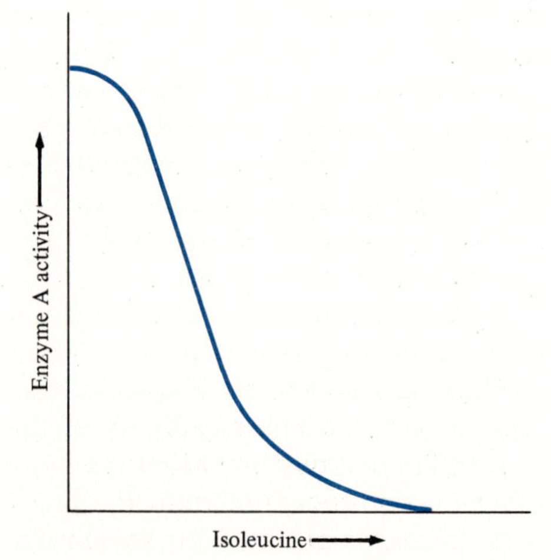

strate. In 1962 John Gerhart and Arthur Pardee at the University of California were investigating the feedback inhibition of the first enzyme in the pathway of pyrimidine biosynthesis. They discovered that the isolated and purified enzyme remained susceptible to inhibition by the end product of the pathway. Gerhart and Pardee learned that their purified enzyme not only had an active site for combining with its substrate, but also had a different site, called the **regulator site,** for combining with the feedback inhibitor. They demonstrated that these two combining sites were quite distinct; abolishing the regulator site for the inhibitor had no effect on the active site for the substrate. The two sites, although on separate parts of the enzyme, could exert their effects on each other; that is, they were functionally linked. To explain this phenomenon, Monod, Jean-Pierre Changeux, and Jacob proposed the hypothesis of **allosteric interactions.** They suggested that a class of proteins **(allosteric proteins)** existed whose three-dimensional structure was altered when combined with certain compounds (inhibitors or activators). In this altered structural form, the enzymatic activity of the proteins was either inhibited or activated. This combination was called **allosteric interaction** (Figure 8.14). For example, during feedback inhibition of an enzyme, it is presumed that the inhibitor combines with the enzyme at its regulator site and changes the appearance of the enzyme. The active site for the substrate is then no longer a perfect "fit" (Figure 8.15B). This alteration in the active site would reduce the activity of the enzyme since it could no longer bind to the substrate at a normal rate. Conversely, in the case of positive control, where enzymes are activated, the combination of a small molecule with the enzyme would change the appearance of the enzyme so that the "fitness" for the substrate is improved (Figure 8.14C).

Allosteric interactions may also be invoked to explain the activation or inactivation of repressors by inducers or end products. The inducer would then act by altering the three-dimensional appearance of the repressor so that the combining site for the operator would be changed. The end product would, on combining with an inactive repressor, improve the "fitness" of the combining site for the operator region.

It should be evident from the foregoing discussion that cells have developed a variety of regulatory mechanisms for coping with changes in their environment. The complex process of growth and reproduction requires highly integrated systems for the production of balanced levels of building blocks for the synthesis of cellular macromolecules. An imbalance of these levels is obviously detrimental. Underproduction of a precursor results in a reduction of macromolecular synthesis, which would result in a reduction in the growth rate; overproduction of a precursor is wasteful since the cell cannot accommodate high concentrations of building blocks and thus excretes them.

The integrated operation of repression and feedback inhibition furnishes microorganisms with a mechanism for adjusting the levels of building blocks in relation to cellular needs and compensates for

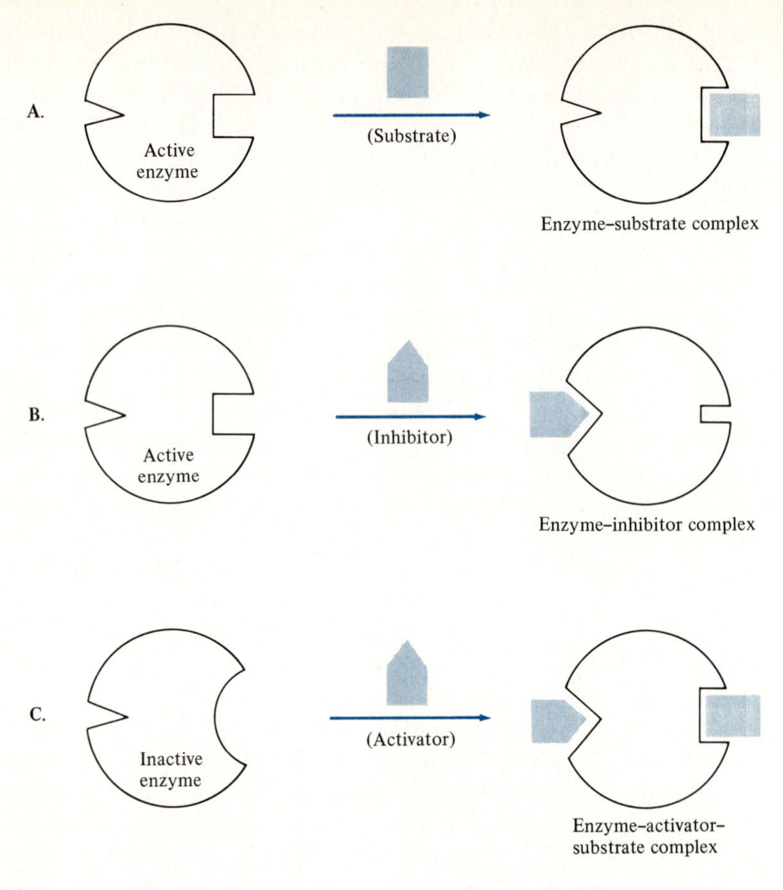

Figure 8.14
Schematic representation of allosteric interactions. (A) Normal reaction between active site of an enzyme and its substrate. (B) Inactivation of enzyme by inhibitor; the inhibitor binds to the enzyme and alters the "fitness" of the active site so that the substrate and enzyme can no longer combine. In (C), the activator improves the "fitness" of the active site for the substrate.

alterations in the environment. By reducing the levels of biosynthetic enzymes via repression, overproduction of building blocks would not be prevented immediately. The reason for this is that the existing enzymes would continue to function and the end product would continue to be made even though it was in excess or no longer needed. This would be wasteful synthesis, a crime that cells attempt not to commit. End product inhibition, on the other hand, would operate immediately, to stop cellular synthesis of the building blocks when they are present in adequate supply.

The generalized control of cellular metabolism by allosteric interactions, then, provides the cell with a mechanism for adjusting the expression of its genetic potential to contend with variations in its environment. An allosteric change in the repressor enables the cell to invoke the synthesis of inducible enzymes when their substrate suddenly appears in the medium. When that substrate has been consumed, an allosteric change in the repressor then shuts off further synthesis of these no longer needed enzymes. At the end of the exponential growth period, terminated possibly because of nutritional inadequacy of the environment, the end products of biosynthetic enzymes would accumulate unless an allosteric interaction with the inactive repressor inhibited the further synthesis of the enzymes. Moreover, it is by still another allosteric interaction with the first enzyme in the biosynthetic pathway that the end product immediately turns off its further synthesis. **It is clear therefore that the relative composition of the environment acts as a switch that triggers the expression of the genetic information of the cell by allosteric interactions.**

The problem of regulation in multicellular organisms is more complex than in bacteria. We know, for example, that in a human being there are liver cells, brain cells, kidney cells, nerve cells, all looking and behaving very differently. Yet to the best of our knowledge, all these cells have identical copies of DNA. Although the (potential) information is the same, the cells turn out to be different. It follows that there must be some sort of switch, something that turns on and off particular genes in appropriate environments. The process by which a single fertilized egg divides into two, then four, then eight, and so on, until an aggregate containing different cell types is formed, is called **differentiation.** At present the underlying control mechanisms for this remarkable process are unknown.

As we have discussed in this chapter, even single-cell organisms, such as *E. coli,* have elaborate control mechanisms. Great variation can exist between the amount of a specific enzyme present when it is needed and when the environmental conditions are such that it would serve no useful purpose. We are just beginning to understand in a few simple cases how control systems operate. It will be interesting to see whether information obtained on bacterial regulation will provide clues to the more complex problem of differentiation.

It is the nature of science that for each question answered, several new and more fundamental problems arise.

QUESTIONS

8.1 In 1969, man landed on the moon and brought back samples from its surface. Assuming that soil analyses revealed the total absence of any organic compounds and combined forms of nitrogen, what type of earthly organism would you expect to grow initially in the soil if it was not kept sterile? What two obligatory reactions would be prerequisite for these organisms to appear?

8.2 Define the following terms: autotroph, psychrophile, diauxic growth, operon, gratuitous inducer, promoter, feedback inhibition, and differentiation.

8.3 Draw a typical bacterial growth curve. Discuss those factors which determine the beginning and end of each phase.

8.4 A fresh medium is inoculated with a single *E. coli* cell. Assuming that growth is logarithmic and that the generation time is 20 minutes, how long will it be until one thousand cells are present in the culture? One million? One billion? Two billion?

8.5 A logarithmically growing culture, having a doubling time of two hours and initially containing 4×10^8 cells per milliliter, uses 4 grams of glucose in growing four hours. How many grams of glucose are used in the first two hours? (It may be useful in solving this problem to make a rough graph of the growth of the culture over the four hours.)

8.6 The following enzyme system is inducible:

$$X \xrightarrow{\text{enzyme 1}} Y \xrightarrow{\text{enzyme 2}} Z$$

Enzymes 1 and 2 are specified by the structural genes *a* and *b*, respectively. Together with their regular gene (*i*) and operator gene (*o*), they occupy the following positions on the chromosome:

i o a b

State which enzymes will be produced, with and without inducer, in zygotes prepared from the following matings:

	Hfr	×	F^-
a)	$i^+ o^+ a^+ b^-$		$i^- o^+ a^- b^+$
b)	$i^+ o^- a^- b^+$		$i^- o^+ a^+ b^-$
c)	$i^+ o^+ a^+ b^-$		$i^+ o^+ a^- b^+$
d)	$i^+ o^- a^+ b^-$		$i^+ o^+ a^+ b^-$

8.7 Explain the diauxic growth phenomenon in terms of induction and catabolite repression.

Suggested Readings

Changeux JP. The Control of Biochemical Reactions. Scientific American, 212: 36, 1965.
A discussion of repression with a more extended discussion of allosteric interactions.

Cohen GN and J Monod. Bacterial Permeases. Bacteriological Reviews, 21: 169, 1957.
An extensive discussion of bacterial permeases, lucidly written (in English).

Monod J. From Enzymatic Adaptations to Allosteric Transitions. Science, 154: 475, 1966.
The Nobel lecture delivered in Stockholm on December 11, 1965, when the author received the Nobel Prize in physiology.

Pastan I and R Perlman. Cyclic AMP. Scientific American, 227: 97, 1972.
A clear account of how cyclic AMP is involved in catabolite repression in bacteria.

Ptashne M. DNA-Operator-Repressor. Scientific American, January 1976.
Modern account of the operon model.

Umbarger HE. Intracellular Regulatory Mechanisms. Science, 145: 674, 1964.
This is an excellent account of feedback inhibition as described by the discoverer.

Environmental Microbiology

Microbes possess a wider range of physiological and biochemical potentialities than do all other organisms combined. Microbes represent forms of life that can persist in nature because they fill particular ecological niches.

C. B. van Niel

Two schools of microbiology began to develop in Europe during the latter part of the nineteenth century. One school, led by Robert Koch, emphasized the importance of studying pure cultures. The other group, influenced primarily by Sergius Winogradsky, contended that a fundamental part of microbiology is how microorganisms grow in nature and bring about changes in the environment.

The Koch school made rapid progress in the isolation and identification of disease-causing microorganisms. Working with pure cultures, they were able to prove (using Koch's criteria, Section 13.1) that specific diseases were caused by specific microorganisms. Furthermore, the study of pure cultures of disease-causing microbes led directly to the immunization procedures that have been so successful in combatting some of the major diseases of humans. What could be more important? The obvious significance of the work of Koch, Pasteur, and their followers was so great that their pure culture approach soon came to dominate the newly emerging field of microbiology and suppressed the growth of the Winogradsky school. Attempts to publish scientific articles that were not concerned with pure cultures were met with scorn. An eminent Dutch microbiologist reflected the general feeling of "serious" microbiologists of the first half of the twentieth century when he said, "The only things that can emerge from studies of mixed cultures are green molds and nonsense."

The pure culture technique not only served the needs of the early medical microbiologists, but also provided the key element for the development of biochemistry and molecular biology. All of the fundamental experiments described in previous chapters that gave rise to our current understanding of the structure and function of cells, gene replication and transcription, protein synthesis, and the regulation of growth were performed with pure cultures. The concept of working with a large number of genetically identical microbes (a pure culture) is so deeply ingrained in the minds of contemporary microbiologists that the worst insult that one scientist can give a colleague is to claim that he or she is working with an impure culture (i.e., contamination).

In Chapter 1 we discussed the usefulness of the scientific method for arriving at a solution to a particular problem. However, the scientific method has very little influence on deciding which problems are important and worth investigating. Here relevance, taste, fashion, availability of research monies, politics, and other non-objective factors play the determining role. Until recently, all these aspects favored the Koch school of pure culture. Recently, three changes have taken place that have begun to elicit a rebirth of the Winogradsky school and bring it into the mainstream of modern microbiology: (1) growing concern about environmental pollution, (2) the energy and food crises, and

(3) the development of new techniques for manipulating microorganisms and measuring their activities under natural conditions.

In this chapter we discuss the fundamental role of microbes in maintaining the balance of chemicals in nature. The enormous metabolic diversity found in the microbial world is largely responsible for the breakdown and recycling of organic and inorganic materials on earth. Furthermore, it is precisely this metabolic diversity that provides mankind with the potential to combat much of the pollution brought about by modern society and to ease the energy and food crises. Several important microbial processes occur in nature only when microbes are in continuous close association with other forms of life. Thus microbial ecology includes the interaction of microorganisms with other living organisms as well as with their abiotic environment.

9.1 The Cycles of Carbon and Oxygen

The most abundant elements in living organisms are carbon and oxygen. Together they make up approximately 80 percent of the weight of biological matter. Central to the cycling of these elements are the complementary phenomena of respiration and photosynthesis. The net result of photosynthesis is the fixation of carbon dioxide (from the atmosphere) into organic matter with the simultaneous production of oxygen gas (from water). In respiration, the oxygen gas and organic materials react to regenerate water and carbon dioxide. The simple view of the coupling of animal respiration and plant photosynthesis is then

$$\text{carbon dioxide} + \text{water} \underset{\text{animals}}{\overset{\text{green plants, light}}{\rightleftarrows}} \text{oxygen} + \text{organic material}$$

There are two major problems with this traditional way of looking at the carbon and oxygen cycles. First, on a quantitative basis, microorganisms play a much larger role than animals and plants combined with regard to both photosynthesis and respiration. Second and more important, most of the organic materials produced by photosynthetic organisms are not digestible by animals. If it were not for the incredible metabolic diversity of microbes, the bulk of plant material would accumulate as organic matter in the soil and water, rapidly depleting the atmosphere of carbon dioxide and making life impossible. Thus, microorganisms are an essential part of our global ecosystem.

Figure 9.1 is a diagrammatic representation of the carbon cycle. The figure can also be used to represent the oxygen cycle by replacing the carbon dioxide with oxygen gas and changing the direction of each of the arrows pointing to carbon dioxide. The atmosphere of earth contains 20 percent oxygen gas and 0.03 percent carbon dioxide. During photosynthesis, carbon dioxide is removed from the air and replaced by oxygen gas. Most photosynthesis on land is performed by green plants, with algae and photosynthetic bacteria playing only minor roles. In the ocean, however, algae and blue-green bacteria are the major contributors to photosynthesis. These free-floating microorganisms of the sea are collectively called **phytoplankton.** It has been

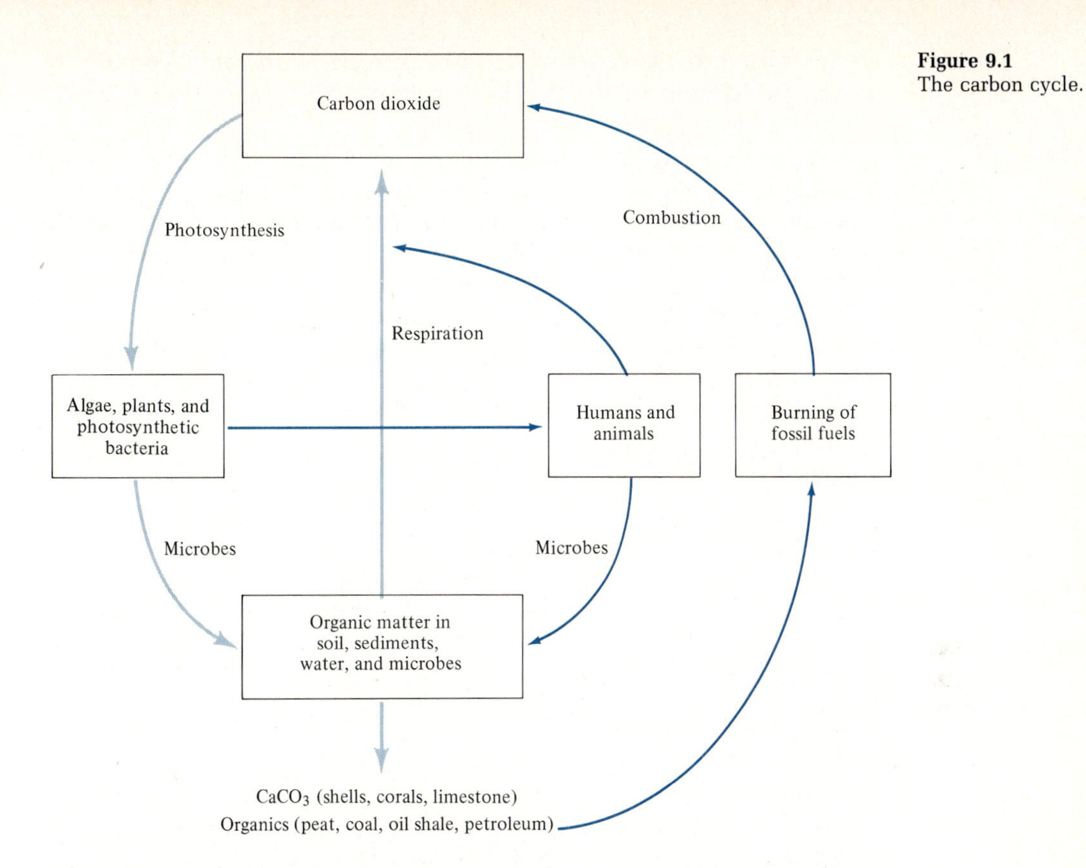

Figure 9.1
The carbon cycle.

estimated that the combined photosynthetic activities of green plants and phytoplankton are sufficient to deplete the atmosphere of all its carbon dioxide within 20 years—if it were not replenished by respiration.

Algae and plants are the primary source of food for other forms of life. Animals eat photosynthetic organisms, either directly (herbivores) or indirectly (carnivores). Part of the organic matter consumed is burned (respiration) with the production of energy (ATP) and carbon dioxide; part is used as building blocks for growth and reproduction; and by far the largest part is excreted as waste material. Keep in mind that the bulk of plant organic matter (cellulose, pectin, and lignin) is not digestible by animals.

Most of the biological conversion of organic compounds back to carbon dioxide is carried out in the soil, sediments, and water by a vast array of microorganisms. The soil is constantly being fed with organic matter in the form of litter from animals and plants, as well as eventually the dead organisms themselves. This organic material is extremely complex, containing thousands of different kinds of organic molecules. In spite of this complexity, there is not a single biologically produced organic compound that cannot be decomposed by some microbe. Although there are a few examples of bacteria that can degrade many different organic compounds, the broad metabolic potential of the microbial world is primarily the result of many different groups of microorganisms working in concert, with each type of microbe specializing in the breakdown of a particular group of compounds under a specific set of conditions.

In considering microbial ecology, one must think "small" in terms of ecological niches, but "large" in terms of numbers and varieties of microbial species. Consider, for example, what happens when a small leaf, weighing approximately 10 grams, enters the soil. Thousands of different kinds of bacteria and fungi attack the leaf; some of them break down cellulose, others digest the leaf protein, still others specialize in the breakdown of plant pigments, and so on. The final result is that the leaf has been consumed, bacteria and fungi have multiplied, and approximately 70 percent of the leaf carbon has been returned to the atmosphere in the form of carbon dioxide. From 10 grams of leaf material, approximately 2 grams of bacteria are produced, corresponding to more than 10 billion microorganisms. All of this takes place in a handful of soil.

What is the fate of the bacteria and fungi that have grown on the leaf? As their population increases, protozoa and other microorganisms begin to feed on them. The protozoa are then preyed upon by mites, and these in turn are devoured by larger animals. At each step in the **food chain,** oxygen is consumed and carbon dioxide returned to the atmosphere. Thus, as you go up the food chain, less and less carbon is available for growth, leading to what is often called a **food pyramid** (Figure 9.2).

A small amount of carbon temporarily escapes the cycling process by being deposited as either calcium carbonate ($CaCO_3$) or organic materials that are highly resistant to further microbial degradation. In the former case, carbon dioxide in the water combines with calcium ions to form a precipitate of calcium carbonate (limestone). The hard outer shells of mollusks and corals are also formed in this manner. Much of

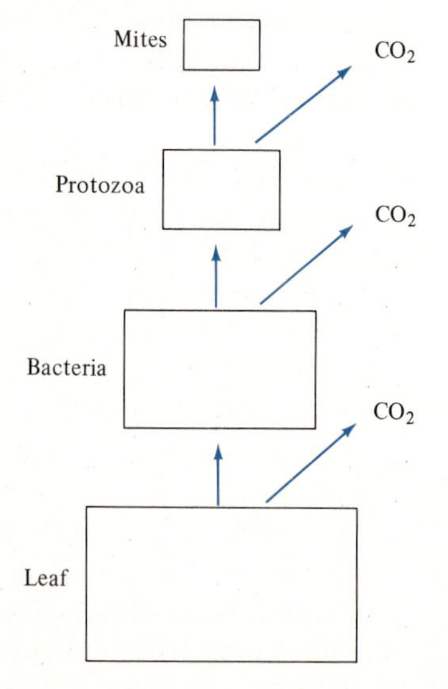

Figure 9.2
A simplified food pyramid: leaf litter → bacteria → protozoa → mites.

this inorganic carbon is eventually released into the atmosphere through the weathering of rocks.

Microbes also play an important role in the formation of fossil fuels, such as peat, coal, oil shale, and petroleum. Some of the organic material that is buried in the soil and ocean sediments is slowly converted by microbes into compounds that are highly resistant to further microbial decomposition. The formation of humus in the soil is an example of such a process. By mechanisms that are not completely understood, these resistant organic molecules are further reduced biologically to hydrocarbon-like materials. Finally, in the course of geological time, the combined action of high pressure and temperature converts the material into a fossil fuel. By harvesting and burning these fossil fuels, man is returning to the atmosphere carbon atoms that were used for photosynthesis millions of years ago.

9.2 The Nitrogen Story

The cyclic transformation of nitrogen is of paramount importance to life on earth. Distortions in the nitrogen cycle, resulting in the accumulation of certain toxic nitrogen compounds, are one of the major causes of pollution. Conversely, the lack of utilizable forms of nitrogen in the soil limits agricultural productivity and is partly responsible for the world's food shortage. Overcoming the problem by adding chemical fertilizers requires enormous quantities of energy, at a time when the supply of fossil fuels is rapidly diminishing. Thus, an understanding of the basic principles of nitrogen transformations in nature is of fundamental concern to all of us.

The nitrogen cycle (Figure 9.3) can be divided into six basic processes: (1) **ammonification,** the formation of ammonia from amino acids and other forms of cellular nitrogen; (2) **nitrification,** the conversion of ammonia into nitrate; (3) **nitrate assimilation,** the uptake of nitrate into cells and its reduction to ammonia; (4) **ammonia assimilation,** the uptake of ammonia into cells and its incorporation into amino acids; (5) **denitrification,** the conversion of nitrate to nitrogen

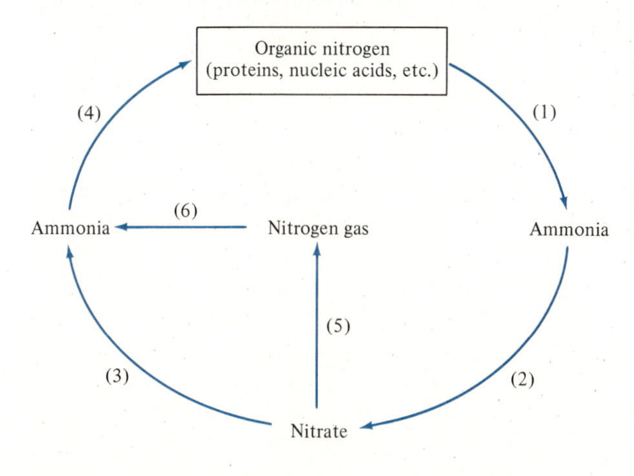

Figure 9.3
The nitrogen cycle. Ammonification (1), nitrification (2), denitrification (5), and nitrogen fixation (6) are carried out primarily by microorganisms. Nitrate assimilation (3) and ammonia assimilation (4) can be performed by both plants and microorganisms. Animals are involved to a limited extent in ammonification.

gas; and (6) **nitrogen fixation,** the biological production of ammonia from nitrogen gas.

What is the role of plants, animals, and microorganisms in the nitrogen story? Plants are able to satisfy their nitrogen requirement by acquiring nitrate and ammonia from the soil. In the case of nitrate assimilation, the plant must first reduce the nitrate to ammonia before incorporating it into amino acids. In general, plants take up only as much nitrogen as they need for growth; there is very little excretion of nitrogen compounds by living plants. It follows that the growth of plants depletes the soil of utilizable forms of nitrogen. How is this nitrogen replenished?

Animals receive nitrogen in the form of protein by eating plants or other animals. The food protein is broken down to amino acids, which then serve as building blocks for the growth, repair, and general maintenance of living cells. If the animal diet contains more protein than is necessary for these functions, the excess nitrogen is eliminated, mostly as urea (by mammals), uric acid (by reptiles and birds), and ammonia (by invertebrates).

Ammonification and Nitrification

The principal role in the nitrogen story is played by the ubiquitous microorganisms. Once an animal or plant dies, many different microbes begin their work of decomposing all of the large and complex molecules in the dead tissues. In the case of nitrogen compounds, the major end product is ammonia (hence the name ammonification). Similarly, the nitrogenous waste materials of animals, urea and uric acid, are broken down to ammonia by specialized groups of microorganisms.* Although the production of ammonia from urea is not an energy-yielding process, these specialized microbes benefit from the reaction in two ways: First, the ammonia serves as the nitrogen source for their growth; second, the ammonia causes the microenvironment to become highly alkaline, thereby inhibiting the growth of competing organisms. The microorganisms that bring about the ammonification are themselves quite resistant to alkaline conditions.

The conversion of ammonia to nitrate (nitrification) proceeds in two steps: In the first step, ammonia (NH_3) is oxidized to nitrite (NO_2^-) by bacteria of the *Nitrosomonas* group; in the second step, nitrite is oxidized to nitrate (NO_3^-) by *Nitrobacter*. Both of these highly specialized groups of bacteria derive all of their energy for growth from the oxidation of the inorganic nitrogen compounds. The nitrate formed by these bacterial processes provides plants with their major source of nitrogen and determines, to a large extent, the fertility of the soil. Even when nitrogen fertilizers, such as manure and

*The smell of babies' diapers that have been left on too long is due to ammonia formed by the microbial decomposition of urea. The ammonia is also the major cause of diaper rash.

ammonia, are added directly to the soil, the combined activities of the ammonifying and nitrifying bacteria must convert the nitrogen into nitrate prior to assimilation by the plants.

Pollution by Nitrogen Compounds

Nitrates are very soluble in water and do not adhere tightly to the soil. Thus, if more nitrate is formed in the soil than can be utilized by the plants, the excess nitrate percolates through the soil into underground water or is washed into rivers, lakes, and seas. The problem of nitrate surplus is especially severe in heavily fertilized agricultural areas. If the concentration of nitrate in drinking water exceeds 45 parts per million, the United States Public Health Service has ruled that the water is unfit to drink, especially for infants younger than four months of age.

What is the basis of nitrate toxicity? The stomachs of infants are much more acidic than those of adults. Under acidic conditions, a population of bacteria develops in the stomach that can convert nitrate to nitrite. It is the nitrite that is extremely toxic. It rapidly destroys hemoglobin in the bloodstream so that the red blood cells can no longer combine with oxygen. When approximately 70 percent of the hemoglobin is inactivated, asphyxia occurs. Furthermore, even very low levels of nitrite are mutagenic and carcinogenic. It is clear that nitrates are a serious hazard to humans and should be kept to an absolute minimum in drinking water.

Denitrification

Sometime nature has a way of repairing damage we inflict upon it. Consider, for example, some events that have taken place in and around the Sea of Galilee† in northern Israel during this century. Because of the year-round warm climate and adequate water supplies, intense agricultural settlements developed in the area. Just north of the Sea, a vast peat bog was drained to alleviate the mosquito problem and provide additional agricultural land. When the land was plowed, buried organic nitrogen compounds came into contact with air, providing ideal conditions for the development of ammonifying and nitrifying bacteria. Large amounts of nitrate were formed and subsequently washed into the Sea of Galilee with the winter rains. The combination of (1) drainage of the peat bog, (2) heavy use of nitrogen fertilizers on the slopes surrounding the Sea of Galilee, and (3) increased pollution from the city of Tiberias and other tourist developments on the shore threatened Israel's major water reservoir. Consequently, the microbial ecologist Moshe Shilo and other water experts brought about the establishment of a laboratory at the Sea of Galilee to study the problem.

†Also called Lake Kinnereth, this small (14 miles long and 8 miles across), deep body of fresh water is mentioned frequently in both Old and New Testaments.

Careful scientific surveys conducted over a period of several years led to the rather surprising conclusion that the level of nitrate in the Sea was *not* increasing significantly. How was that possible if the amount of nitrate entering the Sea was considerably greater than that flowing out? Although Israel is reputed to be the land of miracles, nobody suggested that the Law of Conservation of Mass was being violated. Rather, in the anaerobic depths of the Sea, microorganisms were converting the nitrate to nitrogen gas, which escaped into the atmosphere. This process of **denitrification** is brought about by a number of different microorganisms, but only in the absence of oxygen. As discussed in Section 3.7, certain bacteria can use nitrate in place of oxygen in order to obtain energy from the oxidation of organic or inorganic materials. Since denitrification requires rather specific environmental conditions, including a balanced amount of oxidizable substrate, there is no assurance that further nitrogen pollution of the Sea of Galilee will not lead to the build-up of toxic concentrations of nitrate.

In addition to helping make water in reservoirs available for drinking, denitrification plays an even more fundamental role in the nitrogen cycle. Nitrates are readily leached from the soil and eventually carried to the oceans. Unless there was some mechanism for releasing nitrates from ocean waters and sediments, all of the earth's supply of nitrogen would concentrate in the seas, making life on the land masses impossible. Denitrification in anaerobic layers of the ocean prevents the accumulation of nitrate in the sea.

Nitrogen Fixation

Nitrogen lost to the atmosphere by denitrification is compensated for by another important reaction in the ecology of this planet, **nitrogen fixation,** the conversion of nitrogen gas (N_2) to ammonia (NH_3). Although nitrogen gas makes up approximately 79 percent of the earth's atmosphere, it is chemically inert and thus not a suitable nitrogen source for plants or animals. It is the task of a few groups of nitrogen-fixing bacteria to secure nitrogen from the air, reduce it to ammonia, and then combine the ammonia with other elements to form protein and other nitrogenous organic compounds.

Because of its critical importance to agriculture, nitrogen fixation has been a subject of intense scientific investigation for many years. Recent discoveries about the biochemistry and genetics of nitrogen fixation have stimulated considerable experimentation (and even more speculation) on the possibility of increasing nitrogen fixation by genetic engineering technology. In the next section, the biology of nitrogen fixation is discussed with emphasis on the special symbiotic relationship that is established between bacteria of the genus *Rhizobium* and leguminous plants.

9.3 The Biology of Nitrogen Fixation

The ability of agriculture to feed the growing world population depends on the availability of utilizable nitrogen in the soil. Nitrogen is continuously being removed from agricultural land by incorporation of nitrate into plants (which are subsequently harvested) and, to a lesser degree, by denitrification. If the land is to remain fertile, the nitrogen must be replaced either by addition of nitrogenous fertilizers or by providing conditions in which the amount of nitrogen gas converted to ammonia by nitrogen-fixing bacteria exceeds that lost by harvesting and denitrification.

Traditionally, nitrogen fertilizers were derived from organic wastes, especially animal manures. Today, the vast majority of nitrogen fertilizer is produced synthetically by an industrial process invented in 1914 by two German chemists, Fritz Haber and Karl Bosch. The Haber-Bosch process consists of heating a mixture of nitrogen gas and hydrogen gas at 500°C and 120 atmospheres pressure to produce ammonia. The ammonia can be used directly or converted easily to urea, nitrate, or other nitrogen compounds. Since the hydrogen gas is derived from petroleum or natural gas, the price of chemically synthesized nitrogen fertilizer is closely linked to the price of fuel. Thus, the cost of a barrel of crude oil greatly influences the price of food.

Biological nitrogen fixation is brought about by two groups of procaryotes: those that are free-living and those that live in symbiotic association with certain plants. Free-living nitrogen-fixers can be aerobic *(Azotobacter)*, strictly anaerobic *(Clostridia)*, facultatively anaerobic *(Klebsiella)*, or photosynthetic (blue-green bacteria). The amount of nitrogen harnessed by these free-living microorganisms is quite substantial, especially in warm-wet climates. For example, many of the world's paddy fields receive no artificial fertilizer whatever. Nitrogen fixed in the rice paddies, primarily by blue-green bacteria, is made available to the plants when the blue-green bacteria die and are decomposed by other microorganisms. As shown in Table 9.1, symbiotic nitrogen fixation is even more efficient than nitrogen fixation by free-living microorganisms. Species of the soil bacterium *Rhizobium* that live inside nodules on the roots of leguminous plants such as clover, peas, alfalfa, beans, peanuts, and soybeans can introduce as

TABLE 9.1 Nitrogen Fixation by Microorganisms

Microorganism	Habitat	N_2-Fixed[a] (Kilograms/Acre/Year)
Free-Living		
Azotobacter	Tropical rain forest	35
Blue-green bacteria	Australian desert	2
	Rice paddy in India	15
Total population	Soil under wheat	7
Symbiotic		
Rhizobium	Alfalfa	180
Rhizobium	Soybean	40
Rhizobium	Alder tree	100

[a]Recalculated from M. J. Alexander: Microbial Ecology (see reference at end of chapter).

much as 200 kilograms of fixed nitrogen per acre into the soil each growing season.

Symbiosis and Nitrogen Fixation

Humans have long recognized that the fertility of agricultural land declines when grain crops such as wheat and barley are sown year after year. Farmers soon learned by trial and error that the decline in productivity of the soil could be reversed and the land refurbished by rotating crops, one of which is a legume. A relatively recent example of this practice was the introduction of the peanut plant into southern United States to restore and maintain the fertility of soil depleted of nutrients by years of growth and harvesting of cotton crops. Today, in many of the warmer parts of the world, peanuts and other leguminous plants are grown as economically valuable winter crops, both for harvesting and for enriching the soil.

An explanation for the beneficial action of legumes on soil depended on the development of both chemistry and microbiology. About 150 years ago, the tools of chemistry became sufficiently sensitive to measure the small amounts of nitrogen in soil. Analyses of different soil samples revealed that fertile soils contained significantly higher amounts of nitrogen than did poor soils. Since most plants contain 2 to 3 percent nitrogen, the productivity of the soil could be calculated on the basis of how much nitrogen it contained. For example, a metric ton of soil containing 0.01 percent nitrogen could support the growth of no more than 5 kilograms of plant material. From these kinds of calculations, it became clear that the limiting factor for plant growth in poor soils was generally nitrogen deficiency. Leguminous plants, however, were an exception to this rule. Not only was the growth of legumes not limited by the amount of nitrogen in the soil, but in fact, the amount of soil nitrogen actually increased as a result of their growth. Since atmospheric nitrogen gas was the only possible source for the increase in soil nitrogen, these early soil chemists concluded that leguminous plants were capable of nitrogen fixation. Thus, analytical chemistry provided the first logical explanation for crop rotation.

What is it about leguminous plants that allows them to fix nitrogen? Plant anatomists have known for several hundred years that leguminous plants have peculiar nodular structures on their roots (Figure 9.4). At first these root nodules were thought to be pathological aberrations, analogous to the callus-like tumor tissues, called crown gall, found on the stems of certain plants. With the discovery that leguminous plants could fix atmospheric nitrogen, attention was turned to the possibility that root nodules played a role in the process. The correlation between root nodules and nitrogen fixation was supported further by the occasional observation of a legume plant that lacked nodules on its roots; invariably these rare plants failed to fix nitrogen. Although correlations can play an important part in scientific discoveries, it usually takes a more directed experimental approach to prove cause-and-effect relationships.

Figure 9.4
Root nodule of leguminous plant.

When microscopic examination of fresh root nodules revealed that they were full of bacteria, it was immediately suggested that nitrogen fixation was a property of these bacteria and not of the plant. This hypothesis was attractive, particularly because it was readily testable. First, it was shown that clover seeds, treated chemically to sterilize their surface and then grown in pots of sterile soil, failed to develop root nodules. The growth of these germ-free clover plants depended upon nitrogen in the soil. Second, if crushed root nodules from the same plant species were added to the previously sterilized soil, nodulation occurred and plant growth was no longer limited by soil nitrogen. Third, the famous Dutch microbiologist M. W. Beijerinck succeeded in isolating the nodule bacterium in pure culture in 1888. Addition of the nodule bacterium to germ-free plants led to the formation of root nodules and the ability to carry out nitrogen fixation. Thus, the nodule bacterium was the causative agent of both root nodule formation and nitrogen fixation.

Because of their close association with plant root systems, nodule bacteria came to be called rhizobia (Gr. *rhizo*, "root"). For many years, Beijerinck attempted *unsuccessfully* to demonstrate that pure cultures of rhizobia could fix nitrogen. Unlike the case with free-living nitrogen-fixing bacteria, rhizobia did not grow in culture media lacking a source of fixed nitrogen. Apparently, the ability of rhizobia to fix atmospheric nitrogen depended on their close associations with leguminous plants. Since both plant and microbe benefitted from this interaction, the relationship was referred to as **mutualistic symbiosis.** In recent years, the legume–*Rhizobium* symbiosis has become one of the most exciting areas of biological research, from both theoretical and applied points of view.

The process of infection leading to nodule formation begins when a *Rhizobium* organism in the soil is attracted to the roots of leguminous plants. The *Rhizobium* organism enters the legume through a

root hair, proliferates inside the plant, and eventually the bacteria reach the cortical cells of the root, which then develop into a bulbous enlargement: a root nodule. The nodule consists of giant plant cells packed full with bacteria. Under these conditions, the bacteria are able to carry out nitrogen fixation. Ammonia formed in the process combines with carbon compounds manufactured by the plant during photosynthesis to produce amino acids. These amino acids serve as the raw material for synthesis of both plant and bacterial proteins. The relationship is one of genuine cooperation. Nutrients from the plant derived by photosynthesis nourish the *Rhizobium* organism, while excess nitrogen captured by the bacterium is supplied to the plant; since neither the *Rhizobium* organism nor the plant can fix nitrogen by itself, the value of this symbiosis is clear.

Rhizobia are generally specific to their natural plant partner. The bacteria that form nodules in soybeans, for example, will not infect alfalfa. The specificity of the *Rhizobium*–legume cooperation is of particular practical significance to the farmer. If the appropriate *Rhizobium* population is deficient in the soil, then nodulation will be delayed and the crop will suffer. To avoid this problem, legume seeds are now treated with their distinct matching species of *Rhizobium*. This minor innovation is probably the most important contribution that the science of soil bacteriology has made to agriculture to date.

Although the best studied example of symbiotic nitrogen fixation is the legume–*Rhizobium* association, there are several other extremely interesting cases of symbiotic nitrogen fixation in the biological world. For example, the gut of termites contains several different types of microorganisms. Some of them are able to break down cellulose to glucose. Other bacteria can utilize the glucose as a nutrient and fix nitrogen gas. The combination of these microbes allows the termite to live on a nitrogen-free diet consisting of nothing more than paper or wood. Thus, each of the three organisms benefits—the nitrogen-fixing bacteria by receiving a constant supply of glucose, the cellulose-decomposing microbes by obtaining finely ground cellulose from the termite and amino acids from the nitrogen-fixers, and the termites by growing on a substrate that no other animal can consume. From this three-part symbiosis, we can begin to appreciate the beauty and complexity of symbioses in nature.

Nitrogenase

The enzyme that catalyzes the conversion of nitrogen gas to ammonia is called **nitrogenase.** This enzyme is found in all nitrogen-fixing bacteria. Nitrogenases isolated from widely different bacteria have very similar chemical structures. It does not seem to matter whether the enzyme is isolated from an aerobic *Azotobacter*, an anaerobic *Clostridium*, a symbiotic *Rhizobium*, or a photosynthetic blue-green bacteria. The enzyme always contains the following remarkable properties: (1) It is able to split the extremely stable triple bond in nitrogen gas ($N \equiv N$) at 25°C and one atmosphere pressure. (To break the same

bond chemically, by the Haber process, requires 500°C and 120 atmospheres pressure.) The energy for the enzymatic split of nitrogen gas comes from ATP. (2) Nitrogenase contains two atoms of the biologically rare element molybdenum. Biochemical studies have shown that the molybdenum is essential for the enzymatic activity. This probably explains the fact that nitrogen fixation is always low in molybdenum-deficient soils. Addition of as little as 25 grams of molybdenum per acre has increased the productivity of some poor soils in Australia by as much as ten- to twentyfold. (3) Nitrogenase is very sensitive to oxygen gas; the enzyme is totally and irreversibly destroyed as soon as it comes in contact with air.

If nitrogenase is so sensitive to oxygen gas, how is it possible for aerobic microorganisms to perform nitrogen fixation? Each of the aerobic nitrogen-fixers has evolved a unique mechanism for protecting its nitrogenase. In the case of *Azotobacter*, the oxygen that enters the cell is used so rapidly for respiration that the inside of the cell is anaerobic. The problem of oxygen sensitivity of nitrogenase is an even more serious dilemma for blue-green bacteria because they actually generate oxygen gas during photosynthesis. This seemingly insurmountable predicament has been overcome by evolving a system of compartmentalization. Approximately one of twenty cells in a chain of blue-green bacteria specializes in nitrogen fixation; these differentiated cells do not carry out photosynthesis and must therefore receive their carbon and energy from neighboring sister cells. The photosynthetic cells, in turn, obtain nitrogen in the form of amino acids from the nitrogen-fixing cells. This specialization of metabolic activities is another fine example of differentiation in procaryotes.

Protection of the nitrogenase of *Rhizobium* is an integral part of its symbiotic association with the legume. The plant root tissue synthesizes a special oxygen-binding protein, appropriately called **leg-hemoglobin.** This is the only known form of hemoglobin found in the plant world. The leg-hemoglobin traps the oxygen before it can reach the bacteria. This arrangement may be one of the most important benefits the *Rhizobium* receives from its partnership with the legume.

Nif Genes and Genetic Engineering

The cluster of genes that are responsible for the formation of nitrogenase and nitrogen fixation are termed *nif* genes. Recently, scientists have speculated on the possibility of using modern techniques of "genetic engineering" (Section 5.7) to transfer *nif* genes from nitrogen-fixing bacteria to other organisms. The long-range goal of this endeavor is to construct new food crops that do not require nitrogen fertilizers. The economic significance of a major food crop, such as wheat, barley, or corn, that has the capability of fixing nitrogen would be enormous. It would revolutionize agriculture.

How much of this is futuristic speculation and how much is based on solid scientific data? In 1972, Roy Dixon and John Postgate of the University of Sussex, England, succeeded in transferring the *nif* gene from a nitrogen-fixing bacterium to *E. coli* by sexual conjugation. Prior to the mating, the *E. coli* could not carry out nitrogen fixation. After receiving the *nif* gene, however, the "hybrid" *E. coli* was able to fix nitrogen when grown anaerobically. Since *E. coli* has no special apparatus for protecting the nitrogenase, the enzyme is destroyed and nitrogen fixation turned off when the cells come into contact with air. Sexual conjugation in bacteria is restricted to closely related species. To transfer genes between unrelated species requires the application of genetic engineering.

Two general strategies have been envisioned for "engineering" new plant species with reduced nitrogen demands. One is creating nitrogen-fixing bacterial strains that will provide the fixed nitrogen for non-leguminous plants, either by establishing a symbiotic relationship or simply by being attracted to the leaves or roots of the plant. One way to accomplish this feat might be to transfer the *nif* genes into bacterial species that already live in close association with the particular plant. Alternatively, it may be possible to alter genetically the cell surface of a *Rhizobium* or another symbiotic nitrogen-fixer so that it will form new associations. The second general scheme involves the introduction of plasmids containing *nif* genes directly into individual plant cells. The cloned cells can then give rise to mature plants. This latter approach not only is more difficult to accomplish genetically, but also will probably not lead to nitrogen fixation because of the sensitivity of nitrogenase to oxygen.

As of 1982, the "state of the art" of genetic engineering with regard to nitrogen fixation can be summarized as follows:

1. Genes responsible for nitrogen fixation, *nif* genes, have been shown to be clustered together on the bacterial chromosome.
2. *Nif* genes have been placed on different bacterial plasmids.
3. *Nif* genes can be transferred from one bacterium to another by either conjugation or transformation with plasmids; bacteria receiving *nif* genes gain the ability to perform nitrogen fixation under anaerobic conditions.
4. Although it has not yet been accomplished, techniques are available for the construction of hybrid plasmids containing *nif* genes that can multiply inside eucaryotic cells.
5. There is serious skepticism concerning the possibility of *nif* genes functioning in foreign cells in the presence of air; many scientists feel that much more has to be learned about the stabilization of nitrogenase and the biochemistry of bacteria–plant root interactions before we can improve on nature.

9.4 The Sulfur Cycle

In many respects, the sulfur cycle (Figure 9.5) has a striking resemblance to the nitrogen cycle. Inside living cells, the element sulfur is present, primarily as a constituent of most proteins and a few

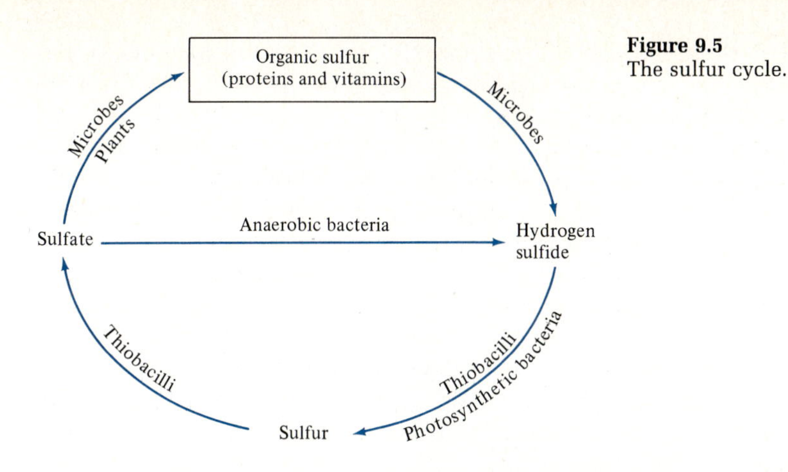

Figure 9.5
The sulfur cycle.

vitamins. Plants and microorganisms derive their sulfur from the soil in the form of sulfate; animals satisfy their sulfur requirement by eating plant or animal proteins. Death and microbial decomposition of organic matter lead to the formation of hydrogen sulfide (H_2S). This process of putrification is similar to ammonification in the nitrogen cycle, not only from the biochemical point of view, but also because in both processes a gas with a disagreeable odor is produced. The hydrogen sulfide is oxidized by specialized bacteria, first to sulfur (S) and then to sulfate ($SO_4^{=}$). In a transformation analogous to denitrification, another group of specialized bacteria can reduce sulfate back to hydrogen sulfide gas.

The conversion of hydrogen sulfide to sulfate can be brought about by either aerobic or anaerobic bacteria. The aerobic sulfur-oxidizing bacteria are called **thiobacilli.** These chemoautotrophs (see Section 3.7) generate ATP by the oxidation of hydrogen sulfide or sulfur to sulfate. The energy is used to convert carbon dioxide into organic material needed for cell growth. Anaerobically, in the presence of light, photosynthetic bacteria use H_2S (instead of H_2O) for photosynthesis and produce S (instead of O_2). In both aerobic and anaerobic production of sulfate, considerable quantities of acid are produced. For this reason, sulfur is often added as a fertilizer to improve alkaline soils.

Bacteria involved in the sulfur cycle are often found in characteristic environments. Bacteria that produce H_2S gas from sulfate, for example, are particularly abundant in the mud at the bottom of ponds, lakes, and seas.‡ In these anaerobic environments, specialized bacteria use sulfate instead of oxygen to decompose organic matter (anaerobic respiration, see Section 3.7). Some of the H_2S gas that is produced reacts with iron to form iron sulfide, giving the mud a distinctive pitch-black color. The remainder of the gas rises to the surface, where part of it escapes into the atmosphere and part is oxidized by the thiobacilli.

‡In some cases, the amount of hydrogen sulfide that is generated is so great that the resulting smell of rotten eggs makes the area practically uninhabitable.

It has been estimated that approximately 100 million tons of hydrogen sulfide gas are released each year into the earth's atmosphere as a result of microbiological activity. In the presence of air, the H_2S is oxidized rapidly to sulfur dioxide (SO_2) and then more slowly to sulfur trioxide (SO_3). In addition, large quantities of sulfur oxides are produced nonbiologically by burning fuels that contain sulfur compounds. The SO_3 combines with H_2O in the air to form sulfuric acid (H_2SO_4). The resulting **acid rain** causes considerable damage to stone and metal sculptures and structures. This problem is most serious in concentrated industrial areas.

Further cycles, which will not be discussed here, include the phosphorus cycle and the iron and manganese cycles. In these days when the concept of recycling has become so important, much can be learned from the natural cycles.

9.5 Human Activities Disturb the Natural Cycles of Matter

Life on this planet depends on the ability of photosynthetic organisms to trap some of the energy of the sun and then use this energy to transform inorganic compounds into biochemicals. Since there is a limited amount of materials from which to construct living organisms, the dead biological matter must be transformed back to inorganic compounds in order to sustain life on earth. (From the point of view of matter, the earth can be considered a closed system—with only minor exceptions, nothing enters, nothing leaves.) All of the natural, biologically driven cycles of matter—the C-cycle, N-cycle, S-cycle, O-cycle, and other cycles not discussed here—work together to ensure a constant turnover of matter. Although the biological world has an enormous capacity to transform matter from one form into another, there are rather well-defined limits that should not be surpassed if the cycles are to function efficiently. When these conditions are not maintained, a chain of events occurs, leading to the disruption of the natural cycles, accumulation of toxic compounds, and the death of important members of the ecosystem.

In this section we discuss how certain products of human activities put stress on microbial communities, often leading to serious pollution problems. In the next section we show how microbes can be exploited to monitor and treat some of these problems.

During the last 10,000 years, the human population has increased from about 5 million to more than 4 billion. This population increase has resulted in the concentration of people into great urban centers, the intensification of agricultural practices, and the creation of high-technology industry. The consequence of each of these developments has been the production of large quantities of material that pollutes and poisons our water supplies. These man-made sources of water pollution can be divided into domestic wastes, agricultural run-off, and industrial refuse (Table 9.2).

Domestic waste materials include a filthy mixture of feces, urine,

TABLE 9.2 Man-made Sources of Water Pollution

Source	Pollutants	Problem
Domestic wastes	Excreta and various discarded domestic items	Waterborne diseases and eutrophication
Agricultural run-off	Fertilizers Pesticides	Eutrophication Recalcitrant chemicals
Industrial refuse	Petroleum Synthetic organic chemicals	Recalcitrant chemicals Toxicity

soapy water, toilet paper, hair, food residues, grease, and innumerable other items that are disposed of through sinks and toilets or washed into sewers by the rain. In agricultural areas, the major pollutants are fertilizers and pesticides. Industrial refuse includes a wide variety of synthetic organic chemicals, petroleum products, and metals. In some cases, large quantities of these pollutants are discharged directly into rivers or lakes; in other cases, pollutants filter through the soils and enter the ground water. The most serious problems arising from these man-made sources of water pollution are (1) transmission of disease by **waterborne pathogens,** (2) **eutrophication** of natural bodies of water, and (3) accumulation of toxic or **recalcitrant chemicals.**

Waterborne Pathogens

Several important human diseases are transmitted from one person to another through the water supply. The mechanism of transmission is rather simple: Disease-causing microbes infect the gut, leave the host in the fecal matter, enter the water supply, and are acquired by the next host by swallowing the contaminated water. Diseases that are transmitted by drinking beverages or eating food containing human excreta include typhoid fever, cholera, bacterial food poisoning, bacterial and amoebic dysentery, and poliomyelitis.

In those parts of the world where community drinking water is disinfected, the incidence of waterborne diseases is low. However, certain diseases, such as viral hepatitis, need not be transmitted directly through drinking water. Approximately 80,000 people in the United States contract hepatitis each year. The vast majority of these cases originate from eating shellfish that were grown in water polluted with feces containing the live virus.

Eutrophication

One of the most important principles in biology is the "Law of the Minimum." The law states that the extent of growth of an organism is limited by that required nutrient which is present in the smallest quantity. Thus, if crops are grown in an iron-depleted soil, it does no good to add more nitrogen or phosphorus. Increased yields can be obtained only by adding the limiting nutrient, in this example, iron.

In the case of oceans and most rivers and lakes, the limiting growth element is nitrogen or phosphorus. Provided with adequate supplies of these two elements, algae and photosynthetic bacteria in the upper layers of water are able to harvest energy from sunlight and fix carbon dioxide into organic compounds. Thus, if it were not for the nitrogen and phosphorus limitation, all bodies of water would soon be covered with a thick green mat of algae. This is precisely what happens during **eutrophication.**

Waste materials from both urban and agricultural areas contain large quantities of nitrogen and phosphorus compounds. When these materials reach natural bodies of water, the following chain of events occurs: (1) Bacteria degrade much of the material, releasing nitrates and phosphates. (2) Algae then multiply rapidly, giving rise to what is generally called an "algal bloom." (3) The algae, in turn, provide food for a greatly increased population of aerobic microorganisms; during this time, a tenuous balance is established between oxygen production by photosynthesis in the upper layers and oxygen utilization by respiration in the bottom water. (4) A shift in environmental conditions, such as a few days of low light intensity, shifts the equilibrium in favor of respiration, thereby causing the water to become anaerobic. Deoxygenation of the water kills many fish and aerobic microorganisms. In lakes and coastal water, the dead algae pile up on the shore in rotting masses, and the oxygen depletion caused by the algal decay further reduces the flora and fauna of the water. These irreversible changes, accelerated by man-made pollution, can occur with drastic suddenness.

Eutrophication has a number of unpleasant consequences. First, only small children and a few microbiologists enjoy wading in the organic ooze of eutrophic waters. Most of us prefer swimming in crystal-clear, unpolluted water. Second, many economically valuable fish, such as trout, require clean, oxygen-saturated water. Third, many algae synthesize potent toxins. A marine dinoflagellate produces a poison almost as potent as the botulism toxin. Consumption of shellfish contaminated with this algae leads to paralysis and death. Toxins produced by blue-green bacteria kill fish and birds, in addition to posing a serious danger of contamination of drinking-water supplies. Finally, blue-green bacteria that proliferate in eutrophic water produce musty tastes and odors in drinking waters.

Recalcitrant Chemicals: Oil Pollution

For practical purposes, materials can be grouped into "biodegradable" and "non-biodegradable," the former being substances which, when introduced into soil or water, are rapidly decomposed by microorganisms. The latter group, referred to in the scientific literature as recalcitrant chemicals, are either not degraded or degraded so slowly that they accumulate in the environment and cause problems.

The dramatic increase in knowledge in science and technology during the last few decades has led to the production of immense quantities and varieties of organic compounds that litter the land and pollute our natural waters. Many of these substances are recalcitrant. This ever-widening group of industrial pollutants includes plastics, hard detergents, coal wastes, insecticides, herbicides, fungicides, textiles, phenolic compounds from the paper and dyestuffs industry, and various other petrochemicals. One of the most widely discussed and serious problems is oil pollution.

During this century, the demand for petroleum as a source of energy and as a primary raw material for chemical industries has resulted in an increase in world production from 150 million to more than 13,000 million barrels per year. This rapid increase in production, refining, and distribution of crude oil has also brought with it an ever-increasing problem of environmental pollution. A great part of this problem arises from the fact that the major oil-producing countries are not the major oil consumers. It follows that massive movements of petroleum have to be made from areas of high production to those of high consumption. It has been estimated that approximately 0.5 percent (or 13 million tons per year) of the transported crude oil finds its way into seawater, largely through accidental spills and deliberate discharge of oily ballast and wash waters from oil tankers.

The toxicity of crude and refined oil to marine ecology and even more directly to humans is well documented. Suffice it to mention that crude oil contains mutagenic, carcinogenic, and growth-inhibiting chemicals, and as little as one ten-thousandth of a gram per liter of certain petroleum fractions can destroy microalgae and juvenile forms of many marine organisms. In short, oil pollution in the ocean in general and in the coastal waters in particular presents a serious problem to commercial fisheries, recreational resources, and public health.

What limits the breakdown of petroleum in the sea? Do microorganisms lack the genetic potential to degrade hydrocarbons, or are the environmental conditions inadequate? The fact that microorganisms can utilize hydrocarbons as food has been known for 75 years. In fact, a wide variety of hydrocarbon-degrading microbes, including many species of bacteria, fungi, and algae, are located in virtually all natural areas. With the possible exception of the asphaltene fraction of crude oil, microorganisms have the enzymatic capability of oxidizing petroleum to carbon dioxide and water.

As mentioned previously, the oceans and most lakes and rivers contain very low concentrations of nitrogen and phosphorus. Hydrocarbons are, as the name indicates, composed almost exclusively of hydrogen and carbon. To utilize hydrocarbons as a food source, microorganisms must be provided with large quantities of nitrogen, phosphorus, and oxygen and lesser amounts of the other elements essential for growth of all organisms. The requirement for oxygen is easily overcome in aquatic environments where the oil floats to the surface and comes into direct contact with air. Thus, insufficient concentrations of nitrogen and phosphorus limit the microbial degradation of petroleum in the sea. In general, balanced nutritional conditions are as important as the genetic potential of specialized microorganisms in considering the biodegradability of any recalcitrant chemical.

An important phenomenon associated with many recalcitrant chemicals is **biomagnification,** the concentration of the substance as it proceeds up the food chain. The classic example of biomagnification is the insecticide DDT. Since this synthetic pesticide is broken down very slowly in soil and water, it accumulates in the sea, reaching concentrations of about one part per billion (ppb) in coastal waters. Microscopic algae growing in these waters contain DDT at concentrations of 10 to 40 ppb. Small fish that feed on the algae may contain as much as 200 ppb residual DDT. Finally, dolphins that eat these fish will contain 1000 ppb DDT residues. The explanation for biomagnification is simple: A dolphin, for example, that weighs 500 kilograms must have eaten more than 5000 kilograms of fish; most of the fish protein and carbohydrate is metabolized for energy, whereas the DDT accumulates in the dolphin's liver. The high concentration of toxic pesticides in fish and birds provides a particular danger to both wildlife and human beings.

9.6 Microbes Can Be Used to Treat Waste Waters

In most modern cities, the waste matter from individual homes and factories flows through a series of collecting sewers to a treatment plant. The purpose of the sewage treatment plant is to remove harmful microorganisms and reduce the concentration of undesirable chemicals to such an extent that the water can safely be discharged into a nearby body of water. Most of the work done in the sewage treatment plant is carried out by microorganisms.

A diagram of a typical municipal sewage treatment plant is shown in Figure 9.6. In the first step, large pieces of solid matter are removed by simply passing the raw sewage through a series of screens or filters. The clarified sewage then enters an **aerobic digestion tank,** where a vast array of bacteria, fungi, and protozoa oxidize most of the organic matter to carbon dioxide and water. To ensure that the maximum amount of waste material is digested, large quantities of air are continuously pumped into the tank. The oxidation of organic substrates is accompanied by the development of a large population of microorganisms. The overall result of the aerobic digestion step, then, is the conversion of nitrates, phosphates, and organics in sewage to carbon dioxide gas and more microorganisms:

$$\text{sewage} \xrightarrow{\text{air}} CO_2 + \text{microorganisms}$$

After the aerobic digestion is complete, the turbid mixture is transferred to a settling tank. The microorganisms and other solids

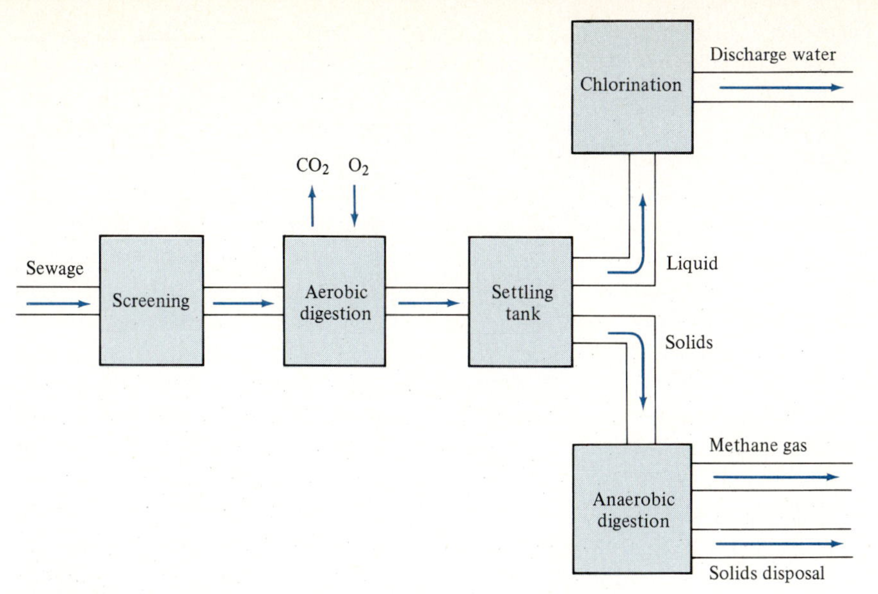

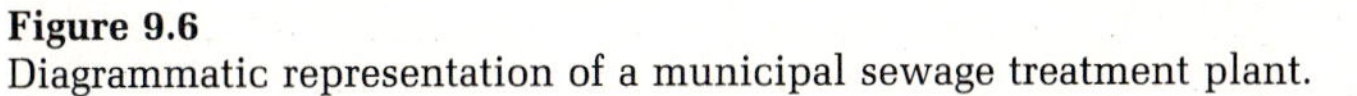

Figure 9.6
Diagrammatic representation of a municipal sewage treatment plant.

slowly fall to the bottom of the tank, leaving clear water that contains less than 10 percent of the organic matter that was originally present in the raw sewage. At this stage, the liquid is disinfected with chlorine gas and discharged into a nearby body of water. This treatment procedure is adequate for coastal cities, which can discharge their effluents some distance out at sea. Inland cities, however, should carry out additional treatments of the water to further reduce the content of recalcitrant chemicals and nitrogen and phosphorus salts. If these additional treatments are not performed, there is a danger of eutrophication of the lake or river receiving the discharged water. A fine example of an advanced waste treatment plant is in South Tahoe, California. More than 5000 gallons of foul sewage per day are transformed by a combination of microbiological, chemical, and physical procedures into sparkling pristine water, safe to drink or discharge into the lake.

Let us now consider the fate of the microbes and other solid materials that fell to the bottom of the settling tank. This sludge is collected and fed into an **anaerobic digestion tank** (Figure 9.7). The small amount of oxygen present in the tank is consumed by the microbes, rapidly causing the system to become totally anaerobic. A wide variety of bacteria then ferment the polysaccharides, proteins, nucleic acids, and other organic matter in the microbial sludge to organic acids, alcohols, and carbon dioxide. These fermentation products are then converted by two distinct bacteria, working in close association, into methane gas (CH_4): One bacterium produces carbon dioxide and hydrogen gases; the second bacterium combines these two gases to form methane. In essence, organic acids and alcohols are

Figure 9.7
An anaerobic sludge digester.

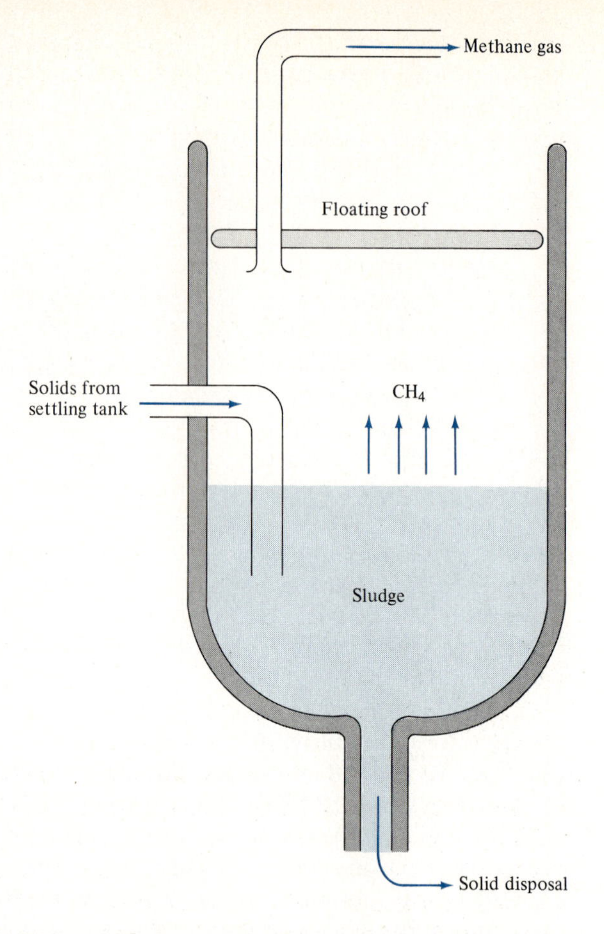

oxidized, carbon dioxide is reduced, and methane gas is formed. The overall process of anaerobic digestion can be represented as follows:

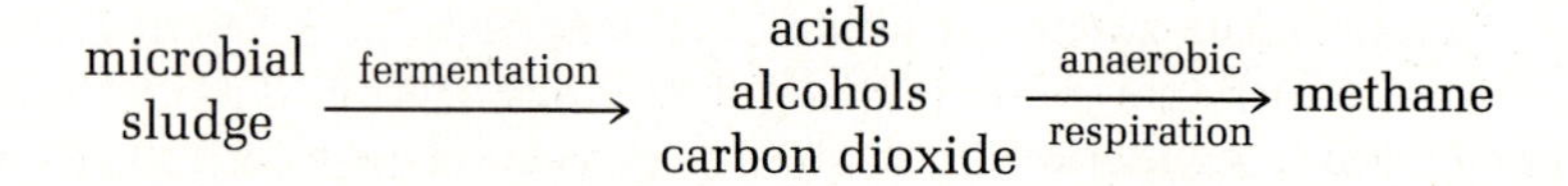

$$\begin{matrix}\text{microbial} \\ \text{sludge}\end{matrix} \xrightarrow{\text{fermentation}} \begin{matrix}\text{acids} \\ \text{alcohols} \\ \text{carbon dioxide}\end{matrix} \xrightarrow[\text{respiration}]{\text{anaerobic}} \text{methane}$$

The methane gas generated during the anaerobic digestion of sewage is stored under a floating roof until it is removed from the top of the tank. Part of the methane is burned, and the energy produced is used to operate the treatment plant. Surplus methane can be incorporated into the city's gas supply. In addition, the solids remaining after anaerobic digestion need not be wasted. Instead of simply burning the spent sludge, many sewage plants now dry it and sell it as fertilizer rich in nitrogen and phosphorus. Milorganite is the trade name of one such fertilizer used widely in the United. States.

The general process of anaerobic digestion of organic material that takes place in sewage treatment facilities can easily be applied on a smaller scale. In recent years it has become popular to convert farm waste materials, such as chicken or cow manure, into methane gas and valuable fertilizer. Regardless of whether the bioconversion takes place in huge tanks in municipal sewage plants or in oil drums in someone's back yard, the same microbial processes are involved—fermentation and anaerobic respiration.

9.7 Testing the Water Supply for Pathogenic Microbes

Modern sanitary practice and an efficient municipal sewage treatment facility minimize the threat of waterborne epidemics. Nevertheless, even in the most advanced parts of the world, human error and accidents occur and lead to fecal contamination of the water supply. The medical literature is full of examples of epidemics that were traced to contaminated drinking water resulting from leaky sewers or faulty chlorination equipment. Thus, the water supply should constantly be monitored for the presence of fecal contamination, the major source of waterborne diseases.

The most widely used indication of fecal contamination of water is the **coliform count.** By a series of easy-to-perform standard procedures (the details of which need not concern us here), the number of coliform bacteria in drinking water can be determined in less than two days. The logic of using *E. coli* (generally not pathogenic) as an indicator of fecal pollution is based on the fact that the bacterium is (1) universally present in large concentrations in human feces, (2) relatively stable in water, and (3) easy to detect. If the coliform count in drinking water is high, it means that the water is polluted with fecal matter and *may* be contaminated with disease-causing bacteria and viruses. The reason why water is not routinely examined directly for pathogenic microbes is that even when they are present in the water supply at concentrations high enough to cause epidemics, they are often difficult to detect and identify. It should be noted that a low coliform count does not prove that the water is free of pathogens. Some intestinal waterborne viruses (e.g., poliovirus) are known to be more stable than *E. coli*.

The United States Public Health Service standards for different types of water are shown in Table 9.3. If the coliform count exceeds ten per liter, the water is not safe for drinking. As mentioned previously, shellfish can cause infectious hepatitis if contaminated with human fecal matter containing the virus. Seawater from which the shellfish are collected should contain less than 1000 coliform bacteria

TABLE 9.3 United States Public Health Service Water Standards

Intended Use of Water	Maximum Permissible Coliforms per Liter
Drinking	10
Collecting shellfish	1000
Swimming	10,000

per liter. Bathing beaches are generally closed by public health officers when the coliform count (from nearshore sewage disposal) in the seawater exceeds 10,000 per liter.

QUESTIONS

9.1 This chapter starts with a quotation from Professor C. B. van Niel, an outstanding scientist and one of the great teachers of microbiology. What "particular ecological niches" in the carbon, nitrogen, and sulfur cycles are filled exclusively by microorganisms?

9.2 What do each of the following terms signify?

Mutualistic symbiosis	Phytoplankton
Eutrophication	Recalcitrant chemical
Nitrogen fixation	Denitrification
Rhizobium	*nif* gene
Food pyramid	Legume

9.3 What is the explanation for the fact that recalcitrant chemicals, such as DDT, increase in concentration as you go up the food chain?

9.4 A recent newspaper article described a new "superbug" that could consume a larger portion of crude oil than any other microbe. Could this organism provide an answer to the problem of oil pollution in the sea?

9.5 Analysis of drinking water obtained from a reservoir showed the presence of high concentrations of nitrate and coliforms. Why is the water dangerous to drink? How do you think each of the pollutants got into the water? How can the water be made safe?

9.6 What microbiological processes are involved in the conversion of chicken manure to methane gas?

9.7 Explain the following statement, made by Louis Pasteur, concerning microbes: "Without them life would become impossible because the act of death would be incomplete."

Suggested Readings

Alexander MJ: Microbial Ecology. New York: John Wiley & Sons, 1971.
A textbook devoted entirely to microbial ecology, by an expert in the field.

Bacom W: The Disposal of Waste in the Ocean. Scientific American, August 1974.

Brill W: Biological Nitrogen Fixation. Scientific American, March 1977.
A well-written article combining the molecular genetics and biology of nitrogen fixation, by one of the leaders in the field.

Brill W: Nitrogen-Fixation: Basic to Applied. American Scientist, July 1979.

Hobson DN: Digesting the Indigestible. New Scientist, 40:142, 1968.

Hughes DE: Towards a Recycling Society. New Scientist, 61:58, 1974.

Janick J, C Noller, and CL Rhykerd: The Cycles of Plant and Animal Nutrition. Scientific American, September 1976.

Porteous A: Sweet Solution to Domestic Refuse. New Scientist, 50:736, 1971.

Scientific American, September 1970. The Biosphere.
The entire issue is concerned with the turnover of matter in nature, with individual articles on the carbon, oxygen, and nitrogen cycles.

Microbiology of Foods and Beverages: Art and Science

CHAPTER 10

"What did they live on?" said Alice, who always took great interest in questions of eating and drinking.

Alice in Wonderland *(Lewis Carroll)*

Many non-biologists believe that microbes are inherently evil. The popular press often refers to them as *germs*—microscopic entities that cause disease, death, and decay. Television advertisements advise us to spray ourselves with the latest antiperspirants and deodorants in order to exterminate the microbes. Why have microorganisms received such a bad press? First, people are generally afraid of things they do not understand. Viruses, bacteria, protozoa, and most fungi are too small to be seen with the naked eye; thus, special equipment and procedures are necessary to become acquainted with them. Second, a small minority of microbes are in fact agents of disease and death. However, as discussed in Chapters 12 and 13, most diseases are caused by man-made circumstances that bring about a temporary imbalance between human beings and their microbial tenants. As for decay, this should be viewed, in general, as a positive contribution of microorganisms to the cycling of matter, without which life on this planet would be impossible.

In this chapter we discuss the direct role that microorganisms play in producing and processing the foods we eat and drink. Meat and milk are available to us only because microbes, in symbiotic relationships with certain animals, are able to digest cellulose. Milk is converted by different microorganisms to a variety of yogurts, cheeses, and butter. Still other microbes are responsible for the preservation of plant materials in the form of tasty products such as pickles and sauerkraut. Alcoholic fermentation by yeast is responsible for bread as well as all alcoholic beverages. Finally, as the human population increases more rapidly than the traditional food supplies, microorganisms are finding increased use as a direct source of protein for animal and human consumption.

There is one recurring theme in this chapter that should be stressed at the outset. Certain microorganisms have the ability of *partially* degrading foodstuffs in the absence of oxygen, thereby producing a variety of alcohols, organic acids, and gases. The importance of these fermentation reactions to humans is that (1) the original food is converted to a product that can be stored because the acids or alcohols inhibit further microbial activity and (2) most of the food value present in the original material is retained in the fermentation products. Ethanol, for example, contains approximately 90 percent of the caloric value of the glucose that was used in its formation.

10.1 Cellulose to Meat and Milk: The Rumen Symbiosis

The ruminants—a group of grazing animals that includes cattle, goats, sheep, deer, camels, and giraffes—exist on a diet consisting primarily of cellulose. However, by themselves ruminants cannot digest cellulose because they lack the enzyme cellulase, which is required for splitting this polysaccharide into glucose. How then can ruminants use grass and other cellulose-rich plant materials as foodstuff? They have gained this valuable nutritional advantage by entering into a symbiotic relationship with cellulose-digesting microorganisms that inhabit their digestive tracts.

Figure 10.1 demonstrates the manner in which ruminant animals are able to grow on an essentially protein-free diet consisting mostly of insoluble cellulose and some inorganic salts. Most ruminants have a stomach that is divided into several compartments. The first and largest section of the stomach is called the **rumen.** In the cow, the rumen is extremely large, having a volume of approximately 100 liters. Food enters the rumen mixed with saliva containing some important minerals. The rumen is teeming with different kinds of microbes, mostly anaerobic bacteria and ciliated protozoa. These microbes attack the cellulose and other insoluble polysaccharides in the food fodder. Initially, the cellulose is split to glucose; subsequently the glucose is fermented to simple organic acids (e.g., acetic acid) and gases (carbon dioxide and methane). The organic acids cannot be further degraded in the anaerobic conditions of the rumen; instead, they accumulate and gradually pass through the rumen wall and enter the bloodstream. Oxidation of the microbially produced organic acids in different parts of the body provides the ruminant animal with its major source of energy for growth. The gases that are formed in the rumen are disposed of by the cow by regular, vigorous belching.

What happens to the microbes in the rumen? In the course of digesting the cellulose, the microorganisms grow and multiply. A

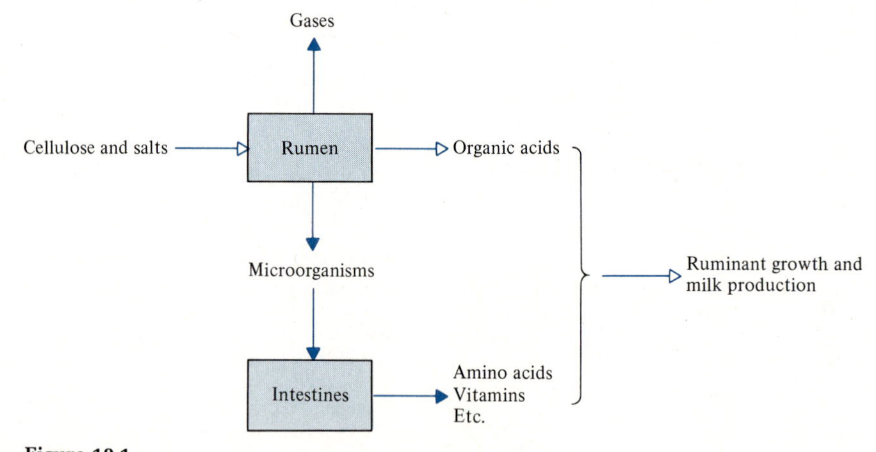

Figure 10.1
Diagram showing how ruminant animals can grow on cellulose and salts.

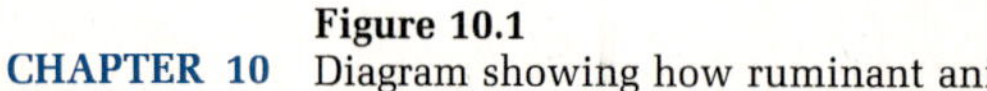

large portion of them are passed into the second part of the stomach, where they are squeezed together with undigested pieces of food into small portions called cuds. The cuds are regurgitated into the mouth, where they are chewed again. When the food is swallowed a second time, it goes down a different route, passing through the small and large intestines. These organs secrete enzymes that digest the microbial cells, producing amino acids, vitamins, and other growth factors. In essence, a ruminant animal satisfies its protein requirement by eating "microbial steak."

The benefit that the ruminant receives from living in close association with microorganisms is clear: Non-digestible cellulose is transformed by the microbes to readily digestible organic acids, proteins, and vitamins. But what do the microbes gain from this partnership? The fact that the rumen of a single cow contains approximately one quadrillion (10^{15}) microorganisms suggests that the growth conditions in the rumen must be close to ideal. Even under the most carefully regulated laboratory conditions, it is difficult to obtain a microbial culture that is as concentrated as rumen fluid. One can consider the rumen as a fermentation vat maintained by the cow for growing anaerobic microbes under virtually perfect conditions. The temperature is kept constant at 39°C; the acidity is carefully regulated; food is constantly fed into the rumen; and waste materials are prevented from accumulating. Although the fate of any individual microbe is to be swept out of the rumen and eventually digested by the cow in the lower stomach, the rumen provides a safe and constant ecological niche for the species. Thus, the interaction between the ruminant animal and the rumen microorganisms is a mutualistic symbiosis.

Certain other non-ruminant herbivores, such as the horse, rabbit, and guinea pig, can also derive some benefit from eating cellulose-rich materials. Again, microorganisms are responsible for the partial breakdown of the insoluble polysaccharide into fermentation products that the animal can subsequently metabolize. However, the process takes place in the large intestines and the cecum, so that most of the microbial cells and partially digested food is lost in the feces. To some extent, rabbits and guinea pigs solve this problem by eating their own feces.

10.2 Milk to Yogurt, Cheeses, and Butter

Milk is an excellent medium for the growth of bacteria. It is rich in protein (casein), carbohydrates (mostly the sugar lactose), fats, vitamins, and minerals. If allowed to stand at room temperature, fresh, unpasteurized milk will undergo a natural souring reaction brought about by lactic acid bacteria. The bacteria ferment the milk sugar, producing lactic acid which in turn curdles the milk protein:

$$(1) \quad \text{lactose} + H_2O \xrightarrow{\text{lactic acid bacteria}} \text{4 lactic acid } (C_3H_6O_3)$$

$$(2) \quad \text{soluble casein} \xrightarrow[\text{calcium ions}]{\text{lactic acid}} \text{insoluble protein curds}$$

Because the acidic conditions brought about by the lactic acid fermentation tend to protect the products from further microbial decomposition, different cultures throughout the world have exploited the lactic acid bacteria to preserve milk in a variety of tasty ways.

Fermented Milks

Buttermilk can be made simply by allowing unpasteurized milk or cream to undergo a spontaneous souring reaction at room temperature. Pasteurized milk cannot be used, since the lactic acid bacteria normally present in milk are destroyed during the pasteurization process. Instead, cultured buttermilk is easily made in the home by combining

1/4 cup cultured buttermilk (from a previous good batch)
1 liter skim milk
1/8 teaspoon salt

Mix well and cover. Let stand at room temperature (about 25°C) until curdled. Stir until smooth and store in refrigerator.

Yogurt is a fermented milk product, originally consumed by Bulgarian tribes but now found in every North American and European supermarket. The longevity of certain Eastern people is reputed to be due to the custom of regularly consuming yogurt and other fermented milk products. The principal bacteria in yogurt are *Streptococcus thermophilus* and *Lactobacillus bulgaricus,* the latter producing the strong final acidity and characteristic flavor. Lactic acid bacteria in fermented milk assist in digestion and contribute significant amounts of vitamin B and other vitamins. Acidophilus milk, containing large numbers of viable lactobacilli, is available commercially for the treatment of gastrointestinal disorders such as diarrhea. Many physicians recommend lactobacilli, either in the form of acidophilus milk or in fermented milk products, especially following antibiotic therapy. Try making yogurt yourself:

1. Prepare one liter of concentrated skim milk. One easy way to do this is to use one and one-half times the suggested amount of powdered skim milk per liter.
2. Inoculate the concentrated skim milk with one tablespoon of "starter" culture. The starter can be from a previous good-quality batch of yogurt, not more than one week old, or can be purchased from a health foods store.
3. Mix well, cover, and store at 45°C until it develops a consistency resembling custard (3 to 12 hours). Small incubators specially designed for making yogurt can be purchased; alternatively, it is often possible to find a warm place (40 to 45°C) in the home (e.g., near a heater). In the absence of a suitable place with controlled warm temperature, the fermentation can be carried out in a well-insulated container. As the bacteria grow, they generate their own heat.

TABLE 10.1 Some Fermented Milks[a]

Fermentation Product	Origin	Principal Microbe	Characteristics
Buttermilk	Bulgaria	*Lactobacillus bulgaricus* *Leuconostoc*	American buttermilk is made from skim milk by the action of at least two different bacteria: *L. bulgaricus* produces lactic acid, and *Leuconostoc* cause the product to become viscous.
Yogurt	Bulgaria and Turkey	*Streptococcus thermophilus* *Lactobacillus bulgaricus*	High-temperature fermentation of concentrated milk. Product has consistency of custard.
Leben	Egypt	*Lactobacillus bulgaricus*	More sour and less firm than yogurt, containing a small amount of alcohol produced by yeast. Usually made from goat's milk.
Kefir	Mountains of Caucasus	*Lactobacillus bulgaricus* *Streptococcus lactis* *Saccharomyces kefir*	Mixed acid and alcohol fermentation; bacteria produce lactic acid; yeast produce the alcohol and carbon dioxide. Since the gas cannot escape, a unique effervescent alcoholic buttermilk is produced.
Acidophilus milk	United States	*Lactobacillus acidophilus*	Produced for medicinal purposes. A pure culture is inoculated into sterilized milk. These bacteria can establish themselves in the lower gut, replacing offensive microbes that are driven off by the lactic acid.
Kumiss	Russian tribe living near the Kuma River	Similar to kefir	Similar to kefir except that mare's milk, which has a higher sugar content than cow's milk, is used. Thus, the yeast produces a higher alcohol content (up to 3%).
Taette	Norway	*Lactobacillus bulgaricus*	Slime-producing bacteria make this product highly viscous.

[a]Fermented milks not shown in the table include curds (Ceylon), gioddu (Sardinia), matzoon (Armenia), dadhi (India), and skyr (Iceland).

4. Divide into appropriate containers and store in the refrigerator. Always remember to save a few teaspoons of yogurt to form the starter for the next fermentation.

A wide variety of fermented milks are made throughout the world (Table 10.1). In all cases, the basic process is the fermentation of lactose to lactic acid, with the resulting curdling of the milk protein. The exact nature of the fermentation product, however, depends on the source of milk (cow, goat, ewe, mare, and so on), the temperature of incubation, and the kinds of microorganisms in the starter culture. In some cases, a mixed fermentation occurs, producing both lactic acid and alcohol. Sour cream is produced in much the same manner as buttermilk except that the bacteria are inoculated into cream rather than into whole or skim milk.

Cheese can be defined, according to the U.S. Food and Drug Administration, as

> the product made from the separated curd obtained by coagulating the casein of milk, skimmed milk, or milk enriched with cream. The coag-

ulation is accomplished by means of rennin or other suitable enzymes, lactic fermentation, or by a combination of the two. The curd may be modified by heat, pressure, ripening ferments, special molds, or suitable seasoning.

Cheese is manufactured from milk in three distinct steps: formation of curd, treatment of curd, and ripening. The first step in cheesemaking is curdling of the milk proteins. This is accomplished either as a result of acid production during lactic acid fermentation or by addition of the enzyme rennin (usually obtained from the stomach of calves, but it can also be isolated from bacteria). In both cases the casein coagulates, forming an insoluble clot (the curd) and a straw-colored liquid (the whey). After the curd has settled into a solid mass, it is separated from the whey by draining, with or without the use of pressure. Drainage without the use of pressure is used for soft cheeses; hard cheeses are heated and pressed with weights until the curd forms a very compact, tough curd. After treatment with salt, the separated curd is molded into the desired shape and is ready for ripening.

Except for a few bland cheeses, such as cottage-type and cream cheeses, the pressed, salted, and shaped curd must be ripened by the action of bacteria and fungi before it is ready to eat. Working under carefully controlled conditions of temperature and humidity, the microbes alter the cheese in a number of different and characteristic ways. During this long ripening period, the cheese changes its texture and develops aroma and flavor that depend on the number and kind of microorganisms present either inside the cheese or on its outer surface.

Cheeses are classified in several ways: on the basis of how the curd was formed (as acid or rennin cheese); on the basis of texture (as hard, semi-hard, or soft cheeses); on the basis of the microbe primarily responsible for ripening (as mold or bacterial cheeses); on the basis of the kind of milk used (as goat, cow, mare, or ewe cheeses); and on the basis of country of origin. There are more than 400 kinds of cheese, with over 800 different names. Some are produced only for local consumption. For example, *Queso de Bola*, a Mexican cheese that is similar to Edam, is available only in the area in which it is made. Table 10.2 summarizes a few of the more tangible characteristics of some common cheeses, classified according to texture.

Cottage cheese is simply the separated fresh curd resulting from the action of lactic acid–producing bacteria on whole or skim milk. Sometimes the curds are salted or otherwise seasoned. When cottage cheese is produced from skim milk, the resulting proteinaceous curds are low in fat content and in calories. Cream cheese is another example of an unripened soft cheese. In this case, however, the final product has a high fat content because whole milk enriched with cream is used as the starting material. Both cottage-type and cream cheeses are highly perishable because of their high water content and because they contain insufficient amounts of acid or salt to prevent microbial spoilage.

TABLE 10.2 An International Cheese Board

Type	Place of Origin	Characteristics
Soft		
Cottage	Europe	Low fat, unripened, slightly salted fresh curd.
Cream	United States	High fat, unripened fresh curd; made from cream-enriched milk.
Limburger	Belgium	Bacteria-ripened for one to two months.
Camembert	France	Ripened for two to five months by enzyme secreted by surface growth of the mold *Penicillium camemberti.*
Semi-hard		
Brie	France	Ripened by a special mold on the outer surface.
Brick	United States	Usually ripened for only about one month; made from pressed curd obtained by rennin treatment of whole milk.
Bleu	Denmark	Inoculated throughout with the mold *Penicillium roqueforti;* the fungal filaments give the blue color.
Roquefort	France	Made from the milk of ewes; ripened by the blue-green mold *Penicillium roqueforti.*
Gouda	France	After ripening with bacteria, the rind is colored with saffron.
Hard		
Cheddar	England	Most popular cheese in the United States; the same bacteria that cause the curdling of the whole milk also are responsible for ripening.
Edam	Holland	Bacteria-ripened; usually made in spherical form, coated with oil, and colored.
Swiss	Switzerland	Bacteria convert the lactic acid in the pressed curd to acetic and propionic acids (taste) and carbon dioxide (holes).
Parmesan	Italy	Ripening with bacteria for over a year, producing a very dry cheese ideal for grating.

Limburger and Camembert cheeses are made soft by the action of microbial enzymes that degrade and dissolve most of the protein casein in the curd. In the case of Camembert, the surface of the curd is inoculated with spores of the mold *Penicillium camemberti*. After the spores germinate, they grow on the surface of the cheese and secrete proteinases that slowly liquify the casein, forming a soft creamy mass at the completion of the ripening period. With Limburger cheese, the solubilization of casein is brought about by bacterial enzymes. During the ripening of these as well as other cheeses, many other complex and as yet unexplained changes occur in the cheese, yielding its characteristic flavor and aroma.

In Roquefort and the other blue cheeses, the curd is inoculated with spores of the mold *Penicillium roqueforti* and then incubated at low temperature (8 to 12°C) and high relative humidity for several months. Since the fungus is aerobic, holes are punched into the curd to facilitate development of the mold throughout the curd. Roquefort cheese has been known for almost a thousand years, originating in the vicinity of Roquefort in southern France. It is made from the milk of ewes, bred particularly for their high milk production. The ripening of the cheese is traditionally carried out in limestone caves; water trickling through crevices in the limestone cools the cave and keeps the air almost saturated with moisture, thus providing ideal natural conditions for the growth of the *Penicillium*. As the growth proceeds, enzymes are secreted that soften the casein slightly, while certain other metabolic products of the mold impart the characteristic aroma and flavor of the cheese. The green-blue veins in the cheese that give it a marbled appearance are due to the spores of the mold, which are colored.

Cheddar cheese, an example of a hard cheese prepared from whole milk, originated in Cheddar, England. Today it is the leading natural cheese manufactured in the United States. The process for making cheddar cheese includes heating and frequently turning the curd ("cheddaring") to force out as much whey as possible. The curd is then salted and formed into large blocks that are carefully wrapped with paraffin or plastic film to prevent growth of contaminating microbes on the surface. During the first few weeks of the ripening process, lactic acid bacteria inside the block increase enormously. These bacteria have very little proteinase activity, so the casein is not liquefied. In the late stages of the ripening process (3 to 12 months), the lactic acid bacteria die off, releasing enzymes and other materials that bring about characteristic changes in the texture and taste of the cheese.

A special group of microbes, the propionic acid bacteria, are responsible for the ripening of Swiss cheese. These bacteria ferment the lactic acid present in the curd to acetic acid, propionic acid, and carbon dioxide gas. The characteristic taste of Swiss cheese is due to the propionic acid, and the holes or "eyes" are produced by the carbon dioxide gas. The appearance of these holes is one of the chief criteria used in judging the quality of Swiss cheese.

Processed cheese is a blend of two or more cheeses homogenized with fresh curds, water, emulsifiers, and preservatives. Before packaging, it is pasteurized to prevent further microbial action. First developed and sold in the United States around 1915, processed cheeses are now used widely throughout the world. Although of minimal gastronomic interest, they have become popular because they keep almost indefinitely under refrigeration and melt smoothly when used in cooking. Nevertheless, most cheese connoisseurs would agree that processed cheeses represent the triumph of technology over conscience.

Although many of the basic processes that occur in the production of cheese are now understood at the biochemical level, the subtle difference that distinguishes an excellent cheese from an ordinary one cannot yet be explained in scientific terms. A first-class microbiologist is not necessarily a good cheesemaker, but a good cheesemaker must be familiar with the principles of bacteriology. Thus cheesemaking is both an art and a science.

Butter

Butter is also a microbiological product. The most popular method of preparing butter consists of adding a "butter starter" culture to pasteurized sweet cream. The butter culture contains two different bacterial species, *Streptococcus cremoris* (or *Streptococcus lactis*) and *Leuconostoc citrovorum*. These two types of bacteria will grow indefinitely together if handled properly. The *Streptococcus* causes souring of the cream by producing lactic acid from the milk sugar. *Leuconostoc* converts citric acid (normally present in small amounts in milk) to a substance called diacetyl, which gives butter its characteristic taste and aroma. Neither microbe by itself yields a satisfactory butter.

After the souring reaction, the buttermilk is churned to separate the fat globules from the other constituents. The resulting butter is washed with a salt solution and "worked" to distribute the water droplets uniformly. The final product contains about 80 percent butterfat, 17 percent water, 2 percent salt, and 1 percent residual curd. Butter should be refrigerated to prevent microbial destruction of diacetyl with accompanying loss of flavor and other undesirable changes.

10.3 Sauerkraut, Pickles, and Silage

Fermented plant materials, such as pickles, sauerkraut, green olives, and silage, are also the work of lactobacilli. In much the same manner that milk sugar is converted to lactic acid in the souring of milk, plant sugars can be fermented to produce lactic acid. The major difference is that the milk sugar is readily available to the microbes, whereas the plant sugar must be "squeezed" out of the plant tissues before the bacteria can utilize them. Consider the following home recipe for sauerkraut:

> Shred fresh cabbage and then sprinkle with salt. About 3 teaspoons (25 grams) salt should be used for each kilogram of cabbage. Fill glass jar with the salted, shredded cabbage. Press down firmly, adding more cabbage if necessary to fill jar. Adjust cover tightly to exclude air and prevent mold growth. Let stand at room temperature (18 to 22°C) for two to three weeks.

The salt serves three purposes: It adds flavor, inhibits the growth of undesirable microorganisms, and brings out the cabbage juices. Bacteria do not have to be added, since ample numbers of the appropriate types are already present on the leaves of the cabbage. After a few days of incubation, anaerobic conditions develop inside the jar as a result of microbial activity. Lactic acid bacteria then begin to proliferate, using sugars extracted from the cabbage as their food supply. Although lactic acid is the major fermentation product, other acids, esters, and diacetyl (the characteristic flavoring agent of butter) all contribute to the taste and aroma of the final product. The combination of acids and salt helps to preserve the vegetable from further microbial decomposition.

The production of pickles from cucumbers takes place in a high concentration of salt (10 to 20 percent), which limits growth of microorganisms to *Lactobacillus plantarum* and other salt-tolerant lactic acid bacteria. Fresh cucumbers contain approximately 90 percent water. When placed in brine, the water is withdrawn from the cucumbers by osmosis. Dissolved in the water are sugars, minerals, and other substances that are fermented by the microbes mainly to lactic acid.

Fermentation is completed at room temperature in approximately two months. Fermented cucumbers in brine are known as "salt stock" and can be kept for years without spoilage as long as the salt concentration exceeds 10 percent. With appropriate seasoning, salt stock can be converted to sweet, sour, or dill pickles.

The following is a tested recipe for homemade, kosher-style dill pickles:

16 small cucumbers	4 small cloves of garlic
2 bay leaves	6 cups of water
a fistful of fresh dill	6 tablespoons salt

Fit the washed cucumbers tightly into a 2-liter jar. Add the bay leaves, fresh dill, and cloves of garlic. Dissolve the salt in the water and pour over the cucumbers. Place the jar near a window that receives direct sunlight. Add water when necessary to keep the cucumbers under the brine. The pickles are ready to eat in two to three days (half done) to a week. Because of the relatively low concentration of salt and presence of garlic, these pickles do not keep for more than a few weeks in the refrigerator.

The tall cylindrical **silo** is one of the first things noticed when driving through an agricultural area. To a microbiologist, the silo is an enormous fermentation vessel for the conversion of fresh green grains, grasses, legumes, and other plant materials into non-perishable **silage.** The same type of anaerobic lactic acid fermentation that is used to preserve milk and plant materials for human consumption is involved in producing silage for animal feed.

Plant materials are first chopped into small pieces and then packed tightly into the silo to ensure anaerobic conditions. Lactobacilli immediately begin to consume the sugars in the plant juices, yielding lactic acid and considerable heat. Other microbes produce acetic acid. After a few weeks, the fermentation slows down and the temperature drops. The acids elaborated during the anaerobic fermentation preserve the silage against further microbial decomposition. When the silage is withdrawn from the silo, it no longer is green, but it contains nearly as much food value as the fresh material from which it was made. Equally important, silage is luscious to animals.

10.4 Alcoholic Beverages and Bread

A loaf of bread, a jug of wine, and thou. . .

Omar Khayyam

The studies of Cagniard-Latour, Schwann, Pasteur, and other scientists of the nineteenth century demonstrated that alcoholic fermentation is a consequence of the growth of living yeast cells on sugar solutions in the absence of air (Section 3.2). The yeast can satisfy all of their energy requirements for growth by fermenting the sugar to alcohol and carbon dioxide:

$$\underset{\text{glucose}}{C_6H_{12}O_6} \xrightarrow{\text{yeast}} \underset{\text{ethyl alcohol}}{2\ C_2H_5OH} + \underset{\text{carbon dioxide}}{2\ CO_2}$$

This microbiological reaction provides the basis for the production of leavened bread as well as all alcoholic beverages. In breadmaking, the carbon dioxide gas made during the fermentation becomes trapped in the dough, causing it to rise, while the alcohol evaporates off during the subsequent baking process. Carbon dioxide also plays an important role in the making of beer and wine. The large quantities of gas that are produced during the fermentation prevent air from entering the liquid, thereby maintaining anaerobic conditions that favor the yeast. Some of the carbon dioxide is retained under pressure in beer and sparkling wines, causing these beverages to bubble and fizz when they are poured.

About Wines

Wine is the fermented juice of the grape or some other fruit (Figure 10.2). When the fruit juice is sufficiently sweet, the fermentation takes place spontaneously. Yeasts that reside on the surface of sweet fruits and nectars of flowers fall into the juice and begin to grow. The small amount of oxygen gas dissolved in the juice is consumed rap-

Figure 10.2
The making of Burgundy wine about 1470, as represented on an old tapestry.

idly and anaerobic conditions soon prevail. The yeasts then carry out an alcoholic fermentation of the sugar.

Since wine production is such a natural process, it is difficult to ascribe its discovery to any particular time or place. Archeological evidence indicates that winemaking was a well-developed art in the Middle East at least 6000 years ago. The Egyptians taught the art to the Greeks, who were responsible for establishing the first great vineyards in southern Italy, France, and Spain. Later, the Romans expanded the vineyards to include Alsace and the valley of the Mosel and organized winemaking into a highly efficient operation in order to supply the needs of their legionnaires. Although the Mediterranean countries have remained the leaders of the world in the manufacture of wine, other parts of the world with Mediterranean-like climates, such as California, have steadily raised the quantity and quality of their wine production.

Converting fruit juice into wine serves two functions. Like most fermentation products, wine is quite stable to further microbial attack and can be stored for years without spoilage. For this reason, in many parts of the world where the water is contaminated, wine is used routinely in place of water for drinking. The second function, of course, is the pure pleasure of drinking a good wine. Throughout the ages, the great poets have sung the praises of wines, referring to them with such endearing terms as friendliness, charm, and character. The Greeks made Dionysus,* the son of Zeus, not only the god of good living but also the god of wine. The drinking of wine is deeply rooted in both the religious and the cultural life of most countries.

Although the subtleties that distinguish a great wine from an ordinary one are currently beyond scientific explanation, the basic principles involved in winemaking are simple enough. First, selected grapes of the proper maturity are crushed, traditionally by treading with bare feet, but nowadays mechanically. The fresh grape juice, or **must,** as it is called, being both acidic and rich in sugars, is an ideal growth medium for yeasts. The first must that flows from the crushed fruit makes the finest wines. Red wines are made by fermenting the must of dark grapes together with their skins. White wines are made from the must of either dark or white grapes, without their skins. Rosé wines are made by allowing only a small amount of the coloring matter in the skin to be extracted into the must.

Most European wines are produced by "wild" yeasts: The dust-like film that appears naturally on the surface of grapes contains many different microbes, including the wine yeast, *Saccharomyces ellipsoideus*. More than a hundred different varieties of *S. ellipsoideus* are known, and the character of the wines from a particular district depends, to a large extent, on the local population of wild yeasts. In California, pure cultures of selected yeasts are used to inoculate the must to ensure the quality of the final product. In either case, the

*Called Bacchus by the Romans.

must is generally treated with sulfur dioxide to inhibit the growth of contaminating bacteria that might otherwise ruin the fermentation. Part of the art of winemaking lies in adding just enough sulfur dioxide to destroy the undesirable microbes without disturbing the taste.

The fermentation is allowed to proceed at a controlled temperature (usually between 22 and 28°C) for a few days, up to a couple of weeks. During this time, yeasts convert the sugars in the must to alcohol and carbon dioxide gas. The fermentation slows down when the yeasts have used up most of the sugar or are inhibited by the alcohol they have elaborated (up to 12 to 14 percent). **Dry** wines are made by allowing most of the sugar to be transformed to alcohol. **Sweet** wines arise from juices that have an excess of sugar.

When the fermentation has reached the desired stage, the young wine is separated from the sediment and transferred to wooden casks for aging at cool temperatures. Two important changes occur during storage in the casks: flavor enhancement and clarification. Flavor, which is a combination of taste and aroma, is developed in the wine as a result of slow chemical changes. The oxidation of aldehydes to acids and the subsequent formation of volatile esters are believed to be of great significance in giving wine its bouquet. Since small amounts of oxygen are needed for the maturation process, young wines stored in airtight bottles cannot age properly. The time required for maturation depends on the original product. Red wines, if properly stored, may get better and better for many years. In general, the higher the alcohol content of a red wine, the better its maturation. Dry white wines are usually ripe in a year or two, and should be drunk before they are ten years old. Although wines may clear naturally over a period of time (especially if the wine is racked into fresh casks occasionally, leaving behind a little sediment each time), filtering aids are frequently used to remove the last traces of turbidity before bottling.

All of the great name wines, such as *Chateau Mouton-Rothschild, Chateau Margaux, Chateau Latour, Chateau Haut-Brion,* to name a few, are made from fermented grape juice. However, excellent fermented beverages can be produced from almost any sweet juice. Cider, perry, and mead, for example, are made from apple juice, pear juice, and honey, respectively. In many parts of Mexico, the cactus *Agave* is tapped to yield a vitamin-rich sweet sap that ferments spontaneously to produce **pulque.** Fortified wines, such as madeira, sherry, port, or vermouth, are wines to which brandy or some other spirit has been added to raise the final alcohol content to 18 to 20 percent.

Champagne and other genuine sparkling wines are produced by trapping the natural fermentation gas inside the bottle. The appropriate clarified wine is mixed with a little sugar, inoculated with a special strain of champagne yeast, and then bottled in extra-strong bottles with specially bolted-on corks. The champagne yeasts carry out a second fermentation, producing more alcohol and carbon dioxide. This time the gas cannot escape and dissolves under pressure in the tightly corked bottle. An ingenious trick, called dégorgement, is used to remove the yeast without losing all the gas. The bottle is gradually turned until it is standing upside down. After the champagne yeast sediments onto the cork, the neck of the bottle is frozen, causing the yeast to stick to the stopper. Next, the dégorgeur removes the cork, momentarily plugging the bottle with his thumb, and quickly inserts a new cork, securing it with wire. Artificial sparkling wines are made effervescent by carbonating wine in the same manner that soda water is prepared. As such, they are considerably cheaper, go flat more rapidly after opening, and have much less body than natural sparkling wines.

Making wine is an enjoyable and economical hobby. The growing availability of fermentation equipment and do-it-yourself books makes home wine-making a wonderful outlet for the creative spirit. Once the basic principles are mastered, it is possible to experiment with new and interesting wines made from any of a wide variety of fruits and flowers. The following recipe, requiring no special equipment, has been used in our laboratory for the last ten years, always yielding a pleasant rice wine:

1.5 kilograms rice
1 kilogram seedless raisins
3.5 kilograms white sugar
0.5 kilogram brown sugar
2 sliced oranges (or lemons)
11 liters water

Pour the boiling water into a large bottle containing the rice, raisins, and sugar. Shake well to dissolve the sugar and adjust the level of the liquid so that it reaches the neck of the bottle. When cool, add the sliced oranges and inoculate with a teaspoonful of brewers' yeast. Plug the bottle with gauze and cotton and allow it to stand in a cool place (around 20°C). Stir daily for about a week. Allow the fermentation to continue for three to five weeks. (The bubbling will cease and the liquid will clear as the yeast settles.) Siphon off the wine from the yeast and other material in the bottom of the bottle into a clean container. At this stage, the wine can be (A) filtered, chilled, and drunk immediately, or, preferably, (B) sealed and stored in a dark place at 10°C for six to eight weeks, then filtered to clarify and bottled.

About Beer, Ale, and Sake

The art of brewing also developed in the Middle East about 6000 years ago, which is rather surprising when one considers the complexity of the process. The fundamental difference between beermaking and winemaking is that beer is made from cereal grains, such as barley, wheat, rye, rice, and corn. Most of the carbohydrate in grains is in the form of the polysaccharide starch rather than simple sugars. Since the yeasts that are responsible for the alcoholic fermentation cannot digest starch, the polysaccharide must be broken down to sugars prior to the yeast fermentation. Thus, from the point of view of biochemistry, beer-making is a two-step process:

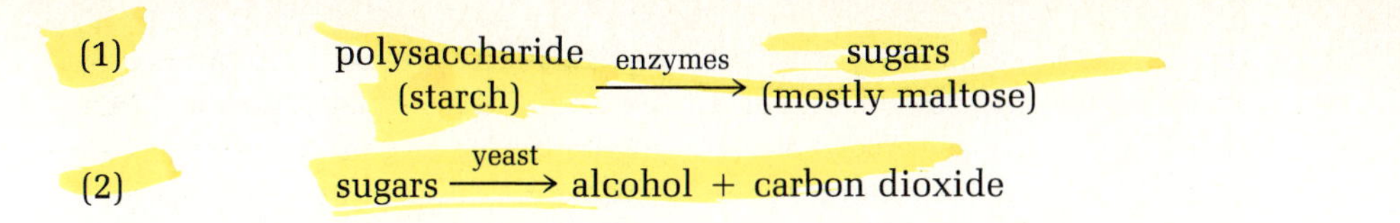

$$(1)\qquad \underset{\text{(starch)}}{\text{polysaccharide}} \xrightarrow{\text{enzymes}} \underset{\text{(mostly maltose)}}{\text{sugars}}$$

$$(2)\qquad \text{sugars} \xrightarrow{\text{yeast}} \text{alcohol} + \text{carbon dioxide}$$

Essentially the process of making beer is this: After the seeds have been steeped for a day or two in water, they are drained and subjected to sufficient temperature to cause the moist grain to sprout and develop enzymes that hydrolyze the starch to sugars; dried germinated barley seeds are called **malt.** The malt is ground, mashed in warm water, and boiled to extract the sugar together with amino acids, minerals, and flavor from the broken cells. **Hops** (the dried female flowers of *Humulus lupulus*) are added when the boiling is almost over, both for their bitter flavor and because they tend to inhibit the growth of contaminating bacteria. The resulting extract, the **wort,** contains a moderate concentration of sugar that has been set free from the starch in the grain. The wort is immediately cooled, and large quantities of yeast are added to it; the yeast is usually obtained from a previous fermentation. Shortly after addition of the yeast, the wort starts the bubbling and frothing from which fermentation derives its name (*fervere,* to boil). After several days, the reaction ceases and a copious precipitate begins to settle to the bottom of the vessel. The clarified liquid (beer) is then stored in vats at low temperatures for several weeks (lagering) prior to bottling. The sediment of yeast (traditionally called the ferment) is used to inoculate a fresh wort.

The selection of an appropriate yeast strain is the most important factor in determining the type and quality of beer produced. Throughout the ages, brewers have selected, cultured, and preserved pedigree strains for their alcohol tolerance, settling properties, and flavor. In general, brewery yeasts fall into two fundamental groups: top yeasts (strains of *Saccharomyces cerevisiae*) and bottom yeasts (strains of *Saccharomyces carlsbergensis*). Top yeasts carry out a vigorous fermentation at 20°C, producing relatively large quantities of alcohol and carbon dioxide. During the fermentation, the yeasts are swept to the surface by the rapid evolution of gas bubbles (hence the name, top yeast). Beers with high alcoholic content, such as English ale, are made by top yeasts. Most breweries in the United States and Europe use bottom yeasts to produce a light beer that keeps well in a refrigerator. The fermentation takes place slowly at 12 to 15°C, yielding less alcohol.

In the manufacture of beer and ale, the enzymes present naturally in the germinating grain are responsible for depolymerizing the starch to sugars. However, these enzymes are not present in some starchy foods, such as rice and corn. Thus, other solutions had to be found for converting the starch in these foods to sugar before they could be fermented. One technique takes advantage of the presence in human saliva of enzymes that depolymerize starch. The fact that such enzymes exist in saliva is readily demonstrable by chewing (but not swallowing) starchy foods, such as bread. The sweet taste that soon develops is caused by the enzymatically catalyzed conversion of starch to simple sugars. In certain parts of Africa, it is common practice to chew starchy roots and then expectorate the mixture into containers, where it undergoes a spontaneous alcoholic fermentation. Similarly, Indians of Central and South America prepare corn beer by chewing the grain and spitting it into vessels containing yeast.

In Japan, a more sophisticated procedure for releasing sugar from starch is used in the production of **sake** from rice. The steamed rice is inoculated with the mold *Aspergillus oryzae* and incubated for several days at 20°C. During this time, the mold grows and produces starch-degrading enzymes. This mixture is then added to a suspension of rice in water and incubated at a low temperature for a few weeks. An alcoholic fermentation ensues involving several yeasts, including *Saccharomyces sake*. The final clarified yellow rice wine contains 14 to 24 percent alcohol. Thus, in sake production, one microorganism is used to convert the starch to sugar and other microbes are used to ferment the sugar to alcohol.

Distilled Alcohol

The alcoholic content of wines and beer is limited by the sensitivity of yeast cells to high concentrations of alcohol. To obtain alcoholic beverages containing more than 15 percent alcohol (30 proof), the fermentation liquid is heated to volatilize most of the alcohol, which is then condensed and collected, a process called **distillation.** Essentially any alcoholic liquid can be distilled, each yielding a characteristic beverage (Table 10.3). Most of the art of distilling is in capturing the right blend of volatile fermentation products together with the alcohol. Aging the distillate and blending procedures also play an important part in producing distilled liquors.

In recent years there has been a growing interest in producing alcohol as a substitute fuel for automobiles. Alcohol is a clean-burning, efficient fuel; it can be used to dilute gasoline (gasohol) or petrol in automobiles and other combustion engines up to 10 to 15 percent with improved efficiency. In countries that produce large quantities of molasses, such as Brazil, sugar is fermented directly by yeast to alcohol and then distilled to yield almost pure alcohol. If the raw material is wood chips or sawdust, the cellulose must first be enzymatically hydrolyzed to glucose before it can be fermented and distilled. A considerable amount of multidisciplinary research is currently being conducted throughout the world to discover rapidly growing plants that can be fermented and distilled in order to produce alcohol at a price that is competitive with fossil fuels.

Bread

Archeological excavations reveal that 4000 years ago in Egypt the brewing of beer and baking of bread were performed in the same building. This is not surprising, because the same yeasts that are re-

TABLE 10.3 Distilled Spirits

Liquor	Fermentable Substrate	Some Characteristics of Distillate
Bourbon whiskey	Mostly corn mash	Originally produced in Kentucky; aged in charred oak containers for at least four years.
Scotch whisky	Barley (pure malt)	From the Highlands of Scotland, bearing the unique flavor of peat fires used to dry the malt.
Brandy	Fruit juice	Many different types; grape brandy distilled in the Cognac region of France is called Cognac.
Rum	Sugar molasses	Manufactured in those countries that grow sugarcane; aged in charred oak barrels.
Gin	Corn or rye mash	Distilled with or over juniper berries; subsequent addition of flavorings.
Vodka	Potatoes (or grain mash)	Filtered through charcoal to remove flavors and color.
Tequila	Cactus juice *(Agave)*	Manufactured in Mexico.
Aquavit	Potato mash	Scandinavian drink, flavored with caraway seeds and citrus peels.
Cordials and liqueurs	—	Sweetened distillates of fruits, flowers, leaves, etc.

sponsible for making beer also are needed to "leaven" or raise the bread. In bread-making, the carbon dioxide gas, rather than the alcohol, is the important product of the fermentation. The following recipe for raisin bread will illustrate the principles:

1 package active dry yeast
1/4 cup warm water
2 cups scalded milk
1/3 cup sugar
1/4 cup shortening
2 teaspoons salt
6 cups sifted flour
2 cups raisins

Soften active dry yeast in warm water (40 to 45°C). Combine hot milk, sugar, salt, raisins, and shortening. Stir in 2 cups of the flour. Add the softened yeast; mix well. Add enough of the remaining flour to make a moderately stiff dough, kneading until smooth and satiny (10 minutes). Shape in a ball and place in lightly greased bowl, turning once to grease surface. Cover; let rise in warm place until double (about 1 1/2 hours). Punch down. Cut and shape dough into 8 small loaves. Place in greased pans. Cover and let rise until double (45 minutes). Bake at 190°C (375°F) for 25 to 30 minutes.

The package of yeast contains billions of living yeast cells (about ten billion per gram). As long as they remain dry and reasonably cool, they retain their viability for months. When moistened and placed in contact with sugar, they begin to grow and ferment the sugar to alcohol and carbon dioxide. The gas expands within the dough, causing it to rise. When the bread is baked, the heat drives off the carbon dioxide, leaving holes within the bread mass that give bread its characteristic fluffy texture. The heat also kills the yeast and evaporates off most of the alcohol. Nevertheless, a fresh loaf of bread can contain up to 0.5 percent alcohol.

Since leavening bread is a biological process, the time it takes for the yeast to cause the dough to rise depends on several factors, such as temperature, amount of sugar, and condition of the yeast. The reaction speeds up as the yeast goes from a resting state to actively fermenting cells. This is the reason that in the raisin bread recipe the second doubling of the volume of the dough occurs faster than the initial doubling.

Dough can also be made to rise by the use of baking powder in place of yeast. The mixture of chemicals in baking powder generates carbon dioxide gas when moistened. However, baking powder produces a highly unsatisfactory bread and has proved useless for that purpose. Obviously, there is more to yeast than carbon dioxide. The yeast adds flavor and changes the texture of the dough in ways that are not clearly understood.

10.5 Microbes As Food: Single-Cell Protein

Human malnutrition, especially protein malnutrition, is widespread today in many parts of the underdeveloped world. Approximately 1.5 billion people in tropical and subtropical areas exist on diets of one staple vegetable crop lacking essential animal protein. Many agricultural economists predict that by the year 2000, when the world population will be close to six billion, the protein problem will be worldwide. What can be done to solve this problem?

One approach is to improve the protein content of cereals and legumes. However, plant scientists have found that hybrid strains that are high in protein or essential amino acids are less productive. A more serious problem is that of amino acid imbalances in plant protein. Legumes are deficient in methionine and cystine, two amino acids that are essential in the human diet; cereals are deficient in the essential amino acid lysine. Diets combining cereals and legumes provide a temporary solution to the problem of protein malnutrition.

Another approach is to use microbes as a direct source of protein for animal and human consumption. Although the idea of eating microorganisms may seem revolutionary, or even absurd, it has in fact been going on for a long time. We have already discussed in this chapter how cows and other ruminants satisfy their protein requirement by digesting the microbes that they propagate in their rumen. A more obvious example is the use of brewers' yeast, a by-product of the beer industry, as a supplement in animal feed. Introduced about 100 years ago, yeast are now used routinely in cattle, swine, poultry, and fish feeds. Yeast are among the most nutritious foods, being an excellent source of protein (including the essential amino acids) and vitamins of the B group. In fact, because they are so rich in vitamins, yeast and yeast extracts have been used by humans for many years to prevent and treat vitamin deficiency diseases, such as beri-beri and pellagra. During the Second World War, food yeasts were produced by the British from molasses and by the Americans and Germans from waste materials of the paper pulp industry. Over 20,000 tons of yeast were produced and incorporated into human food during the war.

After 1945, yeast production plants were established in many parts of the world. Those that were concerned with producing yeast for animal feed were generally successful. The annual world production of feed yeast has now reached 250,000 tons and is still rising. On the other hand, plants that were established to produce yeast to help feed the protein-starved people of East Africa, India, and Indonesia failed, because the human element of the problem was not fully appreciated. People do not necessarily eat what is good for them. Unfamiliarity with a food is a powerful, almost instinctive, deterrent to eating. Thus, by 1950, almost all of the food yeast plants had shut down, and the eating of yeast was restricted to a few health food enthusiasts in Europe and the United States.

Then in the 1960s, the energy and food crises resulting from the expanding world population brought about renewed interest in the production of food from microorganisms. This time, however, considerably more attention was given to the human side of the problem. In a conference held at the Massachusetts Institute of Technology in 1966, the term **single-cell protein** was coined to include all forms of microbial food. The name was chosen because it does not have the unpleasant connotation associated with "bacterial," "fungal," or "microbial." The name is only one part of a large educational effort (or propaganda, depending on how you look at it) to assure the public that microbial food is healthful. On the technical side, research is continuing on methods of improving the taste, smell, and texture of single-cell proteins.

Production of single-cell protein for partially relieving the world shortage takes advantage of several unique properties of microorganisms:

1. **Microorganisms grow rapidly.** The doubling time of microorganisms varies from about 15 minutes to a few hours. As a conservative example, consider yeast growing on molasses with a doubling time of three hours. Within a day, the yeast will double eight times, yielding 2^8 or 256 times the initial mass of cells. To achieve a corresponding increase in the mass of soybeans, poultry, and cattle would require three weeks, three months, and three years, respectively.
2. **Microorganisms have a high protein content.** Depending on

the particular type of microorganism and the growth conditions, the protein content of microbes varies from approximately 30 to 70 percent of the dry weight of the cells. By comparison, legumes, wheat, and rice contain 25, 12, and 8 percent protein, respectively.

3. **Microorganisms can utilize extremely diverse raw materials.** Relatively inexpensive raw materials, including waste products, can serve as the carbon and energy sources for microbial growth. Microbes are available that can grow on sugars, alcohols, cellulose, and petroleum products. Photosynthetic microbes can make protein from carbon dioxide, sunlight, and inorganic salts.
4. **Microorganisms can be easily modified genetically to produce favorable traits.** It takes many years to alter animals and plants genetically. Most strains that are used today for food have undergone continuous strain improvement for centuries and are thus very different from their wild-type ancestors. It should be possible in a relatively short time to improve microbial food strains by mutation and selection as well as by genetic engineering.
5. **Microorganisms can be cultivated on a large scale anywhere in the world, independent of soil or climatic conditions.** Traditional agriculture is at the mercy of natural conditions. Floods, drought, drastic changes in temperature, diseases, and other unpredictable events can ruin crops and lead to widespread famine. Such disasters have occurred periodically throughout history. As recently as 1976, drought caused major crop losses in North America and Europe and resulted in the depletion of the global grain reserves. Conditions for the production of single-cell protein, on the other hand, can be scientifically controlled by the technicians operating the factory. A medium-sized single-cell protein plant, occupying an area of about 3 acres, produces about the same amount of protein as 300,000 acres devoted to soybeans.

Although most single-cell protein processes are still in the development or pilot plant stages, a few large-scale production plants are already in operation. Table 10.4 illustrates some of the processes that are either currently being exploited or under serious consideration. Processes using yeast are the most advanced on a commercial scale. After more than ten years of intensive research and development, the British Petroleum Industry established a large single-cell protein factory in Lavera, France, to produce yeast from gas-oil, a petroleum fraction containing high concentrations of normal paraffins. In the aerobic process, two tons of dry yeast are produced from one ton of the petroleum fraction. Even with increased costs of petroleum products, the yeast protein can be manufactured at a price competitive with other sources of protein. Large yeast-from-petroleum factories are now operating or being constructed in many parts of the world, in-

TABLE 10.4 Processes for Producing Single-Cell Protein

Microorganism	Raw Material	Development Stage	Location
Yeast	Gas-oil	20,000 tons/yr	Lavera, France
Yeast	Ethanol	1000 tons/yr	Czechoslovakia
Bacteria	Kerosene	150 tons/yr	Taiwan
Bacteria	Methanol	1000 tons/yr	Billingham, England
Bacteria	Cellulose	Pilot plant stage	Louisiana State University
Bacteria	CO_2 and H_2	Research project	Germany
Algae	CO_2 and sunlight	300 tons/yr	Lake Texcoco, Mexico

cluding Czechoslovakia, Italy, Russia, Scotland, and Japan. One of the problems of the yeast-from-petroleum process is the danger that small quantities of toxic hydrocarbons will remain in the final product even after extensive extraction and purification. Using ethyl alcohol as the raw material for producing yeast avoids this problem; however, single-cell protein from alcohol is much more expensive than that produced from hydrocarbons.

The major advantage of using bacteria for making single-cell protein is that a wide range of raw materials can be used. In addition to molasses, ethanol, and petroleum fractions, other inexpensive substrates can be used for growing bacteria, such as methane gas, methyl alcohol, starch, and cellulose. Especially attractive is the potential of turning waste materials, which present disposal problems, into usable single-cell protein. For example, only a small fraction of any plant is harvested for food; the bulk of the plant, which is cellulose, is either plowed into the soil or burned. Several research teams are studying the feasibility of converting such waste plant materials into single-cell protein. Both fungi and bacteria are capable of using cellulose as a substrate for growth. A more exotic bacterial process is the use of hydrogen gas as a source of energy for fixing carbon dioxide into cellular material. In essence, any process that converts a cheap, readily available substance into protein is potentially valuable.

Many scientists argue that in the long run, the economically best source of single-cell protein is photosynthetic microorganisms. Algae and photosynthetic bacteria can be grown in fresh water or sea water, supplemented only by carbon dioxide plus inorganic salts, and exposed to sunlight. The major advantages of microscopic algae over higher plants are their more rapid and efficient growth and the fact that the entire microbe can be harvested and used for food—nothing is wasted. The problems that must be solved before mass cultivation of algae becomes economically viable are primarily technical, rather than theoretical: efficient methods of introducing carbon dioxide gas, mixing, harvesting, and preventing contamination by other microorganisms.

The reasons that single-cell proteins have not been more widely accepted by now as a food for humans are partly psychological, partly technical, and partly because of safety considerations. We have al-

ready discussed the psychological problem of introducing unfamiliar foods. In the case of yeast, this problem has already been largely solved. Most of us (usually unknowingly) have learned to appreciate the flavor of yeast in such diverse products as baked goods, breakfast cereals, sausages, baby foods, soups, peanut butter, protein-enriched noodles, and meat dishes. In Japan, considerable progress has been made in introducing single-cell algae into such foods as soups, bread, cakes, yogurt, and ice cream. Consumer acceptance may be more difficult in less developed parts of the world.

The major health hazard associated with single-cell protein is the high nucleic acid content of microorganisms. If the human diet contains large amounts of single-cell protein, there is an enhanced risk of developing gout or kidney stones. For the present, this problem is circumvented by limiting the amount of single-cell protein in the human diet to 10 to 20 percent of the total protein intake. There are reasons to be optimistic that in the future, the combination of general microbiology, microbial genetics, bioengineering, and food technology will lead to new and improved processes for single-cell protein.

In summary, microbial protein is already an established component of animal feeds and a supplement in the human diet. It is likely that single-cell protein will become a larger part of our diet in the future, as technical problems are solved and consumers become more accustomed to the product.

10.6 Food Spoilage by Microorganisms

Although most of this chapter has been devoted to desirable changes in food brought about by microbes, it is impossible to ignore the role of microorganisms in food spoilage. We are all too familiar with bread becoming moldy, vegetables rotting, meat becoming putrid, fats going rancid, cottage cheese going sour, and so on. Most spoilage is due to the development of undesirable changes in the flavor, aroma, texture, or appearance of the food. In these cases, *spoilage* and *undesirable* are subjective terms that depend, to some extent, on the eating habits of the particular individual concerned. For example, a Camembert cheese that becomes very soft and pungent may be considered spoiled by one person and delicious by someone else. On the other hand, types of food spoilage that present health hazards are clearly undesirable—those accompanied by the growth of pathogenic microorganisms or production of toxic products.

To illustrate some of the principles of food spoilage, consider the following series of events leading to a small epidemic of food poisoning in a university dormitory. One summer afternoon, the cook prepared a large chicken casserole and placed it in a hot oven for a couple of hours. The dish was taken out of the oven and served as the main course for dinner while it was still hot. Since a large amount of the casserole was left over, it was taken back to the kitchen, covered, and left unrefrigerated overnight. At 11:00 the next morning, the chef mixed the casserole with freshly cooked rice, warmed it up, and served it for lunch. Three hours later, almost all the students who had eaten the leftover chicken casserole came down with food poisoning.

What happened? When the casserole was served the first time, it contained very few bacteria and was a perfectly safe food. The baking procedure brought out the juices, made the meat tender, and sterilized the entire dish. However, while the casserole was being served, the bacterium *Staphylococcus aureus* fell into it—from either the air, someone's finger, or a dirty serving spoon. The leftover casserole was an ideal medium for the multiplication of the staphylococci. The same baking procedure that made the chicken tender and juicy also converted it to an ideal medium for the growth of bacteria. During the evening, the temperature in the kitchen ranged from 25 to 30°C; under these conditions, *Staphylococcus aureus* has a doubling time of about one hour. If a single bacterium infected the casserole at 7:00 PM, there would be two bacteria at 8:00 PM, four at 9:00 PM, eight at 10:00 PM, and so on. By 11:00 AM, the number of *Staphylococci* would have risen to about 130,000. Except for a slight odor, the casserole appeared normal. Nevertheless, the staphylococci had produced enough toxin to cause food poisoning. Even though the warming procedure used before serving the mixture for lunch killed most of the bacteria that had grown during the night, the heat-stable toxin was not destroyed. Thus, it is possible to come down with a serious case of bacterial food poisoning even though no live bacteria are in the food.

Food Preservation

What can be done to prevent or slow down food spoilage? There are four general methods of preserving food: (1) destroying or greatly reducing the number of microbes initially present on or in the food; (2) treating the food in such a way that it is not a good medium for the growth of microorganisms; (3) packaging and handling the food in such a manner that microbes cannot easily gain entry; and (4) storing the food under conditions that prevent or greatly decrease the growth of spoilage microorganisms.

All natural foods contain viable microorganisms. Fresh fruits and vegetables, for example, have a large number of bacteria and fungi on their surfaces even after extensive washing. The first step in food preservation is killing as many of the microbial food contaminants as possible without harming the nutritional value or taste of the food. The most widely used technique is heat. If a food is exposed to a high enough temperature for a long enough time, all the microbes will be killed; i.e., the food will be sterile. The exact time and temperature needed to sterilize a food item depends on such factors as moisture content, acidity, and how well the heat penetrates the food.†

†The first systematic studies of conditions necessary to preserve foods for canning were performed around 1805–1810 by Nicholas Appert, a French confectioner.

Many foods cannot be heat-sterilized without adversely affecting their taste. Wine and beer, for example, lose both alcohol and taste when heated to a high enough temperature to sterilize them. Pasteur discovered that the main spoilage organisms in wine, acetic acid bacteria, could be killed by heat without severely disturbing the wine. Subsequently, the technique of **pasteurization** was applied to milk and other foods. It should be emphasized that pasteurization is not a sterilization procedure. Pasteurization destroys the major spoilage organisms and all of the disease-producing nonspore-forming microbes. Pasteurized foods keep well under refrigeration because the few microbes that survive the heat treatment generally do not grow well at low temperatures.

Drying food is one of the oldest methods of preserving perishable materials. All living organisms require water for growth. Thus, any procedure that **desiccates** or **dehydrates** food will prevent microbial spoilage. The ancient practice of sun-drying fruits, vegetables, and small pieces of meat is based on this principle. Today, powdered milk and many other dehydrated food products are prepared commercially by a spray-drying process.

Various chemical preservatives can be added to foods to retard spoilage. Part of the success of fermentation processes is due to the production of natural chemical preservatives, such as alcohol, acetic acid, and lactic acid. High concentrations of salt (in pickling) or sugar (in jellies, jams, maple syrup, and honey) also act as preservatives. In recent years, there has been growing concern over the use of artificial chemical preservatives, such as sodium benzoate, sorbic acid, and sodium nitrite. As mentioned previously (Section 9.3), nitrites are potential **carcinogens.**

Storing foods in the refrigerator (around 5°C) prevents the growth of many spoilage organisms. However, psychrophilic bacteria (Section 8.2) and fungi grow slowly at refrigerator temperatures and cause spoilage. During the last 50 years, the technology of quick-freezing foods has been developed to preserve the taste and nutritional value of food. Below −10°C, microbial activity essentially ceases, so that foods kept in a deep freezer remain free from spoilage indefinitely. A recent Russian expedition to the North Pole uncovered meat that had been frozen for thousands of years; close examination of the thawed meat showed no sign of spoilage. Although very low temperatures prevent microbial growth, they do not necessarily kill the microbes. Many bacteria and fungi remain dormant at low temperatures for years. When the food thaws, the microbes begin to multiply and produce toxic products. This is the reason why it is dangerous to refreeze frozen food that has thawed.

QUESTIONS

10.1 Considering the mutualistic symbiosis between cow and microbes

a) How does the cow benefit?
b) How do the microbes benefit?

c) How do humans benefit?
d) What prevents the microbes from converting the cellulose all the way to carbon dioxide and water?

10.2 Explain the fact that the shelf life (time it takes for a food product to become uneatable) of cheese is greater than that of yogurt, which, in turn, is greater than that of milk.

10.3 What do each of the following terms signify?

Malt	Rennin
Processed cheese	Swiss cheese
Cider	Hops
Single-cell protein	Silage
Pasteurization	Cottage cheese

10.4 The first step in the production of alcoholic beverages from grains is the conversion of starch to glucose. How is this accomplished when the grain is barley? Corn? Rice?

10.5 Outline the steps (and microbes) that would be required for the production of alcohol from cellulose.

10.6 What is the difference in the role that yeasts play in the production of bread and wine?

10.7 Explain how it is possible to obtain a serious case of bacterial food poisoning even when the food is shown to be completely free of bacteria at the time of consumption.

10.8 Assume that a particular spoilage microorganism has a doubling time in beef stew of one hour at room temperature (20°C) and ten hours at refrigerator temperature (8°C). How long would it take a single bacterium to reach a population of one million at each temperature?

Suggested Readings

Amerine MA: Wine. Scientific American, August 1964.

A popular account of the wine-making process.

Baron S: Brewed in America—A History of Beer and Ale in the United States. Boston: Little, Brown and Company, 1962.

Peppler HJ and D Perlman: Microbial Technology: Fermentation Technology, volume II. New York: Academic Press, 1979.

An up-to-date professional description of many aspects of food microbiology, including beer brewing (chapter 1), cheese (chapter 2), distilled beverage (chapter 31), and wine (chapter 5).

Wilster GH: Practical Cheesemaking. Corvallis: Oregon State University Press, 1969.

Antibiotics and Other Valuable Microbial Chemicals

Chance favors the prepared mind.

Louis Pasteur

Never neglect any appearance or any happening which seems to be out of the ordinary: more often than not it is a false alarm, but it may be an important truth.

Alexander Fleming

The theme of this chapter is simple: Certain microbes synthesize chemicals that are of great value to humans. The isolation of these microbes and their exploitation for the large-scale production of valuable chemicals have led to the development of **industrial microbiology.** However, the phrase "industrial microbiology" fails to capture the relevance of this subject to all of us. Microbially produced chemicals not only are valuable commercially, but also extend and improve the quality of our lives. The microbial chemicals that are discussed in this chapter are antibiotics, vitamins, amino acids, organic acids, hormones, polysaccharides, enzymes, and insecticides.

11.1 Chemotherapy: The Concept of the "Magic Bullet"

From time immemorial, man has attempted to cure diseases by administering specific substances. Folk medicine is rich in recipes for blending extracts of herbs, roots, shellfish, and many other natural materials. Although in several cases modern science has shown that such folk concoctions actually do contain beneficial substances, very few of the traditional folk medicines were effective in combating infectious diseases. One notable exception is the successful treatment of malaria with cinchona bark (which contains quinine), used by the Peruvian Indians.

In 1909, Paul Ehrlich, a German bacteriologist-chemist (Figure 11.1), coined the term **chemotherapy** and established the logical basis for controlling diseases by specific chemicals. Ehrlich took into consideration two important biological developments of the late nineteenth century: (1) the demonstration by Koch, Pasteur, and other

Figure 11.1
Paul Ehrlich (1854–1915). (Courtesy of Tel Aviv University Medical History Library.)

microbiologists that infectious diseases are caused by microbes and (2) the discovery that certain synthetic dye chemicals specifically adhere to and stain bacterial cells. According to Ehrlich, the fundamental problem in chemotherapy is to find a drug that *selectively* kills the disease-causing microbe. Most chemicals that are toxic to microbes are also poisonous to humans. What is needed is a "magic bullet," a chemical that will penetrate and kill microbes without damaging the host tissues. In modern terminology, Ehrlich was searching for an antimicrobial drug with a **high therapeutic index**—that is, a high ratio of maximum dose at which a drug can be tolerated to minimum dose required to cure infections:

$$\text{therapeutic index} = \frac{\text{toxic dose for humans}}{\text{toxic dose for microbes}}$$

If the value of the ratio is large, the drug is relatively safe to use; if it is small (close to unity), then the dose must be controlled carefully; if the ratio is one or less, the substance is not useful as a medicine.

Ehrlich's strategy for producing chemotherapeutic agents was to combine toxic atoms, such as mercury and arsenic, with organic compounds that stick to microbes. It was hoped that the organic component would impart the necessary selectivity for the microbial cell and that the toxic atom would still be potent. Figure 11.2 shows the chemical formula of Salvorsan, the 606th compound to be tested. Salvorsan, or "Ehrlich 606" as it is sometimes called, proved to be

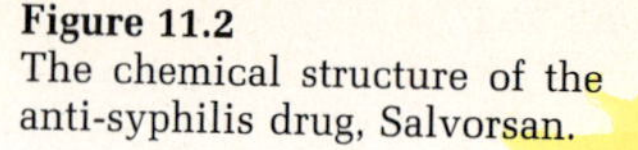

Figure 11.2
The chemical structure of the anti-syphilis drug, Salvorsan.

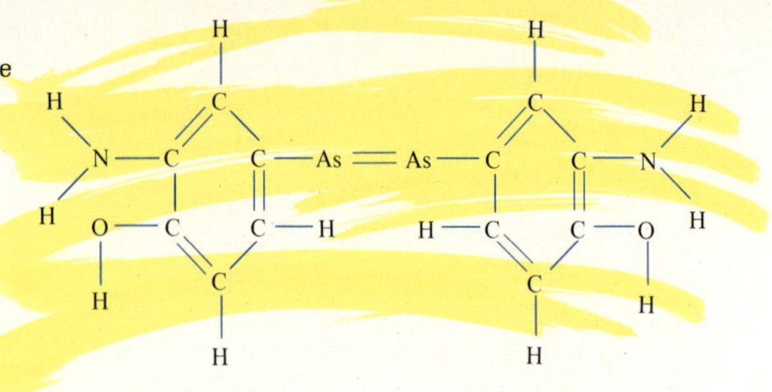

highly effective against the syphilis germ, although it had severe side effects.

The next important advance in chemotherapy was the synthesis of Prontosil in 1935. Once it was shown that Prontosil was active against streptococcal infections, the German pharmaceutical cartel I. G. Farben patented the drug. With an international monopoly on the only drug effective against "strep" infections, the giant Farben industry was in a powerful position to control the world market. "Fortunately for the world," as Sir Alexander Fleming subsequently stated, "Trefouel and his colleagues in Paris soon showed that Prontosil acted by being broken down in the body with the liberation of sulfanilamide and this simple drug, on which there were no patents, would do all that Prontosil could do."

The relatively simple structure of sulfanilamide allowed it to be modified in a variety of different ways to give rise to a class of compounds called **sulfa drugs.** Several of the synthetic sulfa drugs have a high therapeutic index and proved effective in treating a number of infectious diseases, including bacterial pneumonia, rheumatic fever, respiratory infections caused by streptococci and staphylococci, wound infections, and urinary tract infections. During the period 1935–1940, the sulfa drugs had an enormous impact in medicine. Unfortunately, indiscriminate use of the sulfa drugs led to the selection of mutant bacteria that were resistant to these drugs.

How do the sulfa drugs work as chemotherapeutic agents? As shown in Figure 11.3, the sulfa drugs have a striking similarity to a chemical found inside most bacteria, nicknamed PABA. Normally, a

Figure 11.3
The chemical similarity between sulfanilamide, one of the sulfa drugs, and PABA.

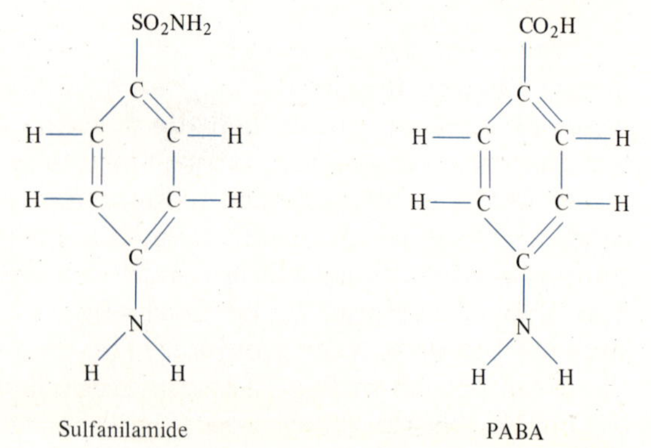

specific enzyme in bacteria converts PABA to one of the B vitamins, folic acid. However, the enzyme cannot distinguish between PABA and the sulfa drugs; in the presence of a sulfa drug, the enzyme combines with the drug and is unavailable for the synthesis of folic acid. Without folic acid, the microbe cannot grow. This explains how sulfa drugs poison bacteria, but it fails to explain why sulfa drugs are not poisonous to humans. The reason is that human beings and other mammals are totally deficient in the enzyme that converts PABA to folic acid. We must obtain folic acid in our diet. (That is, in fact, why folic acid is a vitamin for humans.) Lacking the enzyme system for the fabrication of folic acid, mammals are immune to the toxic effects of the sulfa drugs.

The discovery of the mode of action of the sulfa drugs immediately suggested a rational approach to the development of chemotherapeutic agents—synthesize chemicals that are structurally similar, but not identical, to naturally occurring substances. Such compounds are called **analogs.** Since 1940, thousands of analogs have been synthesized and tested for chemotherapeutic value. Although a few clinically useful drugs have been discovered by this approach, the results generally have been discouraging. Most analogs that "fooled" the microbial enzymes were also toxic to humans.

In science, logic is a necessary but not sufficient basis for success. As it turned out, a chance observation, rather than a logical deduction, led to the major breakthough in antibacterial chemotherapy.

11.2 The Penicillin Saga

The discovery and development of penicillin is one of the great scientific achievements of the twentieth century. Like many events of major historical significance, the contributing factors were varied and complex. Major roles were played by microbiologists, chemists, fermentation engineers, medical doctors, businessmen, industrialists, lawyers, and government officials. The motivating forces ranged from pure scientific curiosity and the desire to alleviate suffering to economic gain and the pressures of the Second World War.

The Initial Discovery: Alexander Fleming

The story begins in 1928 with a fortuitous observation by Alexander Fleming (Figure 11.4), a bacteriologist working at St. Mary's Hospital in London. Fleming was growing a disease-causing staphylococcus in petri dishes containing nutrient agar. An airborne fungal spore fell inadvertently onto the agar and began to multiply. As the contaminating mold grew, a halo, or clear area, developed around the mold colony (Figure 11.5). Such accidents must have happened to hundreds of bacteriologists before Fleming. The ruined agar plates were simply discarded in disgust.

Fleming, however, realized the significance of the clear area around the mold colony. The fungus must have secreted something

Figure 11.4
Alexander Fleming (1881–1955). (Courtesy of Tel Aviv University Medical History Library.)

into the medium that inhibited the growth of the staphylococci. He therefore isolated the fungus and repeated the experiment, this time intentionally adding the fungus to the bacterial culture. Again the fungus excreted a product that killed bacteria in the neighborhood of the mold colony. Fleming went one step further. He grew large batches of

Figure 11.5
Fleming's original culture of staphylococci showing a clear area around the contaminating mold colony. (From Fleming, A.: British Journal of Experimental Pathology, 10:228, 1929.)

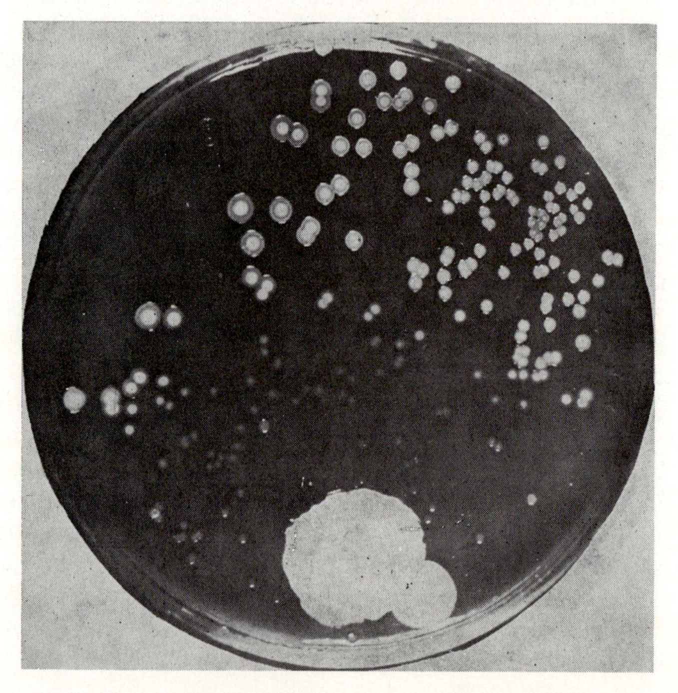

the mold and then separated the mycelial mass from the culture medium. The mold-free juice he prepared was still lethal to bacteria. Since the mold was a species of *Pencillium*, Fleming called the antibacterial material in the juice penicillin.

Fleming published his studies on penicillin in the *British Journal of Experimental Pathology* in 1929. Shortly thereafter he attempted to concentrate and purify penicillin, failed, and abandoned the project. There followed a period of ten years during which there were no significant developments in penicillin research. Several suggestions have been put forth in an attempt to explain this ten-year lag between the discovery of penicillin and its development as a potent chemotherapeutic agent. Although Fleming wrote in 1929, that "it may be an efficient antiseptic for applications to, or injection into, areas infested with penicillin-sensitive microbes," he failed to convince the scientific community of the potential importance of penicillin. Fleming gave up because he was not a good enough chemist to purify the unstable penicillin molecule. Furthermore, he lacked the research budget necessary to hire a chemist to help him. With the millions of dollars being spent today on medical research, it is difficult to comprehend how an established scientist like Fleming could not obtain the $5000 per year that he needed for his penicillin project. Finally, it should be mentioned that scientists are not always the best promoters of their own ideas. Fleming represents the classic example of a scientist whose original findings were subsequently rediscovered, developed, and exploited by others.

Purification and Demonstration of In-Vivo Effectiveness: Chain and Florey

As a consequence of the rise of fascism in the 1930s, many outstanding scientists were forced to flee from Germany. One of these refugees from Hitler was the biochemist Ernst Chain. Shortly after arriving in England, Chain joined with Howard Florey, a pathologist at Oxford, in a systematic search for antibacterial substances. A careful library study led Chain to select penicillin as the target of research. After receiving the penicillin-producing mold from Fleming, the team of Chain and Florey succeeded, in a relatively short time, in developing techniques for the purification of penicillin. The first animal experiment with the partially purified penicillin was extremely encouraging (Table 11.1). Eight mice were used in the experiment. Each of them was injected with 110,000 highly pathogenic streptococci. The four that received no penicillin (controls) became very sick after seven hours and died between 12 and 16 hours after the experiment. The two mice that each received 10 milligrams of penicillin survived four and six days, respectively. One of the mice that received 25 milligrams of penicillin survived 13 days; the other was still alive after six weeks. Subsequent studies demonstrated that penicillin was not toxic to mammals, including humans.

Preliminary tests with humans suffering from incurable bacterial infections showed that penicillin was a miraculous drug. However,

TABLE 11.1 Chain and Florey's First Mouse Protection Test

Mouse	Streptococci[a]	Penicillin[b]	Results
A	+	None	Dead in 12 hours
B	+	None	Dead in 14 hours
C	+	None	Dead in 14 hours
D	+	None	Dead in 16 hours
E	+	10 milligrams	Survived 4 days
F	+	10 milligrams	Survived 6 days
G	+	25 milligrams	Survived 13 days
H	+	25 milligrams	Survived more than 6 weeks

[a]All mice were injected with 110,000 disease-causing streptococci.
[b]The partially pure penicillin was introduced under the skin.

there were some serious disappointments: For example, a London policeman who was showing signs of recovery from bacterial poisoning suddenly died when the supply of penicillin ran out. By 1941, the pressing problem was how to produce enough penicillin for general clinical use. With the outbreak of World War II, the need for drugs effective against battle wounds became even more urgent. Florey and Chain went to the Ministry of Health for help. A decision was made at the highest level of government to assist the Oxford group in making more penicillin. Two different approaches were tried simultaneously: First, penicillin manufacture was to be scaled up in England using whatever manpower and equipment were not essential for the war effort; second, Florey was sent to the United States to try to enlist the help of the Americans.

Production of penicillin was coordinated in Great Britain by the Office of Scientific Research and Development. The technique of producing penicillin at that time consisted of growing the mold on the surface of a shallow layer of culture medium for several days and then pouring the liquid into a large vat, from which the drug could subsequently be extracted and purified. The difficulty with this technique was that enormous numbers of flasks were required to grow the mold. To grasp the magnitude of the problem, consider that (1) about 1,000,000 units of penicillin are required to cure a single person of pneumonia and (2) maximum yield of penicillin was then 1000 units per flask. It follows that at least 1000 flasks were needed to manufacture enough penicillin to treat a single case of pneumonia. There simply were not enough flasks in England to produce even a small proportion of the needed miracle drug. Many different types of vessels were tried. As Florey wrote,

> Apart from the usual laboratory ware, trials were made with various kinds of glass and enamel domestic dishes and utensils, and biscuit and other tins, but it was found that the old-style bed pan with a side-arm and lid was an ideal vessel, providing a relatively large surface area over a shallow layer of fluid.

Unfortunately, there also were not enough bed pans to go around. Instead, researchers turned their attention to one-pint milk bottles, which were produced in large numbers without difficulty by existing glass factories. Giant incubator rooms were fitted with racks to hold the bottles. Several of these so-called "bottle plants" were equipped to hold more than 100,000 milk bottles incubating at one time. After the mold had ceased growing, the cotton plugs had to be removed and the liquid poured into vats for extraction and purification. Most of the labor was provided by schoolchildren. Despite this major effort, England could provide only enough penicillin to treat a fraction of the critical military patients for whom there was no other hope.

Industrial Development: The American Connection

Meanwhile, tremendous advances were being made in the United States. Florey had come to New York in July 1941 with one of his assistants and the penicillin-producing strain. Although the large pharmaceutical companies initially were not interested in the penicillin project because the yields were too low, Florey was sent to Peoria, Illinois, the headquarters of the newly established Fermentation Division of the U.S. Department of Agriculture. The laboratory in Peoria was ideally suited for the project. The proper equipment was available, amd more important, the research staff was experienced in mold fermentations. After a few days of intense discussions, Florey returned to England, leaving his trusted assistant, Norman Heatley, to teach the Americans what was then known about growing *Penicillium* and measuring the quantity of penicillin produced. During the conversations between Florey and the American fermentation experts, a plan emerged that was to provide the basic strategy for the conversion of subsequent laboratory discoveries in microbiology into viable commercial processes:

1. **Media improvement.** Formulation of a diet for the microbe that would be cheap and would allow for maximum production of the desired product. It was already known that the type of food consumed by a microbe can influence greatly the chemicals that it synthesizes.
2. **Development of equipment and conditions for optimal production on a large scale.** In the case of penicillin production, the major problem was converting the process from one that took place only on the surface of the liquid to one that took place throughout the culture fluid. In this way, one large deep vat could replace thousands of small dishes.
3. **Strain improvement.** It was necessary to find or develop more productive microbes. Fleming had discovered by accident one strain that produced the drug. It was likely that other strains existed that produced greater amounts of penicillin.

Progress was rapid on all three fronts. The Fleming strain of *Penicillium*, grown according to the conditions suggested by Florey, produced about four units of penicillin per milliliter. By introducing corn steep (a hot water extract of corn grain) and a few inorganic salts into the growth media, the yield was raised to 20 units per milliliter, a fivefold increase.

The development of the techniques for growing the mold in deep tanks required considerable technical ingenuity. One might ask why they didn't adopt the well-established procedures of breweries for growing yeast in large quantities. The answer is simple: The yeast alcohol fermentation* is an anaerobic process, whereas penicillin production occurs only under aerobic conditions. This is the reason why *Penicillium* initially was grown as a surface culture—in contact with air. To introduce enough air into deep tanks to satisfy the voracious appetite of rapidly growing microbes required the development of a whole new engineering technology. A small tank fermentor (by today's standards) of 10,000 gallons requires about 10,000 liters of sterile air *per minute*. Furthermore, the air must be in the form of small bubbles to enhance the transfer of oxygen from the gas to the liquid phase. By 1943, the engineering problems had been resolved by a cooperative effort of the Peoria laboratory, universities, and industry, and the first stainless steel tank for penicillin was put into operation. A single 10,000-gallon tank (the size of an ordinary room) produced more penicillin than 200,000 surface culture vessels. The bottle plants were phased out and replaced by the more efficient aerated tanks.

The most dramatic results came from studies on improving the microbial strain (Table 11.2). First, many of the *Penicillium* strains

TABLE 11.2 Strain Improvement in Penicillin Production

Year	*Penicillium* Strain	Method of Isolation	Yield (units/ml)
1929	*Penicillium* (Fleming)	Chance contamination	4–20
1941	NRRL-832	Originated in Belgium	60
1943	NRRL-1951 ↓	Isolated from a melon	250
1944	X-1612 ↓	Mutation and selection	400
1945	Q-176 (Wisconsin) ↓	Mutation and selection	900
1949	49-133 (Wisconsin) ↓	Mutation and selection	1750
1953	53-399 (Wisconsin) ↓	Mutation and selection	2700
1980	Commercial strains	Mutation and selection	>20,000

*The term **fermentation** has led to considerable confusion among both scientists and lay persons. Unfortunately, fermentation is used to describe two different processes: (1) generation of energy (ATP) in the absence of air and (2) the microbially catalyzed conversion of food into commercially valuable chemicals. According to the former definition, penicillin production is not a fermentation process; according to the latter definition, it is a fermentation process. To avoid confusion, only anaerobic processes will be called fermentations in this book. However, the student should be aware that newspapers, magazines, and other news media generally refer to both aerobic and anaerobic microbial processes as fermentations.

that already existed in culture collections throughout the world were checked for penicillin production. One of the strains that originated in Belgium, designated NRRL-832, consistently yielded 60 units per milliliter in aerated tanks. For about a year, NRRL-832 was used for the ever-expanding manufacture of penicillin. Encouraged by their success with strain NRRL-832, the microbiologists decided to isolate fresh strains and test them for productivity. Since a good source of this mold is rotting food, they asked everyone to send them samples of bread, fruit, vegetables, or cheese that contained the characteristic blue-green mold. Thousands of samples were checked until, in July 1943, a Peoria housewife walked into the laboratory with a cantaloupe bearing a blue-green mold. After isolation and purification, the melon's *Penicillium* (strain NRRL-1951) proved to be a super-producer—more than 200 units of penicillin per milliliter. In April 1944, the cantaloupe strain was sent to manufacturing companies to help meet the growing demand for penicillin.

At this stage of the strain improvement program, the strategy was altered. Rather than continue the exhaustive search for fresh isolates of *Penicillium* that might yield slightly more penicillin, the scientists decided to concentrate only on the melon strain NRRL-1951. The idea was to produce thousands of mutants of the strain with the hope that one of them would overproduce penicillin. During the period 1944–1953, this technique of strain improvement via mutation and selection was carried out primarily at the University of Wisconsin. The results were remarkable: Strain X-1612, an x-ray–induced mutant of the cantaloupe strain, produce 400 units of penicillin per milliliter; strain Q-176, an ultraviolet light–induced mutant of X-1612, yielded 900 units per milliliter; descendants of strain Q-176, in turn, produced 1750 units (strain 49-133), 2700 units (strain 53-399), and more than 20,000 units (current commercial strains) per milliliter.

What does this applied research program in strain improvement, media optimization, and bioengineering mean to us? During the period 1943–1945, there was only enough penicillin to treat a few privileged individuals; the cost of the penicillin needed to treat a single serious case was at that time more than $1000. By 1951, the United States alone produced 10 tons of penicillin, enough to treat more than 10 million cases at a cost of less than $10.00 per treatment. Today, more than 1000 tons of penicillin are produced at a manufacturing price of less than 20¢ per treatment.

One of the ironic aspects of the penicillin saga is that England, where the drug was first discovered and shown to be effective, had to pay U.S. companies royalties after the Second World War for the technical know-how to produce penicillin. Fleming, of course, never attempted to patent penicillin. It is unfortunate but true that even if he had, the patents would have expired before the drug was produced commercially. The Oxford group, on the other hand, could have obtained considerable financial reward for developing the penicillin extraction procedure, but instead they gave their knowledge freely to the world. The pioneering research on penicillin, however, did not go unrecognized: The 1944 Nobel Prize in medicine was shared by Fleming, Chain, and Florey.

For the immigrant biochemist Chain, there was the added pleasure of knowing that he was a major contributor in developing a drug that saved the lives of thousands of young men who went out to liberate his former homeland—sweet justice.

11.3 Limitations of Penicillin and the Development of Other Antibiotics

Fleming, in his first paper on penicillin, emphasized that the substance was much more effective against certain bacteria than others. In general, gram-positive bacteria are much more sensitive to the drug than gram-negative bacteria. For example, penicillin was a dazzling success when used to treat bacterial pneumonia, rheumatic fever (a heart disease), wound infections, and other diseases caused by streptococci, but it was only slightly effective in treating tuberculosis, gonorrhea, typhoid fever, and meningitis. Furthermore, penicillin has no inhibitory effect whatsoever on viruses or eucaryotic microbes such as fungi and protozoa.

Another important limitation of penicillin is that it is useless when taken orally. The molecule is rapidly decomposed in the acidic conditions of the human stomach. An even more serious problem in the use of penicillin is the development of "resistant strains" of bacteria. As we discussed in Chapter 6, microbial populations are able to adapt rapidly to change in their environments by the mechanism of mutation and selection. In this regard, it is enlightening to examine what has happened to pathogenic staphylococci since the advent of penicillin. Initially, almost all staphylococcal infections could be successfully treated with penicillin therapy. By 1950, most "hospital staphs" had become totally resistant to the drug. Biochemical examination of the penicillin-resistant bacteria revealed that they excreted an enzyme, called penicillinase, which inactivates the drug (Figure 11.6). Later in this chapter we discuss the problem of drug resistance in the more general context of use and misuse of antibiotics.

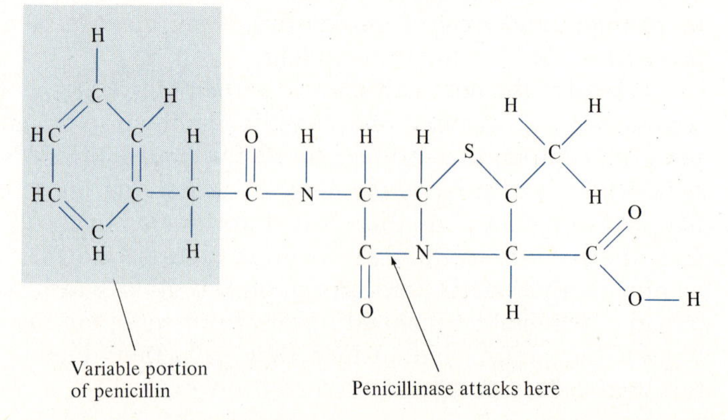

Figure 11.6
The chemical structure of penicillin G. The place where penicillinase splits the molecule is indicated by the arrow. The shaded area of the molecule differs in the various penicillins.

Two approaches have been used to overcome the limitation of penicillin: (1) Chemical derivatives of penicillin have been synthesized that are more resistant to acid and penicillinase and can inhibit a wider range of bacteria than the natural penicillin G and (2) many other different antibiotic† substances have been discovered and developed for clinical use.

The chemical approach was pioneered by research scientists at Eli Lilly Laboratories in the United States and Beecham Laboratories in Great Britain. They found that penicillin could be strengthened by chemically altering the variable portion of the molecule (Figure 11.6). These modified penicillins are called **semi-synthetic penicillins,** since one part of the molecule is synthesized by the microbe and the other part by the organic chemist. Several of these semi-synthetic penicillins have found broad clinical applications. Ampicillin, for example, is a semi-synthetic penicillin that is acid-stable and active against both gram-negative and gram-positive bacteria.

The great success of penicillin encouraged microbiologists to search for other antibiotics. A key figure in this search was the soil microbiologist Salman Waksman of Rutgers University in New Jersey. Between 1939 and 1945, Waksman and his associates examined thousands of microbes for their ability to produce antibacterial substances. A major conclusion that emerged from the Rutgers study was that spore-forming bacteria in the soil, especially the group called *Streptomyces,* were a rich source of antibiotics. Among the thousands of different antibacterial substances produced by different species of *Streptomyces,* Waksman isolated, characterized, and patented several important antibiotics, including streptomycin, neomycin, and actinomycin. Streptomycin, first reported in 1944, quickly emerged as a valuable chemotherapeutic agent for the treatment of tuberculosis and other infections caused by gram-negative bacteria that were resistant to penicillin.

With Waksman pointing his finger to the soil, the "gold rush" was on for new and better antibiotics. Thousands of active compounds have been found, identified chemically, and screened for potential clinical use. Although most of the active compounds were shown to be too toxic for general use, about 50 of them have proven therapeutic value and are currently produced commercially for medical and veterinary use. A partial list of important antibiotics is shown in Table 11.3.

What are the common characteristics of antibiotics? (1) From the chemical point of view, they are an extremely diverse group of organic molecules. The only chemical feature antibiotics seem to share is that they are usually relatively small molecules, containing fewer

---

†The word **antibiotic** refers to a substance produced by one microbe that selectively inhibits the growth of or kills other microorganisms. Thus, penicillin (produced by *Penicillium*) is an antibiotic, whereas the sulfa drugs (chemically synthesized) are not classified as antibiotics. Both penicillin and sulfa drugs are included in the broader term **chemotherapeutic agent.**

TABLE 11.3 Antibiotics Used in Chemotherapy

Antibiotic	Source	Used Medically in Treating	Mode of Action
Penicillin	Fungus *(Penicillium)*	Gram-positive infections	Blocks bacterial cell-wall synthesis
Ampicillin (Penbritin)	Semi-synthetic penicillin	Gram-positive and gram-negative infections	Blocks bacterial cell-wall synthesis
Cephalosporins	Fungus and *Streptomyces*[a]	Gram-positive and gram-negative infections	Blocks bacterial cell-wall synthesis
Streptomycin	*Streptomyces*	Tuberculosis; gram-negative intestinal tract infections	Inhibits protein synthesis in bacteria
Oxytetracycline (Terramycin)	*Streptomyces*	Most bacterial infections	Inhibits protein synthesis in bacteria
Chlortetracycline (Aureomycin)	*Streptomyces*	Most bacterial infections	Inhibits protein synthesis in bacteria
Chloramphenicol (Chloromycetin)[b]	*Streptomyces*	Typhoid fever; Rocky Mountain spotted fever	Inhibits protein synthesis in bacteria
Polymyxin (Aerosporin)	*Bacillus*	Gram-negative wound infections	Destroys bacterial cytoplasmic membrane
Bacitracin	*Bacillus*	Topical infections of eye and skin; burn infections	Blocks bacterial cell-wall synthesis
Novobiocin	*Streptomyces*	As penicillin; also penicillin-resistant staphylococci	Blocks bacterial nucleic acid synthesis
Erythromycin	*Streptomyces*	As penicillin; also penicillin-resistant staphylococci	Inhibits protein synthesis in bacteria
Griseofulvin	Fungus and *Streptomyces*[a]	Fungal infections (taken orally)	Destroys fungal cytoplasmic membrane
Polyenes (nystatin)	*Streptomyces*	Systemic fungal infections	Destroys fungal cytoplasmic membrane

[a]Initially discovered in a fungus, now known also to be produced by *Streptomyces* species.
[b]Initially discovered in a *Streptomyces*, now made chemically.

than 40 carbon atoms. (2) Ecologically, antibiotics are produced almost exclusively by spore-forming soil microorganisms. Furthermore, antibiotics are synthesized at precisely the same time that the cells are undergoing sporulation. Because of these correlations, some microbiologists have hypothesized that the natural role of antibiotics is to regulate the formation of spores in microbes. (3) Medically, the most important trait of an antibiotic is its high therapeutic index, resulting from selective killing and low toxicity. Each antibiotic interferes with an essential and specific microbial function.

Natural and semi-synthetic penicillins inhibit the synthesis of bacterial cell walls. When bacteria grow without concomitant wall synthesis, weak points develop on the cell surface; since the bacterial cytoplasm is at a higher osmotic pressure than the surrounding medium, a "blow out" results at one of the weak points, causing cell lysis. Antibiotics such as penicillin that kill bacteria are called **bacte-**

ricidal.‡ All cell wall antibiotics are bactericidal for growing cells. Using Ehrlich's terminology, penicillin is a "magic bullet" because its target, the peptidoglycan cell wall of bacteria, is completely absent in eucaryotic cells.

The tetracyclines are examples of antibiotics that are **bacteriostatic**§; they inhibit the growth of microbes without necessarily killing them. The target of tetracycline is bacterial protein synthesis. The detailed structure of the bacterial ribosome is sufficiently different from that of eucaryotic ribosomes, so certain antibiotics can selectively inhibit bacterial protein synthesis. Once the growth of the microbe is checked, normal defense mechanisms of the body (Chapter 14) can more easily eliminate the infecting microbe. Antibiotics that block specific steps in bacterial protein synthesis include the tetracyclines, chloramphenicol, streptomycin, and erythromycin. A small group of antibiotics, the most useful of which is polymyxin, has as its target of action the bacterial cell membrane; these drugs behave somewhat like detergents in destroying the cytoplasmic membrane.

There are only a few known antibiotics that are effective against fungal infections. The probable reasons for the paucity of fungal antibiotics are that (1) the fungi tend to bury themselves in tissues in such a way that they are not readily accessible and (2) the fungi are eucaryotic microorganisms and, as such, contain cell structures similar to those of higher animals and plants. Nevertheless, at least two antibiotics, griseofulvin and nystatin, specifically inhibit fungal growth.

The failure to find a single effective antibiotic against viruses has been one of the major disappointments of chemotherapy to date. As discussed in Chapters 2 and 6, viruses grow only inside other living cells, using the normal cell machinery for their own multiplication. It is therefore extremely difficult to find an agent that will inhibit the virus without harming the host. Since, at least in theory, there are a few specific viral functions, such as capsid formation, it is not impossible that some day an effective antiviral antibiotic will be discovered.‖ For the present, antiviral therapy depends largely on strengthening the normal body mechanisms.

11.4 Use and Misuse of Antibiotics

The discovery and development of antibiotics as potent chemotherapeutic agents has resulted in the savings of millions of human lives. Starting with penicillin in the early 1940s and rapidly extending to the semi-synthetic penicillins, streptomycin, the tetracyclines,

‡The suffix "-cidal" refers to a substance that kills. The accompanying prefix indicates the type of organism killed. Thus, we have **bactericidal** agents, **fungicidal** agents, and the more general killing substances called **germicides.**

§The suffix "-static" refers to a material that inhibits growth without killing.

‖A potentially antiviral substance, interferon, is currently undergoing clinical trials. The discovery and properties of interferon are described in Chapter 14.

and many others, antibiotics have been found that can cure each of the important bacterial diseases. In an era of scientific cynicism, it is worthwhile recalling that before the advent of antibiotics, bacterial pneumonia was one of the major killers of humans. Today, death from bacterial penumonia is a rarity. Scarlet fever, once the most dreaded of children's diseases, can be cured easily with antibiotics. The same is true of bone infections, typhoid fever, dysentery, tuberculosis, blood poisoning, syphilis, gonorrhea, typhus, and many other serious diseases. Because of antibiotics, surgeons are now able to perform life-saving operations without the danger that inevitable bacterial infections will kill the patient. Of considerable benefit, although not always appreciated, is the peace of mind that antibiotics give each one of us. We no longer live in constant fear that one of our loved ones will contract pneumonia or another infectious bacterial disease for which there is no cure.

And now for the darker side of the story. . . . Like any material of great value, antibiotics can be (and have been) misused. By employing antibiotics haphazardly when they are not needed and against diseases for which they are impotent, we have greatly reduced their effectiveness. There are at least four general dangers in the indiscriminate use of antibiotics: (1) intrinsic toxicity of the antibiotic, (2) development of allergy, (3) upsetting of ecological balance in humans, and (4) development of antibiotic-resistant bacteria.

Although useful antibiotics must have high therapeutic indices, they all have some toxicity to humans. Prolonged use of streptomycin, for example, gives rise to nerve damage and deafness; in large doses, the tetracyclines cause diarrhea and anal pruritus (itching), polymyxin and bacitracin damage the kidneys, and chloramphenicol causes irreversible reduction of bone marrow, leading to fatal anemia. Several hundred people die each year in the United States because of an allergic response to penicillin. Approximately 7 percent of the patients in U.S. hospitals are strongly allergic to the penicillins.

Human beings live in close symbiotic relationships with beneficial microbes. In Chapter 12 the interaction of humans and their microbial symbionts is discussed in an ecological context. For the present discussion, it is sufficient to point out that antibiotics eradicate the normal bacterial population, making humans weaker and more susceptible to fungal infections and other hostile germs.

Development of antibiotic-resistant bacteria can occur by two different mechanisms: mutation and genetic exchange (plasmid transfer). The method of developing a drug-resistant population by mutation and selection has already been discussed in Section 6.1. Mutations to antibiotic resistance occur spontaneously at a frequency of from one in a million to one in ten billion. In the presence of the antibiotic, there is a strong selective advantage for the mutant; the vast majority of the microbes are killed or their growth is inhibited, whereas the drug-resistant mutant can multiply, giving rise to a progeny population that is drug-resistant. This is the probable mechanism by which pathogenic staphylococci became penicillin-resistant.

The second way in which bacteria develop drug resistance—genetic exchange via plasmid transfer (see Section 6.7)—was first described in Japan in 1959 and has since been thoroughly studied throughout the world. The basic observation was that in a hospital a disease-causing microbe was suddenly found that was simultaneously resistant to all the antimicrobial drugs routinely used at that time. Furthermore, these resistant bacteria could transfer their multiple resistance to other bacteria by the process of conjugation that is discussed in Chapter 6. We now know that the genes for antibiotic resistance tend to collect on plasmids (R factors) that are readily exchangeable between even unrelated bacteria. Thus, bacteria that are part of the normal flora of humans can act as reservoirs of R factors. If invading pathogenic bacteria receive R factors from one of the harmless bacteria, the pathogen immediately becomes resistant to a whole set of antimicrobial agents. The difficulty in treating infections caused by such pathogens is clear.

How should antibiotics be used to increase their effectiveness and minimize the chances of developing resistant strains? To begin with, antibiotics should not be administered for viral diseases or minor ailments. Second, whenever possible, the disease-causing microbe should be isolated and tested for antibiotic sensitivity. The physician will then be in a much better position to recommend the appropriate medication. Third, the antibiotic should be taken in sufficient doses and for a long enough period to eliminate the infection completely. Finally, the non-medical use of antibiotics should be carefully controlled.

It is not generally realized that the largest current market for antibiotics is as growth promoters for poultry, swine, and other farm animals. The exact mechanism by which antibiotics cause animals to grow more rapidly is not completely understood. It has been suggested that antibiotics kill off certain intestinal bacteria that compete with the animal for nutrients. Regardless of the mechanism, farmers know they can produce more meat at a lower price if antibiotics are included in the feed. Balanced against this economic advantage is the fact that massive feeding of antibiotics to farm animals has provided ideal conditions for R factors to flourish and spread. Resistant bacterial strains develop in the animals, leave their host in excreted fecal matter, and find their way into soils, sewers, and polluted rivers; there they can readily transfer their R factors to human pathogens.

11.6 Microbial Production of Vitamins, Amino Acids, Organic Acids, Hormones, and Polymers

Microbes are exquisite chemists, capable of efficiently synthesizing a wide variety of complex organic molecules. Specific strains of bacteria and fungi have been isolated that produce large amounts of certain chemicals that are valuable to humans. In several cases, the amount of product synthesized by the microbe has been further in-

TABLE 11.4 Commercial Chemicals Produced by Microbial Processes

General Product	Specific Example	Producing Microbe	Major Use
Vitamin	Vitamin B_{12}	*Propionibacterium*	Food supplement
	Riboflavin	Fungus	Food supplement
Amino acid	Glutamate	*Corynebacterium*	Flavor ingredient
	Lysine	*Micrococcus*	Food supplement
Organic acid	Lactic acid	*Lactobacillus*	Chemicals, textiles, leather
	Citric acid	*Aspergillus*	Soft drinks, jam, cosmetics
	Gluconic acid	*Aspergillus*	Foods, medicines, cleaning agent
Hormone	Gibberellin	Fungus	Plant growth regulator
	Cortisone[a]	Several	Therapeutic agent
Polysaccharide	Dextran	*Leuconostoc*	Plasma substitute; pharmaceuticals
	Xanthan	*Xanthomonas*	Food stabilizer; oil drilling additive

[a]Many of the steroid hormones, such as cortisone, are produced by a mixture of chemical and microbiological transformations.

creased, initially in the laboratory and later at the pilot-plant stage, by a combination of strain improvement (mutation and selection) and optimization of the medium and growth conditions. Table 11.4 lists some of the important products that are currently manufactured by microbial processes.

Vitamins

Vitamin B_{12} is produced exclusively by microbes. Because it is not present in plant materials, vitamin B_{12} must be obtained by animals and humans either directly or indirectly from microorganisms. Cows and other ruminants satisfy their vitamin B_{12} requirement by living in symbiotic association with microbes that produce the vitamin. Most other animals and human beings must acquire vitamin B_{12} from their diets, either by eating animal products or by specific dietary supplementation. People who are strict vegetarians must occasionally take vitamin B_{12} or they will suffer from pernicious anemia. Large quantities of the vitamin are included in poultry and livestock feed, especially when vegetable protein is used in the feed. Vitamin B_{12} is produced commercially for pharmaceutical or feed supplement use by growing high-producing strains of *Propionibacterium* in large tanks. Another member of the B-vitamin complex, riboflavin, is also produced microbiologically. Riboflavin is added to bread, flour, and other cereal products for human consumption.

Amino Acids

The industrial production of amino acids has been developed primarily by Japanese microbiologists. Over 100,000 tons of glutamate per year are produced from molasses by *Corynebacterium glutamicum.* Marketed as monosodium glutamate, the amino acid is used widely as a flavor ingredient in foods. High-producing mutant strains of *Corynebacterium* that are employed for manufacturing glutamate differ from wild-type bacteria in at least two ways: (1) They have become deregulated; i.e., the mutants keep producing glutamate even when they don't need it for growth and (2) their cytoplasmic membranes are slightly damaged, so they secrete the excess amino acid into the medium.

Several other amino acids, such as lysine and methionine, are produced commercially by microbial processes. All animals and humans require essential amino acids in their diets. These essential amino acids are present in a well-balanced protein diet. However, several economically valuable sources of protein are deficient in one or more of the essential amino acids. Rice, wheat, and corn, for example, are all low in their lysine content. Children raised on an exclusively rice diet fail to grow normally. As the world is forced to turn more to vegetable proteins, the need for specific amino acid supplementation is constantly growing. It is already general practice to include methionine, cysteine, and lysine in breakfast cereals and animal feedstocks.

Organic Acids

Microorganisms produce a number of different organic acids in addition to the amino acids. We have already discussed (in Chapter 10) the process of lactic acid formation (souring of milk) in the context of yogurt and cheese production. When the lactic acid is to be used as an industrial chemical rather than as a food, the process is modified by using the least expensive raw materials available and the highest-producing strains. Lactic acid is used for making plastics, lacquer, varnishes, and polymers, for decalcifying skins in the leather industry, and for finishing silk-rayon fabrics.

Citric acid is used in fruit drinks, jams, candies, cosmetics, and shampoos. Originally produced by extraction from lemons, citric acid is now manufactured more economically by the mold *Aspergillus.* Using carefully controlled conditions, the fungus can convert 100 tons of molasses to 75 tons of citric acid. Approximately 150,000 tons of citric acid are produced annually by this process. Another organic acid of industrial importance is gluconic acid, which is produced in bulk by a different strain of *Aspergillus.* Gluconic acid is used widely in foods and medicines and as a cleaning agent for washing bottles.

Hormones

A microbial product that has found considerable application in agriculture is gibberellic acid and closely related compounds. At concentrations as low as one part in ten million, the gibberellins stimulate the growth rate of plants and can be used to regulate flowering and seed germination. Gibberellins are produced industrially in large aerated fermentation tanks by a fungus that was originally isolated from a diseased rice plant. The optimization of growth conditions and strain improvement techniques followed closely those procedures developed for increasing the yields of penicillin.

In 1950, doctors at the Mayo Clinic in Minnesota discovered that patients suffering from chronic rheumatoid arthritis responded favorably to small doses of the steroid hormone cortisone; within a few days, swelling and pain subsided and frozen joints became mobile. The problem was how to produce enough cortisone for medical use. Extraction of the hormone from the adrenal glands of cattle was not practical; about 100,000 cattle are needed to produce just 1 gram of cortisone. The only other alternative at the time was chemical synthesis, involving 32 separate steps and an overall efficiency of 0.2 percent.

Although microbes do not generally synthesize steroids, several strains can transform relatively inexpensive plant sterols to products that the organic chemist can readily convert to valuable steroids. The combined skills of microbes and chemists are used for the production of cortisone and other anti-inflammatory steroids as well as the steroid oral contraceptives (the birth control pill).

Polysaccharides

Two types of polysaccharides are currently being manufactured on a large scale using microbiological systems. Dextran is a polymer of glucose that is produced in the medium when a special strain of *Leuconostoc* is allowed to grow on sucrose. Dextran is used in the food, pharmaceutical, and chemical industries. Medically, dextran can be used as a plasma substitute in blood transfusions. Xanthan is a complex polysaccharide composed of three different sugars. The high viscosity of xanthan solutions over a wide range of temperatures and acidities makes the substance a valuable food stabilizer (e.g., in French-style salad dressing), an additive to drilling muds, especially in undersea oil exploration, and a viscosity enhancer in cosmetic lotions and toothpastes. There is every reason to expect that in the near future, the production of microbial polysaccharides from cheap raw materials will replace vegetable gums in industry.

11.5 Enzyme Technology

What allows microbes to synthesize efficiently such a wide variety of both simple and complex molecules is, of course, their enzymatic make-up. As discussed throughout this book, enzymes are fantastic catalysts, each one speeding up a specific chemical transformation. In the last few years, an entire industry (with annual sales of more than 150 million dollars) has been built on the concept of isolating specific enzymes and then using them in industry. Table 11.5 presents some of the important microbial enzymes used in industry.

The enzymes amylase, β-galactosidase, invertase, and glucose isomerase all catalyze the conversion of one type of carbohydrate into another. Amylase hydrolyzes starch into maltose and other soluble sugars. This enzyme finds use in the textile industry because threads are routinely prestrengthened with starch before weaving; afterwards the starch is removed (desizing) by soaking the cloth in a warm bath containing amylase.

Lactose, the main sugar in milk, tends to crystallize in the cold; to prevent this from happening in ice cream and ruining the texture, β-galactosidase is added. Another carbohydrate enzyme, invertase, splits the disaccharide sucrose into glucose and fructose. This reaction is used to produce chocolates and other sweets with liquid centers. Small quantities of invertase are mixed with solid sucrose and flavoring ingredients and then covered with chocolate; as the enzyme slowly converts the sucrose to the more soluble glucose and fructose, the center dissolves. Glucose isomerase converts pure glucose into a mixture of glucose and fructose. Since fructose is considerably sweeter than glucose, the enzyme can be used to make very sweet syrups.

Proteases have found a wide variety of applications in industry. In the tanning of leather, protease treatment of the hides produces a softer texture. Bacterial proteases are used in the textile industry to dissolve gummy materials surrounding silk and artificial fibers. Other proteases are used as meat tenderizers and biodetergents. Stubborn stains on clothing and other fabrics can often be removed by soaking

TABLE 11.5 Microbial Enzymes Used in Industry

Enzyme	Reaction Catalyzed	Major Use
Amylase	Breakdown of starch	Desizing in textiles
β-galactosidase	Splitting of lactose (milk sugar)	Ice cream
Invertase	Splitting of sucrose	Candy industry
Glucose isomerase	Conversion of glucose to fructose	Sweet syrups
Proteases	Breakdown of protein	Meat tenderizer; biodetergent; leather and textile industries
Rennins	Splitting and clotting of milk protein	Cheese manufacturing
Pectinase	Degradation of pectin	Clarification of fruit juices
L-asparaginase	Destruction of asparagine	Treatment of leukemia patients
Penicillinase	Inactivation of penicillin	Treatment of penicillin-sensitive patients

in detergents that contain heat-stable proteases. Recently it has become economical to use proteases in the recycling of used photographic film. The protease degrades the proteinaceous gelatin base of the film, making it easier to recover the valuable silver it contains. Rennins are a special class of proteases that split milk protein (casein) in such a way that the casein becomes insoluble in water and forms a clot or curd. Cheeses are formed by further microbial action on this separated curd.

Pectinases are used extensively in preparing concentrated fruit juices. Extracts of many fruits contain large amounts of the gelatinous material pectin. The pectin plays an important role in the preparation of certain semi-solid food products; for example, it causes jams and jellies to set. However, the pectin has to be destroyed in order to obtain concentrated fruit juices that will not solidify when refrigerated. This is accomplished by addition of minute quantities of commercially available pectinases. The enzyme is also used to remove a pectin-like material from the outer surface of freshly picked coffee beans.

It has been reported that neoplastic cells of certain leukemia patients cannot synthesize the amino acid L-asparagine. In an attempt to take advantage of this biochemical difference between cancer cells and normal cells, such patients are treated with the enzyme L-asparaginase. If the specific amino acid is destroyed in the blood by the administered enzyme, and the neoplastic cells, unlike normal cells, cannot synthesize their own, then the cancer cells should starve to death. (There is considerable debate concerning the merits of this therapy.) The major source of L-asparaginase is mutants of *Escherichia coli* that produce up to 1 gram of the enzyme per liter.

The enzyme penicillinase is used to treat penicillin-sensitive individuals who are suffering from an acute allergic or, even worse, anaphylactic response to the antibiotic. The penicillin-splitting enzyme is usually administered by intramuscular injection. Commercially, penicillinase is obtained from penicillin-resistant *Bacillus cereus*.

11.7 Microbial Insecticides

Pasteur's classic investigation in the 1870s on diseases of the silkworm provided direct evidence that insects are susceptible to specific microbial infections. During the following 30 years, a number of disease-causing microbes were isolated from silkworms, honeybees, and other economically important insects. Most of these studies were directed toward controlling epidemic diseases of these valuable creatures.

In 1902, Ishawata in Japan isolated an aerobic spore-forming bacterium from a diseased silkworm. The remarkable feature of this bacillus was that it caused paralysis in silkworms within one hour after they were infected. Hence, Ishawata called the spore-former "sotto-bacillen" ("sudden-collapse bacillus"). During the period 1915–1927, bacterial strains with properties very similar to those of the "sottobacillen" were isolated independently in several countries. This species of bacteria, now referred to as *Bacillus thuringiensis*, has two characteristic features: The organisms are all pathogenic for the larvae (caterpillars) of certain insects, and they all produce diamond-shaped crystals inside the sporulating cells (Figure 11.7).

Until 1956, *B. thuringiensis* with its crystalline inclusion body was only one of many microbiological curiosities. Then, a small group of Canadian microbiologists published the results of a few simple experiments, which changed the entire direction of *B. thuringiensis* research and brought it into the limelight of industrial microbiology. The Canadian scientists showed that the crystal (1) is protein in nature, (2) dissolves only in alkaline solutions, and (3) is responsible for the rapid paralysis of the insect. These data, especially the last point, immediately suggested the possibility of using crystal-containing spores of *B. thuringiensis* as **insecticides.** Thus, research goals were turned completely around—instead of finding ways to prevent epidemics in honeybees, silkworms, and other beneficial insects, the

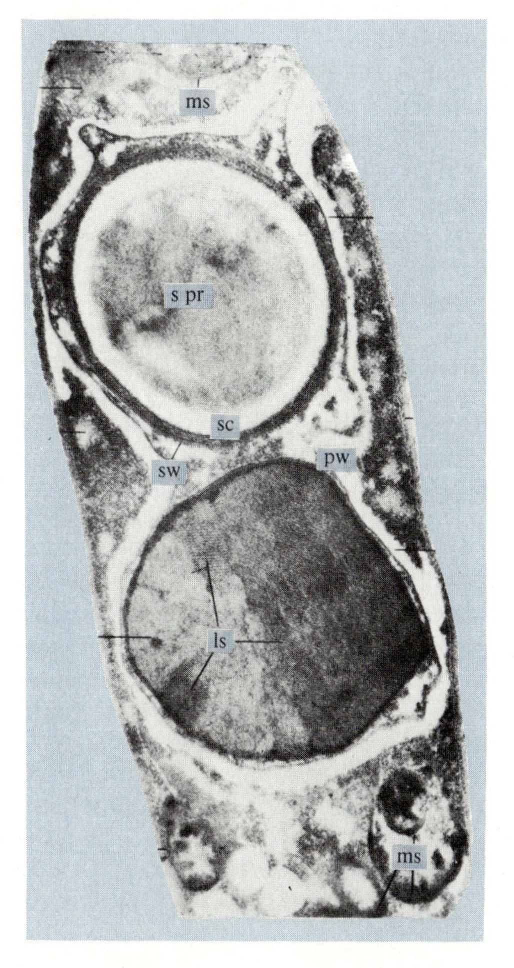

Figure 11.7
Electron micrograph of resting cell of *Bacillus thuringiensis* showing the endospore (spr) and the crystal body (ls).

practical aim became the use of *B. thuringiensis* to kill crop-damaging insects.

How does the microbe paralyze and then kill susceptible insect larvae? While feeding on leaves, the caterpillar ingests a few *B. thuringiensis* cells containing both the spore and the toxic crystal. When the microbial cells reach the highly alkaline gut of the insect, the crystals dissolve and the spores are released. The dissolved protein is then partially hydrolyzed by enzymes in the gut; one of the protein fragments that is formed is extremely toxic, causing small holes to be formed in the wall of the gut. The alkali then leaks out of the gut into the blood of the larva, inducing general paralysis of the insect. As a consequence of the release of alkali, the conditions in the gut become more favorable for spore germination and microbial growth. Within a few hours, the bacilli have multiplied and invaded all of the body tissue. After the bacteria have completely digested the dead larva, they complete the cycle by again forming resting cells containing the characteristic spores and crystals.

Bacillus thuringiensis, or BT as it is referred to in industry, has several advantages over DDT and other chemical insecticides:

1. **Selectivity.** At concentrations of less than one part in ten million, BT is specifically pathogenic to the larvae of a wide variety of lepidoptera. Included in the sensitive group are the following serious insect pests of agricultural crops or forests: gypsy moth (forest trees), tomato and tobacco horn worms, cabbage looper (cabbage, broccoli, potato, lettuce, melon, and cauliflower), cotton bollworm, alfalfa caterpillar, banana caterpillar, cankerworms (forest and shade trees), and the European corn borer. Useful insects, such as bees, are resistant to low doses. Plants and other animals are unaffected, probably because they lack the specific conditions necessary to dissolve and activate the toxic protein. Thus, BT is a selective agent for the control of many serious insect pests.
2. **Resistant mutants to BT have not been found.** One of the major problems with chemical pesticides is the development of insects resistant to the chemical. By the universal biological mechanism of mutation and selection, whole populations of insects have emerged that are resistant to even high concentrations of organochlorines and other chemical insecticides. Resistance of BT has not yet been found either in the laboratory or in the field. To become resistant to BT, insects would probably have to change their basic physiological and biochemical characteristics. Keep in mind that *B. thuringiensis* and insects have lived together in a natural ecological relationship for thousands of years. By preparing BT in factories and dispersing it on certain fields at specific times, human beings simply displace the equilibrium temporarily between insect and microbe.

3. **BT can be produced cheaply and dispersed easily.** Fortunately, *B. thuringiensis* can be grown on inexpensive nutrients and reaches a high cell density. Thus, it can be produced at a price that is competitive with chemical insecticides. Furthermore, both the spore and the crystal are resistant to heat and drying-out conditions. In industry, the bacteria are grown in large aerated fermentors, harvested after crystal formation is complete, spray-dried in vacuum, and incorporated into a dusting powder.

The application of microbial insecticides can be extended to include other insect pathogens and different microbially produced toxins. Recent laboratory experiments indicate that certain bacilli specifically kill flies and mosquitos; other bacteria are active against beetles and termites. As more is learned about the interaction of microbes and insects, it should be possible to decrease the damaging effects of insects without the dangerous ecological consequences of chemical insecticides.

11.8 The New Biotechnology

In the last few years, a new breed of scientists has moved into the field of industrial microbiology. These researchers, many of whom are university professors, are armed with the modern tools of "genetic engineering" (see Section 6.7) and the confidence that this recent development has much to offer of practical consequence. This new biotechnology differs from traditional approaches in one fundamental way: The desired product is manufactured in a strain that has had foreign DNA inserted into it by the technique of **gene cloning.**

Before discussing some of the goals and accomplishments of the new biotechnology, it is useful to clarify the term **cloning.** In current scientific usage, cloning refers to one of three distinct processes. First, since the time of Koch, microbiologists have referred to a population of microbes derived from a single cell as a clone. Thus, an isolated colony on an agar medium is a clone. The second use of the term *cloning* came from biologists who succeeded in growing whole organisms from single somatic cells. For example, many living cells can be isolated from a single tomato plant, each one of them capable of developing into a mature plant. In this way, thousands of genetically identical tomato plants can be produced. This type of cloning was the basis of the novel *The Boys from Brazil.* The newest use of the term, and the one that concerns us here, is in gene cloning—the process of introducing genes into new "host" cells and then selecting those cells containing the desired gene in order to make many copies of the selected gene.

A simple example of gene cloning with potential industrial importance is the transfer of vitamin-producing genes from one bacterium to another. Scientists at Tel Aviv University recently took the

biotin gene from *Escherichia coli* and transferred it via a plasmid to a different strain of *E. coli*. The new strain produced a large excess of biotin because it contained multiple copies of the genetic information needed for synthesizing the vitamin. In principle it should be possible to improve many microbial processes by cloning the relevant genes into microbes that are easier to cultivate and that lack the control genes that normally limit production.

Two of the most spectacular successes in gene cloning are the production of human insulin and the production of interferon in bacteria. To give you an idea of the swiftness with which laboratory results are finding their way to the marketplace, consider the following: In August 1978, Walter Gilbert and his colleagues at Harvard University reported in the *Proceedings of the National Academy of Science (USA)* for the first time the cloning of rat proinsulin genes in bacteria. A few months later, Genentech, one of the new genetic engineering firms based in California, reported at a press conference that they had cloned the *human* proinsulin gene (a precursor of insulin) in *E. coli*. By the end of 1980, Eli Lilly, a large Indiana-based pharmaceutical company, had already demonstrated that human insulin produced in bacteria was active when tested in animals, began clinical trials with human insulin, and applied to the National Institutes of Health (NIH) for a license to scale up production of human insulin using the new cloning technology. One of the reasons that the development of the technology for producing human insulin was allowed to proceed so rapidly was that certain diabetic patients are allergic to animal insulin; the synthetic human insulin is a life-saving hormone for these sensitive individuals. (The hormone is normally produced by extracting the pancreas of animals.)

Interferon (see Chapter 14) is a natural antiviral substance with potential anticancer activity. Many medical scientists consider interferon to be the most important discovery since antibiotics. The major limitation on the development of interferon to date has been the scarcity of the material. Currently, there is not enough interferon even for clinical trials. Thus, the report in 1979 by the Swiss genetic engineering firm Biogen that they had successfully cloned human interferon genes in bacteria is of major potential medical importance. Several large companies, including Hoffman-La Roche, Du Pont, and Schering-Plough, are investigating the possibility of producing large quantities of human interferon by genetic engineering techniques.

These are just a few of the many potential applications of the new biotechnology. Although it is still too early to predict the full impact that genetic engineering will have on medicine and biotechnology, it is clear that we are entering an important era of "applied biology," where discoveries in microbiology, biochemistry, and molecular biology of the last 50 years will greatly influence the future of man. The question is no longer whether or not this new technology will be used, but rather *how* it will be used.

QUESTIONS

11.1 What do each of the following terms signify?

Fungicidal agent
Semi-synthetic penicillin
Essential amino acid
R factor
Vitamin
Gene cloning
Bactericidal agent
Chemotherapeutic index
Strain improvement

11.2 What are the advantages of microbial insecticides over chemical agents for controlling insect pests?

11.3 Present the arguments pro and con for using tetracyclines in animal feed.

11.4 Why do you think many more effective antibiotics have been found against bacteria than against fungi, whereas essentially no useful antiviral agents have been discovered?

11.5 What is the basis of the new biotechnology? How can it be used to obtain higher-producing strains of traditional microbial processes? How can it be used to produce entirely new microbial products?

11.6 The enzyme penicillinase presents a problem in antibiotic therapy. Nevertheless, the enzyme is produced commercially for use by clinicians. Explain.

11.7 What is the advantage of meat over grain as a source of protein? How can vegetable protein be improved using industrial microbiology?

11.8 Why are two or more antimicrobial drugs sometimes given together in treating chronic infections, such as tuberculosis?

Suggested Readings

Aharonowitz Y and G Cohen: Microbial Production of Pharmaceuticals. Scientific American, September 1981.
The entire issue is devoted to various aspects of industrial microbiology.

Davis BD: The Recombinant DNA Scenarios: Andromeda Strain, Chimera and Golem. American Scientist, 65:547–555, 1977.
An analysis of the techniques and risks of cloning DNA in E. coli; *containing a list of the most important references on the subject.*

Dickson D: Patenting Living Organisms—How to Beat the Bug-Rustlers. Nature, 283:128–129, 1980.
The legal aspects of patenting industrial microbes are discussed.

Dixon B: Invisible Allies: Microbes and Man's Future. London: Temple Smith Ltd., 1976.
An excellent popular book with emphasis on the useful functions of microbes. Dixon was a professional microbiologist who abandoned research to become a science writer.

Fox JL: Genetic Engineering Industry Emerges. Chemical and Engineering News, March 17, 1980.
An interesting discussion of how the new biotechnology has moved from the universities to industry.

Hare R: The Birth of Penicillin and the Disarming of Microbes. Reading, MA: Allen and Unwin, 1970.
A popular account of the discovery of Fleming's discovery of penicillin.

Peppler HJ and D Perlman: Microbial Technology, 2nd Edition. New York: Academic Press, 1979.
This two-volume edition discusses many of the topics in this and the previous chapters in greater detail. Particularly relevant are the chapters concerned with making beer, wine, cheese, distilled beverages, single-cell protein, antibiotics, vitamins, and microbial insecticides.

CHAPTER 12

Ecology of Man and His Microbial Guests

A NEW YEAR GREETING

EDITOR'S NOTE

The following verses were written after the poet had read "Life on the Human Skin," by Mary J. Marples (SCIENTIFIC AMERICAN, January 1969).

On this day tradition allots
To taking stock of our lives,
My greetings to all of you, Yeasts,
Bacteria, Viruses,
Aerobics and Anaerobics:
A Very Happy New Year
To all for whom my ectoderm
Is as Middle-Earth to me.

For creatures your size I offer
A free choice of habitat,
So settle yourselves in the zone
That suits you best, in the pools
Of my pores or the tropical
Forests of armpit and crotch,
In the deserts of my forearms
Or the cool woods of my scalp.

Build colonies: I will supply
Adequate warmth and moisture,
The sebum and lipids you need,
On condition you never
Do me annoy with your presence
But behave as good guests should,
Not rioting into acne
Or athlete's foot or a boil.

Does my inner weather affect
The surfaces where you live,
Do unpredictable changes
Record my rocketing plunge
From fairs when the mind is in tift
And relevant thoughts occur
To fouls when nothing will happen
And no one calls and it rains?

I should like to think that I make
A not impossible world,
But an Eden it will not be:
My games, my purposive acts,
May become catastrophes there.
If you were religious folk,
How would your dramas justify
Unmerited suffering?

By what myths would your priests account
For the hurricanes that come
Twice every twenty-four hours
Each time I dress or undress,
When, clinging to keratin rafts,
Whole cities are swept away
To perish in space, or the Flood
That scalds to death when I bathe?

Then, sooner or later, will dawn
The Day of Apocalypse,
When my mantle suddenly turns
Too cold, too rancid for you,
Appetizing to predators
Of a fiercer sort, and I
Am stripped of excuse and nimbus,
A Past, subject to Judgment.

W. H. Auden
(for Vassily Yanowsky)

As we have seen in the previous chapters, microbes are interesting for a variety of reasons. They have provided powerful tools for satisfying our curiosity about the basic genetic and metabolic mechanisms that define life on earth (Chapters 3 and 5 to 8). The history of their discovery and investigation is an exercise in the history and sociology of science and the scientific method. We have also seen how microbes have made important contributions to economics and industry (Chapters 10 and 11). More recently, the new field of genetic engineering is attempting to harness the biological machinery of bacteria and turn them into highly specialized factories for mass producing molecules of immense importance, such as hormones, enzymes, medical reagents, and the like. Thus, an informed person could be motivated to learn about microbes for their scientific, social, or economic importance.

Probably the most compelling reason for gaining some knowledge about microbes, however, derives from the universal experience of each and every one of us of having been sick. Illnesses caused by microbes are a fact of life and call our attention to the importance of microbes in a very personal way.

Medical microbiology and its offspring, immunology, are sciences that have progressed greatly in the last 20 years and are now at the forefront of biology. These sciences have built powerful research tools and produce a vast amount of information that feeds many dozens of scientific journals with thousands of articles a month. Our practical interests in all of this information can be boiled down to three essential questions: How do we keep well, how do we get sick, and, once sick, how do we recover from microbial diseases?

The aim of this and the next four chapters is to supply some answers to these questions. In this chapter we discuss how man and microbes share biological space and dwell together in the same ecological system. The principles of infection and disease are the subject of Chapter 13. In Chapters 14 and 15, we describe the immune system and how it deals with agents of disease. Finally, common and interesting infectious diseases caused by bacteria and viruses, including cancer, are discussed in Chapter 16.

12.1 Man and Microbes Share an Ecological Niche

Our personal worlds are heavily populated by bacteria and viruses. Literally millions of bacteria can be isolated from the skin of each of us. Bacteria are also found on the mucous membranes of our inner surfaces, the nose, throat, and bowels. Most of the weight of fecal material in the human bowel is made up of bacteria. Viruses, too, abound on the skin and mucous membranes of the body. In short, we live our lives carrying a vast microbial load. It is a wonder that most of the time this load does us no obvious harm.

How does this state of affairs maintain itself? How do man and microbes live together? What are the problems that arise in this relationship? How are these problems solved in the least detrimental fashion?

The answers to these questions require us to consider the ecological relationship between man and microbes. Ecology is a word that is derived from two Greek words: *oikos* meaning "home" and *logos* denoting "reason" or "science." The subject matter of ecology is the study of the interactions of an animal species with its home or environment. The sum total of biological and physical factors within which a species lives is called its **ecological niche.** The ecological niche is a description of all the facts of life of a community of creatures—how they compete for or share energy, space, and resources.

The word *niche* calls to mind a tidy recess in a wall suitable for housing some object. An ecological niche also has the connotation of a cozy accommodation to which a species is well adapted to make a living. However, the term *accommodation* is probably a better description of the living quarters of a species than the term *niche*. A niche is static. An accommodation describes lodgings, but it also implies a state of adaptation, an adjustment of differences, the supplying of wants and needs through organized social processes. To accommodate is also to aid or oblige. All these meanings come closer to the dynamic community of living organization that describes man and his microbial parasites.

As we shall see, some of the fundamental problems posed by the relationship between man and microbes can be understood using basic concepts of ecology. Some of the concepts, although important, are subtle in the context of man and his microbes. Therefore, it may be useful to digress for a bit and see such concepts in a more dramatic situation of competitive interaction between creatures. Let us consider lions and zebras.

Lions eat zebras; zebras eat grass. The lions are the predators and the zebras the prey. At first glance it would seem that predator and prey have mutually contradictory self-interests. The lions must eat zebras to survive. In contrast, the survival of the zebra obviously involves not getting eaten and denying the lions their dinner. The lions survive at the expense of the zebras in competition for survival.

From this superficial point of view, the relationship between a parasite and its host might be likened to the relationship between a predator and its prey. The term *parasitism* is usually used to describe a situation in which the parasite is smaller than the host and actually lives on or within the host's body (notwithstanding the exceptional application of the word to some social interaction between people). Nevertheless, parasitism essentially describes the relationship between two creatures in which one, the parasite, always benefits from the other, the host, without paying any compensation. Parasitism is free lunch. The host is the menu.

Predation is the conventional way of describing the relationship between lions and zebras, while parasitism is the term applied to the relationship between microbes and man. It might seem reasonable to believe that, just as lions are out to get zebras, bacteria and viruses are out to get us. If this were true, our reaction to parasitism should be to resist and destroy our parasites. Zebras can only run. We, by learning the secrets of microbes, ought to fight back in all-out war.

This argument, however, is based on false premises. On closer examination it will become clear that the ecological relationship between man and microbes is not simply uncompensated parasitism. The lesson we shall develop is that both prey and predator, and host and parasite, have many common interests, and that these mutual interests can lead to communication and mutual cooperation. The implications are both theoretical and practical.

Let us consider the interests that prey and predator have in common. Both lions and zebras actually have a stake in the survival of the zebras. If the lions were to eat all the zebras, the lions themselves would soon die of starvation. The lions would suffer hunger even if the zebra population were only moderately reduced to the point where zebras became harder to find.

Similarly, the zebras benefit from the lions. Lack of lions leads to an increase in the numbers of zebras, particularly among those segments of the herd that tend to be eaten by lions: the very young, the old, and the infirm. An increase in the numbers of zebras paradoxically can threaten the existence of the entire herd. Survival of all the young produces overgrazing, which can destroy the grass that the zebras require. Epidemics can be generated by overcrowding and by increased numbers of old or weak animals. Starvation and disease are more dangerous to zebras than are lions. The integrity of the herd of zebras benefits from the lions taking the "surplus" of young and old animals.

In short, the prey species derives considerable benefits from the activities of the predator, while the predator depends on the survival of its prey.

This ecological relationship between predator and prey is finely tuned and exists in a delicate equilibrium. Upsetting the ecological balance is dangerous to all involved. For example, the populations of deer in North America are periodically threatened with starvation because their natural predators have been liquidated. It is often necessary to institute hunting seasons to regulate the numbers of deer and ensure their health and survival.

In the Museum of Natural History in Chicago are the stuffed remains of two lions credited with killing and eating a large number of East African railroad workers. These lions had switched their menu from zebras to people because they had become too old to hunt zebras. The response of the people was to kill all the lions they could find, the vast majority of which only ate zebras and were innocent of eating people. All the lions were threatened with extinction because two switched prey.

We have discussed lions and zebras to help dramatize two ecological lessons: (1) predator and prey can have common interests and (2) upsetting a balanced ecological relationship can be costly. These lessons apply also to the interactions between man and his microbes.

The organized interactions between parasite and host and between predator and prey are special cases of a more general ecological organization. Except for plants and other autotrophs that synthesize organic compounds from inorganic materials, most living creatures feed on organic compounds made by someone else. The biosphere, the sum total of living organisms, is composed of an organized chain of the eaters and the eaten. The herbivores prey on plants, the primary carnivores prey on the herbivores, and the secondary and tertiary carnivores prey on other carnivores. All herbivores and carnivores are subject to microbial diseases. Organic molecules are finally recycled by the microorganisms of decay. From this point of view, the definitions of predator and prey and of parasite and host can be seen to be rather arbitrary. We ourselves are both hosts and predators, simultaneously. Of course, parasites tend to have a lifestyle with

common identifying features that allows us to distinguish them from other creatures. However, it is useful to keep in mind that the relationship between parasite and host is part of the great food chain of life. Hence, like all ecological relationships, it is founded on mutual adaptation.

12.2 Normal Human Flora

Bacteria that reside in or on healthy individuals are called normal flora in the medical literature. The term "normal" indicates that the presence of the bacteria usually is not associated with abnormality or disease. They are called "flora" because bacteria formerly were classified as belonging to the plant kingdom.

Normal flora are contrasted with pathogenic bacteria. The term *pathogenic* means disease-producing and refers to bacteria that are capable of causing disease. The term *virulence* is used to describe the relative capacity of a microbe to produce disease. One pathogen may be more or less virulent than another pathogen.

The distinction between normal flora and pathogenic bacteria is important, but it is not absolute. A particular species of bacteria may be part of the normal flora of one individual or population of individuals, whereas the same species of bacteria can be pathogenic in different circumstances. The factors that determine the pathogenicity of a bacterium will be explored in the context of the processes of disease.

The major types of bacteria that make up the normal flora and the particular niches they tend to occupy in our body are outlined in Table 12.1. Normal bacterial flora occupy four main sites: the skin; the cavities of the mouth, nose, and throat; the bowels; and the lower

TABLE 12.1 The Normal Human Flora

Site	Characteristics	Examples
Skin	Gram-positive bacillus	*Corynebacterium*
	Gram-positive coccus	Staphylococci, streptococci
	Acid-fast, gram-positive bacillus	Mycobacteria
	Gram-negative bacillus	See intestine
Nose and throat	Gram-positive bacillus	*Corynebacterium*, lactobacilli, actinomycetes
	Gram-positive coccus	Streptococci, pneumococci, staphylococci
	Gram-negative coccus	*Neisseria*, *Veillonella*
	Gram-negative bacillus	*Bacteroides*, *Hemophilus*, *Klebsiella*
	Spiral bacteria	Treponemes
Intestine	Gram-negative bacillus	*Escherichia coli*, *Bacteroides*, *Pseudomonas*, *Aerobacter*, *Proteus*
	Gram-positive bacillus	Lactobacilli, clostridia
	Gram-positive coccus	Streptococci, staphylococci
Vagina	Gram-positive bacillus	Lactobacilli

TABLE 12.2 Sites of Skin Bacteria

Bacteria	Skin Surface	Epidermis: Cornified Layers	Epidermis: Cellular Layers	Dermis	Hair Follicle	Sweat Gland
Staphylococci		+			+	+
Streptococci		+	+	+		
Corynebacteria					+	
Gram-negative bacilli		+				
Soil bacteria	+					

genital tract. Normal flora are not permitted to penetrate into the body's tissues or internal fluids such as the blood. They are limited to the external surfaces and mucous membranes.

Skin Bacteria

A great many different kinds of bacteria are found on our skin (Table 12.2). The gram-positive aerobic cocci, such as species of staphylococci and streptococci, predominate and are located superficially in the stratum corneum (Figure 12.1). Anaerobic gram-positive ba-

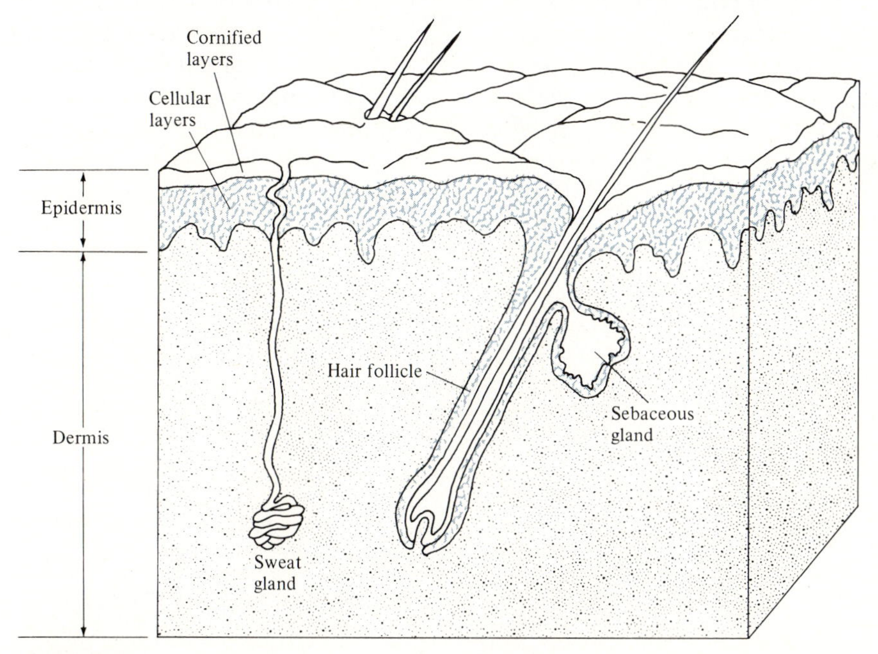

Figure 12.1
Niches of normal bacterial flora in the skin. This schematic diagram shows the skin divided into the more superficial epidermis and the deeper dermis. The epidermis is made up of cornified layers of dead cells, the products of underlying layers of living cells. The dermis is the seat of the sweat glands and the hair follicles (roots). The sebaceous glands secrete an oily material called sebum into the hair follicles. Various bacteria live in characteristic niches in the skin. (Modified from Marples, M. J.: Life on the Human Skin. Scientific American, January 1969.)

cilli, such as species of corynebacteria, live deep within the hair follicles in association with the sebaceous glands. Gram-negative bacilli spill onto the skin from the gut and are found in the anal and genital regions and also on the soles of the feet. However, gram-negative bacteria do not flourish on the skin unless the normal gram-positive bacteria have been reduced by application of antibiotics or other unphysiological factors. Limited areas of the skin can provide highly specific ecological pockets. For example, a species of mycobacteria, *M. smegmatis,* thrives only under the male foreskin. Circumcision abolishes this niche. Yeasts such as *Candida albicans* are also found on normal body skin.

The numbers of skin bacteria are difficult to measure accurately. Their concentrations seem to vary in the different regions of the skin (Figure 12.2). The axilla (armpit) has the densest population of aerobic

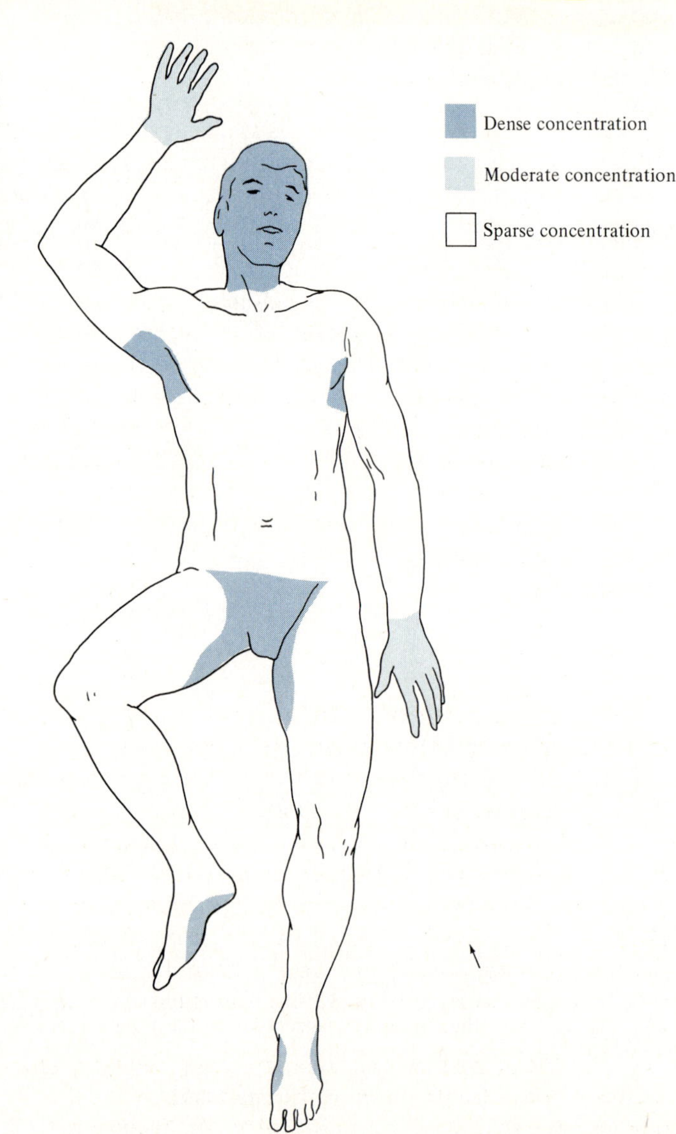

Figure 12.2
Concentrations of bacteria in areas of the human skin. This distorted figure shows the relative concentrations of bacteria on various sites of the body. The heaviest concentrations of bacteria are in the axillae, genital area, scalp, forehead, face, and soles of the feet. The hands have a variable number of bacteria, while the remainder of the skin is sparsely populated. (Modified from Marples, M. J.: Life on the Human Skin. Scientific American, January 1969.)

bacteria, at least 2.4 million bacteria per square centimeter of skin. The scalp with 1.5 million and the forehead with 200 thousand bacteria per square centimeter are also relatively well populated. Other areas of the skin, such as the forearms and back, may have only between a hundred and a thousand bacteria per square centimeter. All of these numbers are minimal estimates of the total number, since they do not include the anaerobes that live deep in the hair follicles.

Compared with other ecological systems, how densely populated is the human skin? The numbers of bacteria and fungi in fertile earth have been estimated to range from 10 million to 10 billion per gram of soil. The numbers of bacteria per weight of scrapings from the cornified skin surface were about one half million bacteria per *milligram* of material in one study. This is the equivalent of 500 million bacteria per gram of surface skin. Hence, some areas of skin support a normal flora as dense as that of good fertile soil.

The numbers of bacteria shed from the skin can be measured more easily than the resident population. In one experiment, clothed subjects were found to shed from 6000 to 60,000 bacteria per minute. Thus, each person moves in his or her own cloud of bacteria.

These normal skin flora include bacteria such as staphylococci or streptococci that are potentially pathogenic. Occasionally, we develop skin infections associated with these bacteria. However, they cause no problems in most people most of the time.

Body odors of any kind are out of favor in western society at present. However, in more robust times and places, some human skin scents have been considered to function as sex attractants. These body odors are produced by the action of bacterial enzymes on the secretions of the large sweat glands located in the axilla and elsewhere. In this way, normal skin flora play some small role, positive or negative according to taste, in human sexuality. The millions of dollars invested in the deodorant industry are a measure of the impact of normal skin flora.

Nose and Throat Bacteria

The upper respiratory tract (nose and throat) harbors countless numbers of bacteria, many similar to species carried on the skin (Figure 12.3). There are also potential pathogens among the normal throat flora, perhaps even more than on the skin. In addition to streptococci and staphylococci, we often carry pneumococci and *Neisseria* in our noses. Pneumococci have the potential for causing pneumonia. Some strains of *Neisseria* can cause meningitis, an inflammation of the covering of the brain. In the vast majority of people, these bacteria are harmless members of the normal flora in the nose and throat. Thus, it would appear to require particular circumstances for them to produce disease.

The lower respiratory tract (trachea, bronchi, and lungs) has no resident normal flora. A variety of mechanisms act to remove bacteria and viruses that stray into the lower respiratory tract from their nor-

Figure 12.3
Normal flora of the upper respiratory tract. Bacteria reside in the nose, nasopharynx, and throat. The trachea, sinuses, and eustachian tubes to the middle ear are normally sterile. Infection of these areas leads to disease.

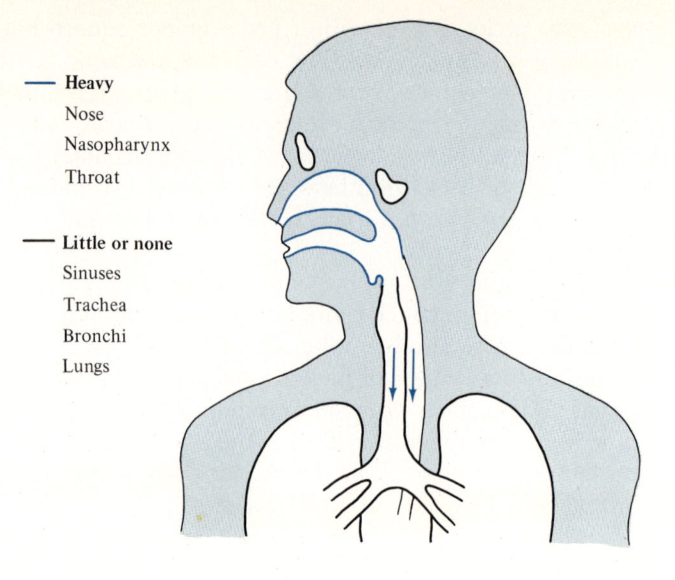

mal niches in the upper tract. These protective mechanisms are discussed in Chapter 13.

Normal Flora of the Intestine

The esophagus, stomach, and upper small intestine (duodenum and jejunum) have relatively few normal bacteria, while the large intestine, or colon, is the home of myriads of bacteria (Figure 12.4). A reliable sample of the solid contents of the colon is found in the stools. One gram of stool contains between 10 and 100 billion bacteria. This high concentration of bacteria indicates that the colon is a very efficient fermentation chamber. The majority of colon bacteria are anaerobic gram-negative bacilli, although other kinds of bacteria are also present (see Table 12.1). The composition of the gut flora is influenced by diet. For example, infants fed on breast milk harbor *Lactobacillus bifidus* as the predominant organism. This bacterial species almost disappears in babies fed other diets. The usefulness of this peculiar ecological specialization is unknown.

Normal Flora of the Vagina

The uterus and fallopian tubes are normally sterile. The lower genital tract, the vagina, has normal flora that vary with the hormonal state of the woman. During a woman's reproductive years, the vagina usually has a weakly acidic pH because of the metabolic activity of the cells that line the vagina. *Lactobacillus* resists low pH and is the predominant normal flora. Before puberty and after menopause, the vagina is not acidic, and gram-positive cocci and gram-negative bacilli predominate over lactobacilli. Yeasts occasionally proliferate after menopause or during pregnancy.

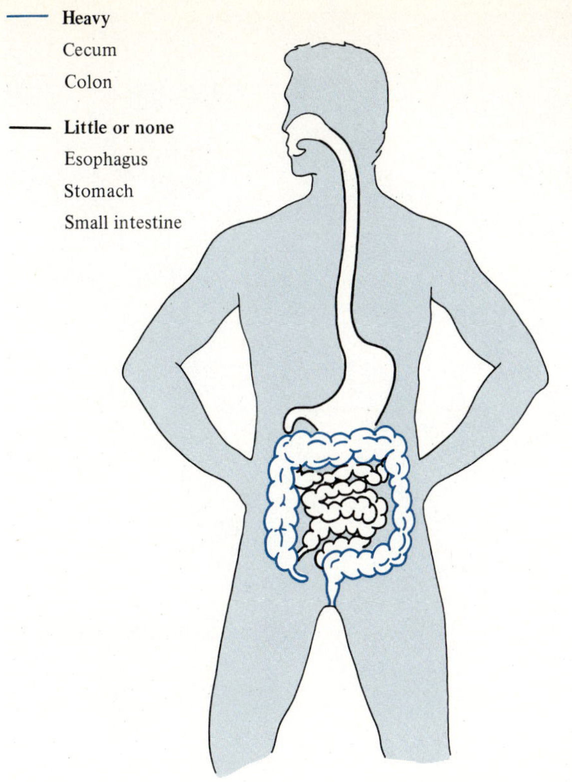

Figure 12.4
Normal flora of the intestines. Bacteria grow in high concentration in the cecum and the colon. The esophagus, stomach, and small intestine are sparsely populated.

12.3 Adherence of Bacteria to Host Cells

We have seen that particular types of bacteria constitute the normal flora of the body and that these strains of bacteria live in certain defined niches. Some, like staphylococci, favor the skin; others, like pneumococci, favor the oral cavity; and still others, like the gram-negative bacilli, favor the bowels. Note that each bacterial strain has its own characteristic niche.

What are the reasons for this specificity of the ecological niche? Why do different bacteria reside in different places on the body? Why are normal flora not crowded out by pathogenic bacteria? The complete answers are unknown. A few years ago, we could only have speculated about nonspecific factors such as the availability of different nutrients in certain sites, possibly favoring one kind of bacterium over another. Today, we have evidence suggestive of a very specific interaction between particular bacteria and certain host cells.

The major finding is that bacteria adhere specifically to host cells (Figure 12.5). Bacteria are equipped with surface molecules called **ligands** that bind to other molecules called **receptors** on the surface of the host cell.

The ability to adhere to a particular host-cell type is a characteristic of the bacterial strain. For example, streptococci obtained from skin adhere better to skin epithelial cells than they do to epithelial

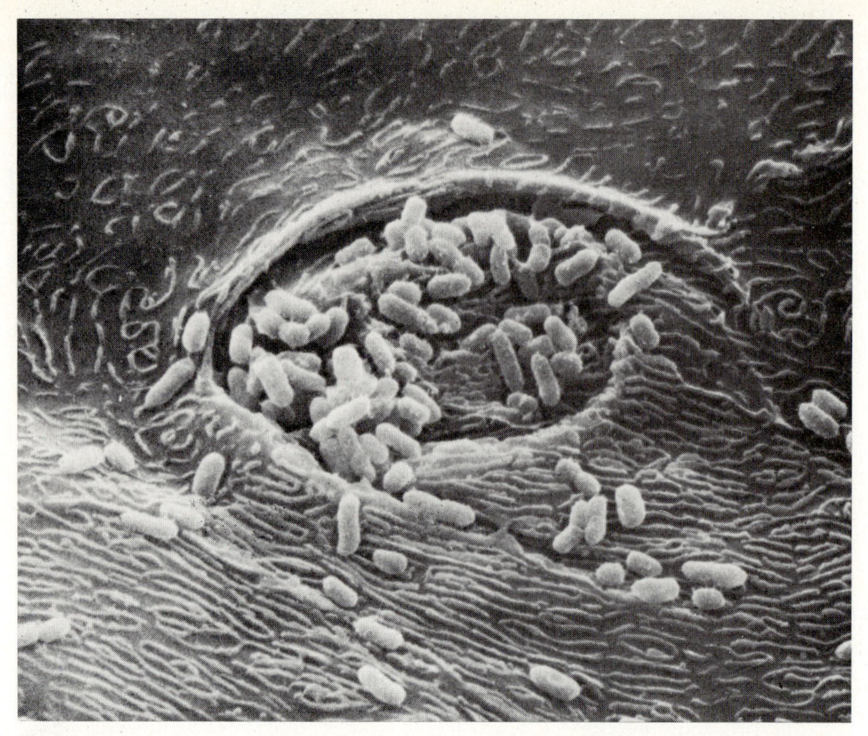

Figure 12.5
Bacteria adhering to cells of a human cheek. This scanning electron micrograph made by Fredric Silverblatt and Craig Kuehn of the Veterans Administration Hospital in Sepulveda, California, shows *Escherichia coli* bacteria, the white cylindrical objects, sticking to cells from the inside of the human cheek, occupying the background. (Courtesy of F. J. Silverblatt.)

cells from the throat. In contrast, streptococci obtained from the throat stick well to throat cells and less well to skin cells. The specificity is even more exact. Streptococci from the mouth, such as *S. salivarius* and *S. mutans,* differ in their sticking to particular surfaces in the mouth. *S. salivarius,* normally found on the gums and oral mucosa, sticks to mucosal cells but only poorly to teeth. *S. mutans,* normally found on the surfaces of teeth, does not stick well to mucosal cell surfaces but sticks to the teeth.

Adherence of bacteria to cells can be studied in the test tube, making it possible to analyze the chemical nature of the binding of the bacterial ligand to the receptor of the host cell.

It appears that the bacterial ligand binds to short chains of sugar molecules attached to the surfaces of the host cells. This binding is based on a three-dimensional fit between the ligand and the receptor, similar to the fit between a lock and a key (Figure 12.6). Different ligands bind or recognize the different types of sugar molecules that characterize various cell types of the host. Therefore, the nature of the ligands made by a species of bacteria can determine the type of cell to which the bacteria adhere. The host cell, too, can regulate the binding of bacteria by controlling its receptors. This principle is illustrated by several examples.

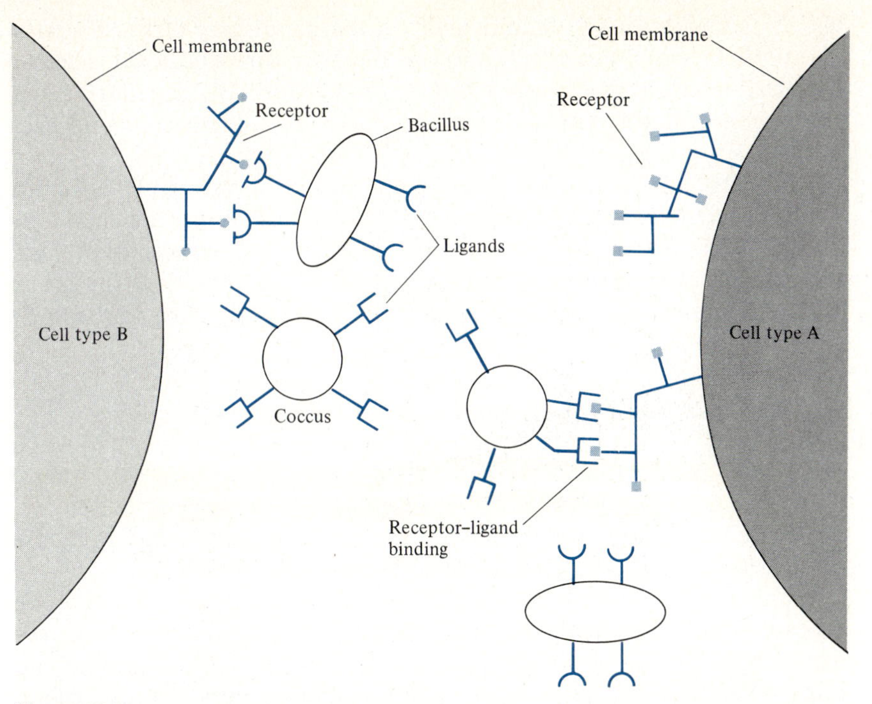

Figure 12.6
Bacterial ligands bind specifically to cellular receptors. This schematic diagram illustrates two types of cells, A and B, each with sugar molecule receptors of a distinct configuration. Two representative bacteria, one a bacillus and the other a coccus, both have ligands whose structure is complementary to the receptors of only cell type A or cell type B. Therefore, each type of bacterium, by virtue of its unique ligand, binds to a distinct cell type. This specific ligand–receptor binding is proposed to function as a mechanism that contributes to the adherence of certain bacteria to specific types of cells.

Mucosal cells from newborn babies have fewer receptor molecules for bacterial ligands than the mucosal cells obtained from the same sites in adults. Hence, the cells of newborns take up many fewer bacteria than do those of adults. This seems to be a reasonable way to introduce the baby gradually into our dirty world and its load of normal flora.

The resistance of the urinary bladder to seeding by gram-negative bacteria from the adjacent large bowel may be strengthened by the fact that healthy bladder cells have relatively few receptors for gut bacteria.

The cells of the vagina have more receptors for lactobacilli, their normal flora, than they do for *E. coli*, a bowel bacterium.

These kinds of findings hint at a molecular mechanism that might regulate the interaction between man and his normal flora. The ligand–receptor binding could serve as a recognition system for communication between our cells and preferred bacteria. However, the experimental evidence is still preliminary and many problems remain. For example, some species of bacteria that are not part of the normal flora of a certain type of host cells may nevertheless stick to such cells. In addition, the binding of bacteria to host cells seems to be influenced by many factors such as the conditions of the bacteria and their phase of growth, the age of the host, and the host's state of health. We need much more information before we can derive the rules governing the natural association between human cells and bacteria.

Nonetheless, it is clear that our load of bacterial parasites is not determined by haphazard chance but, similar to all ecological interactions, is based on an ordered flow of biological information. This information is expressed at least in part by the specificity of the interaction between bacterial ligands and the receptors of particular host cells.

12.4 Benefits of Normal Flora: The Lessons of Germ-Free Life

It is obvious that our normal flora benefit from us; we provide them with a protected environment and a constant flow of nutrients. Do we benefit from them? Is it worth our while to accommodate them? What is the balance of the biological cost-benefit analysis?

The cost of having billions of bacteria on our skin and in our bowels is not insignificant. It is obvious that they rob us of both organic and inorganic nutrients. They also threaten us with disease. The normal flora contain types of bacteria associated with serious illness, such as staphylococci, pneumococci, and *Neisseria*.

The common sense of evolution teaches that to sustain a biological relationship, the sum of the biological cost must be offset by the sum of the biological benefits. The notion is simple. We would in all probability not have survived the constant drain on our resources imposed by normal flora for millions of years unless the loss was replenished by some gain. Do we in fact derive any benefit from our association with normal flora?

When confronted with the problem of detecting the benefit or function of a hitherto unknown organ or process, biologists have often found it helpful to remove that organ or process from the system and observe the consequences. We can learn what the organ or process is good for by observing what happens to the subject when it is lost. The unknown entity can be removed in various ways, each appropriate to the problem we study. A gland in the body of an experimental animal can be removed surgically, or a process can be investigated by isolating a mutant organism that lacks an enzyme, and so on. This strategy would suggest that we study individuals deprived of their normal flora and investigate how they fare without bacterial parasites.

Pasteur asked the question of whether it is possible to rear animals without intestinal bacteria. The German bacteriologists Nuttall and Thierfelder attempted to test this question in 1895. They delivered fetal guinea pigs from their mother's uterus by cesarean section and tried to raise them in sterile isolators supplied with sterile air and food. These early experiments failed for technical reasons. Other scientists attempted to use chicks or baby goats for similar experiments. These early workers were thwarted by a lack of knowledge about vi-

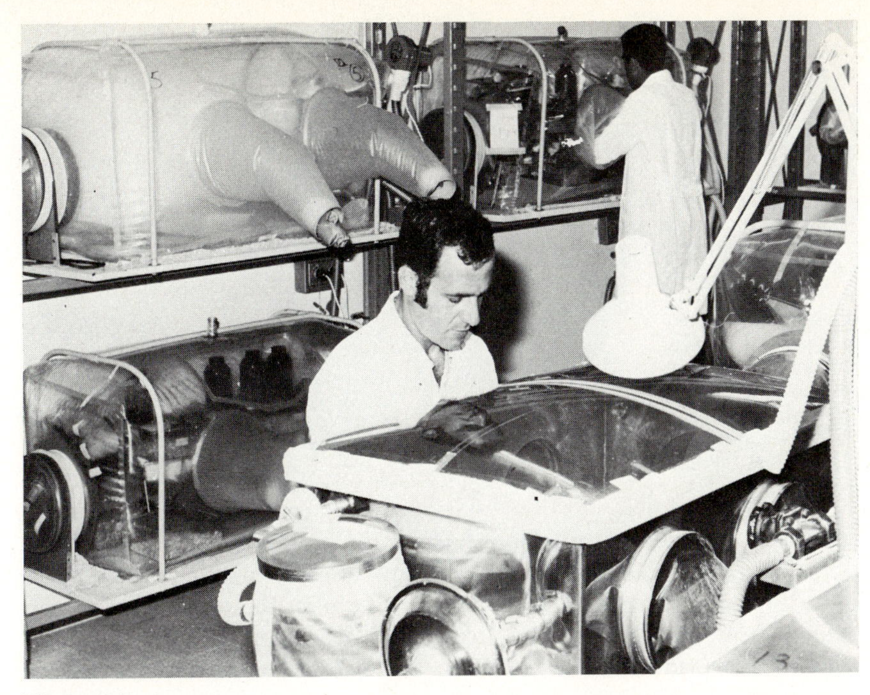

Figure 12.7
A sterile isolator for maintaining germ-free mice. Germ-free animals are born and live out their lives in such isolators. They breathe sterile air and eat sterilized food. The usefulness of normal bacterial flora can be studied by comparing the health and responses of these animals with those of conventional animals populated with normal flora. Access to the mice is obtained by inverting the rubber gloves into the isolator.

tamins and other dietary essentials that were destroyed by the heating used to sterilize the food administered to the animals in isolators.

Only relatively recently has it been possible to breed and maintain experimental animals free of any detectable bacteria or viruses. Such animals are called **germ-free.** Most germ-free animals studied have been mice and rats, because they are the easiest to prepare and maintain. In principle, we still use the experimental design suggested by Nuttall and Thierfelder more than a half-century ago. Germ-free animals are raised in sterile chambers, as shown in Figure 12.7, where they receive sterilized food enriched with vitamins, sterile air, and water and are handled through a sterile barrier to the outside world.

Such germ-free animals have been compared with genetically identical animals, raised in similar chambers but populated with a conventional bacterial flora. The results of these studies demonstrated that germ-free animals differed from conventional animals in at least three ways: nutrition, physiology of the gut, and susceptibility to infection.

Nutrition

It appears that our normal flora synthesize essential vitamins for us. Germ-free mice were much more susceptible to diets low in vitamins K and B than mice carrying normal flora. Furthermore, there is

Figure 12.8
Germ-free animals develop enlarged ceca. The cecum obtained from a germ-free mouse (left) is about five to ten times larger than that obtained from a conventional mouse of the same age and sex. The only difference between these animals is the absence or presence of normal bacterial flora. Enlarged ceca of germ-free animals have been found to contain increased amounts of mucopolysaccharide material and a protein called alpha-pigment. The origin and functions of these materials are not entirely known, but their accumulation may be related to the metabolic and physiological aberrations of germ-free life. The cecum decreases in size and these materials disappear when normal flora are fed to previously germ-free animals. (From Luckey, T. D.: Germ Free Life and Gnotobiology, 1st ed. New York: Academic Press, 1963, p. 346.)

some evidence that our normal flora in the gut are helpful in destroying potentially dangerous materials and preventing their absorption into the body. Gut bacteria also seem to aid in the absorption of nutrients from the gut that are lost in the feces in germ-free mice. In short, the normal flora resident in our bodies derive their nutrients from us, but they also contribute to our nutrition. As described in Chapter 10, ruminant animals are totally dependent on normal flora for their nutrition.

Physiology

Germ-free animals have a large number of metabolic deviations compared with conventional animals. These combine to produce approximately a 25 percent decrease in metabolic rate and a 40 percent decrease in blood flow through the heart. Various enzymes show either decreased or increased activity in germ-free animals. The particular enzymes involved are not important in the context of this discussion. The point is that the absence of a normal flora can have far-reaching effects on the metabolism of the host.

A particularly striking feature of germ-free rodents is enlargement of the cecum, the distal part of the small intestine (Figure 12.8). Some investigators have found that surgical removal of the enlarged cecum tends to restore some of the metabolic deviations of germ-free life. Although the story is as yet incomplete, it seems that the absence of bacteria in the gut can lead to the accumulation of drug-like substances in the cecum and elsewhere. These substances influence the metabolism of the animals as well as the structure and function of the gut itself. Normal gut bacteria apparently destroy these drug-like materials and protect the host from their undesirable effects.

Susceptibility to Infectious Disease

An important reason for putting up with normal bacterial flora is because they may help prevent more serious infection by pathogenic bacteria.

Normal flora help us resist pathogens in at least two ways. First, their physical presence prevents colonization of the skin and gut by pathogenic bacteria. Normal flora simply occupy the sites that other bacteria could use to gain a foothold. In one study, germ-free and conventional mice were fed one billion pathogenic *Salmonella* bacteria, and the numbers of salmonellae in the small intestine were counted one day later. The conventional mice remained with only about a thousand salmonellae, representing a million-fold decrease in the number of bacteria. In contrast, the number of salmonellae increased tenfold in the germ-free mice. Similar findings have been observed using other pathogenic bacteria.

In addition to physically competing for suitable niches, normal flora actively suppress the growth on the skin of more virulent bacteria and fungi. Unsaturated fatty acids present on the skin were known for some time to suppress the growth of skin pathogens. Recently, it has become clear that these protective fatty acids are a metabolic product of human skin secretions released by specific enzymes of normal gram-positive skin bacteria. Thus, the normal gram-positive flora process the host's secretions to generate the unsaturated fatty acids that protect the skin from more virulent gram-negative bacteria and yeasts.

Normal flora also seem to prime the physiology of the body's defense systems. Conventional animals have more cells in a state of readiness to fight infection, and these cells are mobilized to a site of infection more quickly than they are in germ-free animals. Thus, the normal flora may help keep the defense system in a healthy state of alertness.

We cannot control the external environment to rid it of the lurking danger of pathogens. However, by maintaining the normal flora, we actually prevent pathogens from gaining a foothold on our exterior. The normal flora may also keep our defenses in shape and, as a bonus, supply some nutritional and physiological services. In exchange for these benefits, we feed and house them and suffer the danger of their occasional invasions.

12.5 Viruses Carried by Healthy People

The viral analogues to the normal bacterial flora are those viruses that infect our cells chronically but produce no symptoms for long periods. Table 12.3 is a partial list of such viruses. Other viruses, in fact most viruses, infect us without producing disease. For example, during a mumps epidemic, mumps virus can be detected in many people who suffer no symptoms of disease. These asymptomatic infections can occur in approximately 60 percent of susceptible people who have close contact with a person with mumps. These cases of inapparent infection are acute and limited. The virus is quickly eradicated from the body, and very few people become chronic carriers. Such acute symptomless infections lack the protracted host–parasite relationship characteristic of normal flora. The viruses listed in Table 12.3 are more analogous to normal flora in the persistence of infection.

TABLE 12.3 Viruses Carried by Healthy People

Type of Virus	Site
DNA Viruses	
Cytomegalovirus	Salivary glands Kidney White blood cells
Epstein-Barr virus	White blood cells
Herpes simplex I	Skin
Adenoviruses	Gut Upper respiratory tract
RNA Viruses	
Enteroviruses	
Poliovirus, coxsackievirus, echovirus	Gut
Rhinoviruses	Upper respiratory tract
Retroviruses	Many cells

Herpesviruses include viruses that are carried in a wide variety of host cell types (Table 12.3). Cytomegalovirus and Epstein-Barr viruses are usually contracted asymptomatically in childhood and persist in our cells for a lifetime. The presence of such viruses is detected only when the host or the host's cells are treated in special ways, as we discuss in a later chapter. Herpes simplex virus reveals itself in some people by periodically breaking forth as fever blisters around the lips. In this case, expression of the virus can be activated by natural phenomena such as sunburn, emotional stress, or other illnesses. Many people carry herpes simplex virus without showing any recurrent fever blisters.

Adenoviruses and enteroviruses are chronically carried in and excreted from the gut. They may be associated with disease in a small number of infected people. Rhinoviruses and adenoviruses can be carried in the nose.

Retroviruses demonstrate a unique type of chronic infection. These viruses appear to be transmitted from mother to fetus in the uterus or in the milk to the newborn. Retroviruses are characterized by the presence of an enzyme called reverse transcriptase that makes a DNA copy from the viral RNA (Figure 12.9). This RNA-to-DNA reversal of the conventional flow of genetic information allows the viral genes to be integrated into the host's own DNA genetic material. These integrated viral genes can behave in the host eucaryotic cell in a fashion reminiscent of lysogenic infection of bacteria with prophage (see Chapter 4). The integrated viral genome is duplicated along with the host DNA and is passed on to daughter cells without leading to virus replication or cell damage. Occasionally, integrated viral genes are activated, and host cells begin to express viral RNA and viral proteins. Sometimes these components are assembled to form complete virus particles that bud from the host cells (Figure 12.9). From an ecological point of view, retroviruses demonstrate an ultimate relationship between parasite and host. The parasite becomes integrated into

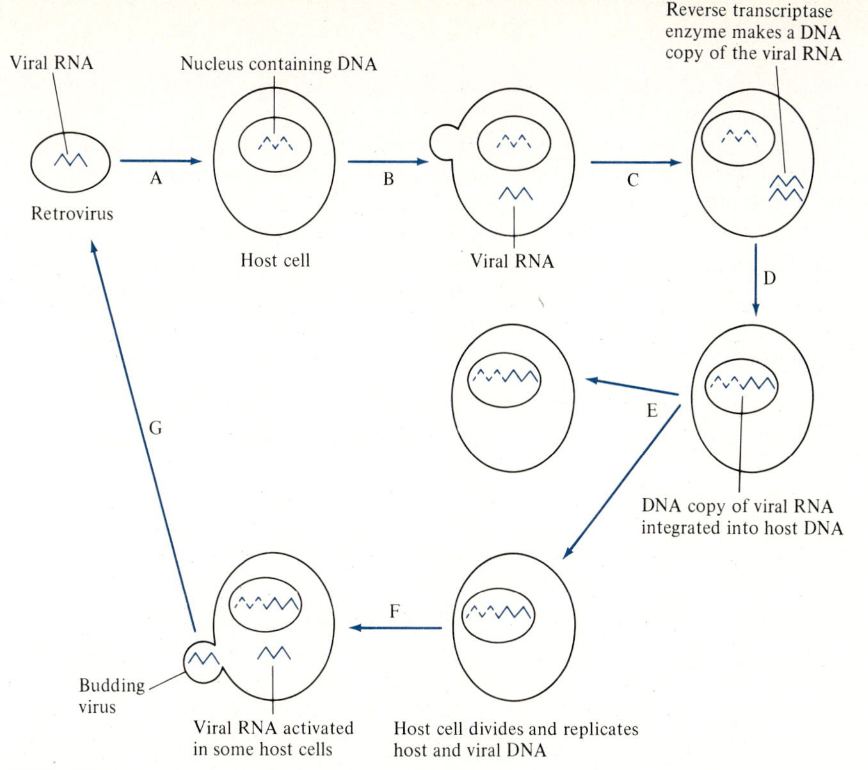

Figure 12.9
Life cycle of a retrovirus. Virus particle containing viral RNA attaches to the host cell (A), and the viral RNA enters (B). The reverse transcriptase enzyme makes a DNA copy of the viral RNA (C). The viral DNA copy can then be integrated into the DNA of the host cell (D). The host cell replicates both host and viral DNA so that daughter cells receive both types of DNA (E). In some cells, the viral DNA may give rise to viral RNA that become part of new viral particles that bud from the membrane of the host cell (F). These viral particles can now infect fresh cells (G).

the host's own genes to the extent that it is often difficult to distinguish between the parasite and the host itself.

QUESTIONS

12.1 How might our normal bacterial flora be affected when we are exposed to antibiotics? What might be the resulting potential dangers or benefits to our health?

12.2 Certain of our cells produce receptor molecules that can bind bacterial ligands (Section 12.3). What benefit do we gain from the binding of bacterial ligands to these receptors? Is it logical to conclude that the only function of these receptor molecules is to bind bacteria? Or is it more logical to suppose that bacteria exploit as receptors molecules of the host cell made for some other purpose?

12.3 List four differences between the predator–prey relationship and the parasite–host relationship. Which of these differences is the most fundamentally important? Give reasons for your choice. What is the most striking difference in mode of parasitism between bacteria and viruses?

Suggested Readings

Costerton JW, GG Geesey, and K-J Cheng: How Bacteria Stick. Scientific American, 238:86, 1978.
This article and the one by Sharon describe how bacteria adhere to host cells and other objects by means of sugar molecules.

Marples MJ: Life on the Human Skin. Scientific American, 220:108, 1969.
The description of our skin as an ecological niche for other living creatures.

Sharon N: Carbohydrates. Scientific American, 243:80, 1980.

Wilson EO, Ed: Ecology, Evolution, and Population Biology. San Francisco: W.H. Freeman and Company, 1979.
Selected readings from Scientific American *that illustrate the concepts of ecology and evolution in their broadest context. An excellent collection of articles.*

Infection and Disease

Pathogenicity is not the rule. Indeed, it occurs so infrequently and involves such a relatively small number of species, considering the huge population of bacteria on earth, that it has a freakish aspect. Disease usually results from an inconclusive negotiation for symbiosis, an overstepping of the line by one side or the other, a biologic misinterpretation of borders.

Lewis Thomas in Lives of a Cell

In the previous chapter, we examined some of the interactions that have developed between human beings and the microbes that live on and in them. We observed that the bacteria and viruses that make their living as normal or persistent residents tend to live in quiet harmony with their host, and even pay something for the cost of their upkeep. In this chapter, we discuss microbes in their role as the agents of disease. Our objective will be to try to understand what is essential to the difference between microbes as harmless tenants and microbes as vicious invaders. We shall develop general concepts of microbial disease and later (Chapter 16) apply these concepts to specific diseases.

The etiology of a disease refers to the assignment of its cause. The term derives from the Greek *aitia* meaning "cause" and *logia* meaning "description" or "science." There are two important factors involved in the etiology of microbial diseases: the etiologic agent and etiologic circumstances.

Among the objectives of the medical microbiologist are to identify the specific microbe that produces a disease and to learn how the microbe, by invading the host, elaborating a poison, or misbehaving in any other noxious manner, causes the disease. The presence of the microbe, the etiologic agent, is a prerequisite for the illness. For example, tetanus is not possible without the work of *Clostridium tetani*, a gram-positive, spore-forming anaerobic rod that releases a powerful nerve toxin in the body. Hence, *C. tetani* is the etiologic agent of tetanus.

However, our understanding of an infectious disease is not complete if we have discovered only the etiologic agent. Our ability to prevent or control the disease requires knowledge of the circumstances of its etiology. What are the environmental factors that have allowed the etiologic agent to misbehave? The presence of *Clostridium tetani* is necessary for the development of tetanus, but it is not sufficient in itself. Spores of the tetanus bacterium are prevalent in the soil and therefore may be present on anything contaminated by soil, such as our skin and our food. For tetanus to develop, spores of *Clostridium tetani* must find their way into damaged tissues that have a very low oxygen content, such as deep wounds with a poor supply of oxygenated blood. If tetanus spores are introduced into healthy tissue, the spores fail to germinate and no disease results. From the ecological point of view, it is insufficient to describe the etiology of tetanus without referring to the circumstances of infection, for example, stepping on a rusty nail. Full understanding of the etiology of microbial disease includes a description of the precipitating ecological factors. Hence, the cause of a microbial disease should be investigated with regard to both the etiologic agent and the etiologic circumstances.

In the following discussion, we describe the etiology of infectious disease, taking into account both the noxious behavior of microbes and the environmental circumstances that abet their implementation. It is worthwhile to think about the environment. Surely, human beings have benefited at least as much from manipulating the etiologic circumstances of microbial disease as they have from discovering and treating specific microbial etiologic agents. Cholera was abolished from the western world by washing hands, treating waste, supplying clean water, and other public health measures, not by identifying and treating the etiologic agent, *Vibrio cholerae*.

13.1 Etiologic Agents

Apprehending the Etiologic Agent: Koch's Postulates

The observations of bacteria by Leeuwenhoek preceded by about two centuries their discovery as causes of infectious disease. Leeuwenhoek himself (1676) described bacteria in pus, but this discovery was ignored. Today, discoveries of much less importance become within months the focal point of intense investigation by international teams of medical microbiologists. Why did it take so long for the bacterial cause of disease to be discerned after bacteria were seen in pus?

There is no simple answer. Scientific observation is not merely seeing; it requires a prepared mind to grasp the potential significance of what is seen. The act of scientific observation is a talent exercised by an individual, but to realize this talent, the individual must be prepared by a cultural heritage. Cultural evolution is faster than genetic evolution, but it still takes time. It took almost 200 years for microbiology to generate and test the hypothesis that bacteria may cause disease. Once the question was raised, progress was rapid.

In 1767, John Hunter, an unfortunate English physician, inadvertently inoculated himself with the etiologic agent of syphilis while attempting to discover in pus the cause of gonorrhea. Three decades later, Edward Jenner (1796) introduced vaccination using material from cowpox as a means of protection against smallpox. It is thought that people in China had been using a form of vaccination against smallpox for generations. These striking empirical observations of the ability of natural materials to transfer disease or to induce protection also failed to stimulate the experiments needed to incriminate microbes as agents of disease. A half century after Jenner, the Viennese obstetrician Semmelweis (1847) discovered by chance that by washing his hands before delivering babies, he could prevent the often fatal puerperal fevers of new mothers. Thus, prevention of infectious disease by manipulating the host and the host's environment through immunization and antiseptic procedures preceded the discovery of microbes as the etiologic agents of disease.

The notion that disease might be caused by microscopic organisms began to be considered seriously in the middle of the nineteenth century. Henle, a German anatomist, in 1840 drew up a statement of the conditions that would prove the microbial etiology of infectious disease. These conditions were first fulfilled in the work of his compatriot, the celebrated physician Robert Koch. Although Koch himself never published a clear, concise statement of the principles of proof, as did Henle, these principles are generally known as **Koch's postulates.**

The original postulates for proving that a given microbe is the etiologic agent of a specific disease can be stated thus:

1. The microbe should always be found in those suffering from the disease and should not be present in healthy individuals.
2. The microbe must be isolated in pure culture from the diseased individual.
3. Inoculation of the pure culture into susceptible laboratory animals should produce the disease process.
4. Pure cultures of the same organism should be recoverable from the experimentally infected animals.

Koch fulfilled these postulates in studying anthrax, a disease of horses, cattle, sheep, swine, and (rarely) humans. *Bacillus anthracis* is a gram-positive, aerobic, spore-forming bacillus that infects farm animals that graze on pastures contaminated with anthrax spores. Human beings are occasionally infected by handling contaminated animal products, as illustrated by an old name for anthrax, "woolsorters' disease."

Koch isolated pure cultures of the anthrax bacterium from the blood of diseased animals and serially transferred the disease to laboratory animals with these pure culture isolates. In this way he established *B. anthracis* as the etiologic agent of anthrax.

There are several problems with the original formulation of Koch's postulates. A trivial but obvious question is why do the textbooks traditionally ascribe to Koch the postulates first put forth by Henle? Although it does not seem fair, this apparent misnomer highlights a fundamental principle of the scientific method. The experiment—the empirical demonstration—takes precedence over the idea. The measures of a good scientific idea are the experiments it generates. The postulates outlined by Henle in a sense became Koch's because he provided them with empirical meaning. Science does not exist where ideas have authority over empirical observations.

A more serious deficiency of Koch's postulates as originally stated is that they are not fulfilled by many microbial diseases. For example, common agents of disease such as staphylococci, streptococci, and pneumococci are carried as normal flora by a great many healthy people, and thus they do not fulfill the first postulate. Many microbial agents of human disease do not infect or grow in laboratory animals, so an experimental model of disease is not possible in these cases. Finally, certain microbes, among them viruses that have been associated with cancer (Chapter 16), may cause different diseases in different individuals, or no disease at all. In short, Koch's postulates are incomplete because they tend to ignore the importance of the environment and the state of the host to the development of infectious diseases. The postulates do not explicitly take into account the etiologic circumstances of infectious diseases. Nevertheless, the postulates were an important contribution because of the emphasis they placed on a rigorous microbiology based on methods of pure culture and exact identification of etiologic agents of disease.

Pasteur also studied anthrax at about the same time as Koch. His earlier studies on fermentation led him to investigate many microbial diseases, including chicken cholera (1880), swine erysipelas (1882), and rabies (1885). Pasteur's approach was different from that of Koch in that Pasteur sought to study the complex interactions of the microbe with the host and the environment. Pasteur developed vaccines to augment the resistance of potential hosts, and he investigated the effects of mixed cultures of organisms on fermentation. He even studied "diseases" of beer and wine. At the risk of oversimplification, we might say that Pasteur's approach took into account the etiologic circumstances, whereas Koch's approach was focused more on the etiologic agent. In complementing the pure culture approach of the Koch school, the followers of Pasteur prepared the foundation for the modern sciences of immunology and microbial ecology.

How Microbes Cause Disease

In the foregoing section we outlined some rules for testing and incriminating the etiologic agents of microbial disease. In this section we attempt to answer the question of how pathogenic microbes damage their host. There is evidence implicating three general mechanisms:

1. Destruction of host cells and tissues by direct microbial invasion.
2. Poisoning by toxic substances elaborated by bacteria.
3. Damage that results from inflammation and allergy generated by the host in response to the microbial invaders.

In discussing these mechanisms of disease, we attempt to discover the essential differences between pathogenic and nonpathogenic microbes.

DAMAGE PRODUCED BY MICROBIAL INVASION Both bacteria and viruses can damage the host by invading and proliferating within the tissues or cells. Microbes can be introduced into the body mechanically through wounds or the bites of animals or insects. Many bacteria and viruses can invade the host on their own by penetrating through the skin or the respiratory, gastrointestinal, and urinary tracts. Such unassisted invasion requires the microbe to adhere to the cells or cell products constituting the barriers of the body to the outside world.

In Section 12.3 we told how normal bacterial flora specifically adhere to cells of the host. Pathogenic bacteria also have ligands that can bind to molecules on the surfaces of host cells. It has been postulated that these bacteria may invade the host by producing enzymes that break down external barriers to penetration. For example, hyaluronidase is an enzyme that digests hyaluronic acid, a polysaccharide that acts as a cement holding host cells together. Bacteria such as staphylococci, streptococci, or pneumococci produce hyaluronidase and other enzymes that could help them spread beyond the barrier of surface cells by breaking down the cement that binds the cells together.

Once bacteria invade the body, the amount of damage they produce is related to their ability to persist and proliferate in the tissues. It is obvious that prompt destruction of bacteria will limit their ability to damage the host. Therefore, pathogenic bacteria often are outfitted with the means of impeding their detection or destruction by the defenses of the host. In the next chapter we discuss in more detail the mechanisms of the host that are designed to detect and destroy invading bacteria. As will become evident, host defense mechanisms usually involve ingestion of the bacteria by types of white blood cells called **phagocytes.** The word is derived from the Greek *phagenin* meaning "to eat" and *cyte* meaning "cell." Ingestion of bacteria or other particles by cells is termed **phagocytosis.** To survive and multiply in the tissues of the host, pathogenic bacteria must avoid being eaten and digested by phagocytes. Some pathogenic bacteria such as pneumococci, streptococci, and anthrax are surrounded by capsules. Encapsulation tends to impede the ability of phagocytes to engulf the bacteria. Variants of these bacteria that lack capsules are readily ingested and destroyed.

Other bacteria such as *Mycobacterium tuberculosis,* the cause of tuberculosis, and *Salmonella typhosa,* the cause of typhoid fever, although they are readily ingested, are not killed, and multiply within the phagocytes. How such bacteria avoid being digested within phagocytes is not known.

Pathogenic bacteria that have invaded tissues of the body rarely proliferate to such a degree that they produce damage by their numbers alone. Bacteria injure the host by elaborating toxins that poison host cells or by provoking an inflammatory response on the part of the host.

Viruses, in contrast to bacteria, can produce damage merely by invading and proliferating. This is because viruses replicate within the cells of the host, subverting the cells' metabolic machinery and often killing them.

The nature of the disease produced by invading microbes can be related to the site in which they settle. For example, the bacterium *Neisseria meningitidis,* if it penetrates the body from its niche in the throat, typically infects the meninges covering the brain and causes meningitis, or inflammation of the meninges. Poliovirus, disseminated from its normal niche in the gut, may enter certain nerve cells. The death of these nerve cells may lead to paralysis.

How do different pathogenic microbes choose their particular target cells? What determines the characteristic site of infection? The answers are obscure at present, but it is likely that the adherence of pathogenic microbes to host cells is directed by ligand–receptor interactions similar to those that bind normal flora to host cells (Section 12.3). Thus, different microbes have a structural affinity for specific molecules on the surfaces of certain host cells. The metabolism of certain host tissues, too, may make them a suitable place for specific types of microbes that require a particular nutrient supplied by the tissue. Other factors may also contribute to the homing of microbes to specific host organs. For example, bacteria entering the blood stream tend to settle out of the blood and invade areas where the blood flow is turbulent, such as in the vicinity of defects in the heart.

TOXINS PRODUCED BY BACTERIA The pathogenic effects of some bacteria can be related directly to the toxins that they produce.

Bacterial toxins are usually divided into two groups: exotoxins and endotoxins. Exotoxins are proteins secreted by bacteria into the external environment. For this reason they are called "exo-" or external. Endotoxins are lipopolysaccharide components of the cell walls of gram-negative bacteria. Although present on the surface, they are "endo-" or internal constituents of the bacterium.

Exotoxins Table 13.1 is a list of some exotoxins. Three kinds of evidence combine to implicate the participation of these exotoxins in their respective diseases: (1) Pathogenicity of the species of bacteria is associated with the ability to produce the exotoxin; variants that do not produce the exotoxin are not pathogenic; (2) injection of the exotoxin into experimental animals mimics the disease; and (3) immunity against the exotoxin protects against the disease. These criteria are fulfilled fairly well by the exotoxins listed in Table 13.1. Other toxic products have been isolated from many kinds of pathogenic bacteria, but we lack similar evidence implicating their role in disease.

Botulism illustrates the extreme example of a bacterial disease in which the etiologic agent, *Clostridium botulinum*, need never enter the host's tissues. The disease can be caused entirely by the action of the toxin produced by the bacterium outside the body of the host. *C. botulinum* is a spore-forming anaerobe whose natural habitat is the

TABLE 13.1 Some Exotoxins Produced by Bacteria

Bacterium	Toxin	Action on Host	Disease
Clostridium botulinum	Neurotoxin	Paralysis	Botulism
Clostridium tetani	Neurotoxin	Paralysis	Tetanus
Clostridium perfringens	Several types	Damage to cells	Gas gangrene
Corynebacterium diphtheriae	Diphtheria toxin	Inhibits protein synthesis	Diphtheria
Vibrio cholerae	Gut toxin	Interferes with regulation of water absorption	Cholera

soil. The bacterial spores do not germinate readily in the tissues, and the bacteria rarely invade the body. Typically the spores survive in canned food that has not been sterilized properly. The anaerobic conditions in the can are suitable for germination of the spores, and the bacteria begin to grow and produce toxin. The activity of the toxin is resistant to digestion by the enzymes of the intestinal tract of the individual who eats the contaminated food. The toxin is called a neurotoxin because it attacks the nervous system, preventing the transmission of signals between cells. Botulinum toxin is one of the most powerful toxins known; 1 milligram is sufficient to kill more than two million adult guinea pigs.

Ecologically, botulism is an accident. *C. botulinum* is not a parasite and apparently gains nothing from poisoning its victim. It kills without an obvious biological motive. However, the logic of evolution insists that *C. botulinum* must use a protein that is toxic to us for some purpose to the advantage of the bacterium; otherwise, natural selection would not have tolerated the metabolic cost to the bacterium of synthesizing a useless protein. The physiological role of botulinum toxin for *C. botulinum* remains a mystery.

Tetanus is another disease produced by a neurotoxin. Tetanus toxin poisons a normal mechanism that prevents overstimulation of the activity of the individual's nerve cells. In the presence of the toxin, orderly signal transmission is blocked by nervous overstimulation. Like *C. botulinum, C. tetani* is an anaerobic spore-former that lives in soil. It requires an environment very low in oxygen to germinate, grow, and produce its neurotoxin. *C. tetani* grows locally in damaged tissues and does not readily spread in the body.

Clostridium perfringens, one of the agents of gas gangrene, produces a great many enzymes that break down the host's tissues. Because of these enzymes, *C. perfringens* spreads within the tissues. Similar to other clostridia, *C. perfringens* is an anaerobic spore-former that normally lives in the soil and may be present in the gut without causing harm. Gas gangrene, historically associated with wars, is usually the result of extensive wounds heavily contaminated by soil that contains spores.

The ecological function of the toxins of the soil bacteria *C. tetani* and *C. perfringens* is as obscure as that of *C. botulinum*. However, the toxins have an unequivocal role in generating the characteristic signs of disease. Tetanus can be prevented by immunization against the toxin alone. *C. tetani* without its toxin is harmless.

Diphtheria is a disease caused by the toxin produced by *Corynebacterium diphtheriae.* In the absence of toxin production, the diphtheria bacterium lives as a member of the normal flora of the throat of many people. Immunity to the toxin itself is sufficient to prevent disease. Thus, disease can be related entirely to the toxin, which kills cells by preventing the synthesis of proteins. The toxin blocks the transfer of amino acids from transfer RNA to growing chains of peptides at the ribosomes (Section 7.3, Figures 7.5 and 7.6).

Diphtheria toxin is produced only by bacteria that are infected with a specific lysogenic phage called β. The toxin is encoded in the phage genome. Non-toxigenic strains of diphtheria become toxigenic and pathogenic upon infection with the β phage. However, even infected bacteria will produce no toxin unless a low concentration of iron is present. It seems that deprivation of iron activates the phage genome within the bacterium, leading to production of the toxin. Thus, diphtheria can arise from the conjunction of a host, a bacterium, a lysogenic phage, and an appropriate concentration of iron.

Cholera is a disease whose symptoms also can be attributed to an exotoxin. *Vibrio cholerae* is a gram-negative bacillus that in nature infects only humans. It is not a member of the normal bacterial flora, and chronic carriers are unknown. *V. cholerae* infects the gut usually by way of contaminated water. The bacteria grow superficially and are unable to penetrate the cells lining the intestines to spread in the body. The disease is caused by the action of an exotoxin that poisons the cellular mechanism regulating the flow of water and salts across the wall of the gut. The result is a massive diarrhea that can kill the host by dehydration.

Endotoxins Many dissimilarities exist between endotoxins and exotoxins (Table 13.2). Exotoxins are produced by gram-positive and gram-negative bacteria. In contrast, endotoxins are produced only by gram-negative bacteria. Endotoxin is the lipopolysaccharide component of the complex of protein and lipopolysaccharide that coats the cell wall of gram-negative bacteria; hence, all gram-negative bacteria produce endotoxins. Lipopolysaccharides are molecules made of lipid and sugar components that can resist heating at 100°C for about one hour. Since exotoxins are proteins, they are much more sensitive to heating and usually denature rapidly. Exotoxins are ten thousand to ten million times more potent by weight than are endotoxins. Exotoxins poison the host by interfering with a single well-defined metabolic process or product, and administration of the toxin alone can mimic many of the important features of the specific disease. Endotoxins, in contrast, produce many effects, and no single process can be identified as the target of the toxic action. Finally, immunization

TABLE 13.2 Characteristics of Exotoxins and Endotoxins

Characteristic	Exotoxins	Endotoxins
Bacterial producers	Mostly gram-positive pathogens	Only gram-negative pathogens and non-pathogens
Molecular class	Protein	Lipopolysaccharide
Response to heating	Sensitive	Stable
Potency	Extremely potent	Relatively weak
Pathogenic effect	Highly specific	Not specific
Injection of toxin	Mimics specific disease	Does not mimic specific disease
Immunity to toxin	Protects host	Does not protect host

against the exotoxin usually protects an individual from disease, whereas immunization against endotoxin offers little protection.

This comparison between endotoxins and exotoxins suggests that the two kinds of toxic bacterial products work their mischief through principally different mechanisms. Exotoxins are metabolic poisons or noxious enzymes. How can we categorize endotoxins?

A useful approach toward characterizing the biological activity of a substance is to inject experimental subjects with the purified material and record the outcome. One of the most sensitive indicators of endotoxin is fever. Injection of as little as one billionth of a gram of endotoxin per kilogram of body weight is sufficient to produce fever. The endotoxin does not stimulate directly the host's temperature control center, but it stimulates the host's white blood cells to release an endogenous **pyrogen** (*pyro-* = "fire"; *gen* = "cause"). This pyrogen travels to the brain and activates a control center in the hypothalamus to raise the temperature of the body.

Endotoxins also influence the ability of an animal to resist infection. We can assay the resistance of experimental animals by injecting them with known numbers of pathogenic bacteria and measuring the minimal number of bacteria needed to kill half the animals. This number is called the LD_{50} (lethal dose for 50 percent). Animals become markedly susceptible to infection during an eight-hour period following injection of very small amounts of endotoxin. However, by 24 hours after injection of endotoxin, the animals become highly resistant to infection; the LD_{50} increases so that 10,000 to 100,000 more bacteria are required to kill 50 percent of the experimental subjects. This heightened resistance lasts for several days. Hence, administration of endotoxin first lowers and then raises resistance of animals to infection. This effect is not specific, in that endotoxin obtained from any bacterium, pathogenic or not, will produce the same transient increase in susceptibility and resistance to challenge with any other bacterium or even a virus. The aggregate of these observations indicates that endotoxin works by evoking a series of responses on the part of the host itself.

DISEASE PRODUCED BY INFLAMMATION AND ALLERGY

Endotoxin also triggers inflammation of tissues. Inflammation is a complex response of an individual characterized, as all of us have experienced, by swelling, redness, and pain. Swelling results from leakage of proteins and fluid from blood vessels into the tissues; redness is a sign of dilation of blood vessels; and pain is caused by the release of chemicals from white blood cells.

Allergy is inflammation produced by the host itself. The microbial stimulators of allergy are called allergens. These allergens need not be toxic or inflammatory on their own account. They may cause inflammation only in people who are sensitive or allergic to them. Hence, allergens are unlike exotoxins or endotoxins, which are intrinsically noxious in all individuals.

How does inflammation cause harm? A large amount of endotoxin released in the blood stream or a sudden, generalized allergic reaction can cause massive dilation of blood vessels and leakage of fluid into the tissue spaces. Such events can cause an individual to die of shock, the lack of a sufficient volume of blood to sustain the cells and tissues. However, localized inflammation can also be harmful. For example, inflammation within the skull, in the brain or the coverings of the brain, can injure nerve cells. The leakage of fluid from blood vessels leads to the build-up of pressure within the confines of the skull. Such a degree of inflammation, which is fatal in the brain, might, in the skin or muscles, produce only an uncomfortable local swelling.

Similarly, minor inflammation of the valves or coverings of the heart can compromise the heart's function and produce serious disease. In this way, the site at which microbes trigger inflammation as well as the magnitude of inflammation can have a critical role in producing disease. The vital function of the host organ involved defines the seriousness of the inflammatory process generated by the host in response to microbial products.

In summary, microbes can cause disease if they penetrate the body and persist and multiply in the tissues. The ability to invade can be a biological property of a pathogenic microbe, or a naturally harmless microbe with little invasive capacity of its own can be introduced into the body mechanically by bite, injury, or even medical treatment. Pathogenic microbes may persist and multiply in the body because, by encapsulation or other mechanisms, they can evade destruction by the host's defenses. However, microbes that are naturally harmless for a healthy host may also persist and multiply in the tissues of a host whose defenses have been weakened by cancer, medical treatment, old age, intoxication, malnutrition, or other factors. Inflammation triggered by endotoxins or allergy may result from invasion of pathogenic microbes or penetration by usually harmless microbes.

In short, except for the diseases caused by exotoxins, there is no absolute distinction in mechanisms of disease between pathogens and normally harmless microbes. The development of disease usually depends on the circumstance of the host–parasite interaction, not merely on the identity of the parasite.

13.2 Etiologic Circumstances

With the exception of a few diseases caused by exotoxins, microbial diseases are infectious or contagious; they are transmitted from one individual to another. Transmission from host to host is the means of livelihood of all parasites. To survive the death of its host, the parasite or its offspring must reach a new host. Microbes that fail the test of transmission must either adapt themselves to living independent of their hosts and abandon parasitism as a way of life, or become extinct. Hence, the process of infection is at the core of a parasite's existence. It is no wonder that parasites, and microbes among them, have evolved ingenious mechanisms for maintaining a chain of infection from host to host.

The host, too, must be concerned about infection. Survival depends on controlling the kinds and numbers of parasites that infect the host, and limiting their activities to particular sites.

Clearly, the characteristics and behavior of both the parasite and the host influence infection. Organisms that exist together in the same ecological environment mutually adapt to each other. They have no choice under the pressure of natural selection. In the previous chapter we discussed how this rule of mutual adaptation is expressed in the maintenance of niches in the host for normal bacterial flora and non-pathogenic viruses. In this section we explore some mutual adaptations and non-adaptations of host and parasite related to the processes of microbial infection and disease. We will see that the etiologic circumstances of serious infectious disease often involve the breakdown of an ecological balance between man and microbes. Prevention of disease requires restoration of this ecological balance based on biological or cultural innovations.

Communicability and Virulence

The ability of microbial parasites to be transmitted from host to host is called **communicability.** Infection connotes a state in which a microbial parasite has succeeded in setting up a home in the host. Diseases caused by microbes are often called communicable or infectious diseases. Communicability and infection, however, do not refer to the behavior of pathogens alone. Normal bacterial flora and harmless viruses are communicable and infectious, but usually they are not virulent agents of disease. On the contrary, it is likely that non-pathogenic normal microbial parasites are more easily communicable in nature than are the most virulent pathogens. For example, rabies is almost always fatal in humans, but it is never transmitted from one person to another. Human beings are infected by the uncommon accident of being bitten by a rabid dog or other animal. For the rabies virus, infecting a human being is a dead end to its communicability.

Another example of divergence between communicability and virulence is illustrated by the pneumococcus. This gram-positive bacterium is a highly communicable member of the normal bacterial flora of a human's nose and throat; between 40 and 70 percent of normal adults carry at least one type of pneumococcus. Only occasionally does an individual suffer from pneumonia caused by the pneumococcus. The same pneumococcus harmlessly carried by humans is wildly virulent for mice. Injection of one or two such bacteria into the abdominal cavity of a mouse is sufficient to kill it within 16 to 36 hours. Yet an experimentally infected mouse will never be able to infect other mice kept in the same cage. The pneumococcus is virulent for mice but in nature is not communicable from one mouse to another.

The behavior of rabies virus in humans and pneumococci in mice illustrates a general rule of the host–parasite relationship. The most dangerous pathogenic microbes are usually those that are not regular parasites of the host species, but infect the individual incidentally. In the previous chapter, we presented the idea that parasites have a stake in the survival of the hosts that they parasitize regularly. For their own good, parasites obliged to live in a particular host cannot wipe out the obligatory host population. Hence, normal bacterial flora and persisting viruses are usually well adapted to living in their obligatory hosts in a way that causes the host a minimum of inconvenience. The corollary to this concept is the notion that microbes need not be so restrained when they infect an incidental host. Rabies virus is maintained by a natural chain of infection that does not include humans. The pneumococcus is part of the normal flora in humans but not in mice. Hence, these microbes do not owe their own survival to the survival of the incidental host, and thus they pay no penalty for causing him disease.

We can sum up this discussion by stating that well-adapted microbial inhabitants of a host population will in general not be markedly virulent for members of that population, but will have efficient mechanisms for communicability and infection within the population. In contrast, microbial parasites that incidentally infect members of a population to which they are not well adapted may be relatively more dangerous. Such microbes may or may not be communicable within the incidental host population.

Therefore, we can expect the etiologic circumstances of serious microbial disease often to involve infection of a host population by relatively new, less well-adapted parasites.

Transmission of Microbes in Host Populations

The study of the communicability of microbes within populations is called **epidemiology**, from the Greek *epi* meaning "among" or "upon" and *demos* meaning "people." Epidemiology does not limit itself to epidemics but deals with all the elements in the spread of microbes through populations.

RESERVOIR OF INFECTION A primary component in understanding the epidemiology of microbial infection is identification of the supply of communicable microbes. The reservoir of infection, the stable source of parasites, is responsible for their continuous propagation and dissemination into the environment.

Man himself is the reservoir of microbial parasites that are natural inhabitants of man. Normal bacterial flora and persistent viruses are usually transmitted by healthy individuals to other healthy individuals without evidence of disease. The occasional appearance of disease produced by normal microbial parasites can also be traced to the reservoir of healthy people. For example, pneumococcal bacteria isolated from patients with pneumonia usually are of the same strains as those prevalent in the healthy local population.

The human reservoirs of certain infections are sometimes people who have recently recovered from the disease but continue to carry and disseminate the etiologic agent. Typhoid fever is caused by *Salmonella typhi,* a gram-negative bacillus. Two to 5 percent of people who have recovered from the disease persist in carrying and excreting the bacterium for more than a year. The bacterium apparently can persist in the gall bladder while causing no symptoms. Thus, *S. typhi* can be excreted continuously in the stools of apparently healthy carriers and spread to others through contaminated food or water.

It was probably Koch who first grasped the significance of healthy or mildly ill carriers in the spread of infectious disease when he studied an epidemic of cholera in Germany in 1892–1893. He noted that some patients with very mild disease could still excrete large numbers of cholera bacteria.

Chronic or convalescent carriers of pathogens who have recovered from acute illness can be distinguished from healthy carriers of normal flora who have never suffered symptoms of disease. However, in principle, both convalescent and healthy carriers are effective human reservoirs of microbes.

Some microbial infections are maintained within a reservoir of acutely ill persons. Measles virus is spread only by contact of susceptible persons with persons who have overt cases of measles. There are no infectious healthy or chronic carriers. Once an individual has recovered from measles he cannot disseminate the virus or be reinfected. Measles, therefore, exists by exploiting a constant supply of susceptible people.

Many microbes maintain a reservoir outside of man. Bacteria such as *Clostridium tetani* or *Clostridium perfringens* exist in the soil. The anthrax bacillus has its reservoir in soil and animals. The etiologic agent of plague, *Pasteurella pestis,* is carried by rodents. A very large number of viruses have reservoirs in animals, including rabies, yellow fever, dengue, western and eastern equine encephalitis, and others. Microbes that persist in animal reservoirs are called **zoonotic;** the term derives from the Greek *zoon,* meaning "animal," and *nosis,* meaning "disease." Since man may be an incidental host to certain zoonotic microbes, there may be no mutual adaptation; thus, many diseases produced by zoonotic microbes can be quite serious. In addition, such microbes usually require special circumstances or intermediary agents to be transmitted to man.

VECTORS OF INFECTION **Vectors** is the term for biological agents that bear parasites from host to host. Arthropods (insects) function as vectors for zoonotic microbes. A special example is the large family (more than 250 members) of viruses called **arboviruses.** The term derives from the description **ar**thropod-**bo**rne **virus.** The virus of yellow fever illustrates some important aspects of arboviruses and vector-borne diseases in general. Yellow fever virus exists in reservoirs of monkeys and other animals living in the tropics. Mosquitoes are the vectors of yellow fever. The virus circulates in the blood of

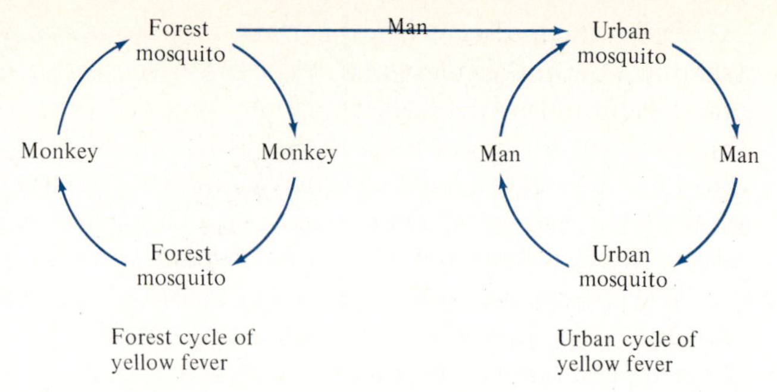

Figure 13.1
Forest and urban cycles of transmission of yellow fever virus. The zoonotic forest cycle of yellow fever virus involves transmission of virus between monkeys and forest mosquitoes. An urban cycle can be initiated when a person ventures into the forest, is bitten by a mosquito, and acquires the virus. Upon returning to the city, the person transmits the virus to urban species of mosquitoes. These in turn transmit virus to humans living in the city.

infected monkeys and is available to blood-sucking mosquitoes, which transmit the virus from monkey to monkey. An interesting feature of arboviruses is that they multiply in the tissues of their vectors. In this way, a few virus particles picked up by a mosquito are sufficient to transmit disease to another host. The number of viruses is increased in the mosquito, and consequently the efficiency of spread to new monkeys is increased greatly.

As we would expect, the species of monkeys that serve as the major reservoir of yellow fever virus are not made seriously ill by the infection. The virus is most virulent for incidental hosts such as other species of monkeys or man. Humans become infected accidentally by venturing into the forest and being bitten by an infected forest mosquito. Upon returning home, the human can infect urbanized species of mosquitoes. These city mosquitoes can infect humans who have never ventured into the forest, and thus an urban cycle of yellow fever infection is initiated (Figure 13.1).

The spread of yellow fever from Africa to the New World, the elucidation of the vector that led to control of the disease, and the subsequent discovery of the etiologic agent illustrate the impact of etiologic circumstances on the course of human affairs.

Yellow fever was imported into the Americas by enslaved West Africans who happened to be infected with the virus. The virus quickly established new reservoirs of infection among populations of suitably resistant New World monkeys.

The humans living in West Africa had a long history of contact with yellow fever virus and appeared to have evolved a relatively high degree of resistance to the disease. Not so the American Indians. The local populations of Indians had no previous exposure and therefore no opportunity to become adapted to yellow fever. They acted as new hosts for whom the parasite was most virulent. It is likely that their decimation by yellow fever, as well as by measles, smallpox, and other new (for them) parasites, hastened the conquest of the Indians by the Europeans.

From its newly established reservoir in the jungles of the Caribbean, yellow fever spread well over subtropical North and South America, even causing epidemics in the southern United States.

In 1900, the American military government in Cuba was scandalized by epidemics of yellow fever in Havana, and Dr. Walter Reed was appointed to investigate the cause. Influenced by the theories of the Cuban physician Dr. Carlos Finlay, Reed and his colleagues succeeded in proving that the disease was transmitted by the stegomyia mosquito. Two of the American investigators, Carroll and Lazear, proved it with their lives.

The American military governor of Cuba, General Wood, declared war on the mosquito, and by 1901 Havana was free of yellow fever. Reed and his fellow investigators had no inkling of the viral identity of the etiologic agent of yellow fever, but discovery of the etiologic circumstances sufficed to bring the disease under control.

This information continued to influence the political and social history of the Americas. The French by 1889 had sacrificed hundreds of lives to yellow fever and finally gave up trying to dig a canal through the isthmus of Panama. The discovery made by Finlay and Reed in Cuba made it possible for the Americans to dig the Panama Canal (opened to commercial traffic in 1914) and to enhance American influence throughout the entire Caribbean area.

The etiologic agent of yellow fever was discovered in 1929. American and English scientists traveled to West Africa to study new epidemic outbreaks of the disease. Stokes, an English member of the team, died of yellow fever shortly after discovering that it was possible to infect rhesus monkeys experimentally. The experimental animal model provided the means of determining that a filterable virus was the agent. Noguchi, an eminent bacteriologist of Japanese origin who worked at the Rockefeller Institute in New York, had announced some years earlier that a spirochete was probably the cause of yellow fever. He hastened to West Africa to re-examine his theories. Both Noguchi and his collaborator died of yellow fever shortly after reaching Africa.

Plague is another historically significant disease with a reservoir in animals. *Pasteurella pestis,* a gram-negative bacillus, infects rats and wild rodents. It is transmitted principally by rodent fleas that acquire bacteria from the blood of infected rodents. Human fleas are very poor vectors of *P. pestis,* and infection of humans requires the unusual circumstance of a person being bitten by rodent fleas.

Plague could never succeed in producing vast epidemics among humans if it was transmitted only by the bites of rodent fleas. However, a change occurs in some infected humans that allows transmission from person to person without a flea vector. *P. pestis* can multiply in the lungs of severely ill persons, and the disease can be transmitted by the small droplets of water produced by coughing. It is this pneumonic, air-borne form of the plague that has been the epidemic scourge of history, wiping out whole human populations (Figure 13.2). Plague lately has become rare; however, the zoonotic agent continues to lurk at the doorstep of man's world. *P. pestis* infects wild rodents even in the western United States.

Figure 13.2
Zoonotic and pneumonic plague cycles of *Pasteurella pestis*. In nature, *P. pestis* is maintained by being transmitted by rat fleas to rats and back to rat fleas. A person bitten by an infected rat flea can develop the pneumonic form of infection, leading to a rapidly spreading plague that is transmitted by coughed-up water droplets from one person to another.

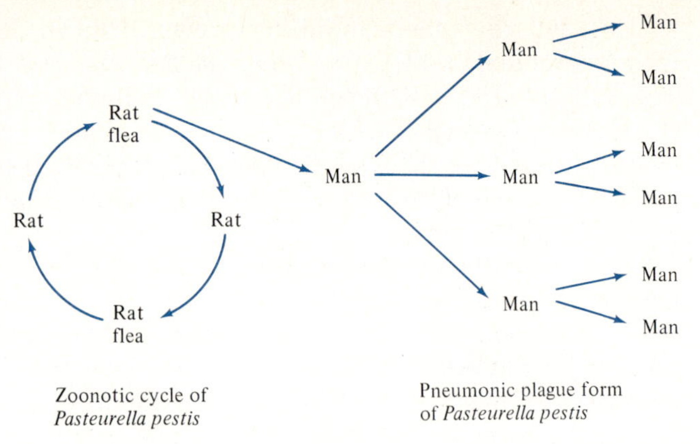

The Site of Infection Influences the Mode of Microbial Transmission

The means by which microbes are transmitted from host to host are suited to the microbe's niche. For example, bacteria and viruses that infect the nose, mouth, or throat are usually transmitted by water droplets in the air. The microbes are carried in droplets expelled by one person and breathed in by the next person. The physical properties of the droplets influence their ability to transmit microbes. Fine droplets are dispersed more widely than coarse droplets, but microbes are in greater danger of drying out in fine droplets than they are in coarse droplets. As a general rule, however, microbes that infect the respiratory tract are adapted to air-borne dissemination and are relatively more resistant to drying than are microbes transmitted in other ways.

Microbes that infect the intestinal tract are usually spread in contaminated water or by contaminated hands. The microbes are ingested and journey from the mouth through the esophagus and stomach to the intestines, in which they reside. The strongly acidic environment of the stomach poses a danger, and intestinal microbes must be well adapted to surviving this passage. They also must resist the digestive enzymes of the host.

Microbes that infect the host by way of the genital tract are transmitted by sexual contact. Hence, these venereal microbes are not called upon to survive outside the body for any length of time. Therefore, venereal microbes lack the adaptations of their less pampered colleagues and are extremely sensitive to environmental hazards such as drying, acidic conditions, and so on.

Microbes that live in the blood of the host require assistance in transmission. Insects that feed on blood can serve as vectors, as we saw in the example of yellow fever. The syringe of the physician, dentist, or drug addict can also inadvertently transmit blood-borne microbes. An important example is serum hepatitis. The virus of this disease circulates in the blood of apparently healthy carriers as well as in patients with hepatitis. It seems to have no way of exiting the body of one host and entering another except by way of injection. Serum hepatitis is usually transmitted by way of blood transfusion or by needles or other sharp medical instruments used on more than one patient. The virus is relatively resistant to procedures of sterilization that readily destroy other microbes. This adaptation is obviously important for such a microbe. The natural origin of the virus of serum hepatitis is a mystery. Blood transfusion and surgical procedures that seem to be the only means of its communicability did not exist a generation ago.

THE ROLE OF SYMPTOMS IN INFECTION **Symptoms** is the term for the conditions of illness that accompany a disease. The symptoms tell you that you are sick. Chills, fever, sneezing, coughing, nausea, vomiting, diarrhea, pain, weakness, achiness, swelling, and pus are some examples of symptoms. They are caused by a combination of the harmful pathogenic mechanisms of microbes and the inflammatory response of the host. The degree and type of symptoms felt by the patient are characteristic for the particular microbial disease.

Symptoms usually appear some time after infection with a pathogenic microbe. The interim between infection and the onset of symptoms is called the incubation period. It reflects the time required for the microbe to proliferate in the tissues to the point of harming the host.

Symptoms, at best, are an inconvenience for the patient; at worst, they can kill him. Does the microbial parasite derive any benefit from causing symptoms? Does the patient derive benefit from experiencing them?

Communicability—the transmission of a parasite from host to host—can be influenced to a great degree by the social activity of the host. The more active the host and the more social contacts made, the greater the chances for the parasite to infect new hosts. In contrast, a debilitated host will be less active in spreading parasites. Furthermore, in the extreme case where the parasites kill the host quickly, the parasites may die with their host before they can spread to a new host. Therefore, less virulent parasites that do not incapacitate the host will be spread and will flourish in the population. Virulent parasites that severely limit the social contacts of the host will spread less well. Hence, less noxious parasites will have a relative advantage over parasites that cause more severe symptoms and will tend to replace them in the obligatory host population.

An example of the success of a less virulent parasite can be seen by comparing the prevalence of parasitism of man by *Escherichia coli* compared with that of *Vibrio cholerae*. *E. coli* is an avirulent member of the normal flora of the bowel and is found in all human populations. *V. cholerae* is also a bacterium that lives in the gut, but it causes marked diarrhea. Although *V. cholerae* has been spread by epidemics to most parts of the world, this species of bacteria has not been able to establish itself as widely as has *E. coli*.

Can we conclude that successful parasites should cause no symptoms at all in obligate hosts? The answer is no. In many instances, the symptoms themselves provide the parasite with a vehicle for spread. For example, coughing or sneezing is a highly efficient way of spreading parasites of the mouth, throat, and nose. Very few microbes are expelled in droplets by talking or breathing. One good sneeze, however, can distribute millions. Parasites such as the viruses of chickenpox or cold sores would have no way of exiting the body of one host and entering another if they did not emerge in rashes. Therefore, it is advantageous to microbes to cause at least some symptoms such as coughing, sneezing, or rashes. Evolution will select against those microbes that cause symptoms severe enough to reduce communicability of the microbial parasite. Adaptation of a parasite to the host population will ensure a balance in which the benefit to the parasite of the symptoms it causes will just outweigh the inconvenience of the symptoms to the host.

In addition to facilitating communicability of the parasite, symptoms may actually be useful to the host. A host who knows that he or she has been invaded by microbes will institute measures to enhance survival and recovery. Bed rest will allow the system to fight the infection. Fever will inhibit the growth of many microbes in the body. Medical treatment will be sought. Since the survival of the host ultimately benefits the parasite, successful parasites should cause the kinds of symptoms that help the host contain infection and control severe disease. Hence, symptoms can be a kind of signal from the parasite to the host, beneficial to both.

The Social and Cultural Circumstances of Infection

Transmission of parasites between hosts is influenced by the behavior of the host population. Therefore, human behavior and human social and cultural institutions dictate many of the features of man's relationship with microbial parasites. Style of living, economic institutions, urbanization, travel, sexual habits, and ritual ceremonies all contribute to the host–parasite interaction. In this section, we present some examples illustrating this principle.

POPULATION SIZE AND INFECTION A major factor in communicability of parasites is the density of the host population. It is obvious that the historical, but now exotic, epidemics of Europe such as plague and cholera depended on the crowding together of large numbers of people under particularly unsanitary conditions. However, even mundane diseases of the civilized world have required a certain concentration of people to maintain themselves. A good example is measles. As we mentioned before, measles virus at present has no known animal reservoir and maintains itself entirely within the human population by causing an acute disease. There are no infectious carriers of measles virus, and people recovering from measles are solidly immune to reinfection. Therefore, the reservoir of measles virus is a chain of susceptible children (almost all adults already being immune) who transmit the virus during the incubation or acute stages of disease. The total period of communicability of measles virus is about two weeks. To keep going, measles virus requires enough people breeding in one place to generate a constant supply of susceptible children. A periodic epidemic occurs when a large number of susceptible persons accumulates. However, measles has to "smolder" in the host population as an acute disease even during nonepidemic years. One can compute the numbers of people needed in a locality to produce susceptible children at a rate sufficient to ensure contact with at least one susceptible host during the two-week period of communicability. It turns out that a population of less than 500,000 people will have very little chance of functioning as a reservoir of measles. A smaller concentration of people could not produce enough fresh susceptible persons. Obviously, virus can be transported from the outside into a small population of susceptible persons, but it takes about half a million people to keep the virus going.

A critical population density is needed to maintain other microbes, such as the viruses that cause the common cold. Small isolated communities cannot support these viruses once all the susceptible persons have recovered and become immune. The common cold disappears from the ice-bound communities of Greenland and Alaska once the winter ice closes the seaports. Colds return only after the spring thaw permits the first boat to arrive with a new batch of cold viruses to which the population is susceptible.

Infectious diseases such as measles and the common cold seem to be well adapted to human beings. Nevertheless, these microbes, dependent on a large population density, must have entered our ecological niche only recently. Table 13.3 presents an estimate of the size of the total human population and its communal aggregates during the last 500,000 years of human evolution. For more than 90 percent of this time (494,000 years), human beings lived in small, semi-isolated bands of hunters numbering less than 100 people. The entire human population of the world for most of this time was probably less than a few million. Only with the development of agriculture did some small cities arise; most people continued to live in village groups of less than 300 people. Cities of 500,000 people or more appeared only about 250 years ago as a result of the industrial revolution.

This leaves us with some very perplexing questions about microbes that are dependent on a high density of the human host population. Where and how did these microbial parasites exist before the cultural evolution of human cities? Are these new parasites? Did they become adapted to humans from an animal reservoir? It is difficult to believe that the viruses of diseases such as measles, mumps, influenza, and some types of common cold have been with us for only a thousand years or so. If these microbes really did evolve so rapidly, what does the future hold for man and his microbes? The ecological history of some very common parasites has yet to be unraveled.

TRANSPORTATION AND TRANSMISSION Earlier we saw how the transportation of slaves from West Africa introduced yellow

TABLE 13.3 The Physical and Cultural Development of the Human Population and Infectious Disease

Number of Years before 1980	World Population (millions)	Number of Generations	State of Human Culture	Type and Size of Human Communities	Characteristic Microbial Infections
500,000	0.1	25,000	Hunting and food gathering	Nomadic bands (<100 persons)	Normal flora; some zoonoses from wild animals; wound infections
10,000	5	500	Beginning of agriculture	Few villages (<300 persons)	Same + zoonoses from domesticated animals
6,000	50	300	Irrigated agriculture	A few cities (≈100,000); mostly villages (<300)	Same + epidemics during wars
250	600	10	Industrial revolution	Some large cities (≈500,000); many smaller cities (≈100,000); many villages (≈1,000)	Same + urban epidemics
130	1000	5	Sanitary reforms; beginning of medical science, modern urbanization	Some cities of 5,000,000; many cities of 500,000; few villages	Control of epidemics; cure of acute disease; beginning of nosocomial diseases

fever into the New World several hundred years ago. Transmission of microbial parasites was in both directions. It appears that an especially virulent form of syphilis was brought back to Europe from the New World by Columbus' sailors. The disease spread as an epidemic among the new European hosts, who had little genetic resistance.

Today, the fast transportation of people, animals, plants, and goods all over the world has introduced previously isolated human populations to all kinds of new parasites. Travelers' diarrhea is merely the result of a visiting host's introduction to the local bacteria. Sometimes more serious parasites are transmitted by modern transportation, as witnessed by the occasional person with cholera or some other exotic tropical disease who lands at an international airport. The wonder is that so few serious problems have arisen from the worldwide exchange of parasites. The alertness of public health organizations is probably more important now than in the past, when travel time was usually longer than incubation time.

HYGIENE CAN BE EXCESSIVE Delay of exposure to certain microbial parasites may be associated with the advance of human culture. This itself may lead to a dangerous ecological imbalance. Paralytic polio illustrates this point. Poliovirus is a member of the family of enteroviruses that infect humans through the intestinal tract. With the introduction of the toilet and the improvement of hygiene in modern times, intestinal diseases such as typhoid fever and dysentery have become much less prevalent. However, until the development of vaccination against polio, the incidence of poliomyelitis had become progressively greater. During the 1930s and 1940s, America was in panic every summer.

The reason for the increase in paralytic polio has now become clear. In the days before the development of toilets and sewers, which is to say for most of human evolution, man's familiarity with fecal material led to infection with poliovirus in early infancy. Thus, children became immune to poliovirus at an age when they were relatively resistant to the virus, either because they had a measure of passive immunity from their mothers (see Section 15.3) or because poliovirus itself is less virulent in the very young. Early infection was universal and paralysis was rare. The wide use of toilets delayed the age of contact with poliovirus until later in life. Hence, more and more people became infected at a susceptible age and without the protection of passive immunity. This led to a large increase in the incidence of paralytic complications. Therefore, an ecological imbalance in the relationship between host and parasite can develop from what appears to be an improvement in public health.

ECONOMIC INSTITUTIONS AND INFECTION The relatively recent (10,000 years ago) development of agriculture and animal husbandry brought human beings into intimate contact with the parasites of domesticated animals after many thousands of generations of evolution as hunters and food gatherers. The diseases anthrax and brucellosis came from contact with animals. Epidemics of plague and

Figure 13.3
The etiologic circumstances of legionellosis. *Legionella pneumophila* inhabits the soil and can be blown into the air on dust particles. Cooling towers of large air-conditioning systems apparently channel *L. pneumophila* into the air-conditioning ducts of the system. The bacteria are then disseminated and can be inhaled by people in the building.

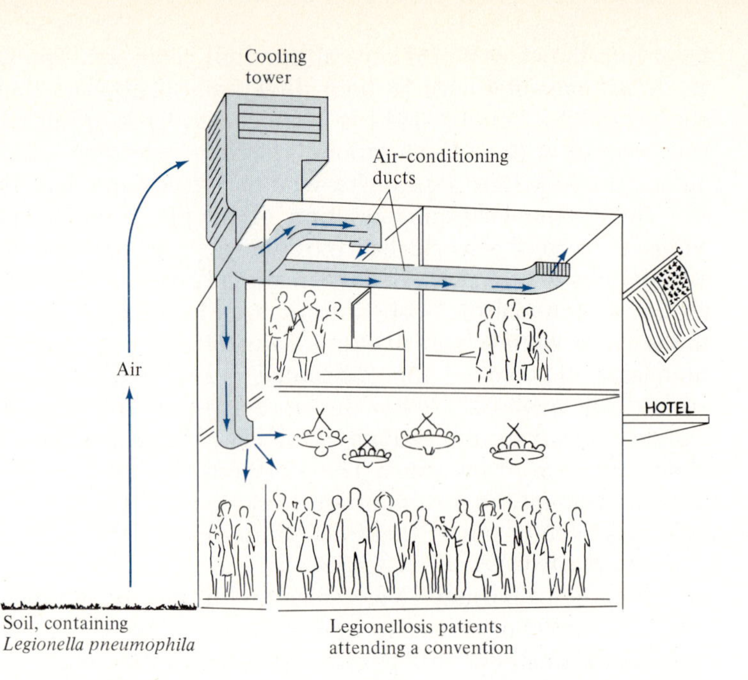

typhus accompanied the gathering of armies for conquest and defense. Tuberculosis increased with mining and the early industrial revolution. The development of the post-industrial world will surely bring us new parasites and new expressions of old parasites. We have only to look for them.

Legionnaires' disease, or legionellosis, is an example of how human cultural institutions and modern technology can influence etiologic circumstances. The 58th annual convention of the American Legion's Pennsylvania chapter, which took place in Philadelphia in July 1976, was marked by a mysterious epidemic that affected 221 people, 34 of whom died of pneumonia. The detection of the etiologic agent, a previously unrecognized bacterium now called *Legionella pneumophila,* is a saga of modern microbe hunting. It appears that *L. pneumophila* is a bacterium that normally lives in soil. However, it seems to be transmissible to human lungs by way of air-conditioning systems. *L. pneumophila* has been detected in the evaporative condensers of cooling towers that are part of these air-conditioning systems. Thus, the social institution of a convention coupled with the technology of air-conditioning gave rise to the etiologic circumstances favorable to the emergence of legionellosis (Figure 13.3).

MICROBIAL PARASITES AND HUMAN SEXUAL BEHAVIOR Our venereal parasites seem to be wedded to us with everlasting fidelity. Most will survive in no other species but human beings, so a major problem in studying *Treponema pallidum,* the agent of syphilis, or *Neisseria gonorrhoeae,* the agent of gonorrhea, has been the difficulty of infecting experimental animals. These parasites have bound their fate not only to our survival but to our reproductive act itself. Futhermore, the biology of human sexuality is necessary but not sufficient to propagate venereal parasites. These microbes require a particular form of social sexuality for their widest dissemination: promiscuity. The "vertical" transmission of venereal parasites from parents to offspring is almost negligible as a factor in the survival of these parasites. They require "horizontal" transmission through venereal contacts between individuals. However, two individuals, or even a group of individuals, whose sexual contacts are limited to their exclusive partners, cannot spread venereal parasites. If two people restrict their sexual activities to each other alone, then even if one were infected from birth, only two people at most would share the infection. The parasites would die with the couple. A chain of infection demands that sexual behavior be relatively unrestricted. As expected, the spread of venereal parasites is a direct measure of the degree of promiscuity in a human population. The total adaptation of venereal parasites to human beings suggests that these parasites have been enjoying human promiscuity for a considerable time in evolutionary history.

RITUAL BEHAVIOR AND INFECTION: KURU AND SLOW VIRUSES Religious ceremonies or rites have served the communicability of microbial parasites by bringing together, in close contact, large numbers of people. The crusades of the Middle Ages were mass movements of people gathered from all the countries of Europe for the religious ideal of conquering the Holy Land for Christendom. The great majority came from relatively isolated small villages and thus were abruptly confronted with new microbes under the stresses of crowding, filth, and travel. It is no wonder that most of the Crusaders died from communicable diseases before ever leaving Europe for the East.

The pilgrimage of Moslems to Mecca brings together people from all of Asia and North Africa. Epidemics of cholera and other microbial diseases have been associated with this pilgrimage even in modern times.

The common feature of pilgrimages is the convocation of large numbers of believers who exchange parasites. Pilgrimages illustrate no specific association between a particular rite and a given microbe. However, we have recently witnessed the discovery of a microbe whose transmission can be related to a distinctive ceremonial rite: cannibalism. Moreover, elucidation of this unusual host–parasite interaction has led to the discovery of what may be a large family of bizarre microbes, some of which may cause diseases for which an infectious etiology was previously unsuspected.

Kuru is a disease characterized by slow degeneration of the cerebellum, the part of the brain responsible for coordinating muscular activity. Its symptoms are loss of balance and shivering-like involuntary movements. Kuru, without exception, progresses to complete failure of nervous control of the muscles and death approximately a year after onset. Kuru attacks mostly women and children and is limited to the Fore tribes who live in the mountains of New Guinea and number approximately 35,000. Kuru, which means "trembling" in the Fore language, used to affect up to 1 percent of the population, but it has decreased markedly over the past 15 years with the decline in ritual cannibalism of dead relatives.

Drs. D. C. Gajdusek and V. Zigas of the National Institutes of Health in Bethesda, Maryland, first described the disease in 1957. Since that time, Dr. Gajdusek and his colleagues have studied kuru intensively in epidemiological surveys in New Guinea and in the laboratory. In 1966, they were able to report that the disease could be transmitted to chimpanzees by injecting them with brain tissue from humans who had the disease.

The investigators demonstrated remarkable insight in seeking an infectious cause for a disease that had all the earmarks of a hereditary degenerative process lacking any pathological signs of infection and inflammation. Their patience was no less commendable; the incubation period (the time elapsed between injection of chimpanzees and appearance of symptoms) was from ten months to almost seven years. The etiologic agent, called a "slow virus," was found to have many of the properties of known viruses, such as the ability to replicate in infected animals and in cell cultures and to pass through filters small enough to exclude bacteria. The agent of kuru also manifested some very atypical properties: long incubation periods of months to years and perhaps decades; failure to trigger in the host an inflammatory or immune response; ability to withstand extremely high doses of ultraviolet irradiation or treatment with formaldehyde; and absence of the diagnostic signs of virus infection detectable by the electron microscope.

It is now believed that kuru was transmitted among the Fore tribe by ritual cannibalism or handling of the contaminated brains of dead relatives carried out as a sign of respect by women and young children.

The finding of a slow virus as the cause of kuru led Gajdusek and his colleagues to attempt to transmit to primates and other animals human diseases with a pathology similar to that of kuru. They have been successful in infecting animals with Creutzfeldt-Jakob disease. This degenerative disease of the brain is rare but is found throughout the world. Until the recent past, it had been considered to be an inherited condition, due to some defective gene and not to an infectious agent.

Transmission from person to person has been inadvertently carried out by surgical transplantation of corneal tissue to the eye of a recipient who later developed Creutzfeldt-Jakob disease. Afterward, the donor of the cornea was diagnosed as having suffered from Creutzfeldt-Jakob disease. The disease has also been observed in brain surgeons and other physicians with access to human brain tissue. Hence, the etiologic circumstances of infectious degeneration of the brain, outside of New Guinea, may be occupational rather than ritual.

Slow viruses grow slowly, and progress in unraveling their mysteries is commensurately slow. However, we must now seriously study the possibility that slow viruses are involved in many other chronic degenerative diseases affecting millions of humans, such as some types of multiple sclerosis, arthritis, and diabetes. Investigation of exotic kuru, with its seemingly limited significance, has uncovered the once hidden trail of human slow viruses. We can only guess where that trail may lead. What are slow viruses? Where have they come from? How are they transmitted? What diseases do they cause?

The notion of undetected slow viruses possibly lurking in the dark corners of our bodies may be disturbing to some. On the contrary, however, the discovery of slow viruses should give cause for optimism. Diseases of microbial etiology offer the hope of control, prevention, or cure once the etiologic agent and circumstances are characterized. It is usually much more difficult to deal with diseases caused by inheritance of faulty genes than it is to control infectious diseases. After all, our genes are intrinsic to ourselves, whereas infectious microbes are extrinsic.

Finally, in introducing us to slow viruses, kuru demonstrates that the work begun by Pasteur and Koch of detecting and characterizing microbial parasites is far from complete.

NOSOCOMIAL INFECTIONS: MEDICAL SCIENCE CAN TURN NORMAL FLORA INTO PATHOGENS The once deadly plagues of infectious diseases are almost gone from the Western world; plague, cholera, smallpox, tuberculosis, and infantile diarrheas have succumbed to toilets, sewers, chlorination, adequate nutrition, vaccination, and antibiotics. The most serious mortality due to infectious agents is now produced by normal bacterial flora.

The term **nosocomial infections** refers to infections related to hospitalization or medical treatment (*nosos*: "disease," plus *komein*: "to take care of;" hence, *nosocomial* means pertaining to a hospital or an infirmary). Nosocomial infections are caused by normal bacterial flora that are pathogenic for the patient because of the type of medical or surgical treatment undergone to remedy some other illness. Very often, the treatment that has made the patient susceptible to nosocomial infection is the very treatment that has saved or prolonged the patient's life. Surgical procedures may unavoidably introduce normal flora into critical areas of the body of a patient whose resistance has been compromised by underlying poor health. People suffering from diabetes are especially prone to infection. Both cancer and its treatment impair the ability of the immune system to fight off even weakly virulent bacteria or viruses. In short, the success of medical science in keeping people alive who might have died of other causes has opened the way for serious nosocomial infections. Normal bacterial flora carried by the patient and all of the patient's contacts prevail in the environment and may exploit any deficiency in resistance.

Some figures can highlight the magnitude of the problem. Nosocomial infection occurs in approximately 3 to 15 percent of all hospital admissions. This means that more than two million Americans, or approximately 1 percent of the total population, are affected by nosocomial infections each year. Invasion of the blood stream by the normal gram-negative bacteria of the gut, occurring about once in each thousand hospital admissions, has a mortality of at least 25 percent. Thus, about 80,000 people die yearly in the United States from invasion of the blood by these normal flora. This is considerably more than the number of people who die from automobile accidents and twice the number who die from cancer of the large bowel and rectum.

We can divide the history of nosocomial infections into three periods: that of sepsis or filth, that of antisepsis, and that of antibiotics. Until antisepsis was introduced by Semmelweis in 1847, a horrendous mortality of approximately 90 percent accompanied all surgical procedures. Gas gangrene was common. Hospitals were death houses for the poor. This situation was improved greatly by the development of techniques for washing and sterilization and the use of rubber gloves and face masks that became standard practice during the second half of the nineteenth century. Aseptic surgery prevented gas gangrene, but the postoperative patient all too often suffered from infection with gram-positive bacteria, notably the streptococcus.

The third and present period was initiated by the discovery and wide use of antibiotics. Treatment with sulfa drugs and penicillin effectively eradicated the threat of infection of surgical patients with streptococci. However, we are now in the era of gram-negative nosocomial infections. Gram-negative bacteria are a serious problem because of their ubiquity in the environment and the ease with which they develop resistance to the most powerful antibiotics.

Paradoxically, the problems of nosocomial infections have become more acute as medical and surgical practices have become more aggressive and more successful in maintaining life. New biological problems are created by the solution of older biological problems. Disease produced by normal flora is part of the price of progress in medicine.

13.3 Adaptation and Ecological Balance

In this chapter, we have attempted to present principles for understanding the essence of infectious disease. We have seen how development of disease is a combined function of the etiologic agent and the etiologic circumstances. The attributes of the microbe and the nature of the host, the host's biology, behavior, and environment, are all important in determining the process of disease.

Microbial parasites that are obliged to exist within a particular host population are exquisitely adapted to maintaining themselves while causing the host as little trouble as possible. This host population in turn is well adapted to the parasite and keeps it confined to a particular niche.

Serious infectious disease often results from changes that upset the ecological balance between host and parasite. The cultural evolution of humans over the past few thousand years appears to have affected this ecological balance. Human beings have vastly changed their way of life in population size and urbanization, in economic activities and travel, in contracting new parasites, and in creating

new avenues for infection through medical and surgical intervention in human biology.

These ecological disequilibria have produced new infectious diseases, plagues, and epidemics. However, humans have continued to flourish despite this ecological upset because they have succeeded in creating a new ecological balance.

This restored balance with the microbial world is based on two kinds of adaptive mechanisms: biological and cultural. Biological adaptation is a consequence of having suitable genes. Such adaptation is common to all forms of life; we call the process evolution. In contrast, cultural adaptation is unique to human beings, who alone accumulate information and pass it on to their fellows as well as to their offspring in the form of symbolic language. Changes in genetic information occur through evolution: the slow and costly process of random mutation of genes and natural selection of those that are more beneficial. Cultural adaptation is direct rather than random and is transmitted horizontally throughout the human population and not just vertically to the next generation. Hence, cultural adaptation is orders of magnitude faster than biological evolution.

Man adapted culturally to the new etiologic agents and circumstances of infectious disease by founding the sciences of microbiology and immunology, by inventing measures and machines for public health and sanitation, clean food and water, by applying vaccination to prevent some diseases, and by using antibiotics to treat others. The evolution of human culture that created circumstances for new infectious diseases also created the means of meeting the challenge. This process is self-evident in the educational experience of us all.

Did man and microbes also adapt biologically to the new situations? The logic of the host–parasite relationship suggests that a virulent microbe should evolve toward lower virulence, while the susceptible host population should evolve increased resistance against the microbe. The science of microbiology is too young to have been able to detect such evolution involving human beings. However, a historical experiment involving rabbits and myxomatosis virus has demonstrated the reliability of this prediction.

RABBIT MYXOMATOSIS A lesson in the biological evolution of the host–parasite relationship was afforded by the introduction of myxomatosis virus into Australia. Myxomatosis virus is a member of the pox virus family of DNA viruses. The virus naturally infects species of wild hares of the *Sylvilagus* genus that are native to North and South America but not found in Europe or Asia. Domestic rabbits have been derived from the *Oryctolagus* genus of Europe and North Africa and not from the American genus.

In 1896 it was noted that a colony of domestic rabbits in South America was wiped out by a markedly virulent disease characterized by swelling and ultimate degeneration of masses of connective tissue over the rabbit's body. About 20 years later a Brazilian scientist, Dr. Aragao, discovered that the disease could be transmitted by insect vectors. Dr. Aragao later learned that the agent of the virulent disease was carried by the local American hares in a very mild skin reaction. Mosquitoes biting an infected hare in the region of the skin lesion pick up the virus and transmit it to the next hare they bite. The infected American hares seemed to suffer no ill effects from harboring the virus. The host and parasite appeared to be well adapted. In contrast, domestic rabbits of European origin were a new, unadapted host for the virus, and infection was fatal in more than 99 percent of these rabbits. This is another illustration of the potential virulence of a parasite for a new host.

Domestic rabbits of the European type had been introduced into Australia by European settlers. These rabbits proliferated unchecked by their natural predators until they became a plague in themselves. The rabbits multiplied so well that they soon competed with sheep for pasture lands and caused great economic losses to Australian sheepherders and farmers.

Dr. Aragao suggested that American myxomatosis virus be used in biological warfare to destroy the rabbit pests of Australia. The Australians embraced the idea, and infected rabbits were liberated in Australia in 1950. At first it appeared that the virus did not take to the new environment. However, myxomatosis suddenly flared up, and within a few years, millions of rabbits died. Mortality was more than 99 percent. The resulting increase in grass made available to sheep led to an increase in the production of wool that was estimated to be worth more than 200 million dollars in one year. The stage was set for observing evolution in action.

Two observations were made: The virus in the field became less virulent, and the wild rabbit host became more resistant. As predicted, the less virulent variants of the virus enjoyed an advantage in communicability; less sick rabbits survived longer and spread the virus better. Hence, the less virulent viruses prevailed, so much so that the Australian farmers had to import fresh stocks of virulent virus to keep the disease going. This of course failed; less virulent viruses always emerged and replaced the virulent ones.

The development of resistant rabbits was also predictable. The more susceptible rabbits were quickly killed off, and only the genetically resistant ones contributed their genes to the next generation. The total numbers of rabbits were reduced, but an equilibrium was established between the attenuated virulence of the virus and the increased resistance of the rabbits. Moderation won the day.

It is very likely that similar experiments in biological adaptation took place during the evolution of man and his parasites, with the same result. The new parasites we contracted and the new ecological circumstances we created led to the appearance of new and terrible infectious diseases. History records how plagues of all kinds wiped out entire populations. These infectious diseases have been reduced in severity by biological adaptation as well as by the development of human culture, although, as in everything else that concerns human beings, culture has probably contributed much more than has genetic evolution.

QUESTIONS

13.1 When the astronauts returned to earth from the moon they were placed into strict quarantine. Explain why this was done.

13.2 How do the agents of nosocomial infections fulfill Koch's postulates as originally stated? How do they fail to fulfill the postulates?

13.3 How would you amend Koch's postulates to bring them up to date?

13.4 When a dog is infected with rabies virus, the virus can be found in two organs: the brain and the salivary glands. How does the presence of the virus in each organ influence transmission of the virus to the next host?

13.5 In general, American Indian or African children infected with measles virus suffer a much more severe form of disease than do European children infected with the same virus. What might you conclude about the evolutionary history of measles virus in these different human populations?

Suggested Readings

Andrewes CH: The Natural History of Viruses. New York: W. W. Norton and Company, 1968.
A readable and interesting survey of the ecology of some virus–host interactions that includes the story of rabbit myxomatosis. The language is not technical.

Camus A: The Plague. New York: Alfred A. Knopf, 1948.
A novel in which the plague bacterium is an important character.

Fraser DW and JE McDade: Legionellosis. Scientific American, 241:82, October 1979.
The fascinating story of Legionnaires disease and its discovery using the methods of modern epidemiology.

Hirschhorn N and WB Greenough III: Cholera. Scientific American, 225:15, August 1971.

Kaplan MM and RG Webster: The Epidemiology of Influenza. Scientific American, 237:88, December 1977.

Zinsser H: Rats, Lice and History. Boston: Little, Brown, 1935.
Classic entertainment based on case histories illustrating the decisive influence of epidemics on human history.

Resistance to Infection and the Immune System

Stimulate the phagocytes.

Doctor's Dilemma, *George Bernard Shaw*

In this chapter we discuss the mechanisms by which we contain our microbial parasites and destroy them when they overstep their bounds. For purposes of understanding and analysis, it is useful to distinguish between two different mechanisms for combatting infection: **constitutive resistance** and the **immune response.** Constitutive resistance refers to barriers to infection that result from normal human anatomy and physiology, barriers that are not specifically formed in response to invading microbes. The immune response, in contrast, includes those mechanisms of resistance that are induced in response to infection. In other words, the immune system receives signals from microbes, while constitutive mechanisms of resistance do not. Hence, a communicative relationship exists between microbes and the immune system but not between microbes and constitutive barriers to infection.

14.1 Constitutive Barriers to Microbial Invasion

The human body possesses a number of physical, chemical, and physiological factors that form a first line of defense against microbial invasion. The intact skin, for example, is impervious to most microbes, so that infection through the skin usually requires a cut, abrasion, or other physical penetration. Skin diseases that break the integrity of the skin can also provide portals of entry for microbes. The mucous membranes, too, resist invasion by many microbes. In addition to physically excluding microbial invaders, the healthy skin actively inhibits the growth of potentially pathogenic bacteria and fungi by secreting sebum in the sebaceous glands. As mentioned earlier (Section 12.4), normal bacterial flora residing in the skin metabolize sebum to produce fatty acids with antibacterial properties. Thus, the skin and its normal flora work together to prevent invasion by pathogens.

Various kinds of chemical substances serve as agents of constitutive resistance. For example, an enzyme called lysozyme is secreted in the tears that constantly bathe the exterior surface of the eyes. Lysozyme breaks down the peptidoglycan of bacterial cell walls (see Section 2.4). This helps keep the eyes free of infection.

The hydrochloric acid secreted by the cells in the stomach is a chemical barrier to many microbes that are inactivated by low pH. Other substances present in the gut, such as digestive enzymes or bile acids, may also neutralize many potential invaders.

In addition to physical and chemical barriers, the movement of fluids and mucus in various organs is an important physiological factor in constitutive resistance. The gastrointestinal, urinary, and respiratory tracts all cleanse themselves by flushing. The constant flow of urine from the kidneys to the bladder and the periodic emptying of the bladder wash out bacteria and viruses. This flushing is critical in keeping the urinary tract essentially free of microbes. Any obstruction to the smooth flow of urine leads to accumulation of bacteria and infection of the urinary tract.

Although the gastrointestinal tract is not sterile, wave-like contractions of the bowels flush the system and keep the numbers of bacteria and viruses within tolerable limits. Obstruction to this flow through the gut allows the number of gut bacteria to increase greatly and decreases the ability of the bowel wall to serve as a barrier to invading microbes. Hence, any bowel obstruction can potentially expose the patient to a massive invasion of bacteria. This is one of the reasons why surgical treatment of obstruction is an acute emergency.

The respiratory tract also flushes itself. As expected, the cleansing process used by the respiratory tract is adapted to the problems of trapping and expelling contaminants borne by inhaled air. The respiratory tract has special cells that secrete mucus. This mucus forms a blanket that lines the air passages and traps microbes or other contaminants of the air. Mucus is constantly propelled out of the lungs and air passages by the wave-like beating of microscopic hairs called cilia attached to the cells of the respiratory tract. Cilia are similar in their structure to the flagella that propel bacteria (Section 2.4, Figures 2.20 to 2.22). Upon reaching the upper airways, the mucus is expelled by coughing or swallowing and the trapped microbes are killed by the acid in the stomach. Fresh mucus is continuously secreted to replace the expelled mucus. As we all know from experience, the amounts of mucus generated in the nose and airways can increase markedly when the mucous membranes are irritated. Cigarette smoke causes stagnation of the mucus by paralyzing the beating of the cilia. This increases the likelihood of respiratory tract infection and contributes to the chronic cough of heavy smokers.

The physical, chemical, and physiological mechanisms that we have described here constitute an important first line of defense.

14.2 Immune Mechanisms of Resistance

The immune system is capable of recognizing invading microbes and responds by destroying the invaders. The branch of science that studies the immune system is called **immunology.** This field has undergone a period of explosive growth in the past 20 years. Immunology is now in an epoch of creative fervor that can be likened to the heroic decades when Pasteur, Koch, and their colleagues built the foundations of microbiology. The recent advances in immunology and their impact on biology and medicine make it important to acquire a basic understanding of the immune system.

The immune system is important and interesting for a number of reasons. Besides performing a critical function in controlling infection, the behavior of the immune system is a key element in the success or failure of organ transplantation; it is implicated in the story of cancer; it contributes to the pathogenesis of common, often serious diseases such as allergies, some forms of diabetes, and arthritic inflammation of the joints; finally, the immune system itself is one of the most fascinating of biological systems, superseded in adaptability and complexity only by the brain. In this section we discuss the mechanisms deployed by the immune system to inactivate or kill invading microbes. In the next chapter (Chapter 15) we complete the story of the immune system by discussing how particular cells called lymphocytes regulate the immune response to infection.

At the outset, we must clarify the term **immunity,** which does not have the same meaning in lay English as it does in microbiology and immunology. In this context, immunity implies resistance to a microbial agent that is capable of infecting the individual. Immunity should be distinguished from lack of susceptibility. Only a potential host is thought of as immune or not immune to a microbial parasite. A human being becomes immune to polio after infection or immunization with the virus. In contrast, humans are defined as not susceptible to tobacco mosaic virus, which can infect only tobacco plants. Immunity describes an aspect of the relationship between a host and potential parasites. Immunity implies resistance to infection that is based on the function of immune mechanisms. We shall concern ourselves with mechanisms of immunity rather than with mechanisms of nonsusceptibility.

14.3 Proof that the Immune System Influences the Host–Parasite Relationship

Does the immune system play a role in regulating the normal relationships between humans and their natural microbial parasites, or does it only occupy itself fighting pathogenic microbes?

Earlier (in Section 12.4), we discussed how it is often possible to identify the function of a biological organ by removing it and then seeing how the organism fares in its absence. Therefore, one way to elucidate the role of the immune system is to remove or inactivate it and see how this procedure affects the host–parasite relationship. Unfortunately, both nature and medicine have provided us with many experimental models for study. Occasionally, individuals are born lacking part or all of their immune system because of a genetic or developmental defect. Other individuals, for their own benefit, have had to have their immune system medically inactivated. For example, recipients of kidney or heart transplants will attack and reject the transplant unless their immune systems are weakened by treatment with drugs or x-rays.

These individuals with depressed or inactive immune systems are found to be in danger of serious infections caused by bacteria or viruses. The important fact to remember for our present discussion is that almost all of these serious infections are caused by the prevalent "normal" bacterial flora or by viruses that are "harmless" for most people.

As we observed when we considered the process of infection and disease (Chapter 13), the virulence of microbes is defined by their behavior in particular hosts under particular circumstances. Pathogenicity is defined operationally; it is not an intrinsic property of the parasite. So-called normal flora and harmless viruses, as well as pathogens, are quite virulent in people who have a deficiency in their immune competence.

From these observations, we may conclude that the immune system plays a critical role in containing normal microbial parasites and confining them to sites on our bodies where they can do little harm. In other words, non-pathogenic microbes are such because they are contained by the immune system. Therefore, the immune system regulates the equilibrium that exists between humans and their normal microbial parasites.

An important conclusion emerges from the fact that the immune system contributes to the stability of the host–parasite relationship. Parasites have an interest in the general well-being of the host species on whom their own survival depends (Section 12.1). Therefore, normal microbial parasites of humans benefit in the long run from a stable, relatively non-pathogenic relationship. Since the immune system of humans contributes to this goal, it is likely that many of our microbial parasites are well served by an efficient immune response. This explains why many microbes actually transmit a variety of powerful signals to the immune system of the host—to ensure that host and parasite get along as well as possible. Therefore, it is useful to view aspects of the immune response as a system of communication between the host and microbial parasites. Coordination of the stimuli transmitted by the microbes with the responses of the host contributes to the stability of the relationship between the host and microbial parasites.

14.4 A Way to Describe the Immune System

The immune system is complicated because it involves communication among many components, including cells of various kinds. Furthermore, the interactions between components are regulated in a very precise manner, so the immune response to a particular microbe is appropriate to the nature of the threat. For example, to be effective, the immune system must use a different mode of defense against an exotoxin circulating in the blood stream than it does against a virus that infects one of our cells (Chapter 13).

A useful way to organize our thinking about a system is to de-

Figure 14.1
Block diagram of the relationship between microbial invaders and the immune system.

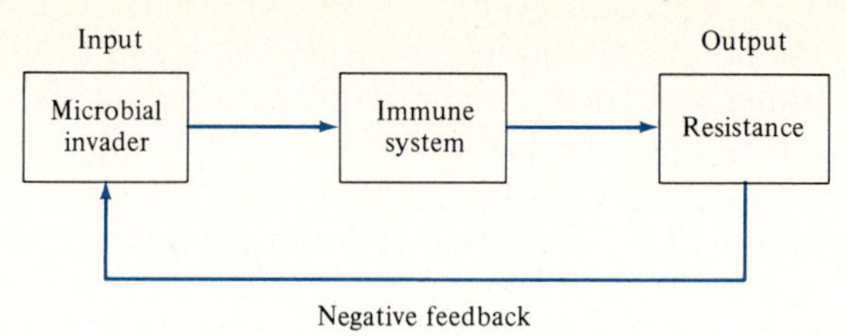

velop a conceptual framework that will aid us in identifying the important parts of the system and in seeing how these parts are related to each other. Engineers sometimes use block diagrams to describe mechanical systems. We shall borrow their approach to explain some aspects of the immune system.

Figure 14.1 shows a block diagram of the relationship between the immune system and invading microbes in very general terms. The diagram is composed of blocks that frame the components we wish to illustrate and arrows that connect the blocks. The blocks delineate entities, processes, or ideas on which we choose to focus our attention. The arrows show how these components are related or influence each other.

In Figure 14.1, we show that a microbial invader signals the immune system. The arrow going from the microbe to the immune system denotes this signal. Although we have yet to discuss the nature of the signal, we can appreciate its function as an "input" into the system.

The immune system receives this information and responds by implementing mechanisms of defense that result in resistance to the microbial invader. Hence, "resistance" is the "output" of the system. The mechanisms constituting resistance, also as yet undefined, in turn "feed back" on the invading microbe and fight it off. The feedback is described as "negative" because the output, the mechanisms of resistance, act to negate or reduce the influence of the microbes. Figure 14.1, therefore, describes the immune response as a closed loop between microbial invasion and the immune response.

Our objective is to characterize each of the blocks and arrows in the figure. We want to know the following: (1) What are the signals by which microbes communicate their presence to the immune system? (2) How does the immune system receive and interpret these signals? (3) What are the defense mechanisms deployed by the immune system and how do they destroy invading microbes? (4) How does the immune system learn from experience and regulate the nature of its response?

14.5 Activation of Effector Mechanisms of Immunity: Signals and Reception

The immune system deploys a variety of chemicals and cells that can attack and destroy microbial invaders. We may call these agents of resistance "effectors" or effector mechanisms, because they pro-

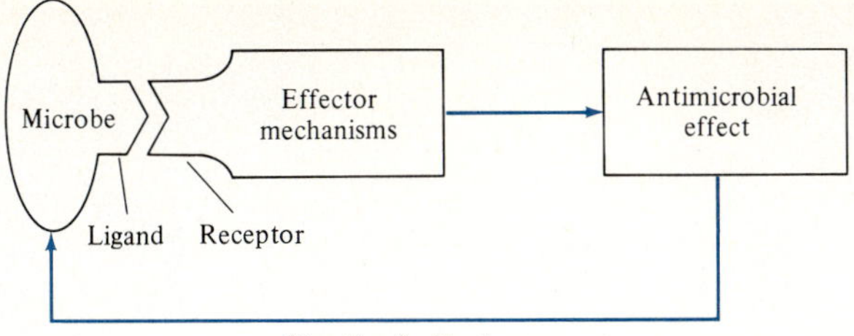

Figure 14.2
An antimicrobial effector mechanism is activated when a receptor recognizes a ligand on a microbe.

duce a noticeable effect on the microbial invader. Effector mechanisms must be able to recognize microbes in order to know when they are present. It appears that effector mechanisms can be signaled by molecular products of the microbes themselves.

How do molecules act as specific signals to the immune system? A molecule is a signal when it is recognized by its target receiver such that the receiver responds to it. In general, molecular signals are recognized by the lock-and-key principle of complementary fit. This principle was illustrated by the adherence of bacteria to specific host cells (Section 12.3 and Figure 12.6), but it is also fundamental to interactions between hormones and cells, or enzymes and substrates. Two molecules specifically bind and interact to the exclusion of unspecific molecules, by the three-dimensional complementarity of particular sites on each molecule. The target molecule that receives the information is called the receptor. The molecular signal that bears the information to the receptor is called by many different names in different circumstances, but in general it may be called a ligand, meaning "to bind." The ligand and receptor bind because of their mutual geometry, as do lock and key, without establishing permanent chemical bonds. This binding comprises the transmission of information and can thus be called recognition.

Effector mechanisms may be activated directly by microbial invaders when the mechanisms have receptors that recognize microbial ligands. Figure 14.2 illustrates this relationship schematically. A microbe is pictured as having a ligand that is recognized by a receptor present on a molecule of a hypothetical effector mechanism. The binding of the ligand to the receptor activates an antimicrobial effect, which in turn attacks the invading microbe.

14.6 Effector Mechanisms of Immunity

In this section, we describe the important effector mechanisms activated directly by microbial ligands.

Phagocytes

The term **phagocyte** (cells that ingest; Section 13.2) was coined by the Russian-born biologist Elie Metchnikoff. Metchnikoff, while studying jellyfish and sponges at Messina, Italy, noted the remarkable behavior of certain cells in the internal cavities of these primitive

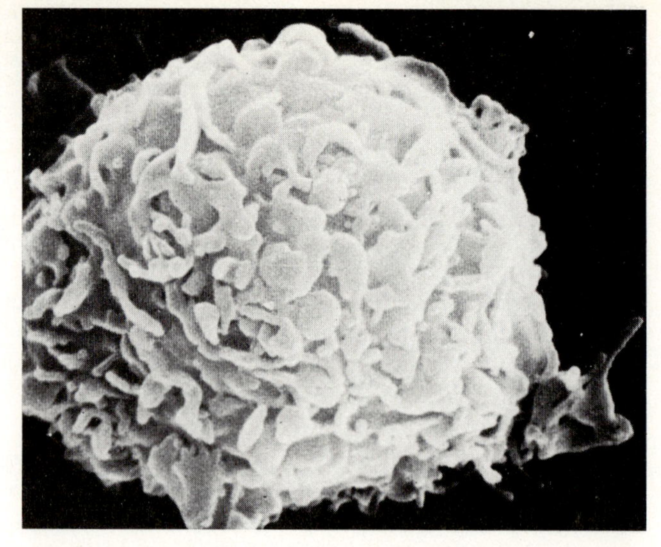

Figure 14.3
Scanning electron micrograph of a macrophage showing the ruffled surface of this cell (magnification 24,000×). (Kindly provided by Dr. S. Cabilly and Dr. R. Gallily of the Department of Immunology, Hadassah-Hebrew University Medical School, Jerusalem, Israel.)

creatures that wandered about ingesting and destroying foreign particles. He proposed that phagocytes functioned to engulf and digest bacterial invaders of the organism (1884).

In 1888, Pasteur invited Metchnikoff to Paris and gave him a laboratory. Metchnikoff soon became one of the leading figures of the Pasteur Institute. His studies of phagocytosis, the engulfment of particles by phagocytes, uncovered an important mechanism of defense.

The initial observation of phagocytes in the most primitive of multicellular organisms was followed by their detection throughout the animal world. Phagocytes, cells specializing in the detection, engulfment, and digestion of microbial parasites and other small foreign substances, are a general mechanism of resistance.

There are two types of phagocytes in human beings: relatively large cells called **macrophages** ("big eaters"; Figure 14.3) and smaller cells called microphages ("little eaters"), more commonly known as **neutrophils.** These cells are produced in the bone marrow along with other blood cells. Both types of phagocyte can adhere to solid surfaces or can wander about. They ingest bacteria (Figure 14.4) and viruses by binding the microbes to a portion of their outer cell membranes.

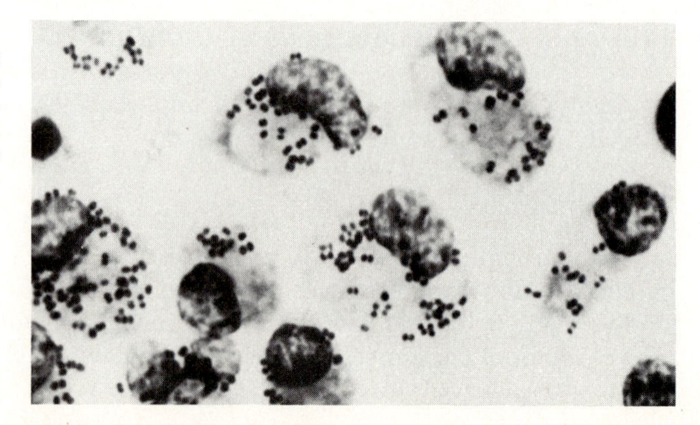

Figure 14.4
Phagocytosis of bacteria by macrophages. This picture, taken by a conventional light microscope (magnification 1200×), shows macrophages filled with staphylococci. The macrophages will proceed to destroy the staphylococci. (Kindly provided by Dr. R. Gallily.)

The phagocytic cell membrane surrounds the microbes, forming a kind of sac, or invagination. This sac is then internalized into the cytoplasm of the phagocyte as a **phagosome** (*-some:* "body or structure"). Other vesicles called **lysosomes** (*lys:* "destroy") contain powerful enzymes capable of breaking down proteins, lipids, polysaccharides, and nucleic acids. The lysosomes fuse with the phagosomes, and their enzymes then digest the trapped microbes (Figure 14.5).

Macrophages and neutrophils differ functionally in several ways. Neutrophils are the predominant type of white cell circulating in the blood: A microliter (one millionth of a liter) of blood has approximately 10,000 white blood cells, of which approximately 60 percent are neutrophils, while only about 3 to 5 percent are monocytes, the term for macrophages in the blood. Neutrophils seem to live in the blood only for several days and arrive relatively quickly, within hours, at the site of an infection. In contrast, macrophages accumulate much more slowly at sites of infection, appearing days later than neutrophils. The lifetime of macrophages is not exactly known but seems to be much longer than that of neutrophils. Macrophages are more prominent in tissues such as lymph nodes, spleen, liver, lung, and skin than they are in the blood stream. Macrophages seem able to digest certain microbes that resist digestion by neutrophils.

Microbial invaders are recognized by both neutrophils and macrophages, leading to phagocytosis and digestion. The microbial signals recognized by these phagocytes are unknown. Most non-pathogenic microbes are easily recognized. Some pathogenic microbes may

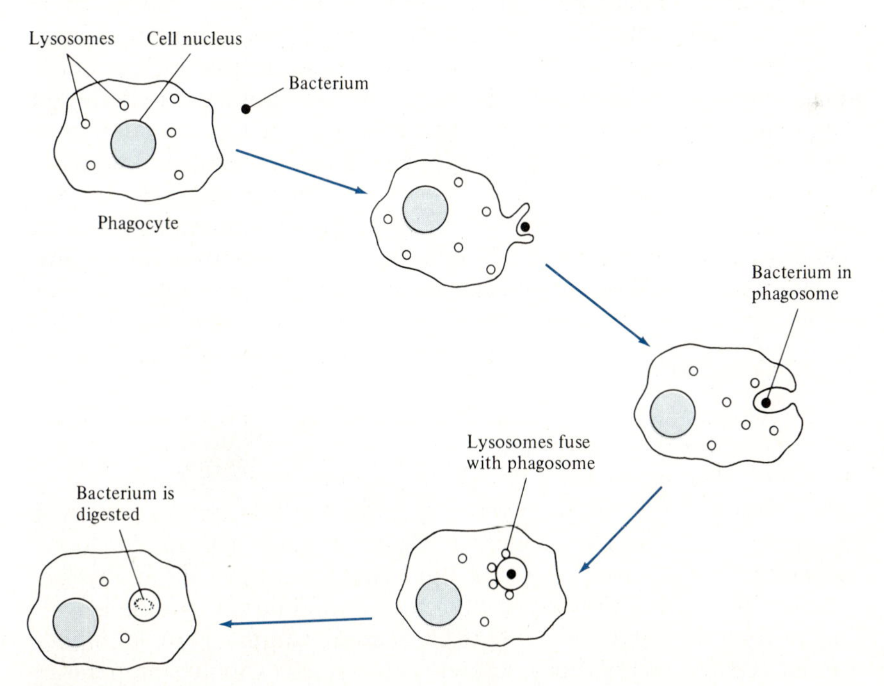

Figure 14.5
Schematic illustration of phagocytosis and digestion of bacteria by a phagocyte. The phagocyte engulfs the bacterium, forming a phagosome. Lysosomes fuse with the phagosome, and their enzymes digest the phagocytosed bacterium.

evade direct phagocytosis by enveloping themselves in capsules (Section 13.2).

To destroy invading microbes, phagocytes must accumulate at the site of infection. How are phagocytes signaled to gather in large numbers at the specific place where they are needed? Attraction of phagocytes to a focus of infection appears to be brought about by a process of chemotaxis (*taxis:* "movement toward"), the movement of cells in the direction of an increasing concentration of a chemical signal. Phagocytes wander apparently at random until signaled by a chemical gradient that focuses their movements in a specific direction. The signals that elicit chemotaxis are products of bacteria or substances produced by the host tissues during inflammation.

We have all experienced the accumulation of phagocytes at the site of infection; we call it pus. Pus is composed largely of phagocytes, many of which have died in the process of ingesting and digesting microbial invaders. However, even dead phagocytes contribute to the destruction of microbes by spilling out their lysosomal enzymes and incapacitating microbes that may have evaded phagocytosis. Pus, because of its low pH and lysosomal enzymes, inhibits the metabolism and hence the proliferation of bacteria. Although this would seem to be to the host's advantage, inhibition of bacterial metabolism may also contribute to the persistence of infection. Poorly metabolizing bacteria are relatively resistant to the action of antibiotics (Chapter 11), and therefore pus can serve as a refuge for bacteria from treatment with antibiotics. For this reason, accumulations of pus often must be surgically drained, even in this era of antibiotics, in order to rid the patient finally of a pocket of invading bacteria.

The bone marrow responds to an increased consumption of phagocytes by pouring out large numbers of these cells to reinforce the battle lines. Thus, one sign of significant bacterial infection is an increased number of neutrophils in the blood, a condition known as an elevated white blood cell count. Infection with viruses usually does not produce pus, and hence the neutrophil count may not be elevated. Thus, an elevated white blood cell count can help the physician distinguish between a bacterial and a viral infection.

Phagocytes and Fever

Phagocytes act to increase the resistance of the host in yet another way: Stimulated by microbial products such as endotoxins, or by products of inflammation, phagocytes secrete pyrogen (*pyro:* "fire"; *gen:* "to generate"). A pyrogen is a substance that stimulates fever. Pyrogens produced by phagocytes travel in the blood stream to the temperature regulatory center in the hypothalamus of the brain. The pyrogen acts to reset the body's thermostat optimum to a temperature higher than 37°C (98.6°F). This triggers an elevation of body temperature above 37°C, a condition known as fever. Elevation of temperature is produced by increased heat production generated by contraction of muscles (shaking and chills) and by decreased heat loss generated by shunting of blood away from the patient's skin (Figure 14.6).

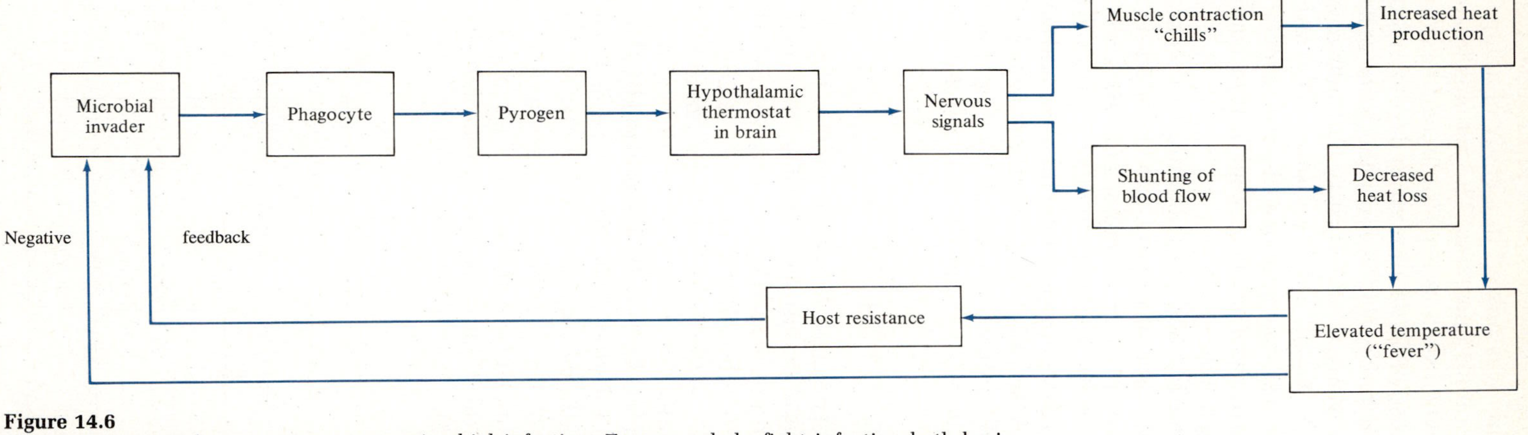

Figure 14.6
Phagocytes trigger fever in response to microbial infection. Fever can help fight infection both by increasing host resistance and by inactivating microbial invaders.

Fever has been shown to increase resistance to microbial infection. This was proved in a study of a large number of patients suffering from mild infections. The patients treated with drugs that reduced their fevers, such as aspirin, tended to suffer longer-lasting infections than those patients allowed to develop moderate degrees of fever. The mechanisms by which fever increases resistance are not well characterized. Many microbes carry out their metabolism best at a temperature of 37°C and are at a disadvantage at higher temperatures. In addition, fever may enhance important metabolic processes of the host. Of course, fever must be kept within bounds. A very high fever can be dangerous.

Elevating body temperature is a tactic employed even by "cold-blooded" creatures, whose temperature is regulated primarily by the environment. For example, animals such as lizards, when infected with bacteria, instinctively expose themselves to direct sunlight and higher environmental temperatures and so raise their body temperatures. A usually mild infection can become fatal if these "cold-blooded" animals are forced to remain in the shade. Hence, fever may be a universal strategy for resisting microbial infection.

Natural Killer Cells

Phagocytes are well adapted to ingesting and destroying free bacterial cells or viruses. However, parasites living within cells of the host are not very susceptible to killing by phagocytosis. One way to kill such intracellular microbes is to kill the host cells that harbor them. To achieve this end, the immune system has evolved **natural killer cells.**

Natural killer cells are able to kill abnormal target cells. Host cells that are infected with certain viruses have been found to be susceptible to killing by natural killer cells. The presence of the viral genes in the infected cell leads to the expression of particular markers on the surface membrane of the cell. These markers are recognized as signals by natural killer cells, which then kill the target cell on contact (Figure 14.7).

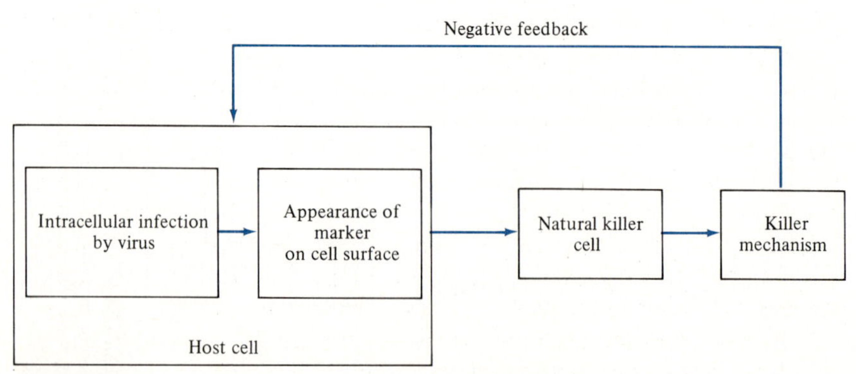

Figure 14.7
Natural killer cells are activated to kill abnormal host cells infected with viruses.

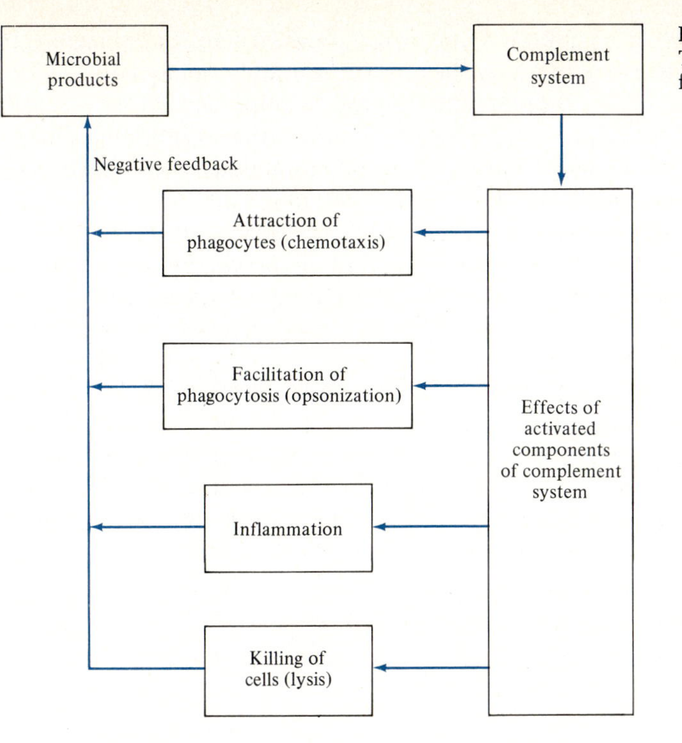

Figure 14.8
The complement system aids the host in fighting off microbial invaders.

Complement

Around the time that Metchnikoff was doing his pioneering studies of phagocytosis, the German microbiologists Nutall and then Buchner published their discoveries that serum, the fluid phase of blood obtained after clotting, had the power to kill bacteria. Buchner showed that fresh serum killed typhoid bacilli but that serum heated at 55°C for 30 minutes did not. Buchner called the substance that killed bacteria *alexine,* meaning "protective substance." He concluded that the resistance of healthy individuals to bacterial infections must be due in part to the action of alexine in the blood serum. Metchnikoff took the side of his phagocytes and rejected the possible existence of a non-cellular or "humoral" agent of resistance. A controversy raged between the proponents of cellular resistance based on phagocytes and the supporters of humoral immunity. The mechanism of humoral immunity was complicated by the observation that the killing of bacteria in serum, in addition to alexine, required a heat-stable substance, later identified as antibody. The heat-labile substance was called "complement" by the French scientist Bordet, because it completed the killing reaction. Alexine and complement were found to be identical and composed of not one but a multitude of proteins that circulated in the blood. Today, complement is understood to include at least 17 different proteins that work together as an integrated defense system.

Figure 14.8 depicts four processes performed by the component proteins of the complement system that defend against microbial invasion.

1. Products of complement direct the movement of phagocytes toward invading microbes in a process of chemotaxis.
2. Products of complement sticking to microbes act as a signal to phagocytes. The complement signal thus stimulates phagocytosis of the microbes. This phenomenon is called **opsonization,** which means "preparation for eating."
3. Products of complement trigger the process of inflammation; that is, they cause the contraction of smooth muscle, increase the leakage of proteins out of the blood vessels, and release from cells various drug-like chemicals. These processes lead to swelling of infected tissues, accumulation of phagocytes and pus, and confinement of microbes to the area of infection.
4. Finally, activated proteins of the complement system can kill microbial or other cells by punching holes in the surface membrane of the cell. Gram-negative but not gram-positive bacteria are susceptible to this form of destruction.

Evidence that complement is important in resisting disease derives from the observation that individuals with defective complement components suffer from repeated bouts of severe infection.

The Story of Interferon

Interferon is a protein produced by cells of the body in response to infection with viruses. The properties of interferon are such that its use in medicine may revolutionize the treatment of viral diseases and possibly some cancers. Some scientists hope that the therapeutic potential of interferon for treating viral diseases will be equivalent to that achieved by antibiotics in bacterial diseases.

Interferon was discovered in 1957 in London by the British virologist Alick Isaacs and a visiting scientist from Switzerland, Jean Lindenmann. During a casual chat over a cup of tea, the two men decided to investigate the observation that infection of an individual with one virus may prevent infection at the same time with another type of virus. For example, a person infected with influenza virus may be resistant to simultaneous infection with poliovirus. The scientists were able to reproduce in the laboratory this phenomenon of mutual interference by viruses. Furthermore, they discovered that the fluid bathing virus-infected cells could transfer to fresh host cells resistance to infection by other viruses. The substance mediating interference was called interferon.

The exact mechanism by which interferon blocks viral infection has yet to be characterized. However, it is certain that interferon itself does not inactivate viruses directly; rather, interferon acts on host cells to make them resistant to infection. Infection of a host cell by virus leads to production of an interferon signal by the infected cell. The signal is taken up by adjacent cells, which respond by making interferon. The interferon then diffuses through the host and can render his or her cells resistant to viruses. Virus cannot reproduce in

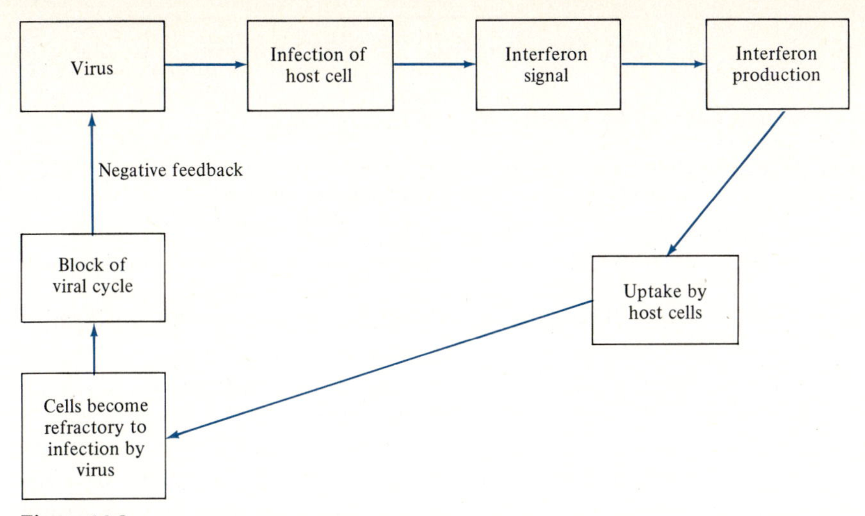

Figure 14.9
Interferon renders host cells resistant to viral infection.

these protected host cells, so the cycle of viral infection and reproduction is broken (Figure 14.9). In other words, interferon makes uninfected cells refractory to productive viral infection. Viruses can attach to and enter interferon-treated cells, but they cannot take over the host cell or reproduce there. The effect of interferon covers a very wide range of DNA and RNA viruses, and interferon stimulated by one particular virus will protect against them all.

The discovery of the effect of interferon on an individual's cells led to study of its effect on cancer. Ion Gresser, an American scientist working near Paris, demonstrated in the 1960s that interferon prevented or arrested the development of several types of cancer in mice. Trials of interferon in some human cancers appear promising; however, it is much too early to forecast the exact situation in which interferon will be most useful as an anti-cancer drug.

Clearly, interferon has been found to cure serious viral infections. However, a major problem in using interferon in human cancer or viral illness is supply; interferon is prohibitively expensive to produce and purify in large quantities. Furthermore, interferon works best in the species of animals that produced it. With the exception of chimpanzee interferon, interferon obtained from animal cells has little effect on human cells. Recently, human interferon genes have been introduced into bacteria, using techniques of genetic engineering (Sections 6.8 and 11.7). It is hoped that these bacteria, grown in large fermentation tanks, will produce the great quantities of human interferon needed for clinical trials.

Recently, interferon has been found to augment the activities of macrophages and natural killer cells. In addition, there is evidence that interferon may regulate the growth of both normal and cancer cells. Thus, interferon has a variety of important biological effects (Table 14.1). The comic strip "Flash Gordon" in 1960 was prophetic in

TABLE 14.1 Beneficial Effects of Interferon

Cells Influenced by Interferon	Beneficial Effect
Potential target cells of viruses	Virus-proof
Normal and possibly cancer cells	Regulation of growth
Macrophages	Increased phagocytosis
Natural killer cells	Increased activity against infected cells

heralding the first, if imaginary, clinical success of interferon (Figure 14.10).

14.7 Integration of the Activities of the Agents of Resistance

In the foregoing discussion, we have seen that many factors contribute to our resistance to invasion by microbes. Constitutive barriers to infection rest on the physical, chemical, and physiological properties of our external surfaces and internal organs. These constitutive barriers are not elicited or induced in response to infection; hence, they do not constitute an immune "response" in the strict sense of the word. Nevertheless, these factors are very important bulwarks against infection.

Once microbes have penetrated this first line of defense and gained entry into the interior of the body, their very presence can trigger a variety of immune mechanisms that neutralize or destroy them. The mechanisms of resistance include (1) phagocytes that engulf and digest microbes and produce fever and pus; (2) natural killer cells that destroy abnormal host cells harboring viruses; (3) comple-

Figure 14.10
Flash Gordon to the rescue! (Copyright © 1960 King Features Syndicate, Inc.)

ment that produces inflammation, facilitates phagocytosis, and punches holes in cell membranes; and (4) interferon that makes normal cells resistant to viral infection. These diverse mechanisms of resistance are not soloists, despite their considerable virtuosity. They combine their talents in a highly orchestrated fashion to produce an integrated response to microbial invaders. The harmonious interaction of the various agents of resistance is directed by a complicated network of signals whose sophistication is just beginning to be appreciated. These signals are produced by a specialized class of white blood cells called **lymphocytes.** Lymphocytes and their products endow the immune response with attributes of memory and adaptability and provide a virtually unlimited repertoire of receptors capable of recognizing any microbial molecule as a signal for action. How they do this is the subject of Chapter 15.

QUESTIONS

14.1 Interferon, which renders cells resistant to infection by viruses, is produced in response to infection. Why do our cells not produce interferon continuously? Would not this prevent all viral infections?

14.2 Why do bacteria make ligands that are recognized by receptors on phagocytes or by the complement system?

14.3 Bacteria have rapidly evolved biochemical processes that make them resist antibiotics. Why have bacteria not evolved as readily the biochemical processes needed to thwart our phagocytes or complement systems of defense?

14.4 Is it very likely, or very unlikely, that "new" microbial diseases of man will emerge?

14.5 What is the essential difference between constitutive barriers to infection and immune effector mechanisms?

14.6 An elevated white blood cell count prompts the physician to search for what type of microbial invader?

14.7 Define phagocyte, interferon, lysozyme, lysosome, opsonization, immunity, macrophage, neutrophil, chemotaxis, complement, natural killer cells.

Suggested Readings

Burke DC: The Status of Interferon. Scientific American, 236: 42, April 1977.

Mackaness GB: The Mechanism of Macrophage Activation. In: Infectious Agents and Host Reactions. Edited by S Mudd. Philadelphia: WB Saunders, 1970, page 61.

Mayer MM: The Complement System. Scientific American, 229: 54, November 1973.

Wood WB Jr: The Pathogenesis of Fever. In: Infectious Agents and Host Reactions. Edited by S Mudd. Philadelphia: WB Saunders, 1970, page 146.

CHAPTER 15

Lymphocytes, Antibodies, and Regulation of the Immune Response

The Greeks, whenever doubt arose with regard to any disease, always thought it best to rely on nature and her doings which were sure, in the last resort, to banish the disease. They based their opinion on the following grounds: Nature, being the servant and provider of all living things in healthy days, helps them also in disease.

Galen, Greek physician of the second century A.D.

We have described (Chapter 14) immune effector mechanisms that help us resist invasion by microbial parasites. Cellular effectors such as macrophages, neutrophils, and natural killer cells, combined with humoral molecules such as interferon and the complement system, are able to destroy invading microbes. However, the adaptability and versatility of the immune system is provided by **lymphocytes.** These cells and their products can help repel microbial invaders. No less important is the ability of lymphocytes to serve as channels of communication between microbes and the host's other effector mechanisms. Lymphocytes receive signals from microbial invaders and are able, in turn, to signal phagocytes or complement into action. Hence, lymphocytes bridge gaps in communication between microbial parasites and host defenses. Two functional characteristics of the body's system of lymphocytes are particularly important: (1) It can recognize almost any microbial product as a ligand, and (2) it can learn from experience.

15.1 Adaptive Immunology: Past, Present, and Future

We have all had the opportunity to observe that the immune system learns from experience. For example, resistance to chickenpox virus is greatly enhanced by previously having recovered from the disease, so much so that a second contact with the virus years later almost never produces overt chickenpox. Based on the behavior of lymphocytes, the immune system is capable of remembering the previous experience of the individual with particular microbes and acquiring the capacity for an augmented degree of resistance specifically directed against these microbes. In short, the immune system is able to adapt itself in the light of its past experience and acquire new capabilities. It was discovered that this adaptation could be induced artificially by immunization.

TABLE 15.1 Vaccinations Routinely Used		
Agent Used for Vaccination	**Protects Against**	**Schedule of Vaccinations**
DPT		
Diphtheria toxoid	Diphtheria	Three injections during the first year of life. Repeat injections at 1½ and 5 years
Tetanus toxoid[a]	Tetanus	
Pertussis antigen	Whooping cough	
Oral poliovaccine	Poliomyelitis	As above
Measles vaccine	Measles	Given once, at about 1½ years
Mumps vaccine	Mumps	Given once, at about 1½ years
Rubella vaccine	German measles	Given once at puberty

[a]To maintain protection against tetanus, booster injections of tetanus toxoid should be administered every 7 to 10 years.

Immunization is the process of endowing individuals with resistance to a specific microbial pathogen by inoculating them with the pathogen or its products. Edward Jenner in England was the first person to deliberately immunize against an infectious disease by using a weakened pathogen. Jenner was intrigued by the claim of a milkmaid that she was no longer susceptible to smallpox because she had already contracted cowpox. Jenner's *An Inquiry into the Causes and Effects of Variolae Vacciniae,* published in 1798, described the successful vaccination of one James Phipps using infectious material from lesions of cowpox. The basic technique of vaccination against smallpox has been used from Jenner's day until the present and has been responsible for the eradication of smallpox. Table 15.1 lists vaccinations that are routinely used in the United States and elsewhere.

The discovery of antibodies occurred almost 100 years after the discovery of vaccination. Roux and Yersin, working in Paris, noted that the blood of animals immunized with diphtheria toxin had the property of neutralizing the effect of the toxin. At about the same time, von Boehring and Kitasato at the Koch Institute in Berlin demonstrated that immunity against tetanus toxin could be transferred from an immunized animal to an unimmunized animal by injection with serum. In 1895, the Frenchman Bordet clearly showed that the killing of certain bacteria by immune serum resulted from the combined action of heat-labile complement and a heat-stable substance. The complement could be obtained from any healthy animal, immune or not, while the heat-stable substance was found only in the serum of immunized animals. The heat-stable antibacterial substance was later called **antibody.**

In addition to collaborating with complement to kill bacteria or to neutralize toxins, antibodies were observed to cause bacteria to stick together (agglutination) and to facilitate their phagocytosis (opsonization). For a number of years, investigators catalogued the many biological effects of antibodies while controversy raged over the proposed molecular mechanism of the interaction of antibodies with their

specific target lectins, called antigens. The great German pathologist Paul Ehrlich, who pioneered the development of chemotherapy (see Section 10.5), propounded the theory that antibodies interacted with antigens by the laws of chemical union.

In the 1930s and 1940s, antibodies were identified as proteins belonging to the class of gamma globulins circulating in the blood, and it was suspected that they were produced by lymphocytes. The chemical structure of antibodies as a class of proteins, the immunoglobulins, was determined in the 1960s. Research in the 1970s uncovered a multitude of classes of lymphocytes, and secretion of antibody molecules was found to be the job of only one of the classes of lymphocytes. Today, immunologists are beginning to investigate the immune system as a complex devoted to processing information. Regulation of the immune response and self-adaptation of the immune system are being studied from different but integrated points of view: genetic, molecular, and cellular. The application of basic immunological knowledge to medical problems of organ transplantation, cancer, and other diseases is a second area of intense research. A third frontier of immunology involves relating the behavior and products of lymphocytes to the organization and expression of DNA in the genes.

15.2 What Defines a Microbial Molecule as an Antigen?

Microbes communicate with the host's lymphocytes through the medium of their antigens. What, then, is an antigen?

An antigen is defined operationally as a molecule that can be recognized by antibodies. Antigens have no common chemical features. A variety of organic molecules may be recognized as antigens, proteins, carbohydrates, or even lipids. Any part of a molecule that binds to an antibody is called an **antigenic determinant.** Similar to other ligand–receptor interactions, an antigenic determinant is recognized by its three-dimensional, lock-and-key complementarity to a part of an antibody called a **combining site** (Figure 15.1). Hence, an antigenic determinant is a geometrical configuration of a part of a molecule that fits the combining sites of antibodies.

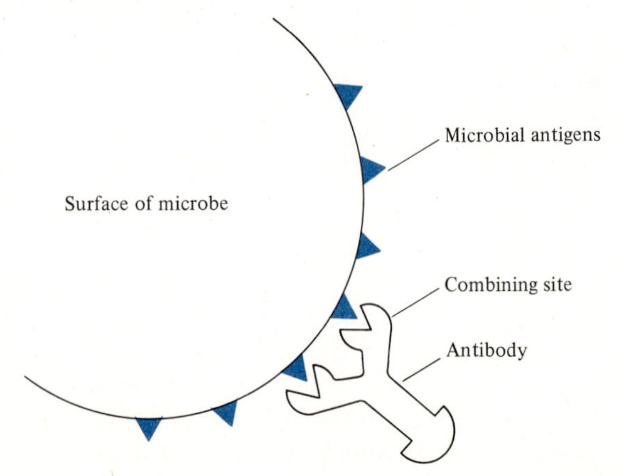

Figure 15.1
Antigen binds to antibody by its three-dimensional fit to the combining site of the antibody. The relative sizes of the molecules drawn in this and in subsequent schematic representations are grossly exaggerated for the sake of clarity.

15.3 The Structure and Function of Antibodies

The function of antibodies requires that these molecules have two almost contradictory attributes: On the one hand, the various antibody molecules must be diverse in structure so that each can recognize different antigenic determinants; on the other hand, antibody molecules must also be uniform in structure, so that any antibody can be recognized by the effector mechanism that it triggers in common with other antibodies. In other words, antibodies that recognize widely different bacterial or viral antigenic determinants must be capable of activating the same phagocytes or the same complement proteins. Hence, antibodies have to be both variable in their structure (to recognize different antigens) and constant in their structure (to activate common agents of defense). Nature solves this dilemma by building antibodies with both variable and constant regions in the same molecule. The variable region of the antibody is the combining site that binds antigen, while the constant region of the antibody can signal other agents of defense.

Figure 15.2 illustrates the unit structure of an antibody molecule. Each antibody molecule is shaped like the letter "Y." The two outstretched arms of the "Y" have variable combining sites that recognize and bind to microbial antigens. This act of binding to antigen modifies the structure of the constant region "foot" of the Y-shaped antibody molecule. The constant region of the antibody in its modified state serves as a ligand recognizable by agents of defense such as phagocytes, complement, or natural killer cells. Therefore, the binding of antibody to a microbial antigen converts the constant region of the antibody into a signal that activates effector mechanisms. In this manner, the antibody serves as a relay or coupler linking the agents of

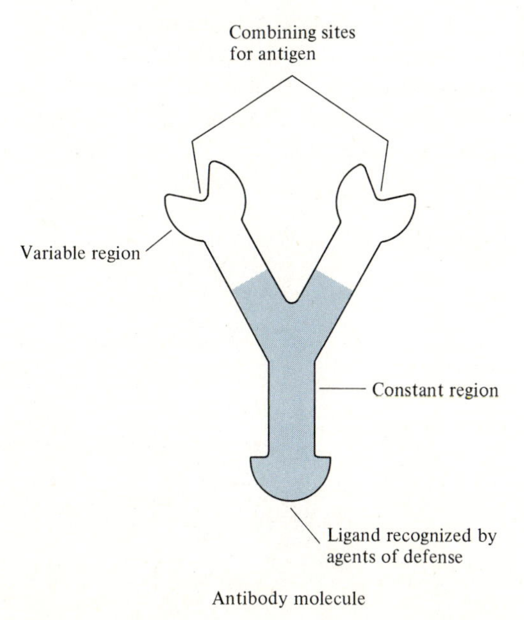

Figure 15.2
Schematic structure of an antibody molecule. The constant region of the molecule is attached to a variable region that contains two combining sites of identical shape.

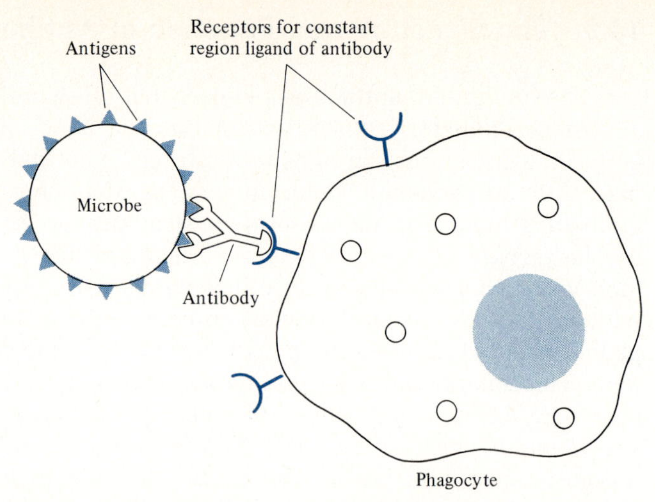

Figure 15.3
Antibodies relay microbial signals. Antibodies binding to microbial antigens can, with their constant region ligands, signal agents of defense such as phagocytes.

defense to the microbe (Figure 15.3). In addition, the binding of antibodies to an antigen may be sufficient in itself to protect the individual. For example, antibodies binding to a microbial toxin can hinder physically the ability of the toxin to damage a host cell, or binding of antibodies to viruses or bacteria may block their penetration into cells or tissues (Figure 15.4).

We now see that agents of defense can recognize microbial invaders by either of two pathways; directly, by recognizing microbial ligands as outlined in Chapter 14, and indirectly, by recognizing antibody molecules binding to the microbe. The indirect pathway mediated by antibodies is highly versatile, because antibodies are virtually unlimited in their capacity to recognize foreign materials such as microbial antigens. Moreover, the indirect pathway is adaptive, because lymphocytes have a memory, as we discuss later.

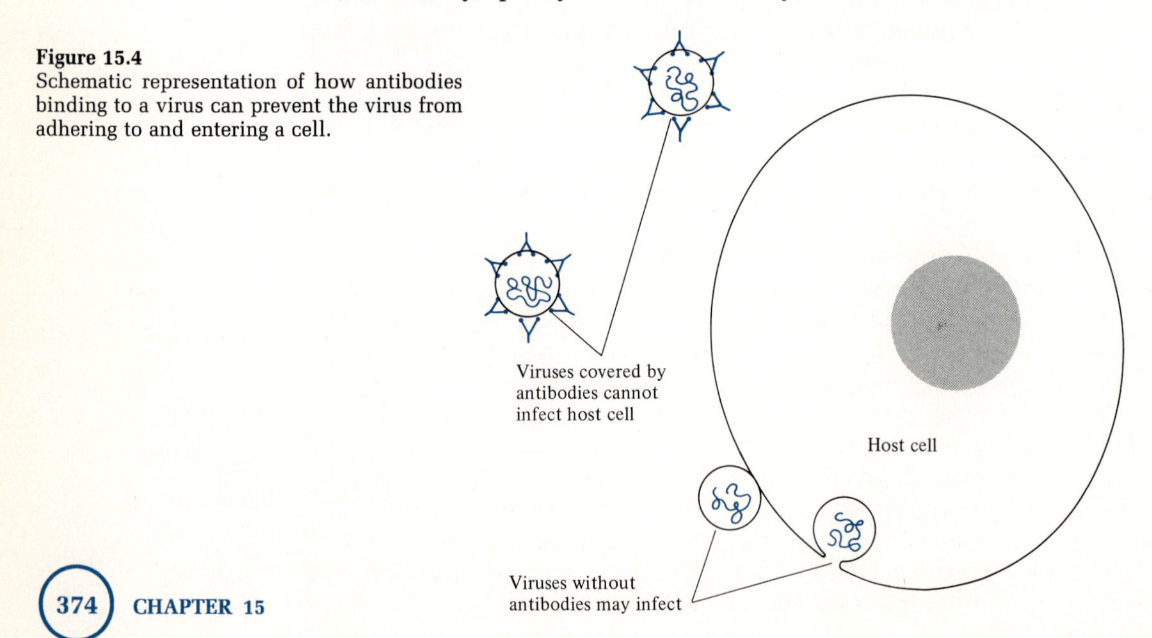

Figure 15.4
Schematic representation of how antibodies binding to a virus can prevent the virus from adhering to and entering a cell.

TABLE 15.2 The Different Classes of Antibodies and Their Properties

Class of Antibody	Properties
IgG	Predominates in blood; relatively long-lived; transferred across placenta to fetus; some subclasses activate complement; opsonizes bacteria; binds to toxins and viruses in blood; increased in memory response.
IgA	Predominates on body surfaces and hollow organs; in tears, saliva, urine, mucus of respiratory and gastrointestinal tracts; strengthens barriers to infection.
IgM	Found in blood; predominates in primary immune response and against some bacterial antigens; activates complement.
IgD	Found mostly as a receptor on certain lymphocytes.
IgE	Causes allergies by activating a special class of white blood cells called mast cells.

Five different classes of antibodies can be distinguished by the biological properties of their constant regions and the effector mechanisms that they specialize in activating. These classes are called immunoglobulins or "Ig" (Table 15.2). **IgG** is the major class of antibodies in the blood. Antibodies of this type are secreted relatively late and to a small degree after the first contact with an antigen. However, IgG plays an important role in adaptive immunity: A second contact with the same antigen leads to a quicker and greatly augmented response of IgG antibodies.

There are various subclasses of IgG that can activate complement or phagocytes (Figure 15.3). IgG antibodies can also bind to toxins and viruses and so interfere with their attachment to host cells (Figure 15.4). IgG is unique in that it is transferred across the placenta from the blood of the mother to her fetus. The infant is born with IgG antibodies of maternal origin and enjoys relative protection from many infections for the first months of life. This allows the newborn a period of grace during which time he or she can slowly accustom his or her own immune system to the challenge of the microbial environment.

IgA is the antibody that appears in the body's external secretions that bathe the surfaces through which microbes may penetrate. IgA is present in tears, saliva, mucus of the respiratory and genital tracts, urine, and bowel secretions. IgA antibodies are produced locally in response to antigens that enter through these surfaces. They are not produced in response to antigens that enter the body by other routes.

A practical example of the importance of localization of antibodies was seen in vaccination against poliovirus. Two types of vaccines developed by American microbiologists have been used: a heat-killed vaccine developed by Jonas Salk and a live attenuated vaccine developed by Albert Sabin. The Salk vaccine, given by injection, was found to induce IgG antibodies circulating in the blood. The Sabin vaccine, taken by mouth, led to the development of IgA antibodies in the gut as well as IgG antibodies in the blood stream. Both vaccines protected against the development of paralytic polio by preventing poliovirus from reaching the nervous system by way of the blood. However, only the oral vaccine also eliminated poliovirus from the gut. The Salk vaccine failed to stimulate IgA in the intestine, and vaccinated individuals could still carry and disseminate virulent virus from their intestinal tracts. Hence, the oral Sabin vaccine, by eliminating the intestinal reservoir, was much more effective in reducing the prevalence of dangerous poliovirus in the community and is the vaccine routinely used today.

The largest antibody molecules are **IgM,** which is the first class of antibodies to be produced in the immune response and activates complement very efficiently.

IgD molecules are found mostly attached to the surface membranes of lymphocytes, where they function as receptors for antigen.

IgE antibodies are notorious in causing allergic symptoms. Although they are secreted by lymphocytes, the foot region of IgE antibodies binds tightly to receptors on a class of white blood cells called **mast cells.** Hence, secreted IgE antibody molecules are rapidly taken up by the surface of mast cells. Inside these mast cells are thousands of vesicles containing histamine and other substances that stimulate inflammation. When an antigen called an allergen binds to the combining sites of the IgE antibody anchored to a mast cell, the mast cell responds by releasing the contents of its many vesicles (Figure 15.5). The released histamine and other materials cause leakage of fluid from blood vessels, contraction of smooth muscle, and increased production of mucus from various glands. Depending on the site of release of histamine and the amount released, the individual can experience an itchy, runny nose, a skin rash such as hives, an attack of asthma, or even sudden loss of consciousness that may be fatal. These symptoms characterize the wide range of allergic responses. Fortunately, the vast majority of allergies are mild, though inconvenient.

The existence of allergies confronts us with many unanswered questions: What is the advantage to the host of an allergic immune response? Why do some individuals but not others have a tendency to produce IgE and hence suffer from allergies? How is it possible for factors such as emotional stress or the weather to affect allergic responses? How do certain antigens such as penicillin or ragweed act as allergens and stimulate IgE in susceptible individuals?

Long before the discovery of IgE and its mechanism of action on mast cells, allergic persons were treated successfully by "desensitization," a technique involving repeated injection of the allergen to the allergic person. Desensitization illustrates how a useful form of therapy may be proposed and applied widely despite an imperfect or erroneous understanding of both the etiology of the disease and the nature of the treatment. Desensitization, rather than decreasing the amount of antibodies to the allergen as originally thought, appears actually to stimulate the production of antibodies. However, the bind-

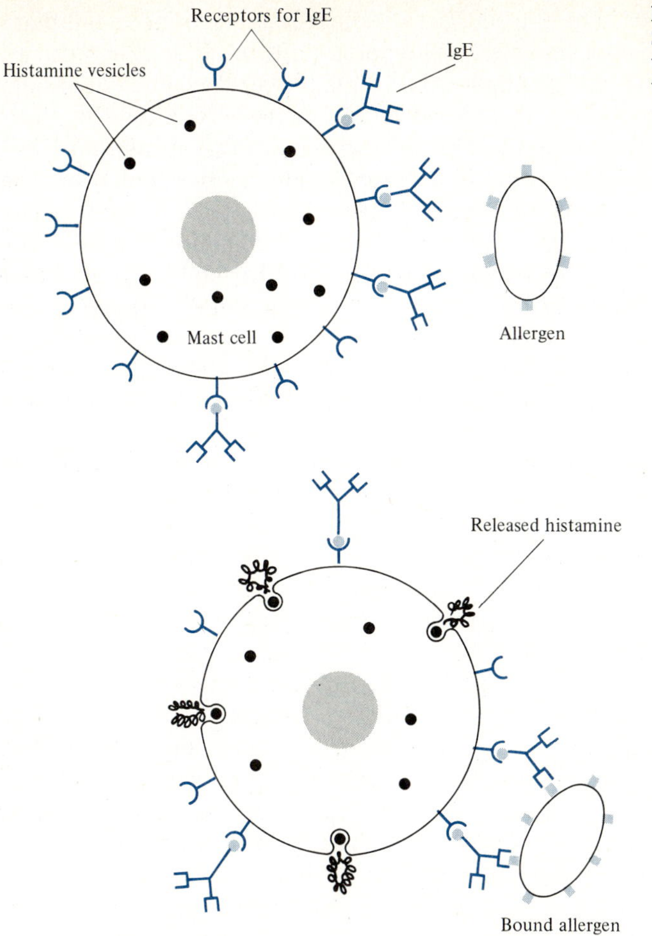

Figure 15.5
Schematic representation of the release of histamine from vesicles of a mast cell triggered by the binding of allergen to IgE antibodies.

ing of antibodies to ragweed pollen in itself is not the cause of hay fever. Antibodies to ragweed of the IgE class alone can release histamine from mast cells. The process of desensitization seems to prevent allergic symptoms by augmenting production of IgG antibodies to the allergen.

A large amount of IgG binding to the allergen preempts antigenic determinants and obstructs the binding of the IgE and, hence, the release of histamine from mast cells. Competition among different classes of antibodies for the same antigen can markedly affect the biological significance of the immune response. The type of antibodies one makes is as important as the quantity.

15.4 How Do Combining Sites of Antibody Receptors Fit Antigens?

One of the most fundamental problems in immunology is how combining sites of antibodies are created to fit the shape of specific antigenic determinants. Two explanations were discussed in the early 1950s: the **instructive** and the **selective** hypotheses. The instructive hypothesis states that the antigenic determinant acts as a template around which combining sites of receptors fold in order to acquire an exact complementary configuration. This theory became untenable when it was discovered that the structure of proteins including antibodies was predetermined by the genetic code of the individual.

Selective theories were first formulated independently by the American David Talmadge and the Dane Niels Jerne. A few years later, in 1959, McFarlane Burnet of Australia expanded these ideas into a "Clonal Selection Theory of Immunity." Burnet postulated that each lymphocyte and its offspring, called a clone, bear antigen receptors with ready-made combining sites of a single configuration. A great many different clones, each with its unique receptor, await the penetration of antigens into the body. An antigen with a certain antigenic determinant will, therefore, combine with and activate to proliferate and secrete antibodies only those clones with complementary receptors. In essence, Burnet suggested that the antigen selects spe-

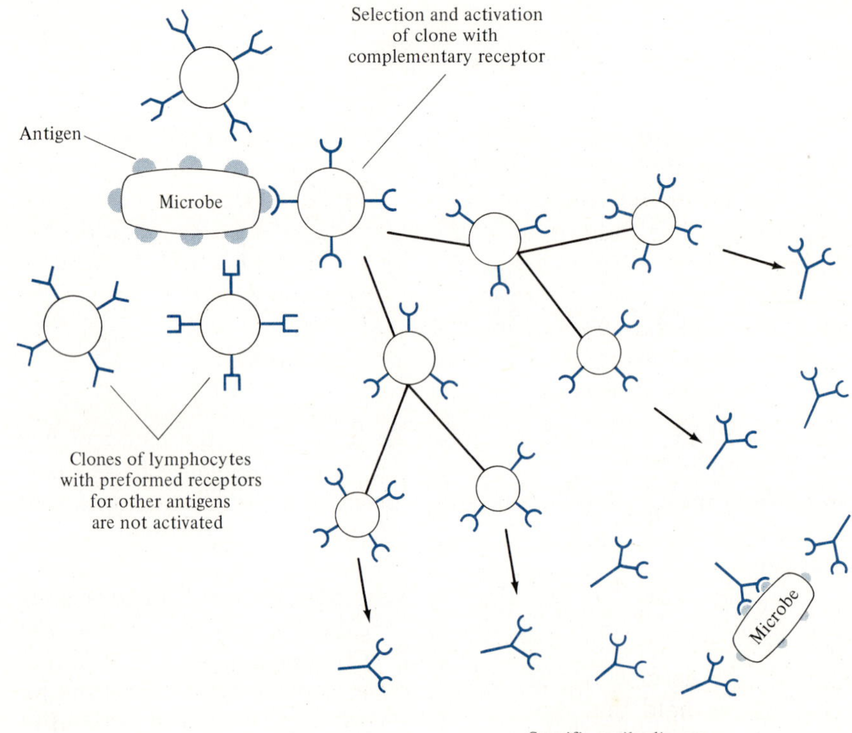

Figure 15.6
The clonal selection theory of antibody specificity. Lymphocytes, each with a preformed receptor, stand ready to interact with antigenic determinants, including those present on microbial invaders. The antigen binding to lymphocytes with complementary receptors triggers the lymphocytes to proliferate and secrete antibodies. The antibodies bear the specificity of the lymphocyte selected to secrete them. Thus, the antigen itself selects clones of lymphocytes and antibodies that fit it.

cific clones of lymphocytes for participation in an immune response. These selected clones of lymphocytes secrete antibodies that have the same combining sites for antigen as do the receptors of the lymphocytes that secrete them. The microbial antigens themselves select the array of antibodies that the host will make to defend against the microbe (Figure 15.6).

The general features of the clonal selection theory proposed by Burnet were later supported by experimental evidence. Clones of lymphocytes were actually found to be committed to recognizing particular antigenic determinants even before contact with these antigens. Furthermore, it was demonstrated that the receptors of lymphocytes are in fact antibody molecules anchored to the cell membrane and that after the lymphocyte meets its predestined antigen and binds to it, the lymphocyte is activated to secrete in free form an antibody with the very same combining site as that found on the lymphocyte's own receptor (Figure 15.6).

15.5 The Tailoring of Genes, or How Our DNA Encodes Millions of Different Antibodies

To recognize the countless number of microbial and other antigens, the immune system has to be able to produce a vast number of different antibodies; their number is estimated to be approximately ten million. The individuality of an antibody molecule is defined by the uniqueness of its variable region that comprises its specific antigen combining site. Although, as we mentioned above, antibodies with different variable regions may share constant regions, the repertoire of antibodies must include millions of different proteins.

Ever since it was discovered that the structure of a protein is determined by a gene, information encoded in the unique sequence of nucleic acids of an individual's DNA (see Sections 7.3 and 7.4), immunologists have attempted to understand the genetic basis of the diversity of antibodies. How does the DNA of each individual encode for the ten million different antibodies?

In the 1960s and 1970s, two theories were put forth to answer this question. The germ-line theory proposed that all the genetic information required to encode ten million different antibodies was carried in the germ cells (the sperm and egg). The somatic (non-germ cell) theory held that the content of DNA in germ cells was insufficient to encode such a multitude of different antibodies, along with the genetic information needed for the rest of our organ systems. Hence, the proponents of the somatic theory considered it more likely that the great diversity of antibodies was produced by processes occurring in the DNA of lymphocytes. This revolutionary idea implied that new genes are manufactured in lymphocytes, genes that were not originally present in the germ cells.

The germ-line and the somatic theories could not be tested experimentally when they were first proposed, because there existed no way to detect differences between germ cells and lymphocytes in the structure of antibody genes. However, the discovery of restriction enzymes and the technology of DNA cloning (see Section 6.8) made it possible to isolate genes of eucaryotic cells and to characterize their DNA sequences and positions on the chromosome.

Experiments carried out in a number of laboratories showed that the sequences of DNA encoding the variable and constant regions of antibody molecules are separated from each other on the DNA molecule in germ cells. Furthermore, there are very many genes coding for different variable regions and relatively few genes coding for constant regions of antibodies. In lymphocytes, the genes for variable and constant regions of antibodies are rearranged and spliced together the same way one would splice film. A particular variable region gene is found to be closer to a constant region gene. Other variable and constant region genes do not appear to change their location in the DNA. This reshuffling of some variable and constant region genes produces new combinations of DNA. These various combinations of DNA give rise to diverse antibodies. Thus, relatively few genes maintained in cells of the germ line, by a process of recombination and splicing, can give rise to a great diversity of antibody proteins. Furthermore, since each lymphocyte makes its own unique DNA recombination, each clone of lymphocytes essentially produces a unique antibody. In short, elements of both the germ-line and somatic theories were confirmed. Most of the information needed to make millions of different antibodies is indeed potentially present in the DNA of the germ cells. However, the realization of this genetic potential occurs in somatic cells, the lymphocytes. Here, new genes are formed by the rearrangement and splicing together of genetic elements inherited from the germ cells.

It seems that rearrangement and splicing of genetic information may occur not only in the making of antibodies. In fact, many proteins have been found to be encoded in widely separated sequences of DNA. These discoveries provide a new way of looking at DNA as a repository of genetic information. The sum total of genes, the genome, is not merely a library of computer programs that are selected and played according to developmental needs. Rather, the genome rearranges and splices its information to generate new relationships between genes that did not exist in the germ cells. Genetic messages are not only read out; genetic messages are composed actively and perhaps rewritten during development of eucaryotic cells. Bacteria, single-celled procaryotes, apparently do not undergo this form of gene tailoring.

These discoveries pose perplexing questions. Except for rare mutations, the genome was held to be an inviolable constitution dictating our way of life; we now have learned that this constitution is regularly amended. Who does the job of cutting and tailoring the genes? What is the mechanism used? Where is the master plan for the job?

15.6 Humoral Immunity and Cell-Mediated Immunity: Differentiation of B and T Lymphocytes

As investigation of the immune response was carried out at the turn of the century by Metchnikoff, Ehrlich, Bordet, Koch, and others, it became apparent that adaptive resistance to microbes could be divided into two distinct types: **humoral immunity** and **cell-mediated immunity.** Humoral immunity was the term used to describe activation of effector mechanisms by antibodies, while cell-mediated immunity referred to immune processes that could not be mediated by antibodies alone but required the presence of living cells. Today, the distinction between humoral and cell-mediated immunity can be related to differences between the function of B lymphocytes and T lymphocytes. B lymphocytes secrete humoral antibodies; T lymphocytes do not. Before discussing the roles of B and T lymphocytes in the immune response, we shall briefly consider the origin and development of these cells.

The developmental process by which a cell specializes to perform unique functions is called **differentiation.** Differentiation involves the relatively irreversible choice of a genetic program that determines the structure and processes needed for a cell to be a specialist. For example, all the cells of an individual are the offspring of a single fertilized egg. However, some of these offspring specialize as brain cells, others as bone cells, muscle cells, skin cells, macrophages, lymphocytes, and other cell types. Each type of cell carries in its nucleus all the genes present in the original fertilized egg, but each expresses only that small fraction of the total DNA that programs its specific tasks. All the cells in the body carry genes coding for production of antibodies, hemoglobin, and so forth; but only B lymphocytes actually express the genetic program for antibody production and only red blood cells make hemoglobin. Division of labor is the essential strategy of multicellular animals and plants. Differentiation of a cell type requires two genetic processes: activation at the proper time of genes needed for specialization, and suppression of the information coded by genes programming for other behaviors that would interfere with the cell's specialized function. For example, to contract properly, a muscle cell must not only make contractile proteins, it must also refrain from producing bone, antibodies, skin protein, and so on. The processes of gene splicing and rearrangement described above are carried out during differentiation of lymphocytes.

The early stages of differentiation of T and B lymphocytes take place in the bone marrow where these cells are continuously propagated along with red blood cells, neutrophils, macrophages, and other cellular elements of the blood. However, freshly born lymphocytes must migrate out of the bone marrow and into other organs to continue the process of their differentiation.

The thymus gland, found in the chest above and in front of the heart, is the finishing school for the functional differentiation of T

Figure 15.7
The sites of differentiation of T and B lymphocytes. Newborn lymphocytes migrate out of the bone marrow. Those destined to become T lymphocytes differentiate in the thymus gland, whereas the B lymphocytes differentiate in other organs, possibly in the spleen or lymph nodes.

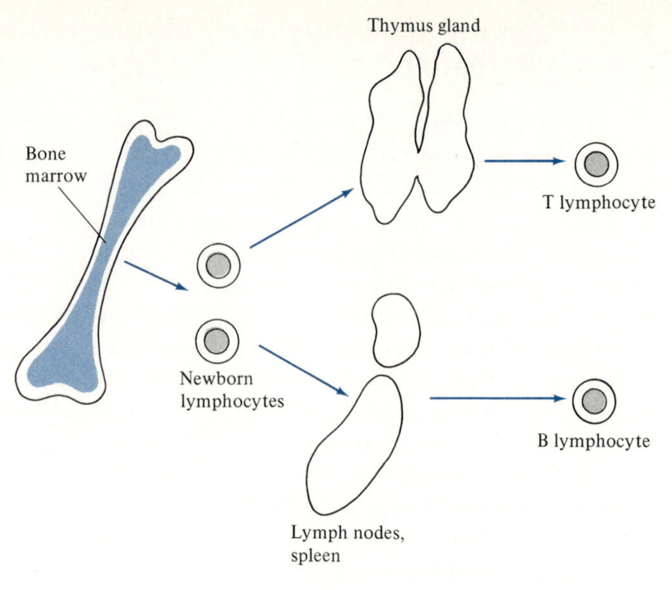

lymphocytes. The name "T" lymphocyte is derived from the thymus (Figure 15.7).

B lymphocytes are named for the bursa, the site of their differentiation in fowl. Mammals do not have an identifiable bursa as birds do, and B lymphocytes in mammals apparently differentiate in other organs serving the function of the bursa, such as lymph nodes and spleen (Figure 15.7).

We have stated that B lymphocytes secrete antibodies, and we have discussed the structure and function of antibodies in activating agents of defense (Section 15.3). What then is the function of the class of lymphocytes called T lymphocytes?

T lymphocytes perform three different roles in the immune system. One class of T lymphocytes has the ability to kill target cells. This class is called cytotoxic T lymphocytes (*cyto:* "cell"). Each clone of T lymphocytes has an antibody-like receptor on its surface, with a unique combining site. Hence, the population of T lymphocytes, similar to the population of antibodies, can recognize untold numbers of different antigens on the surface of the cell. Cytotoxic T lymphocytes also attack foreign cells and tissues transplanted from one individual to another.

A second class of T lymphocytes secretes substances called lymphokines (*lympho:* "lymphocyte"; *kines:* "activators"). Lymphokines activate phagocytes to digest microbes. The secretion of lymphokines can cause a type of inflammation that slowly builds up over several days, as more and more macrophages accumulate at the site of antigen. This type of inflammation is called "delayed" relative to the acute inflammation that is triggered by antibodies that activate complement or mast cells.

A third class of T lymphocytes is involved in regulating the activities of other lymphocytes and will be described in greater detail in the following section.

15.7 To Respond or Not To Respond? The Need for Regulation of the Immune Response

We can conceive of many reasons why a physiological system must have built-in regulatory safeguards. It must start and stop its activities; it must neither overreact nor underreact; it must conserve its resources; and so on. All of the arguments for regulation also apply to the immune system, but two are particularly noteworthy: (1) For the welfare of the host, the immune system must be restrained from attacking the individual himself. (2) To achieve efficiency and effectiveness, the immune system must know whether or not an immune response to a foreign substance is worthwhile.

It is obvious why an immune attack against the individual, a state called **autoimmunity,** is forbidden. As soon as the immune response was discovered, it was clear that autoimmunity was a potential danger and that it had to be prevented by special mechanisms. Ehrlich coined the term "horror autotoxicus," the inadmissibility of self-destruction, to emphasize this idea. The actual danger of autoimmunity is demonstrated by the fact that autoimmune diseases do in fact occur. The immune system of some individuals attacks their normal constituents. Autoimmune diseases may involve organs such as the thyroid gland, the brain, or the joints. The exact processes responsible for autoimmunity have not yet been clarified, but they would seem to include malfunction of mechanisms that regulate the immune response.

The second goal of regulation, that of selecting worthwhile objectives to which to respond, is less obvious. The threat of autoimmunity compels the immune system to know what *not* to attack. Why, however, should the immune system be regulated so as to know what *to attack*?

Both objectives of regulation, in fact, are derived from the same dilemma. The repertoire of antibodies is so large, about ten million different combining sites, that almost any three-dimensional configuration on any biological molecule could conceivably have some chance of fitting at least one clone of lymphocytes. Therefore, some parts of any molecule, including molecules native as well as foreign to the individual, are potential antigenic determinants. Even if we exclude autoimmunity so that the immune system recognizes only antigenic determinants foreign to the host, the system would be paralyzed by the enormous task of responding simultaneously to all the potential antigenic determinants present in the food we eat, the air we breathe, and even on the microbes that populate our environment.

Any system that integrates information must select only a small part of the impinging stimuli for special attention. A person cannot read a book and watch television at the same time, or carry on two telephone conversations, without either becoming totally confused or ignoring one of the sets of information. For our brains to discriminate between information and noise, we must be able to filter the stimuli that enter our senses.

The immune system must also filter out a universe of potential antigens and select for its attention only those antigenic determinants that are potentially dangerous. We can call this process **selection of antigenic determinants.** The immune system must *select against* responses to self-antigens or irrelevant foreign antigens and *select for* efficient responses to important foreign antigens. The immune system achieves this in two ways. First, it is sensitive to the environment in which a potential antigen is recognized; an antigenic determinant present on a bacterium justifiably makes a bigger impression on the system than the same antigenic determinant seen without the bacterium. Secondly, the immune response is modulated by regulatory influences that suppress or enhance the immune response. By balancing the relative strengths of suppressor and helper influences, the immune system can grade the intensity of its response, remember or prevent a future response, or terminate a response when the job is done.

Because of its importance, regulation of the immune response is guaranteed by many different mechanisms. We will briefly outline several of them.

Regulation by Microbial Products

Among the most powerful excitatory signals for lymphocytes are products of microbes or inflammation caused by microbes. This is understandable. The immune system seeks to focus its response against those antigens that are likely to be related to invading microbes because antibodies against these antigens help the host fight off infection. Therefore, microbial or inflammatory products are useful alarm signals to the host; they alert the immune system to take seriously any associated antigens.

Among these alarm signals are units of bacterial cell walls, endotoxin, interferon produced by host cells infected with virus, and products secreted by phagocytes in action. The presence of these materials tells the lymphocytes that trouble is brewing. Some of these materials themselves are antigens. However, they also function as ancillary, adjunct signals that incite lymphocytes to respond to other antigens that may be encountered at the same time.

Feedback Control: Its Use to Prevent Rh Disease

The immune system, like many other biological systems, seems to be turned off by its own products, a process of negative feedback. The presence of antibodies neutralizes antigen and inhibits the production of more antibodies (Figure 15.8).

Medicine has learned to make use of negative feedback to control

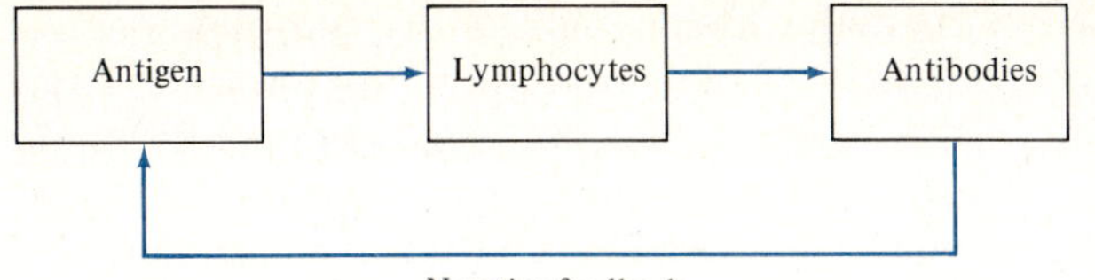

Figure 15.8
Antibodies, products of the immune system, can function as a negative feedback that shuts off an immune response by neutralizing antigen.

the production of undesirable antibodies. An outstanding example is the use of preformed antibodies to prevent a pregnant woman from making antibodies to the red blood cells of her fetus. At the time of birth, blood from the infant usually enters the mother's system. In most cases, this causes no problem. However, a special antigen on red blood cells, called Rh, can be a source of trouble. A mother who is Rh^- (has no Rh antigen of her own) may become immunized against Rh antigens if the red blood cells of her infant are Rh^+. Of course, no immunization will occur and no problem will arise if the infant is also Rh^-. Antibodies against Rh antigen appear only after the first Rh^+ infant has been born, and this child suffers no ill effects. However, the next Rh^+ fetus to develop in the mother is in danger. IgG antibodies that are transferred to the second Rh^+ child across the placenta may recognize and bind to the Rh antigen on the fetus's red blood cells and activate complement or phagocytes to destroy them. Thus, the infant may be born with serious anemia, and in the past, before exchange transfusion became a common procedure, many such infants died. In exchange transfusion, the blood of the newborn infant is slowly withdrawn and exchanged for Rh^- blood of suitable blood type. Thus, the infant is given blood that cannot be destroyed by the mother's antibodies to Rh. In time these antibodies disappear, and the baby makes its own Rh^+ blood cells in safety. However, it is always better to prevent a disease than to treat it.

Immunization of the mother's lymphocytes against her first infant's Rh antigen is now prevented by administering to the mother preformed antibodies to Rh antigen immediately after she gives birth. These antibodies bind to any of the infant's Rh^+ red blood cells that may have entered the mother's system and destroy them before the mother becomes immunized. Moreover, the presence of preformed antibodies to Rh antigen acts as a negative feedback signal that actively inhibits the production of such antibodies by the mother. The passively administered antibodies to Rh disappear from the mother within a few months. The next Rh^+ fetus can develop without danger, because the mother was not immunized by the first fetus and therefore has no memory lymphocytes and no augmented response to the Rh antigens of the second fetus. The mother will never become immunized to Rh antigen as long as she receives antibodies after each delivery.

Histocompatibility Genes and Organ Transplantation

The magnitude of the immune response that an individual can make to certain antigenic determinants is strongly influenced by a complex of genes called the **major histocompatibility complex.** People with some alleles, or variants of genes, in the complex produce large amounts of antibody against a particular antigen, while individuals with other alleles of these genes may produce relatively little or no antibody against the same antigen. Genetic control of the magnitude of the immune response is specific with regard to particular antigens, so that a person who is a genetically low responder to one antigen may be a genetically high responder to another antigen, and vice versa. A theoretically important but as yet unanswered question is how the products of genes in the major histocompatibility complex serve to regulate the character of the immune response.

However, in addition to their theoretical interest, these genes are of great practical importance. They seem to serve as markers by which the immune system of an individual recognizes the "foreignness" of cells originating from another person. The products of foreign genes in the major histocompatibility complex are among the most powerful stimulators of an immune response. Hence, the presence of such foreign genes in a transplanted organ activates in the transplant recipient an immune response against the donated organ. Because of this immune response, the donated organ is destroyed, a process called rejection. Transplantation of organs such as the kidneys, heart, pancreas, or bone marrow is no challenge to surgical technology; transplantation fails because of rejection stimulated by foreign major histocompatibility complex antigens.

The major histocompatibility complex antigens in human beings are termed HLA antigens, an acronym for **h**uman **l**eukocyte **a**ntigens, since they were first discovered on white blood cells. Now, we know that all cells carry HLA antigens on their surface. Each individual has his or her own HLA type, just as his or her own blood type is determined by antigens such as Rh, A, B, or O (H), expressed on the blood cells. To avoid or attenuate the process of immune rejection of transplanted organs, it is helpful to "match" the HLA type of the donated organ with that of the recipient. Typing of blood antigens Rh, A, B, and O (H) is done routinely in blood transfusions, and matching is usually not a problem. HLA typing and matching are much more difficult because of the great numbers of different HLA antigens. The HLA complex of genes is probably the most variable of all gene groups, and the chances of finding a perfect HLA match between unrelated persons is about one in 10,000. Thus, a person's unique set of HLA antigens is a biological marker of individuality. The reasons for the unprecedented genetic variability of the HLA genes are not known, but the diversity of HLA types is a serious obstacle to fulfillment of the goal of exchanging worn-out organs for healthy ones.

15.8 The Immune System Profits from Experience: Memory and Tolerance

A major attribute of lymphocytes is their adaptability: They learn from past experience. In the course of immunization against an anti-

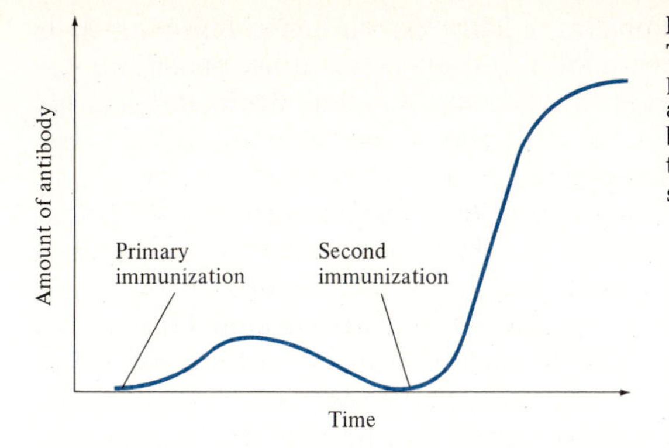

Figure 15.9
The primary and memory responses to an antigen. The primary response to an antigen requires the presence of ancillary signals and takes a relatively long time for antibodies to appear. A second contact with the same antigen triggers memory lymphocytes in the absence of ancillary signals and more antibodies are produced faster.

gen, **memory lymphocytes** are produced that markedly change the nature of the immune response to a second exposure to the same antigen.

As we have stated, in the first contact with an antigen, the immune system hesitates before responding: Is a response necessary? Is it worthwhile to expend resources and energy against the antigen? An unnecessary response is not only wasteful, it could be dangerous. The antigen may be a normal part of the host and a response could produce an autoimmune disease, or the antigen could be on a harmless microbe and a response could divert attention away from a real threat. Therefore, it is to the advantage of the individual that the immune system study the context in which the first contact is made with an antigen before a response is mounted. In this primary response, lymphocytes with receptors for the antigen require additional signals before they are induced to respond by proliferating and producing antibody. The presence along with the antigen of bacterial cell walls, endotoxin, interferon, or inflammation provides an ancillary signal attesting to the association of the antigen with microbial infection. The primary immune response requires time to get started and is usually of relatively low magnitude (Figure 15.9). However, in addition to antibodies, the primary immune response generates memory lymphocytes. These memory lymphocytes are the offspring of the primary lymphocytes that were activated by the first contact with the antigen and proliferated in the primary response. Thus, these memory lymphocytes are more numerous than primary lymphocytes. Hence, they can produce a stronger response than did the primary lymphocytes. In addition, the requirement of ancillary signals is relaxed and memory lymphocytes respond much faster to a second contact with the antigen (Figure 15.9). For example, the first exposure to the virus of chickenpox usually leads to disease because the immune response requires time to overcome its inhibitions and generate a large number of responding lymphocytes. The subsequent memory response to a second exposure to virus is so fast and strong that another bout of chickenpox is almost impossible. Once memory is established, ancillary signals are no longer needed.

The opposite of immunological memory is **immunological tolerance.** This is a state in which the immune system has been induced to remember *not* to respond to a particular antigen. The antigen is ignored even when administered together with powerful ancillary signals. Tolerance is important in protecting the host from attacking his own antigens. The immune system learns that it is forbidden to respond to the specific self-antigen.

Tolerance requires primary exposure to an antigen under special conditions. These conditions mimic the way the immune system sees the individual's own antigens: a sudden high concentration or a prolonged low concentration of antigen in the absence of ancillary microbial or inflammatory signals. Tolerance and memory are outstanding examples of adaptation. Past experience with an antigen modifies the rules of conduct toward that antigen in the future. The mechanisms by which this is done have yet to be elucidated, but they seem to include regulatory helper and suppressor T lymphocytes.

15.9 Immunological Distinction Between Virulent and Avirulent Microbes

Microbes may be classified into two broad groups by the manner in which they interact with the immune system of the host: **virulent microbes** and **avirulent microbes.** Virulence, as we stated earlier (Chapter 13), refers to the relative capacity of a microbe to act as a pathogen and cause disease. Any microbe that invades the host beyond its tolerated niche can cause disease. Since the immune system is charged with repelling invaders, the interaction between a microbial parasite and the immune system is a major determinant of a microbe's virulence. Therefore, operationally speaking, virulence is a property determined jointly by both the behavior of the parasite and the immune response of the host (Chapter 13).

Avirulence, or relatively low virulence, of a microbe can result from easy and open communication between the microbe and the immune system. Avirulent microbes may present ligands recognized directly by agents of defense such as phagocytes or complement without the aid of antibodies. In such a case, the microbe can be destroyed directly (Figure 15.10).

Other avirulent microbes may not have ligands specific for phagocytes or complement, yet they communicate easily with lymphocytes. Such avirulent microbes have antigens that are widely shared throughout the microbial world. Once the immune system has had experience with one of these microbes, memory lymphocytes are gen-

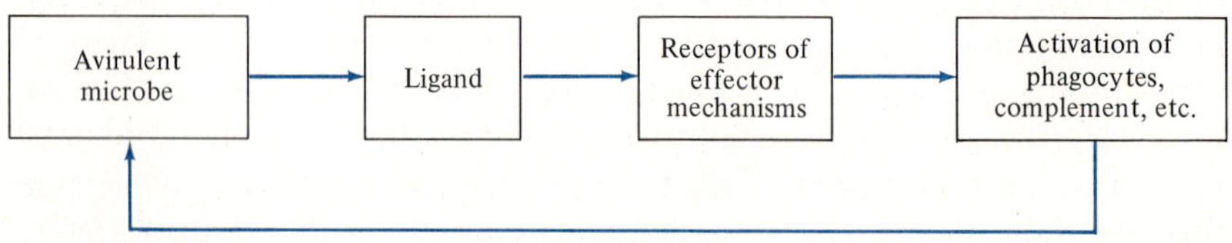

Figure 15.10
Avirulent microbes may present ligands (signals) that are recognized directly by agents of defense.

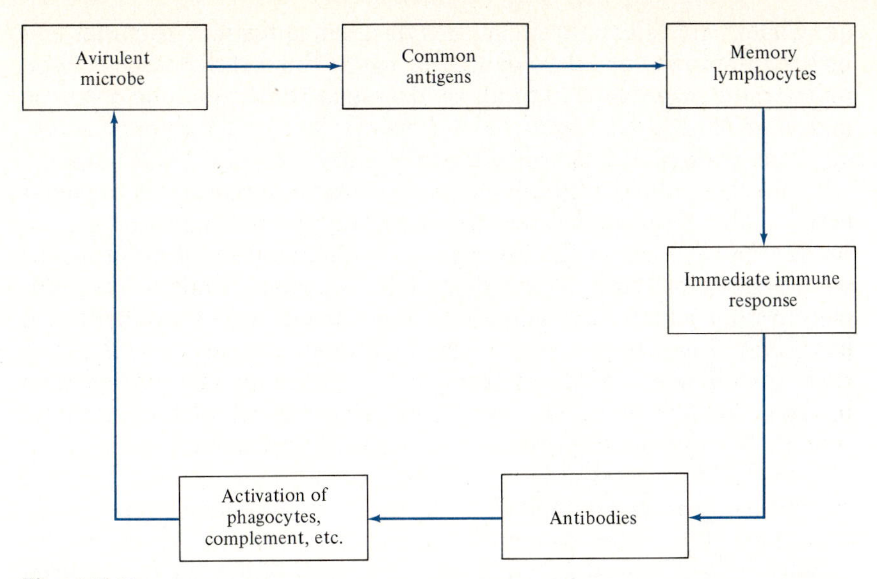

Figure 15.11
Avirulent microbes may present common, easily recognized antigens to the immune system and thus trigger an immediate immune response.

erated that can recognize the same or a similar antigen on a wide variety of other microbes. Hence, an expeditious and powerful "memory" response may be elicited by the first contact with the microbe, because the host has already had experience with similar antigens on other microbes. Figure 15.11 illustrates this alternate channel of communication between avirulent microbes and the host mediated by antibodies.

We know that it is to the long-term advantage of a microbial parasite to be relatively avirulent (Chapters 12 and 13). However, the *microbes* do not know ahead of time what is good for them. Therefore, microbes that do not easily signal immune defenses appear from time to time through genetic changes in avirulent microbes, introduction of new hosts, ecological imbalance, and so on (see Chapter 13). Such microbes are virulent and can cause disease. Ultimately, virulence will be reduced, as we saw in the sample of myxomatosis and Australian rabbits (Chapter 13.3), but the virulent state may require many generations to become attenuated. Virulence is temporary only on the long-time scale of evolution. Given enough time, the microbial parasite and its host will mutually adapt, or one or both will disappear. However, at any one moment, within the environment lurk microbes that are virulent.

Virulent microbes may use several tactics to impede their recognition and destruction by the host's lymphocytes or antibodies. For example, some microbes cover themselves with unique antigens that the immune system has never seen before, and other microbes cover themselves with antigens very similar to those of the host himself.

Confronted with unique antigens, the immune system must take the time to test whether a response is in order. The lymphocytes require ancillary signals from phagocytes, inflammation, bacterial cell walls, interferon, and so on. In the mean time, the microbes have invaded the organism and produce disease until the immune response makes up for lost time and repels them.

The streptococcus provides an example of a virulent microbe with unique antigens. Streptococci of a particular family designated group A that often cause infection of the throat and tonsils produce a surface antigen called M protein. The presence of M protein makes streptococci resistant to phagocytes. Phagocytosis can proceed efficiently only when these streptococci are opsonized by antibodies to M protein (see Chapter 14.6). The M protein antigen is unlike any other microbial antigen, and antibodies recognizing other antigens do not bind to M protein. In fact, more than 50 different types of M protein have been characterized, and antibodies against one type of M protein do not easily recognize other types of M protein. Hence, a previous infection with a streptococcus of one M protein type generates memory lymphocytes and resistance to a second infection with streptococci of the same type, but not against streptococci featuring other types of M protein. For this reason, children are susceptible to repeated streptococcal sore throats, each caused by a different M protein type of streptococcus.

Virulent microbes may also disguise themselves by producing and wearing a cloak of antigens like those of the host. To prevent autoimmunity, the individual's lymphocytes are tolerant of self-antigens and are forbidden to attack. The microbe may exploit this tolerance of the host for self-antigens and gain access to the interior of the body.

The agent of syphilis, *Treponema pallidum*, may possibly use this ploy. In about 1910, the German microbiologist Wassermann sought to develop a blood test that would diagnose patients with syphilis by detecting the presence of antibodies to *T. pallidum*. Since it is impossible to grow *T. pallidum* in culture, Wassermann attempted to extract antigens of *T. pallidum* from the organs of human fetuses that had been aborted because they had been heavily infected with the microbe in their mothers' wombs. Wassermann succeeded in obtaining such an antigen that would bind antibodies from the blood of persons infected with *T. pallidum*. To his surprise, however, he found that the "antigen" of *T. pallidum* also could be obtained from extracts of normal, healthy, uninfected persons. It became obvious that the antibodies that developed in response to infection with *T. pallidum* were able to recognize an antigen that was produced both by the parasite, *T. pallidum*, and by its obligate human host. Thus, *T. pallidum* produces what seems to the immune system to be a normal "self-antigen" of the host. This may be one reason why *T. pallidum* is not effectively removed from the body by the immune response. Fortunately, syphilis is easily cured by treatment with penicillin, and a person does not have to depend on the immune response to get rid of this pathogenic parasite.

A microbe may present to the host his or her own self-antigens in

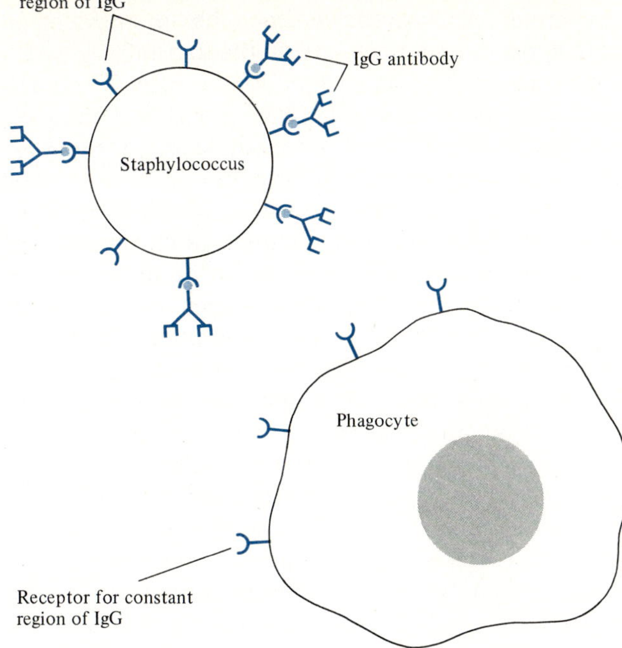

Figure 15.12
A bacterium disguising itself with the host's antibody molecules. This staphylococcus bacterium has receptors that bind IgG antibody molecules by their constant regions. Phagocytes or complement with receptors for the constant region cannot be triggered by the unbound variable regions of the antibody. Compare with Figure 15.3.

yet another way. In this case, the microbe does not produce host-like antigens, but actually borrows the host's own molecules to do the job. To our consternation, microbes can exploit even our antibodies to this end. Staphylococci have surface molecules that act as receptors for the **constant** region of IgG antibody molecules, similar to the way that phagocytes have receptors for the constant region of antibodies. The resourceful microbe, using these receptors, binds the constant, foot end of our antibodies, leaving only the outstretched combining sites available to phagocytes (Figure 15.12). Since our phagocytes do not recognize the combining sites of antibodies, such turncoat antibodies are useless in activating phagocytes to engulf the bacterium.

Another tactic used by virulent bacteria is to thwart digestion by phagocytes. For example, *Mycobacterium tuberculosis*, the agent of tuberculosis, is readily engulfed by phagocytes. The bacterium is able to resist the lysosomal enzymes, however, and thrives inside the phagocytes.

15.10 Recapitulation of Immunity to Microbial Parasites

The essential function of the immune system is to recognize that a microbial parasite has invaded the body and to destroy the invader. In Chapter 14, we outlined a number of effector mechanisms available to the body to destroy or fight off microbial invaders: phagocytes, natural killer cells, and complement. For these defense mechanisms to act, however, they must be capable of recognizing the microbe as a foreign element. They must receive activating signals.

We have seen that agents of defense can be signaled through either or both of two channels: directly by the microbes themselves or indirectly by antibodies, the products of lymphocytes.

Lymphocytes have antibody-like receptors that can recognize three-dimensional forms of molecules called antigens. Upon meeting a specific antigen, B lymphocytes secrete antibodies, and T lymphocytes either secrete lymphokines, become specific killer cells, or function as regulatory helper or suppressor cells. Agents of defense such as phagocytes will be activated to destroy microbes by antibodies binding to microbial antigens. Hence, antibodies can function to create activating signals out of microbial antigens. In this way, the lymphocytes and the antibodies that they produce "translate" or "transduce" microbial antigens into signals recognizable by agents of defense. Antibodies themselves can also protect the host by binding to and inactivating viruses or microbial toxins.

We have also described some of the cellular, molecular, and genetic machinery involved in the behavior of lymphocytes: how lymphocytes differentiate into specialists; how antigens select clones of lymphocytes to respond in a kind of evolutionary "selection of the fittest"; how the activities of this complex immune system are finely controlled and regulated by microbial products, feedback loops, regulator cells, and a unique complex of genes; and how lymphocytes learn from experience and adapt the immune system by remembering the past.

Finally, we saw that avirulent microbes could be defined by the readiness with which they signaled the immune system. In contrast, microbes that avoided detection or destruction by the host's immune system tended to be virulent. Fortunately, the immune system is so resourceful and versatile that the host can succeed in defeating most microbial invaders. Human culture is a powerful ally in the fight. Human beings have learned to use vaccination to endow the immune system with ready-made memory for many microbial antigens. They know how to harness, for their benefit, natural antimicrobial materials such as antibiotics. By knowledge of the ecological circumstances of infectious disease, humans have developed methods for sanitation and quarantine. It is hoped that an awareness and a deeper understanding of the significance of the ecological balance between human beings and their parasites will bring them to use these powerful tools judiciously.

QUESTIONS

15.1 In Section 7.4, we restated the "one gene–one enzyme concept" as follows: "One segment of DNA is responsible for the production of an mRNA which codes for a specific sequence of amino acids." How should this concept now be modified in the light of the genetical basis of the antibody molecule?

15.2 Only vertebrates have lymphocytes and antibodies. The rest of the biological world—insects, worms, plants, and so on—is confronted with the danger of microbial parasites, yet it contains these parasites and survives quite well without the ability to make antibodies. Can you suggest why only vertebrates needed to evolve lymphocytes and antibodies?

15.3 The immune system usually reacts strongly against the histocompatibility antigens on organs transplanted from other persons. What foreign "tissues" are not rejected by a woman's immune system?

15.4 Imagine a microbial parasite that made no antigens. What would be the outcome of infection with such a microbe for the human host? For the parasite?

15.5 What is the biological basis of allergy? How does desensitization help control allergic reactions?

15.6 Define antigen, antibody, variable and constant regions of an antibody, and HLA genes.

15.7 How do antibodies recognize antigens?

Suggested Readings

Burnet FM: The Clonal Selection Theory of Acquired Immunity. New York: Cambridge University Press, 1959.
A readable classic.

Burnet FM, Ed: Immunology: Readings from Scientific American. San Francisco: W. H. Freeman, 1976.
A collection of papers comprising a broad introduction to immunology.

Capra JD and AB Edmundson: The antibody combining site. Scientific American, 236: 50, January 1977.

Cooper MD and AR Lawton III: The Development of the Immune System. Scientific American 231: 59, November 1974.

Lerner RA and FJ Dixon: The Human Lymphocyte as an Experimental Animal. Scientific American, 228: 82, June 1973.

Milstein C: Monoclonal Antibodies. Scientific American, 243: 56, October 1980.
The story of one of the most important discoveries in immunology.

Notkins AL and H Koprowski: How the Immune Response to a Virus Can Cause Disease. Scientific American, 228: 22, January 1973.

CHAPTER 16 Specific Diseases Related to Infection by Microbial Parasites

What checks the natural tendency of each species to increase in number is most obscure.

On the Origin of Species, *Charles Darwin*

In the last four chapters (Chapters 12–15), we discussed the ecological relationship between man and his microbial parasites, the factors involved in the etiology of infectious diseases, and the immune system. In this chapter, we describe specific diseases of general interest. We begin with some diseases of childhood, pass on to representative infectious diseases of the various organs, and end with a section describing a relationship between viruses and cancer.

16.1 Some Diseases of Childhood

Upper Respiratory Tract Infection: The Common Cold

Acute diseases are distinguished from chronic diseases by their relative duration. When a disease develops during a day or two and lasts for only a few days to about a week, it is referred to as an **acute disease.** In terms of the numbers of people affected and the loss of productive work and school days, acute illness of the upper respiratory tract—the common cold—leads all other acute conditions in the toll it exacts from industrialized society. Literally hundreds of millions of work days are lost because of colds yearly in the United States. The cost to the economy is many billions of dollars.

The miseries of the cold are known to all: a stuffy, runny nose, perhaps a sore throat, coughing, sneezing, headache, aches, and pains. The magnitude of the symptoms defines whether it is simply a minor cold or a more severe "flu." Complications, particularly in young children, include ear infections and, less commonly, pneumonia.

By far the most common causes of upper respiratory tract infections are viruses of low virulence. These viruses infect the cells lining the nose and throat and have very little capacity to invade the body. Cold viruses belong to a number of different groups, distinguishable by their size, shape, genetic material (whether DNA or RNA), and biochemical components. Within each group there may be many individual viruses sharing certain antigens but differing from one another by other antigens.

The prevalence of colds is related to the season of the year and to the age, density, and mobility of the host population. There is no apparent reservoir of cold viruses outside the human population. The spread of cold viruses is facilitated by a high density of people in close contact, such as occurs during winter and early spring in Europe and North America, when people tend to remain indoors. Furthermore, the ability of the mucous membrane of the nose to serve as a barrier to the viral infection may be compromised by the dry, artificially heated air of the indoor environment in cold weather.

A relative degree of general immunity to colds develops with age; adults have a much lower incidence of respiratory infections than do young children. As a result of many colds during childhood, the individual's immune system gains experience with the many different families of respiratory viruses and can build up a reserve force of memory lymphocytes. These memory lymphocytes, developing in response to one type of virus, can offer a degree of protection against other viruses that have related antigens. Hence, childhood is a period of adaptation of the immune system to the viruses prevailing in the child's particular environment.

This general immunity to common cold viruses is only relative and can be upset by changes either of the viruses or of the host. For example, introduction into the environment of new types of viruses with previously unrecognized antigens can produce epidemics of respiratory infections in an unprepared population. Epidemics of influenza may be explained in part by the appearance of new and unique surface antigens on influenza viruses.

Alternatively, quiescent viruses surviving in latent form in host cells can be activated by changes in the host's temperature, blood flow, or even emotional state.

Respiratory viruses are resistant to treatment with antibiotics, since like all viruses they use the host's own biochemical machinery for their metabolism. Vaccination of the respiratory tract against the hundreds of different viruses present in the environment might help prevent colds, but such vaccination is not technically feasible at present. Interferon applied locally to the mucous membranes of the nose may prove to be an effective form of treatment, but not enough interferon is available for such use. Furthermore, treatment of colds with interferon, even if feasible in the near future, may be unwise. The use of interferon could prevent an individual from developing active immunity to cold viruses and leave him or her permanently as susceptible to colds as a young child. Barring unforeseen developments, upper respiratory tract infections will be with us for some time.

Measles

In Chapter 13 (Section 13.2), we pointed out how measles virus exists by maintaining a chain of acute infection in human beings. Recovery from a case of measles endows an individual with lifelong immunity, and there are no infectious carriers without symptoms of acute disease. Therefore, persistence of measles in a population seems to depend on a continuous supply of fresh susceptible people who transmit the virus in perpetuity as an acute disease.

In recent years, however, it has become apparent that measles virus has the capacity to persist in certain rare cases. Although this persistence of measles virus is a biological dead end that does not contribute to the cycle of infection, it is enlightening to consider the phenomenon. It was discovered that measles virus can be harbored for long periods in the brain in an extremely small number of children. This persistence of virus is associated with a fatal degenerative brain disease. The finding was unexpected, since the slow onset of brain degeneration was generally distant in time from the acute attack of measles, from which the children all seemed to recover uneventfully. The relationship between brain disease and measles virus was suspected by the finding of high levels of antibodies directed against measles virus in the brain fluid of affected children. Traces of the virus itself were later detected. It is thought that a rare defect in the immune system of these children may be responsible for their failure to eradicate measles virus from the brain. This phenomenon is another example of the role of the host and the immune response in defining the nature of the relationship between a microbial parasite and its human host. Measles is an acute, often severe generalized infection from which almost all healthy, well-nourished children recover completely. Children in underdeveloped parts of the world who are weakened by malnourishment or other causes sometimes experience pneumonia or other acute complications of measles and die. In extremely rare instances of a poorly understood disorder of immunity, measles may behave as a slow virus attacking the brain in a manner reminiscent of kuru (Chapter 13, Section 13.2). Fortunately, vaccination against measles virus promises to eradicate all forms of disease associated with the virus.

Chickenpox

Chickenpox is caused by a virus entirely different from that of measles, although the diseases they cause have certain features in common. Both viruses infect only humans, both cause acute diseases with distinct rashes, and both can hide in the nervous system. However, for chickenpox virus the nervous system is not a biological dead end but is a major factor in the chain of infection to which the virus owes its total existence.

Chickenpox is a highly contagious disease. In young children the disease is usually mild, and recovery is followed by lifelong resistance. In adults chickenpox is usually much more severe; the fever is higher, the rash is more extensive, and pneumonia can occur. Chickenpox in patients with impaired immune systems is dangerous and can be fatal.

After recovery from acute chickenpox, the individual is immune, and the virus can no longer return to produce chickenpox. However, not all of the chickenpox virus is destroyed; some virus is able to persist in nerve cells. After a long time, usually many decades, the chickenpox virus may reappear as the cause of a new disease called shingles, or herpes zoster. Shingles often is associated with the weakening of the immune system characteristic of old age, chronic diseases, or cancer. It appears as a painful rash that is sharply limited to the area of skin serviced by a particular sensory nerve. Hence, it is reasonable to assume that the virus exploits the nerve cell as its lodging and uses the nerve fibers as a means of spread to the skin. The rash of shingles is a source of infectious virus, and susceptible children come down with chickenpox after having contact with an adult suffering from shingles. In contrast to measles, therefore, chickenpox does not require a constant supply of susceptible hosts to maintain itself. By remaining latent in older people, the virus can wait patiently for the birth of a new generation of susceptible children. When grandfather comes down with shingles, he can infect his grandchildren with chickenpox. By vertical transfer from one generation to the next, the chickenpox virus can maintain itself even in a relatively small but stable host population (Figure 16.1).

At present a vaccine capable of immunizing against chickenpox virus is not available. Therefore, natural infection at a young age is the best insurance against experiencing severe complicated chickenpox as an adult. The best way to avoid the appearance of shingles in old age is to be fortunate enough to have a strong immune system that suppresses expression of the virus.

German Measles (Rubella)

In children or young adults, German measles is a mild disease characterized by a rash, low fever, and few complications. Recovery from German measles leads to lifelong immunity to reinfection. But the story does not end there.

In 1941 an Australian physician, N. M. Gregg, noted the occurrence of birth defects, notably of the eyes, ears, and heart, in infants born to mothers who had suffered from rubella during pregnancy. We now know that rubella virus can infect and severely damage a developing fetus while causing very mild, transient disease in the mother. The younger the fetus, the greater the damage. The first four weeks of pregnancy have the greatest risk.

Rubella is an outstanding example of the importance of the state of the host in determining the virulence of a microbial parasite. The same virus that is totally harmless in a nine-month-old infant can be disastrous in a fetus.

Immune mothers do not transmit rubella virus to their fetuses,

Figure 16.1
Transmission of chickenpox virus. (A) A child with acute chickenpox can **horizontally** infect susceptible people (usually children). (B) After recovery from acute chickenpox the child no longer transmits virus, but virus persists in his or her cells in latent form. (C) Latent virus can be activated decades later and reappear as shingles (herpes zoster) in adult life. The patient can now transmit virus to a new generation of susceptible children, a form of **vertical** transmission. These children come down with chickenpox. Other infectious agents that have a cycle of latency and activation can exploit a similar strategy of vertical transmission (see the example of tuberculosis, Section 16.4).

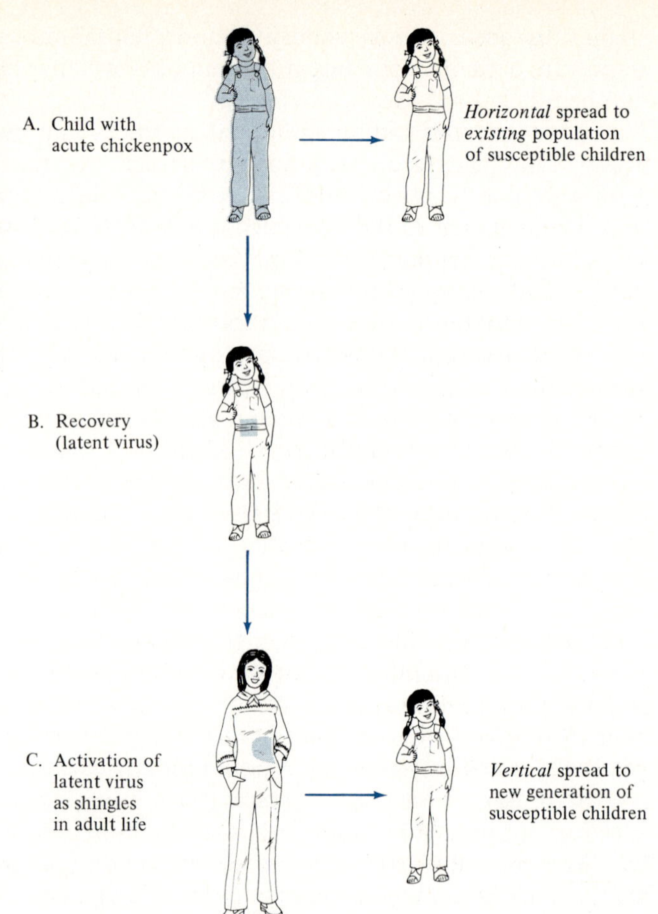

even after direct contact with the virus. Therefore, infection of fetuses can be prevented entirely if girls become immune to rubella before they reach childbearing age. This can be accomplished by immunizing all susceptible girls with rubella vaccine.

Mumps

Mumps is a disease characterized by inflammation of the salivary glands, particularly of the parotid glands located below the ears. Sometimes other organs, such as the brain, are affected. The disease is mild and almost all patients recover without complications. Up to one half of persons infected with mumps never have any symptoms at all. Lifelong immunity is conferred by a single infection with mumps, whether or not symptoms actually have occurred. A vaccine is available.

Streptococcal Infection and Its Complications

Streptococci are a family of gram-positive cocci that are members of the normal bacterial flora of the throat. Streptococcal infections, when they occur, are caused by streptococci that are classified as belonging to a particular subclass, called group A because of an antigen that they all share. However, within group A there are more than 50 different types, defined by variations of another antigen called M protein (see Section 15.9). This antigen, present on the surface of group A streptococci, is associated with resistance to phagocytosis and hence contributes to the virulence of the microbe. Another factor responsible for virulence is a capsule that also impedes phagocytosis. Many people carry group A streptococci as part of their normal flora. How and why these bacteria become pathogenic is not known.

Immunity against one type of streptococcus offers little protection against the other 50-odd types. Hence, an individual may overcome one streptococcal infection and yet remain susceptible to repeated infections with other types of streptococci.

Streptococci can cause infections of the skin, but their chief notoriety derives from their infection of the tonsils and throat, often leading to high fever and swollen lymph nodes in the neck. Sometimes streptococcal infection of the throat is accompanied by a rash, and this disease is called scarlet fever.

Complications of streptococcal infections are of two types: **suppurative,** meaning related to pus and infection, and **non-suppurative.** Suppurative complications refer to spread of the bacteria from the throat to the sinuses, ears, lungs, or coverings of the brain. These complications can be seen as an extension of the invasion of the host by the bacterium.

Non-suppurative complications occur weeks to months after the infection has subsided. *Non-suppurative* refers to states of disease in which there is no evidence of living streptococci in the affected organs. It is suspected that non-suppurative complications are caused by aspects of the immune response of the host triggered by the streptococcus.

The major non-suppurative complication is rheumatic fever. Rheumatic fever can involve the joints in a swollen, painful arthritis that ultimately disappears without a trace. The most serious expression of rheumatic fever is in the heart. The first attack of rheumatic fever usually clears up without leaving permanent damage. However, repeated streptococcal infections followed by repeated bouts of rheumatic fever may leave the heart valves severely scarred and malfunctioning. Diseased valves can compromise the function of the heart in pumping blood.

The incidence of non-suppurative complications of streptococcal infection is low: 1 percent or less following sporadic cases of streptococcal sore throat, and up to 3 percent in epidemics. The severity of the infection is not predictive of the development of non-suppurative complications. Cases of rheumatic fever have followed mild or even inapparent infection. Early and adequate treatment of streptococcal infections with antibiotics, usually for ten days, eradicates streptococci and prevents complications. For this reason, physicians often culture the bacteria obtained from a child's sore throat to determine whether or not group A streptococci are present. Discovery of streptococcal infection necessitates a full course of treatment with appropriate antibiotics to prevent possible development of rheumatic fever.

How streptococci trigger sterile inflammation of the heart or joints is a mystery. Antigens of streptococci have been found to have certain similarities to antigens of the human heart. It is conceivable that an immune response to particular streptococcal antigens can inadvertently trigger effector mechanisms of the host to attack the heart, a form of autoimmunity (see Sections 15.7 and 15.9). Therefore, treatment of streptococcal infection is aimed at avoiding or minimizing the immune response by destroying the bacteria with antibiotics. Diagnosis of the etiologic agent of a sore throat is important because most sore throats are caused by viruses. In such cases, a full course of treatment with antibiotics is unnecessary and potentially harmful.

16.2 Infectious Diseases of the Skin

Acne

The etiologic agents of acne are normal bacterial flora of the skin. The etiologic circumstances relate to the hormone balance of the host and its effect on a constitutive mechanism of defense. During the period of puberty, the nature of the secretions of the sebaceous glands of the skin is influenced by changes in the concentrations of various hormones. The secretions become thickened and tend to plug the pores that normally empty onto the surface of the skin (see Figure 12.1). Such a plugged pore is a "blackhead." Obstruction of the flow of sebum allows normal skin bacteria to accumulate within the pore. Even though these bacteria do not penetrate the skin itself, a small abscess or pocket of pus forms in the plugged pore. The formation of abscesses can be prevented by careful cleansing of the skin, removal of blackheads, and restoration of the flushing process that helps control the numbers of skin bacteria in pores.

Boils or Furuncles

Staphylococci, gram-positive cocci, are normal inhabitants of the skin and the mucous membranes of the nose. However, staphylococci also may invade the skin and cause relatively large pockets of pus called boils or furuncles. The factors in the host and parasite responsible for invasion by staphylococci and subsequent boils are complex, and much remains to be learned. Staphylococcal infection sometimes forms in surgical wounds about one week after a surgical procedure. Staphylococci are very hardy bacteria and often resist treatment with antibiotics. Especially virulent strains of staphylococci are a problem in many hospitals since they can be carried without symptoms in the respiratory tracts of hospital personnel and spread easily to patients weakened by other illness or surgery.

Fever Blisters (Herpes Simplex)

Recurrent sores developing in the skin, often around the lips, are a familiar phenomenon known as fever blisters. One might suspect that fever blisters, similar to colds, are caused by a multitude of different etiologic agents. Actually, fever blisters are caused by one virus, herpes simplex, that persists in the skin despite an immune response. Most of the human population (70 to 90 percent) have antibodies to herpes simplex virus, indicating that infection with the virus is widespread. When susceptible individuals, mostly young children, become infected for the first time, only about 1 percent actually develop symptoms of fever and rash. The other 99 percent never show any signs of acute disease. Whether they suffer symptoms or not, infected individuals develop antibodies to herpesvirus and continue to harbor the virus in latent form in cells in the skin.

Some individuals are prone to express infectious herpesvirus in the form of repeated outbreaks of fever blisters. Others continue to carry the virus without overt signs. Activation of the virus and occurrence of blisters usually requires some additional, often minor stimulus. Sunburn, fever, physical trauma, emotional stress, and menstruation can all elicit fever blisters in susceptible people. Infectious virus present in the blisters is available to infect fresh susceptibles. Thus, the reservoir of herpes simplex virus is the large population of latent carriers, and the vectors (Section 13.2) are the people who tend to develop recurrent overt disease.

16.3 Infectious Disease of the Brain

Meningitis

Infection and inflammation of the membranes covering the brain and spinal cord is called meningitis (*meninx*: "membrane" + *itis*: "inflammation"). The symptoms of meningitis are fever, headache, stiff neck, nausea and vomiting, and general debility. Meningitis can be caused by viruses or by bacteria. A great many different kinds of viruses can cause meningitis, from which recovery is spontaneous and complications are usually absent.

Bacterial meningitis is a much more serious problem. Unless treated vigorously with antibiotics, the disease is fatal. Infection with bacteria, as we have noted (Section 13.1), usually leads to the accumulation of phagocytes and the formation of pus and swelling. Pus within the central nervous system is dangerous because the brain and spinal cord are encased in a rigid frame of bone. The bony covering is inflexible and cannot accommodate changes in volume in response to pressure generated by inflammation and pus. Therefore, the added volume of swelling presses on the brain and spinal cord. Even when the pus is removed, it can be replaced by scar tissue formed in the process of healing. Scar tissue in other organs may be harmless; in the brain it can damage delicate vital tissues. To avoid the development of an inflammatory immune response, pus, and subsequent scarring, it is critical to begin treating bacterial meningitis with appropriate antibiotics as soon as possible. The suspicion of bacterial meningitis requires the physician to use whatever procedures are necessary to confirm a diagnosis and institute immediate therapy.

The etiologic agents of bacterial meningitis are usually bacteria that reside in the nose and throat as part of the normal flora. The nose and sinuses apparently provide access to the coverings of the brain.

Meningitis teaches us the critical importance of the site of infection and the nature of the host's response in determining the outcome of a host–parasite relationship. Pus in the skin is a nuisance; in the brain it is a catastrophe.

16.4 Infectious Disease of the Lungs

Pneumonia

Pneumonia is the term for inflammation of the lungs, usually related to microbial infection. Pneumonia is distinguished from bronchitis, a more common condition in which the inflammatory process takes place in the air tubes (called bronchi) that connect the lungs to the upper airway.

The etiologic agents of pneumonia can be viruses or bacteria. Many of the same viruses that cause a high incidence of upper respiratory tract infections (colds) in children can also cause pneumonia. Therefore, viral pneumonia is more common in children than in adults. The usual case of viral pneumonia develops as a relatively rare extension of an upper respiratory tract infection through the bronchi (bronchitis) and into the lungs (pneumonia). Viruses can also reach the lungs through the blood stream. Hence, viral pneumonia is not always associated with an upper respiratory tract infection, and any severe viral infection can be complicated by pneumonia. In the otherwise healthy host, viral pneumonia resolves without too much trouble.

Pneumonia caused by bacteria is a more serious disease. Most cases of bacterial pneumonia are caused by organisms that are members of the normal flora of the nose and throat, such as pneumococci. These bacteria can reach the lungs by aspiration ("breathing into") from the nose or mouth. This mechanism probably accounts in part for cases of bacterial pneumonia that are not uncommon in alcoholics. Intoxication can anesthetize, or put to sleep, the normal nervous circuits that control swallowing and coughing. Therefore, a person in a state of drunkenness is in danger of aspirating into the lungs infected material from the nose and throat. The cough reflex, whose job it is to expel such aspirated material, functions sluggishly, and aspirated pneumococci are able to establish themselves in the lungs. Thus, an alcoholic suffering from pneumococcal pneumonia is an example of how a social disease provides the etiologic circumstances for penetration into the lungs of bacteria that are harmless in their normal site in the nose. An ecological imbalance in the host–parasite relationship can originate from psychological maladjustment.

Viral infection of the upper respiratory tract can also predispose an individual to pneumonia, as is probably the case in children. Bacterial pneumonia caused by pneumococci usually responds quickly to treatment with appropriate antibiotics.

Tuberculosis

Acute pneumonia would seem to be an unnecessary incident in the life history of pneumococcus and man. This bacterium maintains itself quite well as a member of the normal flora of the nose and throat, and infecting the lungs does not appear to offer it any advantage. The story of *Mycobacterium tuberculosis* is quite different. This bacterium owes much of its existence to its ability to chronically infect the lungs. One of its major modes of transmission is by a special type of pneumonia.

Tuberculosis is a most ancient disease. The earliest mummies found in Egypt show signs of tuberculosis. As expected, the tuberculosis bacterium is well adapted to living with humans and exploits both their biology and their social organization.

Until fairly recently, almost all youngsters growing up in urban environments showed signs of infection by *M. tuberculosis*. However, this primary form of infection was without symptoms of disease in the vast majority of well-nourished children. Children usually acquired the bacterium by inhaling a very few organisms coughed into the air by older people. The inhaled bacteria set up an infection in a small localized area of the lungs. The host's immune response usually prevented any signs of illness by limiting the numbers of living tuberculosis organisms and inhibiting their spread. However, this microbe is able to survive quietly for many decades in the host's own phagocytes, without growing, but without being eliminated. The only signs of infection are evidence of an immune response to mycobacterial antigens (a "positive skin test") and possibly a "spot" on the lungs seen on a chest x-ray. Tuberculosis in its latent state cannot be easily transmitted to another individual.

In the course of time, a few individuals seem to lose their ability to contain their few tuberculosis bacteria. Perhaps the immune system falters in old age or because of stress. The bacteria begin to multiply and spread, and latent tuberculosis becomes active. Individuals with the active form of tuberculosis can cough up living, infectious bacteria and do so for many years. The old grandfather with his chronic cough can become a source of infection for fresh generations of susceptible grandchildren. Thus, tuberculosis bacteria were adapted to survive by vertical transmission from generation to generation of hosts by alternating between an inactive reservoir of primary non-infectious, non-symptomatic latent disease in children, and an infectious, mildly symptomatic, secondary phase in a few older people. Although tuberculosis is very different from chickenpox, in some ways the strategy of vertical transmission in two phases reminds us of the chickenpox-shingles story (Section 16.1 and Figure 16.1).

Tuberculosis has almost disappeared as a problem in medically advanced societies because alert public health agencies have detected and treated infectious carriers of bacteria with appropriate antibiotics.

16.5 Infectious Disease of the Gastrointestinal Tract

Microbial diseases of the bowels are characterized by diarrhea and vomiting, sometimes accompanied by fever and weakness. The most serious complication of acute diarrhea and vomiting is dehydration, a loss of fluid. The vital flow of blood depends on the volume of fluid in the blood vessels, and dehydration can threaten life by compromising the circulation to the brain, kidneys, and other important organs. The human body has a relatively small reservoir of fluid to replace that acutely lost. Hence, a major goal in the treatment of infectious disease of the gastrointestinal tract is to stop the continuing loss of fluids and to replace the fluids that have already been lost. Since most microbial agents of diarrhea and vomiting do not invade the body, measures taken to fight these agents are usually of secondary importance. Agents of microbial infection of the gut are usually spread in contaminated food or water.

Viruses are a major cause of diarrhea and vomiting, particularly in infants and children, although most children infected with these viruses suffer no symptoms.

Normal bacterial flora can be etiologic agents of gastrointestinal disease in people with obstruction of the bowels (Section 14.1). Some newborn infants may also suffer from disease resulting from their first contact with the normal bacterial flora of the population into which they are born.

A number of pathogenic bacteria such as *Salmonella* or *Shigella* regularly produce disease in infected persons, although apparently healthy people can carry and disseminate these bacteria. *Shigella* organisms produce acute and violent dysentery but do not penetrate beyond the walls of the gut. *Salmonella* bacteria may invade the tissues of the body and cause typhoid fever, a chronic disease characterized more by high fever and weakness and less by diarrhea. Cholera, a disease caused by another intestinal pathogen, is described in Section 13.2.

Diarrhea and vomiting may also be caused by ingestion of bacterial toxins (Section 13.1). Thus, the gastrointestinal tract is subject to insult by a wide variety of agents: viruses, normal or pathogenic bacteria, and toxic products of bacteria.

16.6 Infections of the Urinary Tract

The free flow of urine from the kidneys and the periodic emptying of the bladder provide a constitutive barrier to infection of the urinary tract (see Section 14.1). However, this barrier may be broken by any factor that impedes the complete flushing of the urinary system. Even a very minor partial obstruction to the flow of urine can predispose an individual to infection. In such cases, bacteria that are members of the normal flora of the bowel can cause serious infections by gaining a foothold in the urinary bladder or kidneys. Older people and those treated with tubes in the bladder are especially prone to such infections.

Often the etiologic circumstances underlying infection of the urinary tract are not clear and no reason for an impaired flow of urine can be detected. Some young girls tend to suffer from repeated infections. The bacteria associated with these infections, such as *E. coli*, are unremarkable representatives of the normal flora living in the bowel. A problem arises, however, when the infecting bacteria have been selected for their resistance to treatment with antibiotics. The widespread use of antibiotics can create conditions in which antibiotic-resistant bacteria enjoy an advantage over susceptible bacteria with which they compete for ecological niches in the individual. Thus, the person harbors in the gut a reservoir of bacteria that have been selected to resist antibiotics, and infection by these bacteria poses a problem in treatment. Fortunately, many of the young girls who suffer repeated urinary tract infections seem to "grow out" of the tendency.

Urinary tract infections illustrate the importance of constitutive mechanisms of resistance to infection and how these mechanisms can vary with age and sex. Urinary tract infection is much less prevalent in boys than in girls, probably because of fundamental differences in the anatomical arrangement of the urethra, the tube that connects the bladder to the exterior. Urinary tract infections also show how infectious diseases may become more serious as a result of excessive use of antibiotic drugs.

16.7 Infectious Disease of the Generative Organs (Venereal Disease)

The most common venereal diseases are gonorrhea, caused by infection with *Neisseria gonorrhoeae*, and syphilis, caused by *Treponema pallidum*. Both bacteria are obligatory parasites of human beings that depend on promiscuous sexual behavior for their very existence (see Section 13.2). However, the mode of parasitism of each bacterium and the disease each one causes are very different.

Gonorrhea

N. gonorrhoeae is a gram-negative coccus that usually does not invade the body of the host, but maintains its residence at the mucous membrane that lines the tubes of the genital organs. In women the sign of infection may be only a moderate discharge of pus from the genital tract. Occasionally the infection progresses to involve the fallopian tubes, and scar tissue forming in the tubes can lead to sterility by blocking the normal migration of eggs from the ovaries to the uterus. However, a woman can be a symptomless carrier of organisms for a long time, readily transmitting the pathogen to her consorts. Such a woman functions as a reservoir of infection, since she is not compelled by her symptoms to seek medical treatment.

Men who contract *N. gonorrhoeae* almost always suffer acute symptoms of a profuse discharge of pus from the opening of the penis and pain on urination. These symptoms begin several days after infection, and the infected man usually seeks medical care quickly. Therefore, men with acute gonorrhea, in contrast to female chronic carriers, are not very efficient agents for spreading the disease.

Because *N. gonorrhoeae* does not invade the host, the immune system is not activated to produce antibodies in the blood stream. Locally secreted IgA antibodies (see Section 15.3) are produced, but these antibodies are apparently unable to prevent reinfection. Thus, repeated infections with *N. gonorrhoeae* are common. The disease can be cured by prompt treatment with appropriate antibiotics.

Infants born to women harboring *N. gonorrhoeae* are susceptible to infection of the eyes as they pass through the birth canal. Gonococcal eye infections can stimulate violent inflammatory reactions and destroy critical structures, leading to blindness. The eyes of newborn infants are routinely treated to prevent the gonococcal eye infections that once were a significant cause of blindness.

In summary, *N. gonorrhoeae* is not invasive, stimulates little or no immunity, and persists largely in a reservoir of asymptomatic female carriers.

Syphilis

T. pallidum, the cause of syphilis, is a spiral-shaped bacterium that has defied the attempts of generations of microbiologists to culture it outside the body of the host. Accordingly, our knowledge of *T. pallidum* has suffered greatly from the inability to grow the organism in the laboratory.

Syphilis appeared in Europe in 1493 in the form of a terrifying epidemic that was called the "great pox," in distinction to another epidemic scourge, the "small pox." The fact that these two diseases were put in the same class attests to the virulence of syphilis during that epidemic. The coincidence of the arrival of epidemic syphilis with the return of Columbus' sailors from the New World has fostered the belief that syphilis originated in the Americas and was transported by Columbus' crew to Europe. However, there is no proof for this assertion, despite its poetic appeal. It is most likely that syphilis was present in Europe long before the Age of Discovery. Perhaps a native American variant of *T. pallidum* was especially virulent when introduced into the susceptible European population. Whatever the cause, the severity of the "great pox" was recorded vividly in the contemporary literature. In the more than 400 years that have passed since then, the syphilitic germ has lost much of its virulence. It is likely that the organism became attenuated while the population became more resistant (see Section 13.3).

Syphilis manifests three stages of disease. Primary syphilis develops 10 to 90 days after contact with infectious organisms. A local sore or lesion develops on the skin at the site of infection, usually the genital organs. In women, the lesion often is not visible. The lesion is painless and without pus, but it teems with living organisms and is highly contagious. After a variable amount of time, the primary sore vanishes, even without treatment. The germs, however, persist and spread throughout the entire body, often without arousing overt signs or feelings of disease.

Two to 12 weeks after the appearance of the primary sore, secondary syphilis emerges as a generalized skin rash. Despite the presence of a myriad of bacteria in the rash and in other tissues, the host may have little or no fever and few other signs of inflammation and illness.

The subsequent course of the disease is highly variable. The sores tend to clear up, and about 25 percent of the patients become cured of *T. pallidum*, presumably because of their immune response. Another 25 percent of patients continue to harbor living *T. pallidum* but never have symptoms of disease. About half of those infected, however, develop late complications of tertiary syphilis.

The late stage of syphilis may appear many years after the primary and secondary stages. Its manifestations range from minor sores on the skin to debilitating involvement of vital organs such as the liver, heart, or brain. In contrast to the primary and secondary stages of the disease, the lesions of tertiary syphilis are free of any infection with living *T. pallidum*. It would appear that in attacking the bacterium, the host's immune system also inflicts damage on his own tissues. Syphilis is associated with development of antibodies that recognize self-antigens, and it is not inconceivable that autoimmunity contributes to the late stage of disease (see Sections 15.7 and 15.9). This situation reminds us of the non-suppurative complications seen following infections with group A streptococci (Section 16.1).

It is possible that *T. pallidum* avoids destruction by the host's immune system in the early stages of disease by disguising itself with antigens similar to those of the host. Perhaps a variable degree of autoimmunity is the price a host must pay when he or she finally learns to recognize *T. pallidum* as a foreign invader. Fortunately, syphilis is easily cured by treatment with antibiotics, and tertiary manifestations of the disease are now rarely seen.

Our understanding of the relationship between *T. pallidum* and humans is far from complete. Nevertheless, we can appreciate the extent to which this microbe has adapted to both the physiology and the social behavior of its host (see Section 13.2).

16.8 Viruses and Cancer

Is cancer among the diseases of humans caused by microbial parasitism? This question has aroused a great deal of discussion and

controversy as well as a lot of diligent study and ingenious experimentation. The answer is important practically because it guides us to strategies of prevention and cure; it is important theoretically because it involves the rules of organization of multicellular life. The full answer to the question is not yet known. In this section we attempt to clarify some of the issues by describing a viral parasite of man implicated in several types of cancer. We shall see that environmental and ecological factors influence whether infection with this agent will pass without any symptoms of disease, will be expressed as an acute, transient infectious disease, or will be associated with development of cancer.

What Is Cancer?

To approach the question of viruses and cancer, we must first define what we mean by cancer. In biology, good definitions often are operational: We define an entity by what it *does* rather than by what it *is*. Operational behavior is observable and amenable to experimentation. In contrast, what an entity is—its essence—is often a philosophical problem.

Operationally speaking, a cancerous cell is one that grows in defiance of the regulations governing the social behavior of normal cells to a point that threatens the life of the individual. Normal cells are restrained in their proliferation according to the needs of the entire community of cells making up the body. In the test tube, normal cells have a limited life span; they are mortal. In the body, they confine themselves to their appointed sites; brain cells grow in the brain and not in the liver, liver cells grow in the liver and not in the brain, and so on. In contrast, cancer cells are autonomous and ignore the needs of the body as a whole; they proliferate unchecked and invade and colonize tissues that are normally off limits. Cancer cells in the test tube can proliferate without end.

Prohibitions against autonomous proliferation and invasiveness are encoded in a cell's DNA. Since a cancer cell transgresses the rules of social behavior, we can say that cancer can arise from an abnormality of the genetic instructions themselves, or from a defect in the way normal genetic instructions are carried out. In Section 15.6, we introduced the term **differentiation** to describe the specific genetic program expressed by a normal, specialized cell in a multicellular organism. Hence, in the most general sense, cancer can be classified as a disease of cellular differentiation.

Viruses Can Behave as Abnormal Genes

Parasitic bacteria exploit the metabolism of the host while leaving the host's metabolic machinery under the control of the genes. Bacterial parasites merely use the host as their environment. Viruses, in contrast, parasitize the genes of the host. Viruses are packages of genetic information, DNA or RNA, and have no metabolic machinery of their own. They seize control of the metabolic machinery of the host's cells by imposing their own genetic instructions on the cell. In some cases, the viral genes become integrated into the DNA of the host cell, so that the virus can direct the behavior of the cell as if it were a legitimate member of the cell's own governing body of genes. Thus, viral infection may be seen as an infection by abnormal genes. It is easy to imagine that cells under control of such abnormal genes may behave abnormally. Since cells infected with viruses may not carry out a normal genetic program, they can be said to be in a state of abnormal differentiation.

Because cancer also involves abnormal differentiation of cells, many biologists have entertained the notion that some forms of cancer could be related to abnormal genes inserted into the cell in the form of viruses. Is there any evidence for a viral etiology of cancer in human beings?

Burkitt's Lymphoma and Epstein-Barr Virus

In 1962, a British missionary surgeon in Uganda named Denis Burkitt suggested that a certain type of cancer found in children in Africa was caused by a virus transmitted by an insect vector (for discussion of vectors see Section 13.2). This cancer happened to be a lymphoma (a tumor of lymphocytes) most frequent in 6- to 10-year-old children. Burkitt's lymphoma characteristically appears first in the jaws (about 50 percent of cases) but can also start in various organs. Untreated, the lymphoma kills the patient.

Burkitt noted that a high frequency of lymphoma occurred in parts of Africa where there was a specific relationship between annual mean temperature and rainfall. Since a particular form of malaria was transmitted by insects in these areas, Burkitt suggested that insects might also serve as vectors for a virus causing lymphoma.

Burkitt's suggestion stimulated many scientists to search for viruses in the lymphoma cells. Epstein and Barr in 1964 succeeded in detecting a previously unidentified virus in cultured cells from a Burkitt's lymphoma tumor. The Epstein-Barr virus, commonly denoted as EB virus, was found to be a member of the family of herpesviruses and related to the benign virus that causes cold sores (Section 16.2). The association between EB virus and Burkitt's lymphoma was strengthened by the observation that these cancer cells had antigens associated with EB virus and that the lymphoma patients all had antibodies to these EB virus antigens. Moreover, it was shown that genes of EB virus were integrated into the DNA of the lymphoma cells. Finally, there was evidence that EB virus could infect human B lymphocytes in the test tube. Lymphocytes, similar to other normal types of cells, have a limited life span and a limited capacity to proliferate. After a variable number of cell divisions, normal cells stop dividing and die. However, after infection with EB virus, human B lymphocytes became "immortalized" and could divide indefinitely in the test tube, an attribute of many cancer cells. In short, there seemed to be two good reasons for concluding that infection with EB virus

could lead to cancer. Viral genetic material was present in the genes of cancer cells, and infection of normal cells with the virus changed their state of differentiation and made them immortal.

Infectious Mononucleosis and EB Virus

The cogency of the preceding argument was seriously weakened when it was discovered that EB virus was the etiologic agent of a noncancerous acute infection, infectious mononucleosis. The first hint of this fact came when it was reported in 1968 that a laboratory technician developed antibodies to EB virus after she came down with infectious mononucleosis. Her lymphocytes were found to be infected with EB virus during the two months of illness, but not before the illness or after recovery.

Infectious mononucleosis is a disease most frequent among young people of relatively high socioeconomic status, usually college students. It seems to be transmitted in the saliva during kissing. The disease is characterized by fever, sore throat, enlarged lymph nodes, and profound weakness. During illness, the patient's blood shows the presence of strange, atypical lymphocytes. Almost all patients recover completely without complication, and there is no association with cancer of any kind. EB virus can be isolated regularly from patients with infectious mononucleosis, and antibodies to EB virus protect against contracting the disease. Hence, EB virus is undoubtedly the etiologic agent of infectious mononucleosis.

In infectious mononucleosis, too, EB virus infects B lymphocytes and renders them abnormally differentiated immortal cells. However, immortalization or the capacity for unlimited proliferation by itself does not endow infected lymphocytes with the other properties of cancer cells, namely, invasiveness or autonomous, uncontrolled growth within the host. The EB virus–infected, immortalized B lymphocytes developing in infectious mononucleosis are apparently destroyed by the host's own T lymphocytes. These killer T lymphocytes are among the atypical lymphocytes seen in the blood of patients.

Infection with EB virus can also take place without signs or symptoms of infectious mononucleosis or any other disease. Surveys of populations have been made using the presence of specific anti-EB antibodies as evidence of previous infection with EB virus. Newborn infants all possess antibodies to EB virus that they have received from their mothers. These passively acquired maternal antibodies disappear during the first year of life. A major increase in the incidence of antibodies occurs at ages four and five, and by puberty, about 80 percent of children in lower socioeconomic groups have antibodies to EB virus. Individuals raised under the more isolated, sanitary conditions of higher socioeconomic groups miss the opportunity to become infected with EB virus during childhood. These people are the susceptibles in whom infectious mononucleosis develops later, usually in college.

Hence, we may conclude that infection with EB virus in childhood is usually without symptoms of disease and that almost all normal, healthy people in western society have been infected with this apparently harmless virus. Individuals who escape infection during childhood are at risk of developing infectious mononucleosis when infected in early adulthood. This disease, although inconvenient, is usually benign and transient. There is no association between EB virus and cancer in these circumstances.

How can the generally harmless nature of EB virus be reconciled with the observed association of EB virus with Burkitt's lymphoma in Africa? It is possible that EB virus is merely a passenger in the lymphoma cells and has nothing to do with the etiology of the cancer. However, if that were the case, it is difficult to understand why only Burkitt's lymphoma and not all lymphocyte tumors express EB virus. We are led to conclude that the association between EB virus and Burkitt's lymphoma is not due to chance alone. But how can we charge EB virus with responsibility for Burkitt's lymphoma in equatorial Africa, when it is present the world over and is almost universally harmless?

It has been postulated by Berenblum in Israel that the induction of cancer involves several stages and that one factor may serve to initiate a process whose termination in cancer requires additional factors. Burkitt, in 1969, proposed that chronic infection with malaria is the additional factor that converts an otherwise benign infection with EB virus into a cancer.

Recently, it has been found that the cancer cells in Burkitt's lymphomas all have a characteristic abnormality in the appearance of particular chromosomes. George Klein, a cancer biologist working in Stockholm, proposed in 1979 that the cancerous development of Burkitt's lymphoma (and possibly of other forms of cancer) occurs in two stages. Infection with EB virus initiates a process of abnormal differentiation, creating immortal lymphocytes. Additional factors cause a second change in the genes of these cells expressed as a characteristic modification of the cell's chromosomes. This second change completes the transformation of the normal cell into a cancer cell with the added properties of autonomous growth and invasiveness.

Despite the circumstantial evidence cited above, we cannot yet incriminate EB virus as an etiologic agent of cancer. Many more facts must be uncovered by observation and experimentation before we can begin to understand the relationship between EB virus as a viral parasite and as a possible initiator of a cancerous change in cells of the host. The connection between other viruses and human cancers is much more tenuous. In fact, it is not at all certain that the EB story has any general applicability; it is possible that it may be an exceptional case of a unique relationship between viral infection and human cancer. Nevertheless, the EB story is important because it illustrates some of the many possible faces of the interaction between man and his microbial parasites; it forcefully demonstrates the importance of etiologic circumstances in determining the outcome of this interaction; and it provides some insight into the complexities involved in understanding cancer.

Many questions remain to be answered: What information does the EB virus insert into the cell's DNA, and how does this information change the cell's state of differentiation? What additional factors combine with viral infection to accomplish the transformation of a normal cell into a cancerous one? Is the cancerous change reversible so that cancer cells may be returned to a state of normal differentiation? Could we prevent cancers by preventing viral infections? Questions such as these will keep biologists busy for many years to come.

Despite our ignorance of many of the critical aspects of the relationship between infection with EB virus and cancer, the knowledge that has accumulated hints at a fundamental role for microbial parasites, particularly viruses, in the most intimate aspects of the genetic programs expressed by cells of the host.

QUESTIONS

16.1 What is EB virus and what does it cause?

16.2 What is the evidence that EB virus is the etiologic agent of Burkitt's lymphoma? What is the evidence against this notion?

16.3 How might B lymphocytes become "immortalized" by being infected with EB virus?

16.4 Name the reservoirs of the agents of the following diseases: the common cold; measles; chickenpox; acne; fever blisters; meningitis; tuberculosis; gonorrhea.

16.5 Why is the accumulation of pus most dangerous in the brain?

16.6 Why do we suffer only once from measles, whereas streptococcal sore throat may recur many times?

16.7 What are the suppurative and the non-suppurative complications of infection with group A streptococci?

16.8 What is the difference between *horizontal* and *vertical* transmission of an infectious disease? Provide examples.

Suggested Readings

Burnet FM: Natural History of Infectious Disease. Cambridge: Cambridge University Press, 1966.
A philosophical approach toward understanding microbial disease.

Kaplan MM and RG Webster: The Epidemiology of Influenza. Scientific American, 237: 88, December 1977.

Klein G: Lymphoma Development in Mice and Humans: Diversity of Initiation Is Followed by Convergent Cytogenetic Evolution. Proceedings of the National Academy of Science (USA), 76: 2442–2446, 1979.
A theory describing how injection with a virus can lead to development of cancer, provided that additional events take place. Written in technical language.

Rafferty KA Jr: Herpesvirus and Cancer. Scientific American, 229: 26, October 1973.

The First and Second Laws of Thermodynamics

In 1847, the German physicist Hermann van Helmholtz formulated the law of **the conservation of energy.** To date there is no known exception to this law.* It states that the quantity of energy remains constant during the numerous changes that occur in nature. Energy can exist in a number of different and interchangeable forms, such as electrical, chemical, gravitational, light, mechanical, and thermal energy. The important point is this: For any quantity of one form of energy that is lost, an equivalent amount of other forms must be generated. The law of conservation of energy is also called the **first law of thermodynamics.**

First law: The energy of the universe is always constant. Considering only the first law of thermodynamics, it should be possible to build a perpetual-motion machine. For example, we can imagine an isolated machine of the following type: Electrical energy is converted into mechanical energy (for example, a garbage disposal unit); the mechanical energy thus produced is utilized to generate electricity (a dynamo). Such an imaginary machine could then operate indefinitely, with no additional source of energy. In reality, however, the machine would cease to function after a few cycles. This would seem to contradict the first law, since there is an apparent loss of energy in the system. The explanation, of course, is that there is less than 100 percent efficiency in the interconversion of electrical and mechanical energy. Other forms of energy such as heat are also produced. However, even if we could salvage the thermal energy and other forms of usable energy and "feed" them back into the machine, it would still cease to function after a short time. It is for this reason that physical chemists defined a new state of energy called **entropy,**† which is characterized by its inability to perform work. Using modern statistical

*With the discovery by Albert Einstein in 1905 that mass and energy are interconvertible ($E = mc^2$), the law was generalized to include the conservation of both mass and energy. However, for the biological systems that we discuss, the interconversion of mass and energy can generally be disregarded.

†To be more precise, it is entropy multiplied by temperature that has the dimensions of energy.

physics, it is possible to demonstrate that entropy is analogous to randomness or lack of order; the greater the randomness, the greater the entropy.

The **second law of thermodynamics** cannot be expressed in such concise form as the first law; it is stated in various ways according to the type of problem under investigation. To emphasize the irreversibility of real processes, the second law is stated as follows: In any real physical, chemical, or biological process, there is a net increase in entropy.

The formulation of the second law arose as a consequence of a *reductio ad absurdum*:

1. The sum of all energies in the universe is constant (first law of thermodynamics).
2. Entropy is the unique state of energy that cannot be utilized to perform work (by definition).
3. If entropy either remains constant or decreases in any series of real processes, there would be no loss in the capacity to perform work. This would lead directly to a perpetual-motion machine.
4. Perpetual-motion machines are contrary to experience.
5. Therefore, **the entropy of the universe always increases** (second law of thermodynamics).

Since it is experimentally difficult to determine entropy directly, chemists and biologists frequently use the concept of **free energy** to predict whether or not a reaction is energetically feasible. The free energy is that part of the total energy that can do work at a fixed temperature. Since the energy unavailable for work (entropy) increases in any real reaction, the energy available for work (free energy) must decrease. It may be said, then, that the **direction** of a chemical reaction is determined by the sign of the free energy change; the reaction can proceed spontaneously only in the direction that results in a decrease of free energy.

The second law has certain philosophical, as well as scientific, implications. The statement that the entropy, or the randomness, of the universe is always increasing points directly to a relationship between time and energy. The further we go back in history, the greater the amount of energy that was available to do work. As the philosopher Sir Arthur Eddington succinctly expressed it, the second law of thermodynamics is "time's arrow." It follows, then, that the ultimate fate of the universe is to reach a state of complete randomness or maximum entropy, which has been called **entropic doom.**

Does the second law of thermodynamics apply in biology? A fundamental characteristic of all living organisms is the ability to reproduce their highly organized nature. This would seem, at first glance, to contradict the second law; i.e., it would lead to an increase in orderliness and a decrease in entropy. It must be emphasized, however, that the second law states explicitly that the entropy *of the universe* always increases. Thus it is necessary to think not only about the living organism, but also about its surroundings. Consider, for example, a microorganism growing in an aqueous medium containing sugar and inorganic salts. The growth of the microbe is dependent on the breakdown of sugar molecules into simpler (less ordered) substances. Careful analyses have demonstrated that the total decrease in entropy of living matter during the growth process is more than offset by the increase in entropy of the nutrient molecules. Although a certain part of the system has become more highly organized, the net result is an increase in entropy. Therefore, the second law of thermodynamics has not been violated.

GLOSSARY

Exactness cannot be established in the arguments unless it is first introduced into the definitions.

Henri Poincare

Abiogenesis. See *Spontaneous generation.*

Actinomycetes. A large group of filamentous bacteria that are common inhabitants of the soil. Most of the clinically important antibiotics are produced by strains of Actinomycetes.

Activated sludge. The mixed microbial population used in sewage treatment processes.

Active site. That part of a given enzyme molecule into which the substrate fits.

Adenine. A purine found in both RNA and DNA (see Figure 1.13).

Adenosine triphosphate (ATP). A molecule consisting of adenine, ribose and three phosphate groups. (For chemical formula, see Figure 3.4.) It is the key energy-rich compound in the cell.

Aerobic cells. Those that utilize oxygen.

Agar. A relatively inert polysaccharide derived from seaweed; it is used widely in microbiological laboratories as a solidifying agent.

Algae. Eucaryotic protists that obtain their energy by photosynthesis and possess one or more chloroplasts.

Alkaptonuria. A hereditary disease in which a person lacks the enzyme for degrading homogentisic acid, thus excreting it in the urine.

Allosteric protein. Proteins whose three-dimensional structure and biological properties are altered by the binding of specific small molecules at sites other than the active site.

Ames test. A relatively simple procedure for examining whether a substance is potentially carcinogenic.

Amino acid. The building block from which proteins are constructed. (For general chemical formula, see Figure 1.11.)

Ammonification. The reactions leading to the production of ammonia (NH_3) from nitrogen-containing organic compounds.

Anaerobic cells. Those that can live without oxygen.

Analogs. Chemicals that are structurally similar, but not identical, to naturally occurring substances.

Angstrom (Å). A unit of length equal to 10^{-8} cm.

Antibiotics. Chemical substances produced by certain microorganisms that, at low concentrations, can inhibit growth of or kill other microbes.

Antibody. Proteins produced by lymphocytes in response to foreign substances. An individual can make millions of different types of antibodies, each with a unique combining site that recognizes and binds to a particular foreign substance.

Antigen. The structure on a foreign substance that can be recognized by an antibody molecule. Antigens have no characteristic chemical features other than their shape, which fits the combining site of an antibody.

Antiseptic. A material that inhibits or kills microbes but is generally not harmful to human tissues. *Antiseptic condition* implies cleanliness and reduced microbial populations but does not signify sterility.

Atoms. The smallest units in which elements can exist and still have their characteristic properties.

Attenuated. Modified by genetic mutation or by

chemical treatment so as to be incapable of causing disease under ordinary circumstances.

Autoradiography. A method used for locating specific chemicals within the cell. A photographic emulsion is placed in contact with cells or sections of cells that have been made radioactive with specific components. The silver grains on the emulsion that develop point out where that chemical is concentrated.

Auxotrophs. Organisms that use carbon dioxide as their primary source of carbon for growth.

Azotobacter. A gram-negative bacterium that is unique in its ability to use energy obtained from oxidation of organic compounds to carry out nitrogen fixation.

Bacillus. (1) Any cylindrical-shaped bacterium. (2) A specific group of gram-positive, endospore-forming bacteria.

Bacteriophages. Viruses that infect and multiply in bacteria.

Base-pairing. The stereochemical binding of adenine with thymine (or uracil) and guanine with cytosine.

Biochemistry. The cross-discipline in which chemical principles are used to study biological problems.

Biosynthesis. The process by which cells build large and complex molecules from small and relatively simple components.

Blue-green algae. See *Blue-green bacteria.*

Blue-green bacteria. Procaryotic microbes that carry out photosynthesis in a manner similar to higher plants. Blue-green bacteria, formerly referred to as blue-green algae, are sometimes called cyanobacteria.

BT (Bacillus thuringiensis). Gram-positive bacteria that infect and kill larvae of a wide variety of lepidoptera. BT is produced in large amounts for use as an insecticide.

Calorie. A measure of energy; the calorie is defined as the amount of energy required to raise one milliliter of water from 14.5°C to 15.5°C.

Cancer. A group of diseases characterized by uncontrolled cellular growth.

Capsule. A layer of variable thickness surrounding a bacterial cell wall.

Carbohydrate. Organic molecules having the general formula $C_n(H_2O)_n$, including sugars and polysaccharides (see Figure 1.14).

Carcinogen. Any substance that induces cancer.

Casein. The most abundant protein in milk.

Catalyst. An agent that increases the rate of a reaction without being consumed.

Cell. The basic organizational unit of life.

Cell biology. An aspect of biology in which the emphasis is placed on the cell, rather than the total organism.

Cell wall. A rigid structure found outside the cytoplasmic membrane; it is characteristic of procaryotic cells and plants.

Cell-free extract. A liquid that contains most of the ingredients of cells, prepared by breaking open cells and removing any remaining intact cells.

Cellulase. An enzyme capable of splitting cellulose into glucose residues.

Cellulose. A polysaccharide consisting of glucose units joined together in such a manner that only few organisms have the enzymatic ability to split the polymer. Cellulose is the most abundant organic molecule in the world.

Cheese. The product made from the separated curd obtained by coagulating the casein of milk. The coagulation is accomplished by means of rennet or other enzymes, lactic acid formation, or by a combination of the two. The curd may be further modified by heat, pressure, ripening ferments, special molds, or suitable seasoning.

Chemical evolution. The theory, independently proposed by Oparin and Haldane, which states that life arose gradually from non-living material. The theory emphasizes the need for a long series of chemical changes as a prerequisite to the formation of life.

Chemoautotroph. An organism that gets its energy from the oxidation of inorganic chemicals and its carbon from carbon dioxide.

Chemosynthetic bacteria. The group of bacteria that obtain energy by the oxidation of inorganic molecules and then utilize the energy to fix carbon dioxide into cellular material.

Chemotaxis. Directed movement of cells in response to a chemical gradient. Thus, positive chemotaxis is the movement of cells toward a higher concentration of a specific chemical signal.

Chemotherapy. The use of chemicals to combat infectious diseases. The chemical can be either an antibiotic or a chemically synthesized substance.

Chitin. A nitrogen-containing polysaccharide found in the cell walls of fungi and exoskeleton of arthropods and other invertebrates.

Chlorophyll. A green pigment found in chloroplasts that is utilized to convert light energy into chemical energy.

Chloroplast. Chlorophyll-containing cytoplasmic structures found in green plant cells; they are the sites of photosynthesis.

Chromosome. A structure containing nucleic acid and protein that is present in the nucleus of all eucaryotic cells during cell division.

Cilia. Short hair-like appendages present on the surface of some cells, usually in large numbers and arranged in rows. Cilia are responsible for motility in certain protozoa.

Citric acid cycle. See *Krebs cycle.*

Clone. A population of cells all derived from a common ancestor.

Cloning. (1) In classical microbiology, the technique of obtaining a pure culture by obtaining a colony of cells all derived from a single ancestor. (2) In the context of higher animals and plants, the technique of generating a whole organism from a single somatic cell; for example, the production of hundreds of genetically identical tomato plants from one parent. (3) In the terminology of modern genetic engineering, the cloning of foreign pieces of DNA (genes) in microbes.

Clostridia. Gram-positive, spore-forming anaerobic rod-shaped bacteria. One species, *Clostridium tetani,* is the causative agent of tetanus.

Coacervate. A type of colloid that Oparin postulated played an important role in the origin of life. A coacervate is formed when a group of large molecules, such as protein, associate to form microscopic droplets in a liquid medium.

Coat proteins. The outer proteins of a virus.

Codon. A sequence of three adjacent nucleotides that code for an amino acid or chain termination.

Coliforms. Gram-negative rods, including *Escherichia coli,* which normally inhabit the colon.

Colony. A population of cells growing as a compact mass on a solid surface, usually descended from a single ancestor.

Communicability. The ability of a microbial parasite to be transmitted from one host to another.

Complement. A system of many different proteins that circulate in the blood and work together as an integrated defense system against infection.

Compounds. Combinations of two or more different atoms held together by chemical bonds.

Conjugation. A physical association and exchange of genetic material.

Constitutive enzymes. Those produced by the cell at all times, irrespective of growth medium.

Cosmozoa. The theory that life on earth arose from "seeds" (spores) that constantly bombard our planet.

Crossing over. In genetics, the process of exchange of genetic material between homologous chromosomes.

Cyclic AMP. A small molecule produced by a wide variety of cells that regulates metabolic activity.

Cytology. The study of cell structure.

Cytoplasm. That part of the cell inside of the cytoplasmic membrane, but excluding the nucleus.

Cytosine. A pyrimidine found in both RNA and DNA (see Figure 1.13).

Dark reactions. Those parts of the photosynthetic process that can proceed in the absence of light; the formation of glucose from carbon dioxide and water.

DDT. An insecticide that accumulates in fatty tissues and is toxic to humans and other animals when swallowed or absorbed through the skin.

Death phase. That part of the growth cycle in which the number of viable cells decreases sharply.

Degenerate codons. Two or more codons that specify the same amino acid.

Deletion. In genetics, loss of a section of genetic material from the chromosome.

Denitrification. The anaerobic conversion of nitrate (NO_3^-) to nitrogen gas (N_2) brought about by a number of different microbes.

Deoxyribonuclease (DNase). Enzyme that catalyzes degradation of DNA.

Deoxyribonucleic acid (DNA). The chemical substance that is the genetic material of all cells. It is a polymer of deoxyribonucleotides. (For chemical formula, see Figure 5.4.)

Deoxyribonucleoside. One of the nitrogen bases joined to deoxyribose (deoxyribose + base).

Dextran. A polysaccharide consisting of glucose that is produced when a bacterium, leuconostoc, grows on sucrose.

Diatoms. Algae that contain hard, silica-containing cell walls.

Diauxic growth. The phenomenon in which there are two distinct exponential growth phases separated by a short lag period.

Differentiation. The developmental process by which a cell in a multicellular organism becomes a specialist; involves implementation of a characteristic genetic program.

Diploid. A cell that contains two copies of each type of chromosome.

DNA ligase. The enzyme that links DNA fragments.

DNA polymerase. Enzyme that catalyzes the formation of DNA from deoxyribonucleoside triphosphates utilizing existing DNA as a template.

Dominant. In genetics, referring to a gene that is always expressed phenotypically.

Ecological niche. The sum total of biological and physical factors within which an organism lives.

Ecology. The study of the relationships of organisms to one another and to their environment.

Electron microscope. An instrument that uses beams of electrons to visualize material.

Embryo. An early developmental stage of an organism produced from a fertilized egg.

Endoplasmic reticulum. An extensive system of membranes found in the cytoplasm of eucaryotic cells, often coated with ribosomes.

Endotoxins. Cell-surface components of bacteria that are toxic to animals; the outer membrane (lipopolysaccharide) of gram-negative bacteria is the major endotoxin.

Enterics. A medically important group of gram-negative bacteria, many of which can grow inside the intestine of human beings and other warm-blooded animals. The best studied example, *E. coli,* is part of the normal flora of a healthy person.

Entropy. A measure of randomness (see Appendix A).

Enzymes. Proteins capable of accelerating chemical reactions.

Epidemiology. The science dealing with all the factors involved in the spread of infectious diseases through populations.

Episome. Plasmids that alternate between a state in which they are integrated into the chromosome and a state in which they multiply autonomously.

Escherichia coli (E. coli). A generally non-pathogenic bacterium found in the intestines of humans; it is easy to grow and manipulate in the laboratory and thus has been used widely in the study of molecular biology.

Etiology of disease. The study of the causes of disease; this involves both the etiologic agent (the microbe) and the etiologic circumstances.

Eucaryotic cell. A category that includes all cell types except bacteria and blue-green bacteria, characterized by a well-defined nucleus separated from the cytoplasm by a nuclear membrane and a structurally differentiated cytoplasm.

Eutrophication. The enrichment of a body of water with nutrients, often leading to overproduction of algae.

Exotoxins. Poisonous proteins produced by microbes and released into the external environment.

Exponential phase. That part of the growth curve during which the rate of increase is maximum and constant.

F^+ cell. A bacterium that possesses a small circular piece of DNA called the F(ertility) factor; the F factor can be transferred to a recipient F^- cell.

F^- cell. A bacterium that does not contain an F(ertility) factor, but can receive an F factor from an F^+ cell.

F(ertility) factor. A small extrachromosomal piece of DNA containing genes for the production of extracellular appendages, called sex pili. The F factor mediates conjugation in bacteria.

Feedback inhibition. The inhibition of the activity of the first enzyme of a biosynthetic pathway by the end product of that pathway.

Fermentation. (1) The energy-yielding enzymatic breakdown of nutrients in the absence of oxygen. (2) In industrial microbiology, any microbially catalyzed conversion of food into a commercially valuable chemical.

First law of thermodynamics. The energy of the universe is always constant (see Appendix A).

Flagella. Cell appendages that are responsible for motility in many microorganisms.

Fossil. Any remains of an organism or evidence of its presence.

Free energy. That part of the total energy that can perform work at a fixed temperature (see Appendix A).

Fungi. A heterogenous group of non-photosynthetic eucaryotic protists that contain rigid cell walls and are generally immotile.

β-Galactosidase. An enzyme that catalyzes the splitting of lactose into glucose and galactose; it is the best-studied example of an inducible enzyme.

Galactoside permease. An enzyme located on the cell surface that regulates the entrance of lactose and related sugars into the cell.

Gasohol. A mixture of gasoline and alcohol that can be used in automobile engines.

Gelatin. A protein obtained from animal skin, bones, tendons, and so on. It is soluble in boiling water; on cooling, it solidifies to form a transparent gel.

Gene. Part of the hereditary material located on the chromosome. In modern terms, a gene can be considered a segment of DNA carrying the information for a single protein.

Genetic code. The linear sequence of bases in DNA that determines the sequence of amino acids in a protein.

Genetic engineering. An artificial procedure for the modification of the genetic properties of an organism by deliberately introducing pieces of DNA (genes) from one organism into another.

Genotype. The genetic makeup of an organism.

Germination. Resumption of growth by spores or other resting cells.

Glucose. A simple sugar having the formula $C_6H_{12}O_6$; also called dextrose and grape sugar.

Glycogen. A large polysaccharide composed of glucose units; it is the major storage product in animal cells.

Glycolysis. One type of fermentation in which glucose is broken down to lactic acid.

Gram-negative. Bacterium that is stained red by a certain dye mixture (Gram stain). Gram-negative bacteria have a thin multi-layered cell wall.

Gram-positive. Bacterium that is stained purple by a certain dye mixture (Gram stain). Gram-positive bacteria have a single, thick cell wall.

Gratuitous inducer. A substance that provokes the synthesis of specific enzyme(s), but which itself is not a substrate.

Growth. The orderly increase of all cellular constituents, leading to an accurate duplication of the existing pattern.

Growth curve. A plot of the number of cells as a function of time; the general shape of the curve is characteristic of all living systems.

Guanine. A purine found in both RNA and DNA (see Figure 1.13).

Haploid. A cell that has only one copy of each type of chromosome.

Hemoglobin. The iron-containing protein in the blood that transports oxygen.

Heterotroph. An organism that requires organic material as its major source of energy for growth.

Heterozygous. In diploid cells, the situation in which the two members of a pair of genes located on homologous chromosomes are different.

Hfr males. Bacteria that have the F(ertility) factor integrated into the bacterial chromosome.

Histamine. A small molecule that when released by cells into the body fluids causes inflammation.

Homozygous. In diploid cells, the situation in which the two members of a pair of genes located on homologous chromosomes are identical.

Host cell. A cell that is used for the growth and multiplication of a virus.

Hypha. A single tubular filament of a fungus; the hyphae together comprise the mycelium.

Immunization. Inoculation of an individual with a pathogen or its products with the aim of endowing the person with resistance to the specific microbial pathogen.

Immunoglobulins. Antibody molecules.

Immunology. The branch of science that studies the immune system. The immune system is capable of recognizing foreign substances, such as microbes, and responds by neutralizing the invaders.

Inducer. A substance that induces the production of a specific enzyme or group of enzymes by the cell.

Inducible enzymes. Those produced by the cell only in response to specific chemicals in the medium.

Inorganic. Non-living; in chemistry, molecules that lack carbon (except CO_2 and CO).

Interferon. An antiviral substance produced by cells of the body in response to infection with viruses that stimulates the defenses of the body and augments resistance to infection. It does not attack viruses directly.

Inversion. The process of breakage and reunion such that whole segment of the chromosome is replaced in reverse order.

In vitro. (Latin: "in glass"). Pertaining to experiments performed with cell-free extracts.

In vivo. (Latin: "in life"). Pertaining to experiments performed with intact living organisms.

Koch's postulates. A series of experimental conditions that must be fulfilled in order to prove that a particular microbe is the etiologic agent of a specific disease.

Krebs cycle. A cyclic series of cellular reactions in the breakdown of pyruvic acid to carbon dioxide and the formation of hydrogen for oxidative phosphorylation.

Lactobacilli. Gram-positive bacteria that can convert sugars into lactic acid in the absence of air.

Lag phase. That part of the growth curve which comes prior to the exponential phase.

Legume. A group of plants (including clover, peas, alfalfa, beans, peanuts and soybeans) that develop symbiotic relationships with nitrogen-fixing bacteria.

Lethal dose (LD_{50}). The amount of a substance which kills 50% of the test organisms.

Ligand. In general, that part of one molecule that binds to another molecule (receptor) by virtue of a three-dimensional lock-and-key fit (see Figure 12.6). Antigens are one type of ligand. Some bacteria and viruses have ligands on their surface envelopes that can bind and anchor these microbes to receptors on specific cells of the host.

Light reaction. The part of the photosynthetic process that requires light; i.e., the first stage in which the radiant energy is captured and transformed into chemical energy.

Linkage. In genetics, the location of two or more genes on the same chromosome so that they are passed on together from parent to offspring.

Lymphocytes. Cells capable of recognizing antigens and mediating the immune response.

Lysis. The bursting of a cell by destruction of its cytoplasmic membrane.

Lysogenic bacterium. One that contains a prophage.

Lysosomes. Vesicles found in the cytoplasm of certain eucaryotic cells, such as phagocytes, which contain potent degrading enzymes.

Lytic cycle. The sequence of events that leads to the multiplication of viruses and the lysis of the host cell.

Macromolecule. A large molecule, usually built up from small units.

Macrophage. A type of phagocyte that ingests and destroys microbial invaders.

Meiosis. The process of cell division without chromosome duplication so that the daughter cells have half the number of chromosomes.

Mesophiles. Organisms that grow best at intermediate temperatures (between 20° and 45°C).

Messenger RNA (mRNA). RNA manufactured in the nucleus, which moves into the cytoplasm, attaches to ribosomes, and serves as a template for protein synthesis.

Metabolism. The series of biochemical reactions by which the cell is able to acquire energy from nutrients and then utilize the energy for the synthesis of cell material.

Micron (μ). A unit of length equal to 10^{-6} meters, 10^{-4} cm, or 10^{4} Å.

Minimal medium. A medium containing only those components essential for growth, and which the organism cannot synthesize itself.

Mitochondrion. A structure found in the cytoplasm of aerobic eucaryotic cells. It is the major site of ATP production.

Mitosis. The process of cell division leading to two genetically identical daughter cells.

Molds. Refers to those fungi that contain a mycelium.

Molecular biology. An aspect of biology in which the emphasis is placed on the relationship between the structures of molecules and their biological functions.

Molecules. Combinations of two or more atoms held together by chemical bonds.

Mucopolysaccharide. Jelly-like, slippery substance that provides intercellular lubrication and acts as flexible cement between cells.

Mutagen. Mutagenic agent; any substance that increases the frequency of mutation.

Mutagenic agent. Any substance that increases the rate of mutation.

Mutation. An inherited alteration of the genetic material.

Mycelium. A tangled, mat-like aggregate of fungal hyphae. The cottony growths of molds on decaying fruit and bread are commonplace examples of mycelia.

Mycoplasmas. A medically important group of bacteria that lack cell walls.

Neurospora crassa. The common bread mold; it is especially useful for biochemical and genetic studies.

Neutrophils. The predominant type of phagocyte circulating in the blood.

nif genes. A cluster of genes that are responsible for the formation of nitrogenase and other proteins necessary for nitrogen fixation.

Nitrification. The microbial conversion of ammonia (NH_3) to nitrate (NO_3^-).

Nitrogen bases. Molecules composed of rings of carbon and nitrogen atoms. Important nitrogen bases in the cell are the purines and pyrimidines. (For chemical formula, see Figure 1.13.)

Nitrogen fixation. The biological conversion of atmospheric nitrogen gas (N_2) into ammonia (NH_3).

Nitrogenase. An enzyme that catalyzes the conversion of nitrogen gas (N_2) to ammonia (NH_3).

Nosocomial infections. Diseases caused by normal (usually non-pathogenic) bacteria or viruses as a result of the medical or surgical treatment that a patient receives to remedy some other illness.

Nucleic acid. A polymer of nucleotides. (See *Deoxyribonucleic acid* and *Ribonucleic acid.*)

Nucleoside. One of the nitrogen bases joined to a sugar (base + sugar).

Nucleotide. One of the nitrogen bases joined to

a sugar that is also connected to a phosphate group (base + sugar + phosphate).

Nucleus. That part of the cell that contains the genetic material. In eucaryotic cells, the nucleus is bounded by the nuclear membrane.

Operator. A site on the chromosome that regulates the expression of a group of genes; when the repressor combines with the operator region, transcription of the structural genes is inhibited.

Operon. A segment of the chromosome composed of a group of genes whose expression is regulated by a common operator.

Organic. Living; in chemistry, molecules that contain carbon (except CO_2 and CO) are called organic compounds because they are characteristic of living organisms.

Osmotic work. The energy-requiring process by which cells are able to transport and concentrate molecules into the cell from the environment.

Oxidation. A chemical reaction involving the loss of electrons. In some cases, the electrons are accepted by oxygen gas.

Oxidative phosphorylation. The process by which oxygen utilization is coupled to ATP production from ADP.

^{32}P. A radioactive isotope of phosphorus with an atomic weight of 32; it is commonly used for labeling nucleic acids in biological experiments.

Paper chromatography. A technique used for separating small quantities of organic molecules.

Parasitism. Traditionally defined as the relationship between two organisms in which one, the parasite, always benefits from the other, the host, without providing any compensation. Upon closer analysis, it becomes apparent that the host can also benefit from such parasites.

Pasteurization. A heat treatment that kills many of the microbes that are responsible for spoilage or for causing diseases without destroying the taste of the food.

Pathogenic microbes. Microorganisms capable of causing disease.

Permease. A general name for any enzyme which regulates the entrance of molecules into the cell.

Phages. See *Bacteriophages*.

Phagocytes. Types of cells (white blood cells) found in the blood that are able to detect, engulf, and digest foreign particles such as bacteria. The process of engulfment is called phagocytosis.

Phenotype. A recognizable feature of an organism (e.g., eye color, length of ears, and so on.)

Photosynthesis. The enzyme-catalyzed conversion of light energy into useful chemical energy and use of the latter to form sugars and oxygen from carbon dioxide and water.

Photosynthetic phosphorylation. The process by which light energy is utilized to produce ATP from ADP.

Pili. Filamentous extracellular appendages on bacteria; different pili play roles in sexual conjugation, adherence to surfaces, and a variety of cell-to-cell interactions.

Plaque. The clear area that results when viruses cause the lysis of cells in a localized area.

Plasmids. Extrachromosomal genetic elements found in many microbes. Plasmids consist of circular, double-stranded DNA molecules that can replicate independently of the bacterial chromosome, and often contain genes that are not absolutely essential for the growth of the microbe.

Polymer. A large molecule made up of regular subunits (monomers).

Polysaccharide. Large carbohydrate molecules composed of many sugars joined together: for example, starch and glycogen (see Figure 1.14).

Procaryotic cell. A category that includes bacteria and blue-green bacteria, characterized by the absence of both a nuclear membrane and a structurally differentiated cytoplasm.

Promoter. Region of a DNA molecule at which RNA polymerase binds and initiates transcription.

Prophage. The state of a bacterial virus in which it is integrated into the host chromosome.

Protein. A polymer of amino acids.

Protista. A third kingdom of living organisms, distinct from higher plants and animals. Members of the kingdom Protista include bacteria, algae, protozoa, and fungi. Protists have a relatively simple biological organization.

Protozoa. Non-photosynthetic single-cell eucaryotic microbes that obtain their energy by metabolizing organic matter.

Pseudomonads. A diverse group of motile, gram-negative bacteria that are found abundantly in soils and waters.

Psychrophiles. Organisms that grow best at low temperatures (below 20°C).

Pure culture. A population of cells that contains a single kind of microorganism and that has originated from a single cell.

Purine. Molecule consisting of two fused rings of five carbon and four nitrogen atoms. Two of the purines, adenine and guanine, serve as bases in RNA and DNA. (For chemical formula, see Figure 1.13.)

Pyrimidine. Molecule consisting of a single ring of four carbon and two nitrogen atoms. Two of the pyrimidines, cytosine and uracil, serve as bases in RNA and two, cytosine and thymine, serve in DNA. (For chemical formula, see Figure 1.13.)

Radioactive isotope. An isotope that has an unstable nucleus, causing it to emit ionizing radiation.

Recalcitrant chemicals. Substances that are degraded slowly (or not at all) by microbes in natural bodies of water or soil. These substances accumulate in the environment and cause problems.

Receptor. A molecule to which a ligand binds. For example, a virus receptor is a molecule on the surface of a host cell to which the viral ligand binds. An antibody on the surface of a lymphocyte functions as a receptor for antigen. If the ligand is viewed as a key, the receptor can be seen as the lock.

Recessive. In genetics, referring to the lack of phenotypic expression when a gene is in the presence of its dominant allele.

Recombination. In genetics, the production of a new genotype in an offspring by independent assortment of genes.

Reduction. The gain of electrons by a chemical substance.

Rennin. An enzyme that splits casein (milk protein) in such a manner that it precipitates ("curdling"). Traditionally, rennin is obtained from the stomach of calves; however, recently rennins isolated from bacteria are also being used in cheesemaking.

Replica plating. A technique for the simultaneous transfer of many separated colonies from one medium to another.

Repression. The inhibition of the formation of enzymes of a biosynthetic pathway by the end product of that pathway.

Repressor. The protein product of the *i* gene that prevents transcription of specific regions of chromosomes. (See *Operator*.)

R(esistance) factors. Extrachromosomal pieces of DNA (plasmids) containing genes for resistance to antibiotics and other chemicals.

Respiration. The complete breakdown of nutrient molecules to carbon dioxide and water by aerobic cells.

Restriction enzymes. DNases that cut DNA only at specific sequences. The DNA fragments produced by restriction enzymes are readily incorporated into foreign DNA, thus making genetic engineering possible.

Reverse transcriptase. An enzyme that catalyzes the formation of DNA utilizing RNA as a template.

Rhizobia. Bacteria that can fix nitrogen when they live in symbiotic association with leguminous plants.

Ribonuclease (RNase). Enzyme catalyzing degradation of RNA.

Ribonucleic acid (RNA). A linear polymer of nucleotides.

Ribosomal RNA (rRNA). The nucleic acid component of ribosomes. It is the major type of RNA in the cell.

Ribosome. Cytoplasmic particle consisting of RNA and protein; it is the site of protein synthesis.

RNA polymerase. Enzyme that catalyzes the formation of RNA from nucleoside triphosphates utilizing DNA as a template.

Rumen. The first and largest section of the stomach of a group of animals, called ruminants. The ruman is teeming with microbes that convert cellulose to glucose and then to organic acids.

Ruminants. A group of grazing animals that include cattle, goats, deer, sheep, camels, and giraffes. These animals contain a rumen full of microbes responsible for the breakdown of cellulose.

^{35}S. A radioactive isotope of sulfur with an atomic weight of 35; it is commonly used for labeling protein in biological experiments.

SCP (single-cell protein). All forms of microbial food, including yeast, algae, and certain bacteria.

Second law of thermodynamics. The entropy of the universe always increases (see Appendix A).

Semi-synthetic penicillins. Chemotherapeutic agents in which part of the substance is produced microbiologically and part is synthesized by an organic chemist.

Spirillum. (1) Any spiral or curved bacterium. (2) A specific group of spiral, unicellular bacteria that swim with the aid of polar flagella.

Spontaneous generation. The theory that living organisms develop from non-living matter.

Spore. A thick-walled cell capable of surviving adverse environmental conditions.

Staphylococci. Bacteria which, when examined in the microscope, appear as clusters of spherical cells. One species, *Staphylococcus aureus*, is one of the most important causes of skin infections in human beings.

Starch. A large polysaccharide composed of glucose units; it is the major storage product in plant cells.

Stationary phase. That part of the growth curve in which the population has reached the maximal level that the environment permits; in this phase, there is no net increase or decrease in the number of cells.

Sterilization. A process that *completely* destroys or removes all living organisms.

Strain improvement. The largely empirical procedure of selecting variants of strains which have enhanced properties for which you are concerned. For example, selection of strains that produce more antibiotic.

Streptococci. Bacteria which, when examined in the microscope, appear as chains of spherical cells.

Substrate. (1) Substance acted upon by an enzyme. For example, DNA is the substrate for DNase. (2) Substance used for the growth of a microbe. For example, glucose is a good substrate for the growth of *E. coli*.

Sugar. Simple carbohydrate molecule, generally having a sweet taste. (For some examples, see Figure 1.12.)

Taxonomy. The branch of biology that is concerned with the description, nomenclature, and classification of living organisms.

Tetracyclines. A group of antibiotics produced by Streptomyces that are active against both gram-positive and gram-negative bacteria.

Thermophiles. Organisms that grow best at high temperatures (over 45°C).

Thiobacilli. Chemoautotrophic bacteria that can obtain energy by the oxidation of hydrogen sulfide (H_2S) or sulfur (S) to sulfate (SO_4^{2-}).

Thymine. A pyrimidine, generally found only in DNA (see Figure 1.13).

Toxoid. A toxin modified so that it no longer is toxic but is still able to induce immunity.

Transcription. A step in protein synthesis in which the DNA is copied by base-pairing to produce a complementary RNA chain.

Transduction. Transfer of heritable characters (genes) from the cells of one strain of an organism to the cell of another strain by means of a bacteriophage.

Transfection. A process by which DNA purified from phages enters bacteria that are able to be transformed. Once the viral DNA enters the cell it then reproduces and gives rise to many new complete phages. The process is identical to infection except for the manner in which the DNA penetrates the bacterium.

Transfer RNA (tRNA). One of a group of small RNA molecules which combines with a specific amino acid and ensures that the amino acid lines up correctly on the messenger RNA.

Transformation. A conversion from one form to another. (1) In bacteria, the genetic change in an organism induced by incorporating free DNA. (2) In the context of animal virology, the process by which viruses convert normal cells to malignant cells.

Translation. A step in protein synthesis in which the information in mRNA is utilized to direct the synthesis of a specific protein.

Translocation. In genetics, a type of mutation in which a segment of one chromosome becomes attached to a non-homologous chromosome.

Trypsin. An enzyme that breaks down protein.

Tyndallization. A sterilization technique, consisting of heating to 100°C for 30 minutes on each of three successive days.

Uracil. A pyrimidine, generally found only in RNA (see Figure 1.13).

Vaccination. The process of introducing a vaccine into a human or an animal. Vaccines are materials, such as killed or attenuated microbes or innocuous toxoids, which stimulate production of antibodies and immunity without causing the disease.

Vegetative phage. The state of the phage after the DNA has penetrated the susceptible cell.

Virulence. The degree to which a microbe is pathogenic, i.e., causes disease.

Virulent phages. Those bacteriophages that invariably lyse the cells they infect during the process of producing progeny phages.

Virus. An infectious agent that lacks cell structure and which can only multiply inside a host cell; viruses contain either DNA or RNA.

Vitamin. Chemical substances that are required in the diet in relatively small quantities.

Yeasts. Unicellular fungi that do not form hyphae.

Yeast extract. Prepared by extracting dead yeast cells with water and evaporating the liquid to dryness. It is a rich source of vitamins and other growth factors.

Zygote. The diploid cell that is formed by the union of male and female sex cells.

INDEX

Note: Page numbers in *italics* indicate illustrations; those followed by (t) indicate tables.

A

B

C

D

E

F

G

S

T

U

V

W

X

Y

Z